TEUBNER-TEXTE zur Informatik   Band 1

J. Buchmann, H. Ganzinger, W. J. Paul (Hrsg.)

Informatik
Festschrift zum 60. Geburtstag
von Günter Hotz

Informatik

Festschrift zum 60. Geburtstag
von Günter Hotz

Herausgegeben von
Prof. Dr. rer. nat. Johannes Buchmann
Prof. Dr. rer. nat. Harald Ganzinger
Prof. Dr. rer. nat. Wolfgang J. Paul
Universität Saarbrücken

B. G. Teubner Verlagsgesellschaft
Stuttgart · Leipzig 1992

Die Deutsche Bibliothek – CIP-Einheitsaufnahme

Informatik : Festschrift zum 60. Geburtstag von Günter Hotz /
hrsg. von Johannes Buchmann... – Stuttgart ; Leipzig :
Teubner, 1992
   (Teubner-Texte zur Informatik)
   ISBN 978-3-8154-2033-1   ISBN 978-3-322-95233-2 (eBook)

   DOI 10.1007/978-3-322-95233-2

NE: Buchmann, Johannes [Hrsg.]; Hotz, Günter: Festschrift

Gesamtherstellung: Druckhaus Beltz, Hemsbach/Bergstraße
Umschlaggestaltung: E. Kretschmer, Leipzig

Vorwort

Dieser Band erscheint aus Anlaß des sechzigsten Geburtstags von Günter Hotz. Er enthält Arbeiten seiner Schüler, Freunde und Kollegen.

Günter Hotz ist seit 1969 Professor für Numerische Mathematik und Informatik an der Universität des Saarlandes. Er hat am Aufbau des Fachbereichs Informatik der Universität des Saarlandes großen Anteil, und er hat die Entwicklung der Informatik in Deutschland wesentlich mitgeprägt. Dies wird durch die Vielfalt der hier erscheinenden Arbeiten eindrucksvoll belegt.

Mit den Beiträgen im vorliegenden Buch möchten die Autoren bei Herrn Hotz einen Teil des Dankes, zu dem sie aus unterschiedlichen Gründen verpflichtet sind, abstatten.

Saarbrücken, im November 1991
J. Buchmann, H. Ganzinger, W.J. Paul

Inhaltsverzeichnis

On the Physical Design of PRAMs

Ferri Abolhassan
Reinhard Drefenstedt
Jörg Keller
Wolfgang J. Paul
Dieter Scheerer

Computer Science Department
Universität des Saarlandes
6600 Saarbrücken
Germany

Abstract

We sketch the physical design of a prototype of a PRAM architecture based on RANADE's Fluent Machine. We describe a specially developed processor chip with several instruction streams and a fast butterfly connection network. For the realization of the network we consider alternatively optoelectronic and electric transmission. We also discuss some basic software issues.

1 Introduction

Today all parallel machines with large numbers of processors also have many memory modules as well as a network or a bus between the processors and the memory modules. The machines however come with two radically different programming models.

The user of multicomputers is given the impression, that he is programming an ensemble of computers which exchange messages via the network. The user has to partition the data, and exchange of data between computers is done by explicit message passing. A very crude model of the run time of programs on such machines is: as long as no messages are passed, things are obviously no worse than on serial machines. As soon as messages are passed, things can become bad, because of the network.

The user of shared memory machines is given the impression, that he is programming an ensemble of CPUs which simultaneously access a common memory. This is much more comfortable for the user but there is a catch. Because the underlying machine has several memory modules (and/or several large caches) there is of course message passing going on (e.g. by transporting cache lines). Again this message passing can cause serious deterioration of performance, but because the message passing is hidden from the user it is very difficult for the user to figure out, under which circumstances this effect can be avoided.

In spite of this drawback the ease of programing provided by the shared memory model is considered such an advantage, that one tries to provide this view even for machines, which were originally designed as multicomputers.

The best of both worlds would obviously be provided by a shared memory machine whose performance is highly independent of the access pattern into the shared memory. In the theoretical literature such machines are called PRAMs [9]. An impressive number of ingenuous algorithms for these machines has been developed by theoreticians, and simulations of PRAMs by multicomputers were extensively studied. Among these simulations [16] was generally considered the most realistic one.

In [13] a measure of cost–effectiveness of architectures was established, where hardware cost is measured in gate equivalents and time in gate delays. In [1, 2] the simulation from [16, 17] was developed into an architecture which according to this measure is surprisingly cost–effective even if compared with multicomputers under a numerical workload.

This paper describes a possible physical realization of a 128 processor prototype of the machine described in [1, 2]. Roughly speaking the paper deals with those aspects of the hardware, which are not captured by the model from [13]: pins, boards, connectors, cables etc. We also treat two basic software issues: synchronization and memory allocation.

2 The Fluent Machine

The Fluent Abstract Machine [17] simulates a CRCW priority PRAM with $n \log n$ processors. The processors are interconnected by a butterfly network with n input nodes. Each network node contains a processor, a memory module of the shared memory and the routing switch. If a processor $\langle col, row \rangle$ wants to access a variable stored at address x it generates a packet of the form (destination,type,data) where destination is the tuple $(node(x), local(x))$ and type is READ or WRITE . This packet is injected into the network, sent to node $node(x) = \langle row', col' \rangle$ and sent back (if its type is READ) with the following deterministic packet routing algorithm.

1. The packet is sent to node $\langle \log n, row \rangle$. On the way to column $\log n$ all packets injected into a row are sorted by their destinations. The reason for the sorting is the fact that two packets with the same destination have to be combined.

2. The message is routed along the unique path from $\langle \log n, row \rangle$ to $\langle 0, row' \rangle$. The routing algorithm used is given in [16].

3. The packet is directed to node $\langle col', row' \rangle$ where memory access is handled.

4. – 6. The packet is sent the same way back to $\langle col, row \rangle$.

RANADE proposes to realize the six phases with two butterfly networks where column i of the first network corresponds to column $\log n - i$ of the second one. Phases 1,3,5 use the first network, phases 2,4,6 use the second network. Thus the Fluent Machine consists of $n \log n$ nodes each containing one processor, one memory module and 2 butterfly networks.

3 Improved Machine

In RANADE's algorithm the next round can only be started when the actual round is completely finished, i.e. when all packets have returned to their processor. This means that overlapping of several rounds (*pipelining*) is not possible in the Fluent Machine. This disadvantage could be eliminated by using 6 physical butterfly networks. Furthermore the networks for phases 1 and phase 6 can be realized by n sorting arrays of length $\log n$ as described in [2]. The networks for phases 3 and 4 can be realized by driver trees and OR trees, respectively. Both solutions have smaller costs than butterfly networks and have the same depth.

The processors spend most of the time waiting for returning packets. This cannot be avoided. But we can reduce the cost of the idle hardware by replacing the $\log n$ processors of a row by only one physical processor (pP) which simulates the original $\log n$ processors as virtual processors (vP). Another advantage of this concept is that we can increase the total number of PRAM processors by simulating $X = c \log n$ (with $c > 1$) vP's in a single pP. VALIANT discusses this as *parallel slackness* in [19]. The simulation of the virtual processors by the physical processor is done by the principle of *pipelining*. A closely related concept is *Bulk Synchronous Parallelism* in [19].

In vector processors the execution of several instructions is overlapped by sharing the ALU. If a single instruction needs x cycles, pipelined execution of t instructions needs $t + x - 1$ cycles. Without pipelining they need tx cycles.

Instead of accelerating several instructions of a vector processor with a pipeline, we use pipelining for overlapped execution of one instruction for all X vP's that are simulated in one physical processor. To simulate X vP's we increase the depth of our ALU artificially. The virtual processors are represented in the physical processor simply by their own register sets. We save the costs of $X - 1$ ALU's.

The depth δ of this pipeline serves to hide network latency. This latency is proved to be $c \log n$ for some c with high probability [16]. If $\delta = c \log n$ then normally no vP has to wait for a returned packet. This c increases the number of vP's and the network congestion. But network latency only grows slowly with increasing c. Thus there exists an optimal c.

When the last of all vP's has injected its packet into the network, there are on the one hand still packets of this round in the network, on the other hand the processors have to proceed (and thus must start executing the next instruction) to return these packets. CHANG and SIMON prove in [7] that this works and that the latency still is $O(\log n)$. The remaining problem how to separate these different "rounds" can easily be solved. After the last vP has injected its packet into the network, an *End of Round Packet (EOR)* with a destination larger than memory size m is inserted. Because the packets leave each node sorted by destinations, it has to wait in a network switch until another EOR enters this switch along its other input. It can be proved easily that this is sufficient.

One problem to be solved is that virtual processors executing a LOAD instruction have to wait until the network returns the answer to their READ packets. Simulations indicate, that for $c = 6$ this works most of the time (see [2]). But this is quite large in comparison to $\log n$. We partially overcome this by using delayed LOAD instructions as in [15]. We require an answer to a READ packet being available not in the next instruction but in the next but one. Investigations show that insertion of additional 'dummy' instructions happens very rarely [15]. But if a program needs any dummy instructions, they can easily be inserted by the compiler. This reduces c to 3 without significantly slowing down the machine.

Our machine will consist of 128 physical processors (pP) with 32 virtual processors (vP) each. The vP's correspond to the different pipeline steps of a pP.

4   The Processor Chip

The instruction set of our processor is based on the Berkeley Risc processor [15]. The basic machine commands are quite similar to this processor except the special commands for handling the several instruction streams. Instead of register windows we have the register sets of the virtual processors. The processor has a LOAD–STORE architecture, i.e. COMPUTE instructions (adding, multiplying, shifts, logarithmical and bit oriented operations) work only on registers and immediate constants. Memory access only happens on LOAD and STORE instructions. All instructions need the same amount of time (one cycle). We do not support floating point arithmetic but the

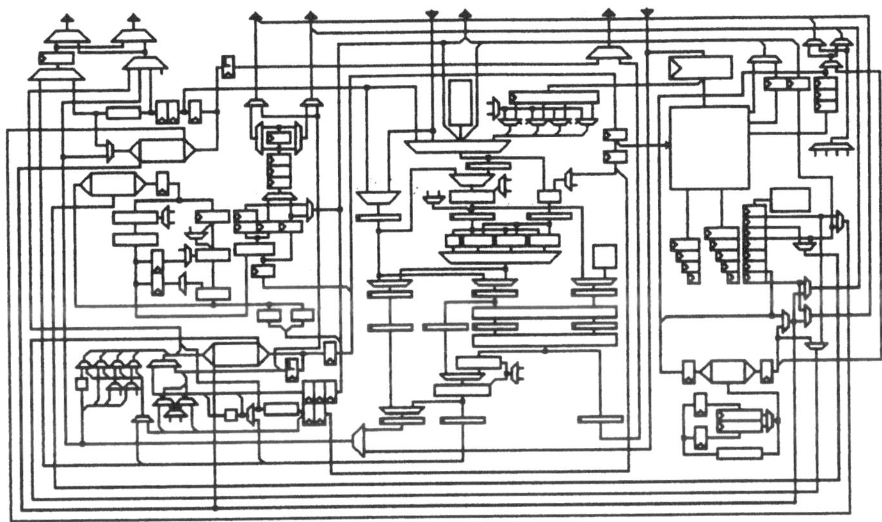

Figure 1: Data paths of the processor chip

addition of a commercial coprocessor is possible.

Because of the LOAD–STORE architecture one multiplier can be used for multiplications in COMPUTE instructions and for hashing global addresses with a linear hash function in LOAD and STORE instructions. This means that hashing does not require much special hardware.

The processor will be located in a 299 PGA and will consist of about $50,000$ gate equivalents. Figure 1 shows the data paths of the processor.

Each virtual processor is represented by its own register set consisting of 32 registers $R_0 - R_{31}$ each 32–bit wide. R_1 of each register set is the program counter, R_2 the local stack pointer and R_3 the global stack pointer. The register sets are held in a static RAM outside the chip. The vP's are handled in pipeline in a round robbin manner. Each cycle of a vP corresponds to a step in the pipeline. The cycle time of the pipeline will be $120ns$ in $2\mu m$ CMOS technology. One step of all 32 vP's takes $32 \cdot 120ns = 3840ns$ time. Additionally each vP can support up to 32 contexts which we will also call logical processors (lP) later on. Therefore the programmer can handle $32 \cdot 32 \cdot 128 = 131,072$ contexts without any software overhead.

5 Several Processes

Each vP is able to simulate 32 lP's without any software overhead. In the following we describe the hardware support of this simulation. In this section we call the work of a logical processor a process. The vP's needs machine commands to "create", "switch" and "terminate" processes. A process (lP) is represented by the values of its register set including the program counter, stack pointers and status register. We call these values the "context" of a process. If a vP switches from one process to another it has to switch the context, i.e. the current lP has to save the value of its register set somewhere and has to load the value of the register set of the next lP from somewhere. This is a complex operation and a fast mechanism to realize that is needed. However the execution time of the commands to switch, terminate and create contextes should be as fast as the other machine instructions. Because it is impossible to hold $32 \cdot 32$ register sets on chip, the register sets are located in a $32K \times 32$ static RAM outside of the processor. Access time to the large static RAM is not the critical part of computation and therefore does not slow down processor speed. To switch from one lP to another one has only to compute the base address of the new register set.

An arbitrary number of processes has to be emulated in software. This could be done e.g. by using a FIFO queue of process descriptions that is located in global memory. Parallel management of that queue needs constructs similar to parallel storage management as given in section 8.1.

A new process can only be created by a process on the same vP. A process can only terminate itself. A switch of processes can only activate the next inactive process (lP). The control of the different lP's is handled for each vP by a 32 bit wide mask b (the reason for the upper bound of lP's per vP). The 32 masks are held on chip. The value of b_i indicates, whether the i–th register set contains a process ($b_i = 1$) or not ($b_i = 0$). At the beginning $b = (0, \ldots, 0, 1)$, i.e. only the first process (lP_0) of every vP is active.

If a process lP_i wants to create a new process one looks for the smallest j with $b_j = 0$ and $i < j < 32$, if this exists. If that does not exist, one looks for the smallest j with $0 \leq j < i$. One changes the bit ($b_j = 1$) and sets the program counter of the j–th register set. The status register has an additional bit which indicates whether further process can be created ($b = (1, \ldots, 1)$). If there is no free register set nothing can be done.

If lP_i switches the process, one is looking for the smallest j with $i < j < 32$ and $b_j = 1$. If that does not exist, one looks for the smallest j with $0 \leq j \leq i$. This exists (e.g. $j = i$). The "actual" process is now lP_j. If a process lP_j is terminated, one sets the corresponding bit b_j to 0 and switches the process. The last process of a vP can not be terminated. The status register contains a flag, that is set if and only if b contains exactly one 1, i.e. if only one process is active.

The additional commands for the support of the different processes are the following: **CREATE** R_x, R_y, R_z creates a new process (if possible). The program counter of the new process is loaded with the value R_x of the current process, register R_y of the new process is loaded with the value of R_z of the current process, **SWITCH** switches a process, **KILL** terminates a process.

6   Network Design

As already mentioned, the prototype uses a butterfly network for processor–memory communication. It consists of 8 stages with 128 network nodes per stage. Packets from processors to memory modules consist of a 32 bit address, 32 bit data and 6 control bits specifying modus and operation. Packets on the way back consist of 32 bit data and 1 control bit. In each direction of a link there exists a bit specifying whether the input buffer of the node at the end of the link is already filled up or not. One link between two network nodes has to be $32 + 32 + 6 + 32 + 1 + 2 = 105$ bits wide (71 forward, 34 backward).

We have to decide how to partition network nodes on VLSI chips, how to partition these chips on printed cicuit boards (PCB's) and how to arrange the boards in racks. Clearly these decisions are not independent of each other. A chip is restricted by maximum numbers of gates and pins available. A PCB is restricted by its area and by the number of connections that can leave it. An arrangement of boards is restricted by the form of the available racks. The wires should not be too long because length of a wire restricts transmission speed and increases delay. The wiring should allow removal of boards.

6.1   Mapping Network Nodes to Chips

A network node that realizes RANADE'S routing algorithm and is able to perform multiprefix operations [17] needs the data paths shown in figure 2. It needs about $15,000$ gate equivalents and a total of 420 pins plus power supply. The largest commercially available ASIC VLSI chips have about $70,000$ gates and 300 pins (HDC105) or $48,000$ gates and 240 pins (HDC064) [12]. This means that we have enough gates to implement several network nodes on one chip but not enough pins to realize the links for only one network node. Distributing a network node on several chips does not solve the problem because all parts of the node are connected by wide busses which lead to a lot of additional pins.

If we half the width of the links and send packets in two parts, we loose a factor of 2 in speed of the network but can implement one network node on one chip HDC064 — but we waste two third of the chip area. Further reduction of the links' widths is not useful because it would slow down the network too much. Thus the links have width $w = 53$ bits, $w_1 = 36$ in forward and $w_2 = 17$ in backward direction.

Fortunately RANADE'S routing algorithm allows to increase the gate/pin ratio by a factor 2 without increasing the number of links. One network node can be cut in two halves such that only $w + 2$ bits cross the cut if w denotes the width of a link. The cut can be seen in figure 2. We implement in a chip a 2×2 butterfly but take only the last part of the nodes in one stage and take only the first part of the nodes in the following stage. Figure 3(a) shows the partitioning and 3(b) shows the implementation with 4 chips. The resulting butterfly network contains 7 stages with 64 chips per stage. One chip now contains 4 half network nodes or 2 nodes and 4 links.

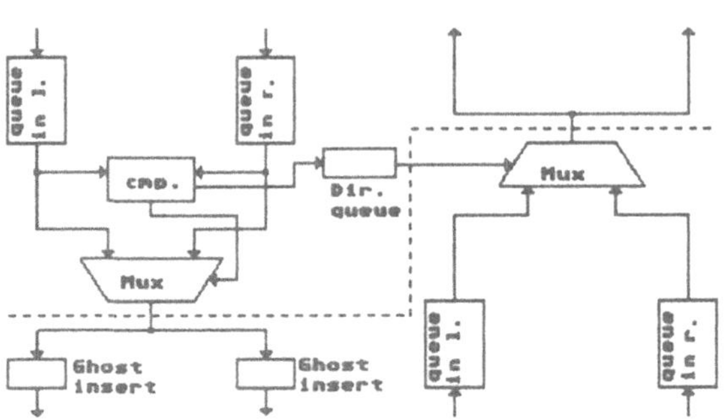

Figure 2: Cut of network nodes

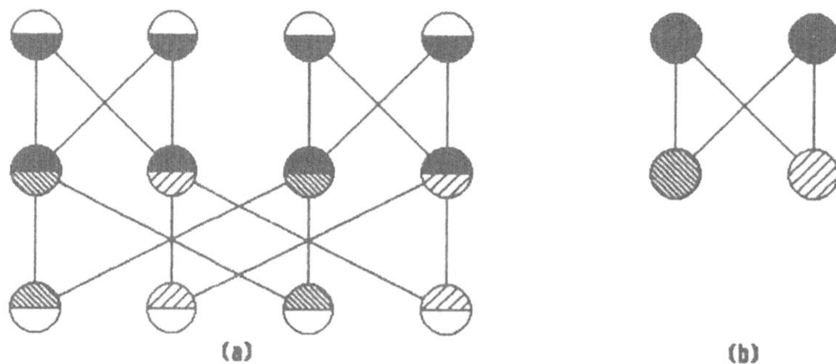

Figure 3: (a) Partitioning of the butterfly nodes. (b) Implementation of the nodes on chips

part k	stage i	board
1	0,1	$\left(1, \left\lfloor \frac{x}{4} \right\rfloor\right)$
2	2,3,4	$\left(2, 4\left\lfloor \frac{x}{16} \right\rfloor + x \bmod 4\right)$
3	5,6	$(3, x \bmod 16)$

Table 1: Board for node $\langle i, x \rangle$

The first half of network nodes in the first stage and the second half of network nodes in the last stage can be deleted because in RANADE'S algorithm they only have one input (output).

We will denote chip $x \in \{0, \ldots, 63\}$ of stage $i \in \{0, \ldots, 6\}$ with $\langle i, x \rangle$. For $i < 6$ chip $\langle i, x \rangle$ is connected to chips $\langle i+1, x \rangle$ and $\langle i+1, x \oplus 2^i \rangle$ where $a \oplus b$ here denotes the number which has a binary representation that is obtained by the bitwise exclusive or of the binary representations of a and b. Because we will only talk of the network of chips we will call the chips also nodes.

6.2 Mapping Chips to Boards

Available Printed Circuit Boards with standard size have an area of $366mm \times 340mm = 124,440mm^2$ [5]. An HDC064 chip has an area of $2237.3mm^2$ [12]. If we consider that wiring on the board and connectors also consume a large amount of the board's area, the chips can only cover about 30% of the board, resulting in at most 16 chips per board. In order to reduce the number of links between boards, one board should contain a butterfly of appropriate size. In this case this is a butterfly with 3 stages and 4 chips per stage. The board then has 12 chips and 16 connectors.

Because of the 7 stages we have to install at least 3 network parts. We choose to design two kinds of boards. The first kind looks like sketched above, for the second we delete the third stage and obtain a board with two 2×2 butterflies. If we cut the network after the second and after the fifth stage we obtain a number of small butterflies that exactly fit on the boards designed above. The first and the third part are made of boards of the second kind, the second part is made of boards of the first kind. Each part consists of 16 boards. Board $j \in \{0, \ldots, 15\}$ of part $k \in \{1, 2, 3\}$ is called (k, j) .

The following tables shows how the nodes $\langle i, x \rangle$ are distributed on the boards. Table 1 gives for each node the board on which it is mapped. Table 2 gives for each board the nodes that it contains.

The 256 links between boards are the most critical ones because they traverse the longest distances. We have to take care of them when arranging the boards.

part k	stage i	nodes
1	0,1	$\langle i, 4j \rangle , \ldots, \langle i, 4j+3 \rangle$
2	2,3,4	$\langle i, \tilde{j} \rangle , \langle i, \tilde{j}+4 \rangle, \langle i, \tilde{j}+8 \rangle, \langle i, \tilde{j}+12 \rangle$ $\tilde{j} = 16 \left\lfloor \frac{j}{16} \right\rfloor + j \bmod 4$
3	5,6	$\langle i, j \rangle, \langle i, j+16 \rangle, \langle i, j+32 \rangle, \langle i, j+48 \rangle$

Table 2: Nodes for board (k, j)

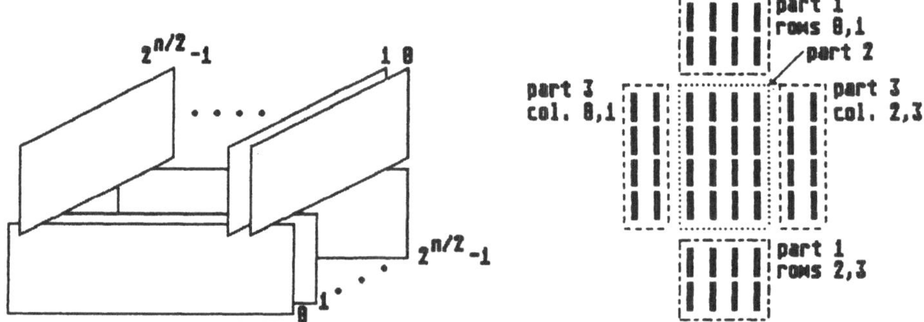

Figure 4: (a) Wise's arrangement of boards. (b) New arrangement of the boards

6.3　Arrangement of Network Boards

Arrangements of butterfly networks normally assume in contrast to reality that the implementation has a homogenous area of sufficient size, e.g. a VLSI plane or a large PCB [4]. WISE proposed in [20] a 3 dimensional arrangement of boards to implement a butterfly. This is the only paper known to us which addresses the problem. Assume that we have a butterfly with n stages and 2^{n-1} nodes per stage. Assume further that n is even. WISE makes a cut after $n/2$ stages and obtains boards that contain butterflies with $n/2$ stages and $2^{(n/2)-1}$ nodes per stage. Each of the two parts contains $2^{n/2}$ of these boards. One can prove that each board of the first part is only connected to all boards of the second part. WISE suggests the following arrangement: all boards stand vertical, the inputs of a board are on its top, the outputs on its bottom. The first part stands on top of the second part. The arrangement looks like given in figure 4(a).

This arrangement has the advantage that the parts ideally can be connected directly without any cables. The longest wire is on one of the boards, that means it is relatively short. The arrangement unfortunately has several disadvantages.

- Because each board can only hold a 3 stage butterfly (see subsection 6.2), the arrangement is only suitable for up to 6 stage butterflies, i.e. butterflies with 64 inputs.

- A direct connection of the boards with standard connectors requires using rectangular connectors which have a length of $8cm$ [5] for a 64 bit connection. Thus a board connected with 8 other boards would have a minimum length of $64cm$. Furthermore removal of single boards would require a large physical force due to the number of connectors.

- If the boards are directly connected they do not fit in standard racks, because in racks only connections in the front and back of boards are usually allowed. If one puts the boards in two racks one on top of the other and preserving the order of the boards as given in figure 4(a) one has to use cables to connect them. This offers the possibility to place all connectors in a way that the boards can have reasonable size but the cables have to be longer than one board (minimum length about $60cm$). Otherwise the boards cannot be removed anymore. These racks still do not fit in standard cabinets because there the front of all racks has to be on one side of the cabinet.

  If we turn the upper rack to use standard cabinets we have an arrangement similar to that in the DATIS–P machine [18]. But there the cables between network boards have length $150cm$. This is not too long for the DATIS–P machine which works at 16 MHz, but it might be too long for a frequency of 25 MHz needed here.

We will use a different arrangement based on an observation how the boards are connected if we cut the network in three parts as described in 6.2.

Theorem 1 *If the boards of each of the three parts are numbered with*

$$\varphi : \{0,\dots,15\} \to \{0,\dots,3\} \times \{0,\dots,3\}, \varphi(x) = \left(\left\lfloor \frac{x}{4} \right\rfloor, x \bmod 4\right)$$

then boards $(1,(i,0)),\dots,(1,(i,3))$ *of the first part are only connected to boards* $(2, (i,0)),\dots,(2,(i,3))$ *of the second part for* $0 \le i \le 3$ *and boards* $(2,(0,i)),\dots,(2,(3,i))$ *of the second part are only connected to boards* $(3,(0,i)),\dots,(3,(3,i))$ *of the third part for* $0 \le i \le 3$.

Proof: Board $(1,j)$, $j \in \{0,\dots,15\}$ contains nodes $\langle 1,4j\rangle$ to $\langle 1,4j+3\rangle$ (see table 2). These nodes are connected to nodes $\langle 2,4j+l\rangle$, $l \in \{0,\dots,3\}$ because node $\langle 1,x\rangle$ is connected to nodes $\langle 2,x\rangle$ and $\langle 2,x\oplus 2\rangle$ for all $x \in \{0,\dots,63\}$. Node $\langle 2,4j+l\rangle$, $l \in \{0,\dots,3\}$ belongs to board $(2,4\lfloor j/4\rfloor + l)$ (see table 1). Thus the first part of the claim holds.

Board $(3,j)$, $j \in \{0,\dots,15\}$ contains nodes $\langle 5,j\rangle$, $\langle 5,j+16\rangle$, $\langle 5,j+32\rangle$, $\langle 5,j+48\rangle$ (see table 2). These nodes are connected to nodes $\langle 4,j+16l\rangle$, $l \in \{0,\dots,3\}$ because node $\langle 4,x\rangle$ is connected to nodes $\langle 5,x\rangle$ and $\langle 5,x\oplus 16\rangle$ for all $x \in \{0,\dots,63\}$. Node

$\langle 4, j + 16l \rangle$ belongs to board $(2, 4l + j \bmod 4)$ (see table 1). Thus board $(3, j)$ is connected to boards $(2, 4l + j \bmod 4)$, $l = 0, \ldots, 3$ and the second part of the claim holds. ∎

The theorem indicates the following arrangement: the boards of the each part are arranged in a 4×4 square, the square of part 1 on top of the square of part 2, and the square of part 3 on the right of the square of part 2. Then all connections are horizontal or vertical. In order to have the arrangement symmetric, the first and the third square are split in two rectangles: the boards of the first part are arranged in two rectangles on the top and on the bottom of the second square. The upper rectangle holds rows 0,1 the lower holds rows 2,3. The boards of the third part are arranged in two rectangles on the right and left of the second square. The left rectangle holds columns 0,1 the right holds columns 2,3. The arrangement is shown in figure 4(b). It has several advantages.

- The boards can be put in standard racks and cabinets.

- All wiring between boards is horizontal or vertical.

- The arrangement can even be extended for butterflies with 9 stages when for all parts the boards with 3 stage butterflies are used.

A complete geometric design has not yet been worked out. If electrical wiring is too long, one can consider using optoelectronic transmission.

7   Optoelectronic Transmission of Signals

To realize the network in the prototype it is necessary to transmit data across long distances. Therefore we check whether optoelectronic transmission should be used.

7.1   Components for Optical Point–to–Point Connections

Data transmission by fiber optic operates sequentially. This is in contrast to the demand for parallel transmission between the network nodes. To improve the total throughput one can use several channels of the same kind. Figure 5 shows a schematic outline how to build up the point–to–point connection by optical components. In each cycle of the network clock w_1 bit data are injected into the network. They pass the parallel/serial converter on board 1, the optical transmitter, the fiber, the optical receiver and the serial/parallel converter on board 2.

The necessary transmission speed can be computed as follows: Like mentioned in section 6 the number of multiplexed electric lines w involves $w_1 = 36$ (forward) and $w_2 = 17$ (backward) for each link (see section 6). Let t be the period of the network clock and p_f and p_b the number of parallel channels in forward and backward direction. Thus the necessary transfer rate d_r in the optical medium for the way

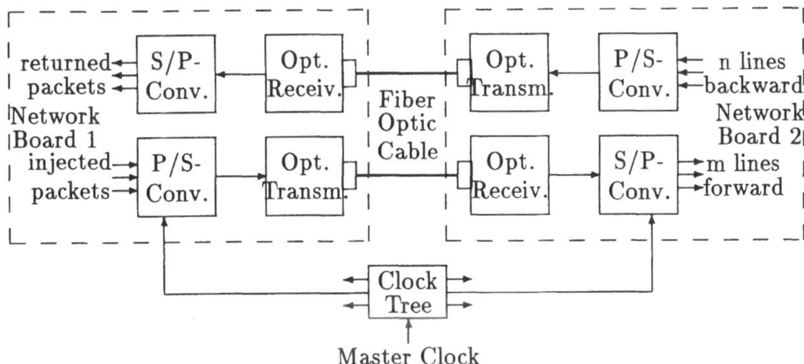

Figure 5: Components for optical data transmission

forward is $d_r = 2 \cdot m/(t \cdot p_f)$. For the way back is $d_r = 2 \cdot n/(t \cdot p_b)$. For given transfer rates we can use this equation to obtain the number of necessary channels.

To realize optical channels with these transfer rates we have studied two possible solutions:

First: Separate optical transmitters and receivers for *data communication* are available up to 1.2 Gbit/s [6]. For a reasonable price we can get compact and small modules (e.g. $15 \times 15 \times 60mm^3$) with a transmission rate of $d_r = 266$ Mbit/s, but an external P/S–converter made from ECL–chips is needed. If $t = 50ns$ (the clock cycle time of the network switches) then the above equations yields $p_f = 5$ and $p_b = 3$. Thus we need a total number of 8 sets of fiber optic cables, transmitters and receivers for each link.

Second: The optical unit and the P/S–converter chip are mounted together in a metal cover. The **T**ransparent **A**synchronous **X**mitter–receiver **I**nterface–chip set (TAXI) [6] provides a high performance transparent fiber optic 8 bit interface. Data transfer rates are up to 125 Mbit/s and the transfer is performed with error detection. Because of the integrated P/S-converter there is a lower bound of $t = 80ns$ for the clock period. This increases the number of bits to be transferred in parallel by a factor of 1.6 because of the $50ns$ clock of the network switches. In this case we obtain $p_f = 9$ and $p_b = 5$. Thus 14 pairs of the TAXI–chip set and logic are necessary to realize one link. The logic includes an interface between network link and the TAXI chip set, the details are not yet worked out.

7.2   Network design based on optical links

We assume that the components of one link occupies a reasonable area of PCB (e.g. $22 \times 7cm^2$), the transceiver board. But the costs of one pair of transceiver boards representing one link are still high (up to 9500,– DM). Therefore we use the advantages of fiber optic only to substitute the set of longest wires. This however

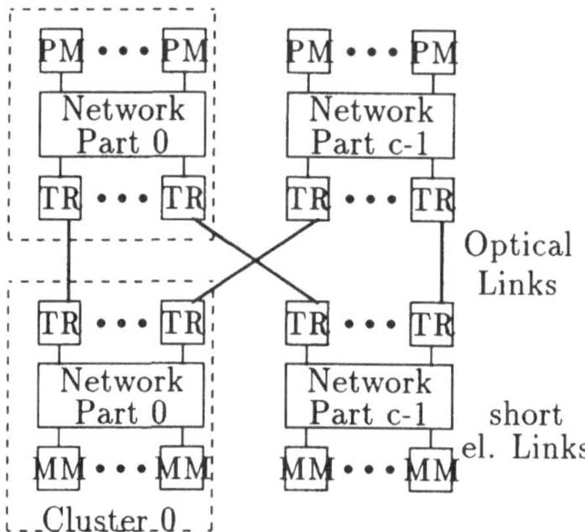

Figure 6: Optical links connect clusters

changes the network.

Fig. 6 shows the situation if we make a cut through the network horizontally. It is divided into $2 \cdot c$ clusters of boards. A cluster is a number of functionally associated boards. In our case it includes one network part of depth $s \in \{3, 4\}$, $2^s \in \{8, 16\}$ processor or memory boards and the same number of transceiver boards. Because the distances within the clusters are short we avoid long electrical links. But then we have one set of long optical links to connect the clusters with each other.

In the following table we compare cost and time for transmitting data by electrical lines and by fiber optic (using the TAXI chip set). Electrical lines may be single ended lines or twisted pair cables. Two commonly used types of driver/receiver combinations are listed. The length of electrical links may be increased using high speed trapezoidial bus drivers [3].

Costs for electric lines are computed as follows: $4 \cdot 128$ electrical interfaces are necessary to connect the three parts network boards with the PM's and MM's (refer to the whole design in section 6).

Because of the more compact layout shown in figure 6 additional electrical links may be avoided by using fiber optic. In this case the costs of 128 pairs of transceiver boards are listed. The technical data are taken from several data sheets. In general each link between two boards adds one or two delay units of network clock. If it is possible to use special low voltage swing drivers the power dissipation can be decreased [11] — however they are not yet commercially available.

	single line	twisted pair	fiber optic
driver/receiver type	AS1034B/F14	26LS31/32	DL6000
max. prop. delay of IC's	12.5 ns	47 ns	80 ns
power dissipation per link	15 W	19 W	43 W
number of links	$4 \cdot 128$	$4 \cdot 128$	128
total power dissipation [kW]	7.7 kW	9.7 kW	5.5 kW
IC's per link	2×10	2×11	2×14
area $[cm^2]$	50	70	154
relative costs	1	2.9	25

The table reveals that there is a large trade–off between costs on the one hand and power dissipation and wiring overhead on the other hand. In the near future we will test some types of electrical links and optical channels in order to decide which type of link provides the necessary throughput considering the real distances between the boards.

8   Basic Software Issues

A new architecture does not only has to have new efficient and powerful hardware, it also has to support hardware by suitable software, especially by an operating system with efficient resource management and by a compiler for a high–level language. A high–level language called FORK has been proposed [10] that is suited for a PRAM. The work on a compiler already has started. The operating system still has to be developed.

As examples of the problems that have to be solved we will present solutions for parallel storage management and for synchronization of multiple instruction streams. Both synchronization and memory management take advantage of the multiprefix (MP) [17] and SYNC (MP without return values) commands that are supported by hardware [2].

8.1   Parallel Storage Management

In a parallel machine that presents its user a shared global memory handling storage management is much more complicated than in a distributed machine where each processor has its private local memory on which it acts (and allocates memory) just like a sequential computer. In a shared memory machine several processors could try to allocate storage at the same time.

First we consider a simple solution for parallel storage allocation without worrying about freeing memory. Let $s(i)$ be the content of memory cell i in global memory and let cell 0 contain a pointer to the first cell of free global memory. If several processors $P_i, i \in I$ want to allocate memory of sizes $m(i)$ they execute a multiprefix command MP $0, +, m(i)$. As a result each processor P_i receives $s(0) + \sum_{j \in I, j < i} m(j)$ and the content of cell 0 is $s(0) = \sum_{j \in I} m(j)$. Each processor thus receives a pointer to its requested memory block and cell 0 contains a pointer to the new begin of free memory. Correct freeing is only possible if the program guarantuees that the freeing operations are performed in reverse order to the allocating operations.

If we choose a fixed block size and allow processors only to allocate a number of not necessarily subsequent blocks, the problem becomes a little bit easier. Let cells 1 to max contain pointers to free memory blocks and let cell 0 contain a pointer to max. If processors $P_i, i \in I$ want to allocate $m(i)$ blocks they execute MP $0, -, m(i)$. Each processor P_i receives a pointer to a cell $x = s(0) - \sum_{j \in I, j < i} m(j)$. Cells $x, \ldots, x - m(i) + 1$ contain pointers to the $m(i)$ memory blocks for P_i. The pointers have to be copied. If processors want to free memory blocks they execute MP $0, +, m(i)$. Each processor receives a pointer to a cell x and writes to cells $x, \ldots, x + m(i) - 1$ the pointers to the $m(i)$ memory blocks it wants to free.

This stack mechanism only works correctly if we prevent the machine from freeing while others are allocating and vice versa. This can be done by programming a semaphore. One can get rid of this by using a FIFO queue. Allocate and free operations now take time proportional to the number of blocks. However allocation of memory with subsequent addresses larger than one block is not possible.

8.2 Synchronization Primitives

FORK provides synchronous execution of high level language commands. The runtime of this code however is often not predictable at compile time. In that case synchronization is necessary. A simple synchronization could be realized in hardware. But several synchronizations can happen simultaneously. Thus one has to provide several synchronizations within "groups" of processors.

Suppose each processor of a group that has to be synchronized later on knows the address a of a cell in global memory. First all processors store 0 in that cell. Then each processor executes SYNC $a, +, 1$. The cell then contains the number of processors in the group. The processors now execute the code for the high level language command after which they have to be synchronized. Each processor that has finished execution of that code executes MP $a, -, 1$. After that it reads the content of a until this content is 0. Then all processors of the group are synchronized again.

A problem that has not been mentioned is what happens if a conflict occurs because of one processor executing an MP on a and one loading the content of a at the same time step. We avoid that by only executing MP commands in time steps with even numbers and LOAD commands in time steps with odd numbers when synchronizing. Our processor design supports this by a modulo flag in the processor status word that is flipped after each machine instruction[1] and with a conditional jump on the value of that flag. If the address a is stored in register R_a the code for synchronizing has the form

    1. ... all processor have reached this point at the same time

---

[1] Remember that all machine instructions take equal amounts of time.

2. STORE $R_a, R0, R0$

3. SYNC $R_a, +, 1$

4. ... code for high level language command

5. ... processors could reach this point at different times

6. JMP modulo clear PC,PC,R0

7. MP $R_a, -, 1$

8. LOAD $R_a, R0, R0$

9. JMP zero set $PC, PC, \# - 1$

10. ... all processors reach this point at the same time

If all processors reach line 5 at the same time, the synchronization overhead is 7 commands: (2), (3), (6), (6), (7), (8), (9).

Reality is somewhat worse because the LOAD in line 8 is delayed and therefore at least one NOP has to be performed before the content of a and thus the correct zero bit is available. In order to have line 8 executed always in an odd time step we have to include two NOP commands after line 8.

It now can happen that a part of the processors in the group reaches line (10) at the same time and the rest of the group reaches line (10) two steps later. This gap can be closed by a second synchronization part where the processors perform a SYNC on a cell with previous content zero, then load the new value and check whether all processors have reached this point. The check will fail for the "faster" processors and succeed for the "slower" ones. If it fails processors execute two NOP's. The final synchronization needs 7 commands: SYNC, LOAD, NOP, CMP, conditional JMP, 2NOP's. The overhead for the synchronization is then 9 for the first part of code and 7 for the second part, in total 16 commands.

One can still argue that 16 commands is too much for synchronizing when compared to about 10–20 commands needed as code for one high level language command. However static analysis of programs allows compilers to reduce the number of necessary synchronizations. Synchronization is only necessary at points where several instruction streams are split and merged later on and where runtime of different streams is not predictable. The FORK compiler uses this kind of analysis [10]. The exact factor of reduction however still has to be determined in practice.

Acknowledgements

We would like to thank Helmut Seidl for helpful discussions about low level software topics.

Bibliography

[1] F. Abolhassan, J. Keller, and W. J. Paul. On the cost–effectiveness of PRAMs. In *Proc. 3rd IEEE Symp. on Parallel and Distr. Processing.* IEEE, Dec. 1991.

[2] F. Abolhassan, J. Keller, and W. J. Paul. On the cost–effectiveness and realization of the theoretical PRAM model. FB 14 Informatik, SFB-Report 09/1991, Universität des Saarlandes, May 1991.

[3] R. V. Balakrishnan. Weniger Störungen auf Mikrocomputerbussen. *Elektronik*, 4:106–112, 1989.

[4] R. Beigel and C. P. Kruskal. Processor networks and interconnection networks without long wires. In *Proc. 1989 ACM SPAA*, pp. 42–51. ACM, 1989.

[5] Bicc Vero Electronics. *Elektronik Handbuch*, 1988.

[6] BT&D Technologies Ltd. *Data Communication Products*, 1991.

[7] Y. Chang and J. Simon. Continuous routing and batch routing on the hypercube. In *Proc. 5th ACM Symp. on Principles of Distr. Comp.*, pp. 272–281, 1986.

[8] The Influence of Technology on the Choice of a Multiprocessor Interconnection Network. *Proc. 2nd Workshop on Parallel and Distr. Processing*, pp. 91–110, 1990.

[9] S. Fortune and J. Wyllie. Parallelism in random access machines. In *Proc. 10th ACM STOC*, pp. 114–118, 1978.

[10] T. Hagerup, A. Schmitt, and H. Seidl. FORK: A high–level–language for PRAMs. In *Proc. PARLE 91*, 1991.

[11] T. F. Knight and A. Krymm. A Self-Terminating Low-Voltage Swing CMOS Output Driver. *IEEE Jounal of Solid-State Circuits*, 23(2), 1988.

[12] Motorola, Inc. ASIC Division, Chandler, Arizona. *Motorola High Density CMOS Array Design Manual*, July 1989.

[13] Silvia M. Müller and Wolfgang J. Paul. Towards a formal theory of computer architecture. In *Proceedings of PARCELLA 90, Advances in Parallel Computing*. North–Holland, 1990. 94.

[14] C. A. Neugebauer. Materials for High-Density Electronic Packaging and Interconnections in the Higher Packaging Levels. *Journal of Electronic Materials*, 18(2), 1989.

[15] D. A. Patterson and C. H. Sequin. A VLSI RISC. *Comput.*, 15(9):8–21, 1982.

[16] A. G. Ranade. How to emulate shared memory. *J. Comput. Syst. Sci.*, 42(3):307–326, 1991.

[17] A. G. Ranade, S. N. Bhatt, and S. L. Johnson. The Fluent Abstract Machine. In *Proc. 5th MIT Conf. on Advanced Research in VLSI*, pp. 71–93, 1988.

[18] D. Scheerer. Entwurf und Realisierung eines Verbindungsnetzwerkes als Teil des Multiprozessorsystems DATIS-P-256. Master's Thesis, Universität des Saarlandes, FB Informatik, 1989.

[19] L. G. Valiant. General purpose parallel architectures. In J. van Leeuwen, (Ed.), *Handbook of Theoretical Computer Science, Vol. A*, pp. 943–971. Elsevier, 1990.

[20] D. S. Wise. Compact layouts of banyan/FFT networks. In H. T. Kung, B. Sproull, G. Steele, (Ed.), *Proc. CMU Conf. on VLSI Syst. and Comput.*, pp. 186–195, 1981.

Synthesis for Testability:
Binary Decision Diagrams

Bernd Becker

Computer Science Department
Johann Wolfgang Goethe-Universität Frankfurt
6000 Frankfurt
Germany

Abstract

We investigate the testability properties of Boolean circuits derived from (Reduced Ordered) Binary Decision Diagrams. It is shown that BDD-cirucits (or at least) BDD-like circuits are easily testable with respect to different fault models (cellular, stuck-at and path delay fault model). Furthermore the circuits and the test sets can be constructed efficiently.

Keywords VLSI structures, algorithms and data structures, synthesis, (complete, full) testability, fault model

1   Introduction

Binary Decision Diagrams as a data structure for Boolean functions were introduced by Lee [17] in 1959. They have been intensively studied in several differing areas of computer science and electrical engineering: in theoretical computer science, especially in complexity theory (e.g. [25, 19]), in electronic design automation, especially in design verification (e.g. [1, 9, 8]). Here a restricted form of BDD's, namely reduced ordered BDD's, have gained a whitespread use. In this paper we point out that BDD's are not only well suited for design verification which is necessary to prove or at least give some evidence of the correctness of the design, but also may be used to synthesize circuits whose testability can be guaranteed in advance. At first, we want to emphasize the practical relevance of testing problems by making some general remarks. For a detailed treatment of the topic see e.g. [27].

As a result of technological improvements, VLSI electronic circuitry may nowadays contain hundreds of thousands of transistors on a single silicon chip. Even if the chips are correctly designed, i.e. design verification has been done successfully, a non negligible fraction of them will have physical defects caused by imperfections occurring during the manufacturing process (e.g., open connections induced by dust particles). Therefore, there has to be a test phase in which "production" verification is performed, i.e. in which the "good" chips are sorted from the "bad" ones.

Because of the variety of possible defects restrictions on a subset of the possible faults are necessary; these simplifying assumptions based on the experience of many years are manifested in fault models. Since tests are generated to test for the fault mechanisms described by the assumptions the reliability of the chip is at least partly determined by the accuracy and effectiveness of the fault model (measured e.g. in detected physical failures). The classical stuck at fault model (SAFM) is well-known and used throughout the industry [6]. It assumes that a defect causes a basic cell input or ouput to be fixed to either 0 or 1. Thus all failures with this effect will be detected by tests for stuck-at faults. It has been observed by many authors that a large amount of defects typical for today's VLSI technologies are not covered by stuck-at faults [4, 3, 15, 16]. Other fault models are necessary to overcome this problem, e.g. flexible fault models which verify the correct static behaviour of a combinational circuit based on inductive fault analysis [18, 12] or dynamic fault models which allow to model stuck open faults or timing issues [23, 26, 22]. The strongest cell-based fault model to control the correct static behaviour of a combinational circuit is the cellular fault model (CFM), which tries to completely verify the function of each basic cell in the circuit [13, 10]. A fault model considering timing issues is the path delay fault model (PDFM), which checks the correct timing of any path from a primary input to a primary output [22]. Correctness in PDFM is a very strong result, it especially implies the non existence of stuck-open faults and gate delay faults. Of course, it is desirable to prove the correctness of a chip under these fault models. However, due to the large costs of the test phase, in practice often only SAFM is considered. Furthermore the investigations are usually based on the single fault assumption, i.e. one assumes that there is at most one fault (according to the considered fault model) in the circuit.

With increasing complexity of VLSI circuits, the costs for the test phase have risen dramatically, at least 25% and up to 60-70% of the total product costs count on testing [20, 24]. Thus, the necessity of new methods for the test phase is evident. Specialists in the field of testing agree that testability issues have to be considered from the very beginning of the design process to control the test costs and to guarantee the testability of the circuit at the end of the manufacturing process. One step in this direction is the use of synthesis algorithms to generate logic designs that not only meet area and/or speed constraints but also are easily testable. For this, it is important to understand the relation between testability and structural properties of Boolean functions and circuit realizations. There has been a lot of work in this direction starting with investigations of testable realizations of Boolean functions in the seventies and leading to the incorporation of testability constraints in commercially available logic synthesis systems today. For an overview see e.g. [6, 14, 11].

In this paper we focus on the synthesis of circuits derived from BDD's and their testability aspects. As pointed out before BDD's are often used in design verification. In this case they are available for free in the subsequent phases of the design process. In addition, they can be easily transformed into multiplexer-based Boolean circuits which can for example be used as a first realization during the synthesis process for irregular logic like control circuitry. Thus, it is natural to investigate the testability of these circuits and to ask for synthesis algorithms which enhance testability, if necessary. The relation between BDD's and testability has so far only been studied in the context of functional testing based on BDD's [2]. We show that BDD's lead to circuits with good testability properties. Criteria used in the following to measure testability are

- number of different fault models considered
- single or multiple fault assumption
- complexity of complete test set construction
- number of redundancies and complexity of their detection
- possibility and complexity of circuit modifications to obtain a fully testable circuit

A little bit more in detail the results of this paper can be sumarized as follows: The basis of our investigations is a characterization and (efficient) computation of the testability properties of BDD-circuits with respect to the cellular fault model. Testable and redundant faults can be easily classified. A sufficient condition is given, which either implies the non-existence of redundant faults or at least allows the removal of the redundancies. In this case the resulting BDD-like circuit is fully testable in the cellular fault model.

It follows easily that in a BDD-circuit any (single) stuck-at fault on an input- or output-line of a multiplexer is testable, i.e. a BDD-circuit, as a circuit realized with multiplexers, is fully testable in the (single) stuck-at model. This must not necessarily be the case, if the multiplexers are replaced by a standard *AND, OR, INVERTER*-gate based realization: there may be redundant stuck-at 1 faults at the inputs of the

AND-gates. However, it is shown that they correspond to redundant cellular faults, which can be removed in any BDD-circuit, and thus result in a BDD-like circuit, which is fully stuck-at testable on the gate-level. While the above results rely on the usual single fault assumption, for the stuck-at fault model we can even construct a test set, which detects any multiple stuck-at fault. Thus we obtain full testability with respect to multiple faults.

We then consider the path delay fault model to verify the correctness of the circuit with respect to timing constraints. The set of paths which is (robustly) testable with respect to path delay faults is determined. This implies a characterization of the BDD-circuits, which are fully path delay testable.

The remaining part of the paper is structured as follows: in section 2 we provide further definitions from the field of testing including our model for combinational logic circuits. Then we introduce BDD's and BDD-circuits in section 3. In section 4 the testability properties of BDD- and BDD-like circuits are derived. We finish with a discussion of the results and mention open problems.

2   Testing and Fault Models

In this section we give a short introduction to basic notions and concepts of testing which are important for the understanding of this paper.

A *combinational logic circuit* (CLC) over a fixed library is modeled as a directed acyclic graph $C = (V, E)$ with some additional properties: each vertex $v \in V$ is labeled with the name of a basic cell or with the name of a primary input (PI) or primary output (PO). The collection of basic cells available is given in advance by the fixed library. Very often the library STD consisting of the 2-input, 1-output *AND, OR* gate and the 1-input, 1-output inverter *NOT* is used. In this paper we consider circuits constructed with help of basic cells from STD $\cup$ MUXLIB $\cup$ $\{XOR\}$. MUXLIB is a library, which merely consists of several types of multiplexers with (at most) three inputs and one output, *XOR* is a cell with two inputs and one output. In general basic cells with arbitrary complexity, especially with an arbitrary number of inputs and outputs, are possible. The inputs and outputs of each basic cell are ordered. There is an edge (u, v) in E from vertex u to v, if an output pin of the cell associated to u is connected to an input pin of the cell associated to v, i.e. edges contain additional information to specify the pins of the source and sink node they are connected to. Vertices have exactly one incoming edge per input pin. Nodes labeled as PI (PO) have no incoming (outcoming) edges. The *size* of the circuit $|C|$ is as usually given by $|E| + |V|$.

Now assume further, that for each basic cell (n inputs, m outputs) in the library the functional behaviour of the cell is defined by a Boolean function from $\mathbf{B}^n \to \mathbf{B}^m$. (For STD the functions are well-known, for the multiplexer cells in MUXLIB the functional behaviour is defined in section 3 (see figure 2 and 3). An *XOR* cell realizes the exclusive or operation.) A CLC with N PI's and M PO's then defines a Boolean function from

$\mathbf{B}^N \to \mathbf{B}^M$ in the obvious way. For an example of a CLC over the library STD and MUXLIB see the circuit C_{MUXLIB} in figure 4 and the circuit C_{STD} in 5, respectively. These examples also demonstrate that a circuit over the library MUXLIB can easily be "expanded" into a circuit over STD with identical behaviour: the multiplexer cells have to be substituted by a standard cell realization of a multiplexer over STD. In general, let C_1 be a CLC over a fixed library L_1. Assume that some basic cells in C_1 can be substituted by some subcircuits over a library L_2 without changing the overall behaviour of the circuit, then C_2 is called an *expansion* of C_1. (For a detailed definition of this notion see e.g. [7, 5].)

As mentioned above, even if a CLC is correctly designed, a fraction of them will behave faulty because of physical defects caused by imperfections during the manufacturing process. Fault models which cover a wide range of the possible defects are defined and tests for faults in the fault models are constructed. We give a short description of the fault models considered in this paper.

The Cellular Fault Model (CFM)

In the CFM ([13, 10] it is assumed that a fault modifies the behavior of exactly one node v in a given CLC C and that the modified behavior is still combinational. Since this fault can be detected by observing the incorrect output values of v for one suitable input combination, it suffices to test for faults of the following kind. A *cellular fault* in C is a tuple $(v, I, X/Y)$, where v is the faulty node (= fault location), I is the input for which v does not behave correctly, and X (Y) is the output of the correct (faulty) node on input I. A test for the cellular fault $(v, I, X/Y)$ is an input to C that generates I on the input lines of v and propagates the difference X/Y to at least one PO of C.

The Stuck-at Fault Model (SAFM)

A fault in the SAFM ([6]) causes exactly one input or output pin of a node in C to have a fixed constant value (0 or 1) independently of the values applied to the PI's of the circuit. A *stuck-at fault* with fault location v therefore is a tuple $(v[i], \epsilon)$ or $([i]v, \epsilon)$. $v[i]$ $([i]v)$ denotes the i-th input (output) pin of v, $\epsilon \in \{0, 1\}$ is the fixed constant value. An input t to C is a test for a stuck-at fault f, iff the output values of C on applying t in the presence of f are different from the output values of C in the fault free case. Usually the stuck-at model is only considered on the gate level, i.e. for circuits over the library STD. We have given here an obvious generalization to circuits over arbitrary libaries.

We finish the discussion of CFM and SAFM with some general definitions and remarks on the relation between both fault models. For this let C be any CLC over a fixed library and FM a fault model as defined above. A fault in FM is *testable*, if there exists a test for this fault. The goal of any test pattern generation process is a *complete* test set for the circuit under test in the considered fault model FM, i.e. a test set that contains a test for each testable fault. It follows easily from the definitions that, given a fixed circuit C, a complete test set in CFM is also complete in SAFM. Thus, the cellular fault model is stronger than the stuck-at fault model. In any case, the construction of complete test sets requires the determination of the faults which are not testable (= *redundant*), even though it is easily seen that in general the detection of redundancies is *co-NP complete*. Redundancies have further unpleasant properties: they may e.g. invalidate tests for testable faults and often correspond to locations of the circuit where area is wasted ([6]). All in all, synthesis procedures which result in non redundant circuits or at least circuits with known redundancies are desirable. A node v in C is called *fully testable* in FM, if there does not exist a redundant fault in FM with fault location v. If all nodes in C are fully testable in FM, then C is called *fully testable* in FM.

Now consider a circuit C_2 which results from a circuit C_1 by expansion (see figures 4 and 5). Then one can easily show that a complete test set for C_1 in CFM is also a complete test set for C_2 in CFM (and SAFM). Thus the CFM is more powerful, if the size of the basic cells increases. We call this property the *completeness property* of CFM. Notice, that in SAFM there is a trend in the opposite direction. This is the reason why in general the strongest version of SAFM, i.e. SAFM for circuits over STD is considered.

The Path Delay Fault Model (PDFM)

Whereas the CFM and SAFM are fault models to verify the static behaviour of a circuit, the purpose of delay testing is to ascertain that the circuit under test meets its timing specifications. In the PDFM ([22, 21]) it is checked whether the propagation delays of all paths in a given CLC are less than the system clock intervall.

For our discussion of the PDFM assume that the considered CLC C is defined over the library STD. A *path* π is given by an alternating sequence of nodes and edges $(v_0, e_0, v_1, \ldots, v_n, e_{n+1}, v_{n+1})$ starting in a PI v_0 and ending in a PO v_{n+1}. Inputs of nodes on the path where no edge e_i of the path ends are called *off-path inputs*.

A *transition* $(0 \to 1 = rising$ or $1 \to 0 = falling)$ *propagates along* π, if a sequence of transitions $t_0, t_1, \ldots, t_{n+1}$ occur at the nodes $v_0, v_1, \ldots, v_{n+1}$, such that t_i occurs as a result of t_{i-1}. π has a *path delay fault*, if the actual propagation delay of a (rising or falling) transition along π exceeds the system clock intervall. For the detection of a path delay fault a pair of patterns (I_1, I_2) is required rather than a single pattern as in the CFM and SAFM: The *initialization vector* I_1 is applied and all signals of C are allowed to stabilize; then the *propagation vector* I_2 is applied and after the system clock intervall the outputs of C are controlled. A two-pattern test is called a *robust test* for a path delay fault on π, if it detects that fault *independently* of all *other delays* in the circuit and all *other delay faults* not located on π. A two-pattern test is called a *non-robust test* for a path delay fault on π, if it detects that fault under the assumption that the off-path inputs of all nodes on π stabilize to their final (non-controlling) values prior to the time, at which the transition propagated along π. (A *controlling* value at the input of a node is the value that completely determines the value at the output (e.g. 1 (0) is the controlling value for OR (AND) and 0 (1) is the non-controlling value for OR (AND).)

In this paper we concentrate on the robust testing of path delay faults. It turns out that the construction of tests with the following property is possible: for each testable path delay fault there exists a robust test (I_1, I_2) which sets all off-path inputs to the non-controlling values on application of I_1 and remains stable during application of I_2, i.e. the values on the off-path inputs are not invalidated by hazards or races.

3 Binary Decision Diagrams

In this section we shortly review the essential definitions and properties of BDD's and introduce BDD-circuits. For a detailed treatment see [9].

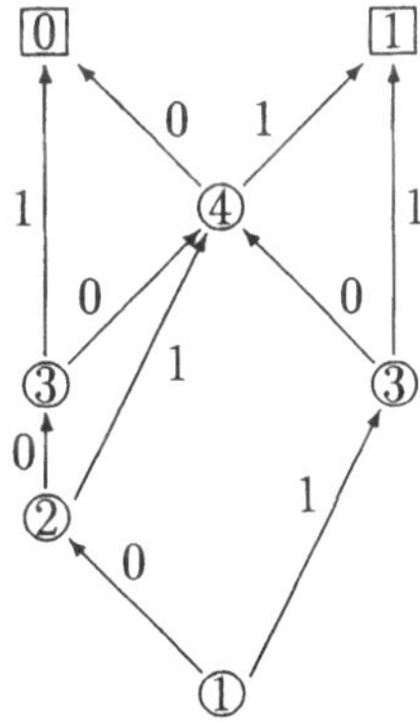

Figure 1:   The reduced, ordered BDD for the Boolean function $f(x_1, x_2, x_3, x_4) = x_1 x_3 + x_2 x_4 + \bar{x}_3 x_4$

A *Binary Decision Diagram* (BDD) is a rooted directed acyclic graph $G = (V, E)$ with vertex set V containing two types of vertices, *nonterminal* and *terminal* vertices. A nonterminal vertex v has as label an argument index $ind(v) \in \{1, \ldots, n\}$ and exactly two outgoing edges with sink nodes $zero(v), one(v) \in V$. A terminal vertex v is labeled with a value $val(v) \in \{0, 1\}$ and has no outgoing edges. For an example see figure 1. There the edges $(v, zero(v))$ $((v, one(v)))$ are marked with a 0 (1).

There exists the following correspondence between BDD's and Boolean functions: A BDD having root vertex v denotes a function f_v defined recursively as:

> If v is a terminal vertex and $val(v) = 1$ ($val(v) = 0$), then $f_v = 1$ ($f_v = 0$).
> If v is a nonterminal vertex with $ind(v) = i$, then f_v is the function
> $f_v(x_1, \ldots, x_n) = \bar{x}_i \cdot f_{zero(v)}(x_1, \ldots, x_n) + x_i \cdot f_{one(v)}(x_1, \ldots, x_n).$

(Vice versa, it follows directly that for each Boolean function f there exists a BDD denoting f.) BDD's that are used as data structure in design automation normally fulfill the following two additional properties:

- The BDD's are *ordered*, i.e. for any nonterminal vertex v, if $zero(v)$ is also nonterminal, then $ind(v) < ind(zero(v))$, and similarly, if $one(v)$ is also nonterminal, then $ind(v) < ind(one(v))$.

- The BDD's are *reduced*, i.e. there exists no $v \in V$ with $zero(v) = one(v)$ and there are no two vertices v and v' such that the subgraphs rooted by v and v' are isomorphic.

Henceforth we only consider reduced, ordered BDD's. The example BDD in figure 1 is reduced and ordered. Reduced, ordered BDD's have the following important properties:

- Let G be an ordered BDD for f with size $|G|$. Then the satisfiability problem for f can be solved in time $O(|G|)$.

- The reduced ordered BDD of a given function f is uniquely determined and can be computed in time $O(|G| \log |G|)$, given an ordered BDD G for f.

- Let G_i be a (reduced, ordered) BDD for f_i $(i = 1, 2)$ and $\diamond$ any binary operator. Then the BDD for $f_1 \diamond f_2$ can be computed in time $O(|G_1||G_2|)$.

It is well-known, that BDD's directly correspond to multiplexer based Boolean circuits called BDD-circuits in this paper. More exactly: BDD-circuits are CLC's over the library MUXLIB or STD. MUXLIB consists of a *multiplexer cell MUX* and *degenerated multiplexer cells DegMUXi*. A *MUX* is defined in figure 2 by its standard *AND-, OR-, INVERTER*-based realization. The western input is called *control input*, the left (right) northern input *0-input (1-input)*. *DegMUX$_i$* cells result from a

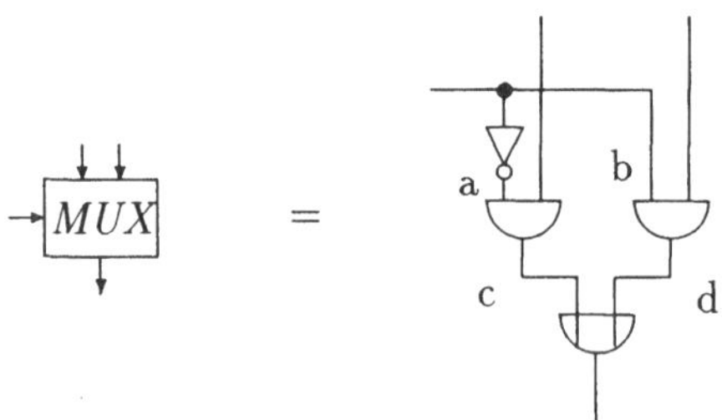

Figure 2: multiplexer cell *MUX*

MUX, if one (both) of the northern input lines has a constant value 0 or 1 (have different constant values 0 and 1). It follows directly that there are 6 different cells $DegMUX_1, \ldots, DegMUX_6$. Two of them are given in figure 3. The remaining ones are defined analogously.

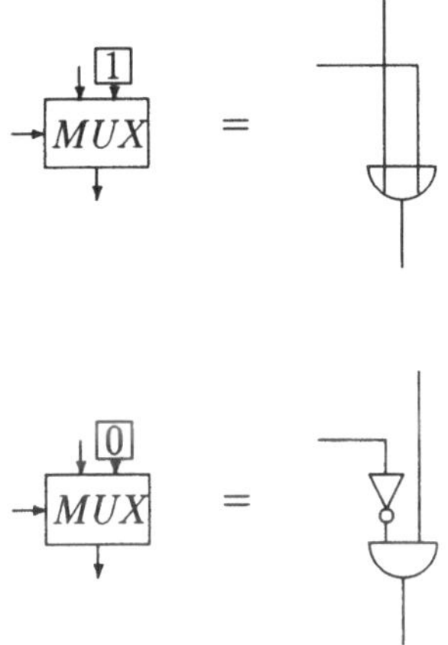

Figure 3: two degenerated multiplexer cells

The *BDD-circuit C_{MUXLIB} of a BDD G over the library* MUXLIB is now obtained by the following construction: Traverse the BDD in topological order and replace each nonterminal node v in the BDD by a MUX cell, connect the control input with the PI $x_{ind(v)}$, connect the 0-input to $zero(v)$, the 1-input to $one(v)$. At the end replace *MUX* cells which are connected to a terminal node by a corresponding *DegMUX$_i$* and connect the output of the multiplexer which replaced the root node with a PO. Figure 4 shows the BDD-circuit of the BDD given in 1. For simplicity we often identify a node in a BDD G with the corresponding multiplexer node in its BDD-circuit C_{MUXLIB}. Also $f_{zero(v)}$ $(f_{one(v)})$ is used to denote the function computed at the 0-input (1-input) of the multiplexer v. We now define BDD-circuits over STD. For this take the BDD-circuit C_{MUXLIB} of a BDD G and substitute the multiplexer cells by the standard realization of the *MUX* and *DegMUX$_i$* cells over STD. The resulting CLC C_{STD} is called the *BDD-circuit of G defined over* STD. The BDD-circuit over STD for the example BDD is given in figure 5.

4   Results

In this section we derive the testability properties of BDD-circuits (over MUXLIB and STD) with respect to the fault models introduced in section 2. Furthermore we discuss possibilities to remove redundancies from BDD-circuits, if occurring, and thus synthesize circuits with enhanced testability properties.

We start with some general remarks concerning the controllability of a multiplexer cell in a BDD-circuit over MUXLIB: The control input of any multiplexer (degenerated or non degenerated) is directly controllable. Furthermore, the values of the northern

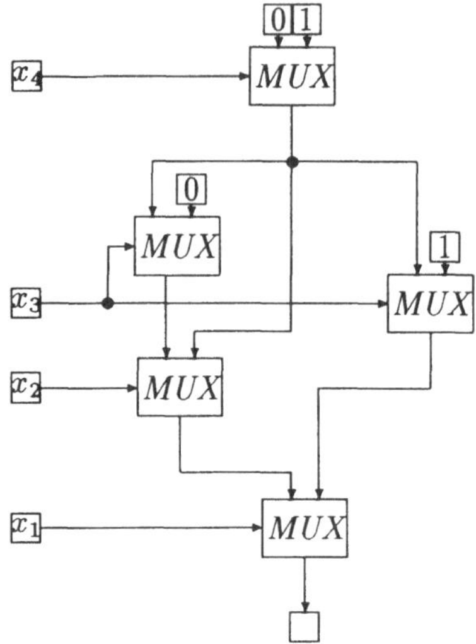

Figure 4: BDD-circuit C_{MUXLIB} over MUXLIB for the example BDD in figure 1

inputs of a multiplexer cell do not depend on the value of the control input, since we consider ordered BDD's. It follows that the control input can be set to 0 or 1 independently of the values of the northern inputs and that the controllability of the cell is completely determined, if we know which of the four possible input combinations $00, 01, 10, 11$ are applicable to the northern inputs. (The first (second) bit represents the value of the 0-input (1-input)).

In a first theorem we now show that the determination of redundancies and the computation of complete test sets with respect to CFM can be done in polynomial time for any BDD-circuit.

Theorem 1 (CFM testability) *Let C be a BDD-circuit over* MUXLIB. *The redundancies of C and a complete test set for C in* CFM *can be computed in time* $O(|C|^3)$.

Proof: Let v be any multiplexer node in the BDD-cirucit C. The set of applicable inputs is determined as follows: Compute the BDD's of the functions

$$\bar{f}_{zero(v)} \cdot f_{one(v)}, \quad f_{zero(v)} \cdot \bar{f}_{one(v)}, \quad \bar{f}_{zero(v)} \cdot \bar{f}_{one(v)}, \quad f_{zero(v)} \cdot f_{one(v)}$$

This can be done in time $O(|C|^2)$, since the BDD corresponding to C is ordered and reduced. Then solve the satisfiability problem for the resulting BDD's (in linear time) to determine the set of applicable inputs and to obtain (partially defined) primary input combinations to C which generate the applicable input values at the northern

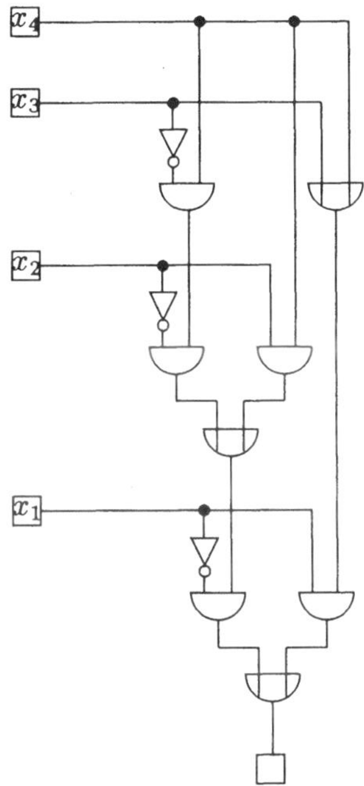

Figure 5: BDD-circuit C_{STD} over STD for the example BDD in figure 1

inputs of v. (Notice that e.g. $\bar{f}_{zero(v)} \cdot f_{one(v)}$ is satisfiable, iff 01 is applicable to the northern inputs of v.)

Now consider any cellular fault $(v, I, X/Y)$ (with v a multiplexer node, a PI or a PO). The difference X/Y can be propagated to the PO by setting the control inputs on a path from v to the PO appropriately. (Notice that these control inputs have not to be set to generate any input at v and are set exactly once on the path.) So, propagation is no problem and we get: a cellular fault $(v, I, X/Y)$ is redundant, if and only if I is not applicable. This finishes the proof. ∎

A test set constructed according to the above lemma is the strongest test set which can be obtained for a circuit derived by expansion from a given BDD-circuit over MUXLIB in any static combinational fault model. This is formalized in the following corollary whose prove follows directly from the completeness property of CFM and the relation between CFM and SAFM pointed out in section 2.

Corollary 1 *Let C be a BDD-circuit over* MUXLIB *and T a complete test set for C in* CFM. *Then T is a complete test set of any expansion of C for CFM and SAFM.*

Theorem 1 and the corollary give evidence to our claim that circuits synthesized from BDD's are easily testable. Nevertheless there may be redundancies, as indicated by the example in figure 4: it is easy to see that in the example exactly the cellular faults with northern input combination 10 at the non degenerated multiplexers are redundant. We now turn to the general question how redundancies in BDD-circuits can be classified and possibly removed.

Firstly we show that degenerated multiplexers never have redundant faults. Then we partition the non degenerated multiplexer nodes of a BDD-circuit into disjoint *redundancy classes*. A redundancy class contains all multiplexer nodes in the BDD-circuit that have identical input combinations applicable to the northern inputs of the multiplexer. Following the proof of the above theorem the redundancies at a node v are given by the set of non applicable input combinations at the northern inputs and thus two nodes in the same class have the same types of redundant faults. At a first glance one might think that there are 15 different cases for the set of applicable input combinations at the northern inputs of a non degenerated multiplexer node resulting in 15 redundancy classes. The subsequent lemma shows that the number of classes can be reduced to 6. This turns out to be the key to many of the testability properties shown in the remainder of the paper.

Lemma 1 *Let C be a BDD-circuit over* MUXLIB, v *a node of C. Then the following holds:*

i) *If v is a degenerated multiplexer with one northern input, then each value is applicable to this input, i.e. a degenerated multiplexer node has no redundancies.*

ii) *If v is a non degenerated multiplexer, then v belongs to one of the following 6 redundancy classes:*

red. class	input combinations applicable at northern inputs			
1	00	01	10	11
2		01	10	11
3	00	01	10	
4		01	10	
5	00	01		11
6	00		10	11

Proof: IF we consider a northern input of any node in the BDD-circuit, then the function computed at this node is neither the 0-function nor the 1-function. (Otherwise the BDD would not have been reduced.) So, given a northern input of a *MUX* or *DegMUX$_i$* cell, there exist primary input combinations to the BDD-circuit, which generate a 0 and a 1 at this input. For degenerated multiplexer nodes we directly get the statement of the lemma.

Now, assume that v is a non degenerated multiplexer node. One of the combinations 01 or 10 must be applicable to the northern inputs of any *MUX* cell. Otherwise the functions corresponding to the 0-input and 1-input would be identical in contradiction to the reducedness of the BDD. If 01 and 10 are both applicable, then v belongs to one of the classes 1-4. If 01 is and 10 is not applicable, v belongs to class 5 because of the following: from above we know that there exists a primary input combination which yields a 0 on the 1-input. This input combination cannot generate a 1 at the 0-input according to our assumption. So 11 is generated at the northern inputs. Similar reasoning shows that 00 can be applied. The arguments for class 6 are analogously. This proves the lemma. ∎

Nodes in class 1 are nodes with no redundant cellular faults. Nodes in class 2-6 have redundancies. We consider possibilities for the removal of these redundancies. The overall strategy is as follows. Perform a local transformation of the BDD-circuit by replacing a node with redundancies by a "simpler" cell or combination of cells which has the same behaviour for the applicable input combinations and at the same time is fully testable in CFM with the applicable input combinations.

We start with the discussion of redundancy class 5, i.e. input combinations $00, 01, 11$ are applicable to the northern inputs. Consider the circuit over STD given by figure 6. It is not difficult to check that the function of this circuit is identical

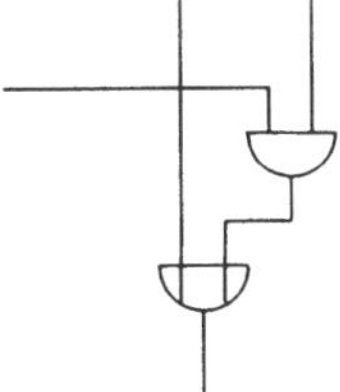

Figure 6: removal of the redundancies for nodes in class 5

to the multiplexer function up to the redundant input values (= input combinations with 10 at the northern inputs). Furthermore the circuit (as a circuit over STD) has no redundant cellular faults, even if we only allow input combinations unequal to 10 at the northern inputs. Therefore all multiplexer nodes with redundancy class 5 in a BDD-circuit C can be replaced by the circuit given in figure 6 to remove this type of redundancy without changing the functional behaviour. The resulting circuit is a *BDD-like circuit* over STD ∪ MUXLIB denoted with $C_{\{5\}}$. If this transformation is done for the BDD-circuit C in figure 4, the resulting BDD-like circuit C_5 is fully testable in CFM. Analogous arguments allow the removal of the redundancies for class 6. Thereby a node in class 6 is replaced by the circuit of figure 7.

Now consider a node v in redundancy class 4. The function computed at the 0-input is the complement of the function computed at the 1-input, i.e. one of the northern inputs, say the 1-input, is superfluous. We therefore delete the edge connected to the 1-input. If the source node of this edge has no remaining outgoing edges,

                                     Bernd Becker

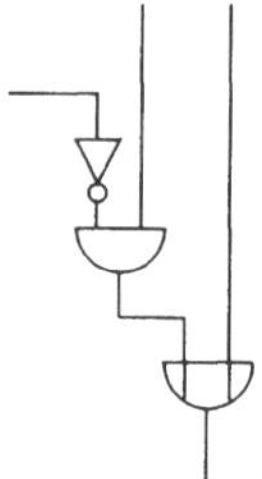

Figure 7: removal of the redundancies for nodes in class 6

the source node is also deleted. The deletion process is continued until neither edges nor nodes are superfluous. The multiplexer at node v is now replaced by an *XOR* cell. This transformation can be executed for each node in class 4. The resulting BDD-like circuit, denoted with $C_{\{4\}}$, is a CLC over MUXLIB $\cup$ $\{XOR\}$. Again the functional behaviour has not changed and redundancies have been removed.

It is easy to see that these transformations can be combined and the resulting synthesized BDD-like circuit which we denote by $C_{\{4,5,6\}}$ is a well-defined CLC over STD$\cup$MUXLIB$\cup\{XOR\}$. Again the functional behaviour has not changed compared to C. All redundancies for nodes in classes 4,5,6 have been removed. It can be shown that there unforunately is no way to remove the remaining redundancies for nodes in the classes 2 and 3 by using a strategy as indicated above. All in all, this leads to the following theorem.

Theorem 2 (Full CFM-testability) *Let C be a BDD-circuit over MUXLIB with empty redundancy classes 2 and 3. Then a BDD-like circuit over STD $\cup$ MUXLIB $\cup$ $\{XOR\}$ which computes the function of C and is fully testable in CFM can be efficiently constructed.*

We now come to the testability properties of BDD-circuits with respect to SAFM and PDFM. It turns out that most of the results can be easily deduced from the analysis of the CFM-testability. We start with the results for SAFM.

Theorem 3 (SAFM-testability) *Let G be a BDD, C_{MUXLIB} the BDD-circuit over MUXLIB, C_{STD} the BDD-circuit over STD.*

i) *C_{MUXLIB} is fully testable in SAFM.*

ii) *C_{STD} is fully testable in SAFM, if and only if C_{MUXLIB} has empty redundancy classes 5 and 6.*

iii) *A BDD-like circuit over STD which realizes the function of G and is fully testable in SAFM can be efficiently constructed.*

Proof: We give a sketch of the proof:

to i)

Since a node which is fully testable in CFM is fully testable in SAFM, it suffices to check the full SAFM-testability of multiplexer nodes in the redundancy classes 2-6. For each stuck-at fault at a node v in class 2-6 prove that there exists an applicable input combination to the inputs of v which makes the fault visible at the output of v.

to ii)

Consider a multiplexer node v in C_{MUXLIB} and the corresponding subcircuit in C_{STD}. If the multiplexer is degenerated, it follows directly that all stuck-at faults in the subcircuit are testable. Therefore consider a subcircuit which corresponds to a non degenerated multiplexer. It follows from i) that stuck-at fault on the input and output lines of the subcircuit are testable. The testability of stuck-at faults on the internal lines a, b, c, d (compare figure 2) has to be checked. A stuck-at 1 fault on c or d is equivalent to a stuck-at 1 fault on the output line of the subcircuit and therefore is testable. Stuck-at 0 faults on a, b, c, d are testable by setting the northern inputs of v to $1x$ and $x1$ ($x \in \{0, 1\}$) and combining this values with 0 and 1 at the control input. These combinations are applicable independently of the redundancy class of v. The stuck-at 1 faults on lines a and b remain to be discussed: check figure 2 to see that a stuck-at 1 fault on a (b) is testable if and only if v is not in redundancy class 5 (6).

to iii)

For the proof of theorem 2 we have constructed the BDD-like circuit $C_{\{4,5,6\}}$ over $STD \cup MUXLIB \cup \{XOR\}$. This construction removed all redundant cellular faults at nodes in redundancy classes 4,5,6. Replace each XOR cell and each cell from MUXLIB in $C_{\{4,5,6\}}$ by a realization of this cell over STD to obtain a circuit over STD. The resulting circuit is a BDD-like circuit over STD which is fully testable in SAFM. (Make sure that full CFM-testability of an XOR cell implies full SAFM-testability of the equivalent realization over STD.) ∎

Until now our investigations were based on the single fault assumption, that exactly one fault is in the circuit. In contrast to this the *multiple stuck-at fault model* (MSAFM) considers a collection of single stuck-at faults as a fault. Due to the large number of possible multiple faults (exponential in the size of the circuit) testing for multiple faults is usually not performed, even though the model is the more realistic one. We point out in the following theorem that synthesis for testability may yield circuits which are fully testable in MSAFM.

Theorem 4 (MSAFM Testability) *Let G be a BDD and C the corresponding BDD- or BDD-like circuit over STD. Then full SAFM-testability of C implies full MSAFM-testability.*

Proof: We outline the proof. Show that a multiplexer cell (as a circuit over STD) and the circuits of figures 6 and 7 are fully multiple testable in MSFAM. Then consider the root v in G. Let C_0 (C_1) be the part of C that computes $f_{zero(v)}$ ($f_{one(v)}$). Assume inductively that test sets T_0 (T_1) exist, which detect any multiple fault in C_0 (C_1). Then one can show that the following test set detects any multiple fault in C:

> Combine any test in C_0 (C_1) with a 0 (1) on the control input of v. Compute two inputs which generate 01 and 10, respectively, at the northern inputs of v and combine both inputs with 0 and 1 at the control input of v. (Notice that the existence of the inputs follows from the full SAFM-testability of C.)

∎

We finish this section by a discussion of BDD-circuits under the path delay fault model (PDFM). Let G be a BDD, C_{STD} its BDD-circuit over STD and π a path in C_{STD} from a PI x_i to the PO. Consider the subcircuit of C_{STD} corresponding to the first multiplexer v met by π. Assume that v is degenerated. Then π is uniquely determined in the subcircuit and the off-path inputs in the subcircuit can be easily set to non-controlling values by setting PI's unequal x_i appropriately. Assume that v is non degenerated. Then x_i is connected to the control input of v. There exist exactly two paths to the output of the subcircuit. On each path two transitions have to be checked. A close look at the subcircuit shows that the off-path inputs in the subcircuit can be sensitized in all cases if and only if 10 and 01 are applicable at the northern inputs, i.e. if and only if v is not in redundancy class 5 or 6. The remaining part of the path is uniquely determined by the specification of a sequence of nodes in the BDD. The path is sensitized by setting the control inputs of these nodes appropriately. The propagation vector then differs from the initialization vector exactly at position x_i. (Notice that a change in the value of x_i does not influence the off-path inputs.) Thus, we have proved the following theorem.

Theorem 5 (PDFM-testability) *Let G be a BDD, C_{MUXLIB} the BDD-circuit over MUXLIB, C_{STD} the BDD-circuit over STD. Then C_{STD} is fully testable in PDFM with robust tests, if and only if C_{MUXLIB} has empty redundancy classes 5 and 6.*

5 Conclusions and Open Problems

We have characterized the testability of BDD-circuits with respect to the cellular fault model (CFM), the single stuck-at fault model (SAFM) and the path delay fault model (PDFM).

A detailed investigation for CFM turned out to be essential for the proof of the remaining results. We showed that the computation of a complete test set T in CFM is possible in time cubic in the size of the circuit. T is universal in the sense, that it is also complete for the circuits constructed in the remaining part of the paper for CFM and SAFM as well. For a classification of the redundancies of a BDD-circuit in CFM we defined 6 disjoint classes. This classification was used to determine redundancies which are removable by a simple linear time substitution process. We finally obtained a sufficient criterion for the existence of a full testable BDD-like circuit.

The redundancy classes were further used to characterize the full testability in SAFM and to finally obtain fully testable BDD-like circuits for SAFM and MSAFM. We finished by showing that the results for CFM can also be applied to characterize the BDD-circuits which are fully robustly testable.

Our results demonstrate that BDD-circuits are easily testable and in many cases can be modified to be fully testable for different fault models which together cover a large variety of possible physical defects.

Nevertheless some questions are left open in this paper: No modification was given to obtain full testability in CFM at nodes in redundancy classes 2 and 3. Is there any possibility to obtain full testability with different methods of modification? In the stuck-at model we proved the full testability even in the presence of multiple faults. A similar result for cellular faults is not given. The existence of nodes in redundancy classes 5 and 6 implies the existence of non testable faults in PDFM. A modification that yields full robust testability in PDFM is not given. One might think that the BDD-like circuit over STD, which results from the BDD-circuit by modification of the nodes in classes 5 and 6 and which is fully testable in SAFM, is also fully testable in PDFM. A look at the circuit derived from the example BDD proves that this is not the case.

The circuits considered in this paper resulted from **reduced, ordered** BDD's. These restrictions may lead to BDD's of large size (possibly exponential in the number of variables). It is an interesting question if the results can be generalized (in modified form) to other classes of BDD's. It is clear for example that many arguments given in this paper remain valid if the BDD's are *read once only*, i.e. if the indices on any path in the BDD are pairwise disjoint.

Bibliography

[1] S. B. Akers. Binary decision diagrams. *IEEE Transactions on Computers*, C-27(6):509–516, June 1978.

[2] S. B. Akers. Functional testing with binary decision diagrams. In *9th Int. Symposium on Fault Tolerant Computing*, pages 75–82, 1979.

[3] P. Banerjee and J.A. Abraham. Generating tests for physical failures in MOS logic circuits. In *Proc. IEEE Int. Test Conference*, pages 554–559, 1983.

[4] C.C. Beh, K.H. Aray, C.E. Radke, and K.E. Torku. Do stuck-at fault models reflect manufacturing defects? In *Proc. IEEE Int. Test Conference*, pages 35–42, 1982.

[5] B. Becker, T. Burch, G. Hotz, D. Kiel, R. Kolla, P. Molitor, H.G. Osthof, G. Pitsch, and U. Sparmann. A graphical system for hierarchical specifications and checkups of VLSI circuits. In *Proc. of the 1st European Design Automation Conference (EDAC90)*, pages 174–179, 1990.

[6] M.A. Breuer and A.D. Friedman. *Diagnosis & reliable design of digital systems.* Computer Science Press, 1976.

[7] B. Becker, G. Hotz, R. Kolla, P. Molitor, and H.G. Osthof. Hierarchical design based on a calculus of nets. In *Proceedings of the 24th Design Automation Conference*, pages 649–653, June 1987.

[8] K. S. Brace, R. L. Rudell, and R.E. Bryant. Efficient implementation of a BDD package. In *Proc. 25th Design Automation Conference*, pages 40–45, 1990.

[9] R.E. Bryant. Graph-based algorithms for boolean function manipulation. *IEEE Transactions on Computers*, C-35(8):677–691, 1986.

[10] B. Becker and U. Sparmann. A uniform test approach for RCC-Adders. In *Proc. of the 3rd Aegean Workshop on Parallel Computation and VLSI Theory, Lecture Notes in Comput. Sci. 319*, pages 288–300, 1988.

[11] S. Devadas, H.K.T. Ma, R. Newton, and A. Sangiovanni-Vincentelli. Optimal logic synthesis and testability: two faces of the same coin. In *Proc. IEEE Int. Test Conference*, pages 4–12, 1988.

[12] F.J. Ferguson and J.P. Chen. Extraction and simulation of realistic CMOS faults using inductive fault analysis. In *Proc. IEEE Int. Test Conference*, pages 475–484, 1988.

[13] A.D. Friedman. Easily testable iterative systems. *IEEE Transactions on Computers*, C-22:1061–1064, December 1973.

[14] H. Fujiwara. *Logic Testing and Design for Testabilty*. The MIT Press, 1985.

[15] J. Galiay, Y. Crouzet, and M. Vergniault. Physical versus logical fault models MOS LSI circuits: impact on their testability. *IEEE Transactions on Computers*, C-29(6):527–531, June 1980.

[16] P. Lamoureux and V.K. Agarwal. Non-stuck-at fault detection in NMOS circuits by region analysis. In *Proc. IEEE Int. Test Conference*, pages 129–137, 1983.

[17] C.Y. Lee. Representation of switching circuits by binary decision programs. *Bell Systems Technical Journal*, 38:985–999, 1959.

[18] W. Maly. Realistic fault modeling for VLSI testing. In *Proc. 24th Design Automation Conference*, pages 173–180, 1987.

[19] C. Meinel. *Modified Branching Programs and Their Computational Power. volume 370 of LNCS*, Springer Verlag, 1989.

[20] B. Milne. Testability, 1985 technology forecast. *Electronic Design*, 10:143–166, 1985.

[21] M.H. Schulz, F. Fink, and K. Fuchs. Parallel pattern fault simulation of path delay faults. In *Proc. of the 26th Design Automation Conference*, June 1989.

[22] G.L. Smith. A model for delay faults based upon paths. In *Proc. of IEEE Int. Test Conference*, pages 342–349, September 1985.

[23] R.L. Wadsack. Fault modeling and logic simulation of CMOS and MOS integrated circuits. *The Bell System Technical Journal*, 57, 1978.

[24] T.W. Williams, W. Daehn, M. Grützner, and C.W. Starke. Comparison of aliasing errors for primitive and non-primitive polynomials. In *Proc. IEEE Int. Test Conference*, pages 282–288, 1986.

[25] I. Wegener. *The Complexity of Boolean functions*. Wiley Teubner, 1987.

[26] J.A. Waicukauski, E. Lindbloom, B. Rosen, and V. Iyengar. Transition fault simulation by parallel pattern single fault propagation. In *Proc. IEEE Int. Test Conference*, pages 542–549, September 1986.

[27] H.J. Wunderlich. *Hochintegrierte Schaltungen: Prüfgerechter Entwurf und Test*. Springer Verlag, 1991.

Ähnlichkeit von Grammatiken -
Ansätze und Erfahrungen

Eberhard Bertsch

Praktische Informatik
Ruhr – Universität
4630 Bochum
Germany

Zusammenfassung

Es wird erörtert, unter welchen Umständen die Ersetzung der Syntax einer Sprache durch eine äquivalente Syntax zu praktisch nutzbaren Ergebnissen führen kann. Die betrachteten Verfahren werden insbesondere anhand der Ausdrucks-Syntaxen von PASCAL und MODULA-2 konkretisiert. Unter Rückgriff auf frühere theoretische Ergebnisse werden Perspektiven für den Ausbau der einfachen Ansätze aufgezeigt.

1 Überblick

Dieser Beitrag beschreibt in den ersten Abschnitten ein einfaches Konzept zur Umformung vorgegebener Grammatiken, theoretische Eigenschaften dieses Konzepts, an denen grundsätzliche Möglichkeiten und Grenzen deutlich werden, sowie praktische Erfahrungen und Meßergebnisse.

In einem weiteren Abschnitt wird der Bezug zu Arbeiten von G. Hotz und Mitarbeitern aus den sechziger und siebziger Jahren aufgezeigt, die in allgemeinem Sinne die Verwandtschaft äquivalenter Grammatiken zum Gegenstand hatten. Die Ausrichtung auf Effizienzfragen ist in diesen frühen Arbeiten bereits vorhanden; es läßt sich jedoch ausführen, daß ein konkreter Einsatz der dort angegebenen Techniken erst bei großzügigem Umgang mit dem Abtausch von Platz und Zeit sinnvoll wird. Anders ausgedrückt: Erst die heutigen Rechnersysteme gestatten die Ausnutzung dieser seit langem bekannten Ansätze für Zwecke der Compiler–Generierung oder a fortiori der Behandlung natürlicher Sprachen.

Die zunächst ausschließlich manuellen (durch interaktive Editing–Routinen unterstützten) Transformationsverfahren auf Grammatiken legten die Forderung nach geeigneten Heuristiken zur automatischen oder halbautomatischen Transformation nahe.

In einer Diplomarbeit wurden solche Heuristiken entwickelt und mit erstaunlichem Erfolg angewendet. Dabei ist anzumerken, daß die verfügbaren Zeit- und Platz–Ressourcen erheblich gesteigert werden müßten, um diese Verfahren beispielsweise für übliche Programmiersprachen–Syntaxen voll auszufahren. Wir geben hier die wesentlichen Merkmale der behandelten Vorgehensweise an.

Beim experimentellen Arbeiten mit Grammatiken wird erkennbar, daß die Verwendung von Standardtechniken für die syntaktische Analyse eine Einschränkung eigener Art darstellt. In welcher Weise und in welchem Umfang das Variieren geeigneter Parameter der Analysatorprozedur – bei gleichbleibender Syntax einer vorgegebenen Sprache – ebenfalls experimentelle Erkenntnisse gestattet, wurde bisher kaum untersucht. Einige Vorschläge hierzu werden andeutungsweise diskutiert.

In dem vorliegenden Artikel wird der Versuch unternommen, bis an den Rand des Vertretbaren durchgängig informell zu argumentieren. Ein Grund hierfür ist ganz einfach das Bestreben, die Bereitschaft zum tatsächlichen Durchlesen bei Nichtspezialisten etwas zu fördern.

2 Umformung von Regeln

Zu jeder kontextfreien Grammatik G_1 lassen sich beliebig viele andere Grammatiken G_2, G_3,... angeben, deren erzeugte Sprachen $L(G_2)$, $L(G_3)$,... mit der Sprache $L(G_1)$ übereinstimmen. Man spricht von (schwacher) Äquivalenz zweier Grammatiken G_1 und G_2, wenn $L(G_1) = L(G_2)$ gilt.

Von besonderem Interesse sind solche Grammatiken G, die folgende Forderungen erfüllen:

G ist äquivalent zu G_1.

G ist auf einfache Weise aus G_1 zu gewinnen.

G gestattet deterministische Syntaxanalyse, genügt also beispielsweise den LALR(1)-Kriterien.

G führt in allen oder zumindest in vielen Fällen zu schnelleren Analysen als G_1.

G erfordert bei der Syntaxanalyse weniger oder zumindest nicht viel mehr Speicherplatz als G_1.

Diese Forderungen stellen einerseits eine willkürliche, starke Einschränkung der in diesem Umfeld insgesamt möglichen Problemstellungen dar, sind jedoch andererseits in der vorliegenden Formulierung ausgesprochen unscharf. Zu brauchbaren Erkenntnissen wird man nur kommen, wenn die erforderlichen näheren Präzisierungen dieser Begriffe den praktischen Gegebenheiten und Möglichkeiten gerecht werden.

Zunächst wenden wir uns der Einfachheits-Forderung zu. Eine einfache Gewinnung von G liegt sicher dann vor, wenn im wesentlichen eine einzige Regel von G_1 geändert werden muß.

Wir betrachten deshalb Veränderungen einer Grammatik G_1 von folgender Art:

Sei $A \to w\,B\,v$ eine Regel, wobei A , B Zeichen aus dem nichtterminalen Alphabet und w , v beliebige (eventuell leere) Strings seien.

Andererseits seien $B \to b_1$, $B \to b_2$, $\ldots$, $B \to b_n$ (abgekürzt: $B \to b_1 \mid b_2 \mid \ldots \mid b_n$) alle diejenigen Regeln, die B als linke Seite haben.

Dann läßt sich eine zu G_1 äquivalente Grammatik $G^{(A,w,B,v)}$ durch Entfernen der Regel $A \to w\,B\,v$ und Hinzufügen neuer Regeln $A \to w\,b_1\,v \mid w\,b_2\,v \mid \ldots \mid w\,b_n\,v$ konstruieren.

Das Entfernen von $A \to w\,B\,v$ dient der Erhaltung der Eindeutigkeit der Grammatik. Der wichtige Fall $A = B$ (bei einer direkt rekursiven Regel) bedarf keiner definitorischen Sonderbehandlung.

Da nur eine Regel weggenommen wird, aber n neue hinzukommen, wird die Grammatik bei Anwendung solcher Veränderungen offenbar umfangreicher. Welche Auswirkungen auf die Größe zugehöriger Analysetafeln vorliegen, ist weniger offensichtlich.

Grundsätzlich führt das Ersetzen von $A \to w\,B\,v$ und $B \to b_i$ durch $A \to w\,b_i\,v$ zur Verkürzung von Pfadlängen in Ableitungsbäumen. Dieser Effekt hat enorme praktische Konsequenzen, wenn die betroffenen Regeln in sinnvoller Weise ausgewählt werden.

Der Aspekt des Verkürzens gewisser Pfade gibt Anlaß, solche Operationen als "Fusion" der Regeln und darüber hinaus der Grammatik zu bezeichnen. In [5] wird der Begriff der Fusion (unter Verwendung einer anderen graphischen Intuition) mit "Flattening" wiedergegeben.

3 Deterministische Analysierbarkeit

Für die nun eingeführte Art der Veränderung von Grammatiken stellt sich die Frage nach der deterministischen Analysierbarkeit:

Falls G_1 wünschenswerte Eigenschaften hinsichtlich der Analysierbarkeit besitzt, was läßt sich dann über die Gültigkeit dieser Eigenschaften bei der abgeänderten Grammatik G aussagen ?

Der Artikel [5] behandelt diese Frage bezüglich der LL(1)- und LR(1)-Eigenschaften. Wir wollen nicht nur die Resultate, sondern auch einige der zu ihrer Begründung erforderlichen Schritte wiedergeben, verzichten hier jedoch auf eine formale Beweisführung.

Die Klasse der LL(1)-Grammatiken ist nicht abgeschlossen unter Fusionen.

Wenn an der Stelle eines nichtterminalen Zeichens, das nicht am linken Ende einer rechten Regelseite steht, die verschiedenen Alternativen zu dessen Ersetzung neue rechte Regelseiten bilden, dann ist die bei der LL(1)-Analyse erforderliche Eindeutigkeit der Auswahl einer Regel zur weiteren Ableitung nicht mehr gegeben.

Andererseits läßt sich zeigen, daß jegliche Fusion an linkester Stelle (leftmost flattening nach [5]) die zuvor vorhandene LL(1)-Eigenschaft nicht zerstört. Unter Berücksichtigung der üblichen Fallunterscheidungen für leere und nichtleere Anfangsworte erkennt man, daß die eindeutige Auswahl aufgrund eines einzigen vorausgelesenen Zeichens in der vergrößerten Regelmenge weiterhin möglich ist.

Auch die Klasse der LR(1)-Grammatiken ist nicht abgeschlossen unter Fusion.

In folgender Weise läßt sich ein Gegenbeispiel angeben: Zwei verschiedene rechte Seiten zu einem nichtterminalen Zeichen der Ausgangsgrammatik seien in einem Anfangsstück gleich. Bei der Analyse entscheide sich somit erst im Weiterlesen, um welche Regel es sich handelt. Wenn nun ein nichtterminales Zeichen in dem gemeinsamen Anfangsstück nur in einer der beiden betrachteten Regeln durch zugehörige rechte Seiten ersetzt wird, führt die entstehende Grammatik zu einem Shift–Reduce–Konflikt.

Bei der Untersuchung dieser Erhaltungseigenschaften bestand die Hoffnung, zeigen zu können, daß das angegebene Gegenbeispiel die einzige Art der Verletzung der LR(1)-Eigenschaft kennzeichnet, daß also sozusagen der Raum aller möglichen Gegenbeispiele durch einen Grundtypus aufgespannt wird. Die Erwartung hat sich als falsch erwiesen, da auch anders geartete, wenngleich wesentlich kompliziertere Fälle zu einem Verlust der eindeutigen Analysierbarkeit führen.

Diese Überlegungen haben unmittelbare praktische Relevanz.

Bei den weiter unten beschriebenen Experimenten mit Varianten der MODULA–2–Syntax war die Gewinnung einer optimierten Regelmenge davon abhängig, ob die experimentell als notwendig erkannten Fusionen ohne Verlust der eindeutigen Analysierbarkeit tatsächlich durchführbar sein würden. Erfreulicherweise – und erstaunlicherweise – hat sich sogar die noch stärker einschränkende sLR(1)-Eigenschaft unter den vorgenommenen Veränderungen als äußerst robust erwiesen. Diese Tatsache bietet eine gewisse Entsprechung zu der Feststellung, daß es schwierig war, auf der rein theoretischen Ebene Gegenbeispiele zu finden.

Weiterhin gilt, daß die Klasse der LR(1)–Grammatiken abgeschlossen ist unter Fusion an rechtester Stelle. Es hat den Anschein, daß eine Art Dualität zur Abgeschlossenheit der LL(1)–Klasse unter linkesten Fusionen vorliegt. Allerdings verlangt der Beweis eine völlig andere Vorgehensweise. Überhaupt ist der definitorische und beweistechnische Aufwand bei der Betrachtung von LR–Grammatiken durchgängig größer als der für LL–Grammatiken erforderliche.

Es ist zu zeigen, daß jede Verletzung der LR(1)–Eigenschaft in der fusionierten Grammatik auf eine schon zuvor bestehende Verletzung in der ursprünglichen Grammatik zurückgeht. Anhand der allgemeinen (nicht an das Erzeugungsverfahren für LR–Zustände geknüpften) Knuth'schen Definition der LR(k)–Eigenschaft [14] unterscheiden wir die Möglichkeiten des Shift–Reduce– und des Reduce–Reduce–Konflikts, wobei die unübersichtlichste Situation darin besteht, daß in einem Reduce–Reduce–Konflikt eine alte Regel und eine neue, durch Fusion entstandene an gleicher Stelle angewendet werden können.

Schwierig ist hier nun der Fall, daß gerade die alte Regel auch zur Fusion diente. Die genauere Fallunterscheidung verlangt eine Betrachtung der letzten Schritte der Rechtsableitung vor dem Auftreten des Konflikts. Sowohl für das nichtrekursive wie für das rekursive Fusionieren läßt sich dann leicht einsehen, daß alle Konfliktsituationen auch schon in der alten Grammatik vorlagen. Die Argumentation hängt wesentlich davon ab, daß Fusionen nur am rechten Ende von rechten Seiten auftreten. Die oben angegebene Konstruktion von Gegenbeispielen macht ja gerade von der Nichterfüllung dieser Forderung Gebrauch.

4 Experimente zur Ausdrucks-Syntax von PASCAL

Wir wollen nun die Effizienzforderung an veränderte Grammatiken näher untersuchen. Knapp ausgedrückt besagt sie, daß eine fusionierte oder sonstwie unter Erhaltung qualitativer Eigenschaften gewonnene neue Grammatik schnellere Parsing–Vorgänge gestattet als die zuvor vorhandene und daß sich der eventuell zusätzlich nötige Speicherplatz in erträglichen Grenzen hält.
Es dürfte klar sein, daß die so vorgegebene Fragestellung nur bei extremer Spezialisierung des betrachteten Sachverhalts zu Antworten führen kann.

In einer 1985 veröffentlichten Untersuchung [4] wählten wir einen kleinen, aber wichtigen Ausschnitt der PASCAL-Syntax, nämlich die Regeln für

`factor`, `set`, `element`, `term`, `simple-expression` und `expression`.

`expression` dient als Startsymbol.

Da die Syntax des PASCAL-Standards mit geringfügigen Anpassungen dem Recursive–Descent–Verfahren zugänglich ist und diese Eigenschaft auch nach strukturellen Veränderungen behalten soll, bieten sich die LL(1)–erhaltenden linkesten Fusionen unmittelbar an.

Dabei lassen sich durch Iteration des Konstruktionsverfahrens Grammatiken G_1, ...,G_6 erzeugen, deren Speicherbedarf bis zum vierfachen des zuvor erforderlichen ansteigt. Wegen der extremen Einfachheit der Ausgangssyntax handelt es sich bei üblicher Darstellung hier immer noch um wenige Kilobytes, sodaß wir hinsichtlich des Merkmals Speicherbedarf keine Überlegungen zur Optimierung anstellen mußten. Für die Experimente stand ein Sirius/Victor–Mikrorechner unter MS–DOS zur Verfügung.

Wie sollen unterschiedliche Grammatiken bezüglich ihrer Analysezeiten verglichen werden?

Man stellt sehr schnell fest, daß brauchbare Aussagen nur bei genauer Kenntnis der tatsächlichen Nutzung grammatischer Konstrukte möglich sind. Im Hinblick auf PASCAL–Programme war hierbei das Buch von Carter [7] eine Hilfe, in dem statistische Daten über die äußere Form von PASCAL–Programmen aus Hochschulen und Industrie zusammengestellt sind.

Die folgenden Arten von Ausdrücken wurden als typisch angesehen:

(a)   v * v * ... * v

(b)   [v , v , ... , v]

(c)   v + v + ... + v

(d)   (v < v) and ... and (v < v)

(e)   not v or not v or ... or not v

(f)   ((... ((v + v) * v + v) ...) * v + v)

Ein naheliegender Ansatz zur Messung von Analysezeiten ist nun, für die Grammatiken G_0 bis G_6 (G_0 ist die unveränderte PASCAL–Syntax) mit den Ausdrucksarten (a) bis (f) und verschiedenen Iterationsstufen innerhalb jeder Art Analyseläufe durchzuführen. Die Größenordnung von hundert Mikrosekunden wurde als ausreichende Genauigkeit der Durchschnittswerte für jede solche Messung angesehen.

Um die Grammatiken zu vergleichen, ist weiterhin eine Vorschrift für die Gewichtung der Messungen erforderlich. Es bietet sich an, alle Iterationsstufen der Ausdrucksstruktur bis zu einer festen oberen Grenze entweder gleich hoch oder gleichmässig abnehmend zu gewichten. Die letztere Methode läßt sich durch Carters Daten zur Häufigkeit arithmetischer Ausdrücke in Abhängigkeit von deren Länge stützen.

Wir verzichten hier auf die präzise Angabe der in [4] verwendeten Berechnungsvorschriften. Die absoluten Zahlen sind ohnehin außer durch die bisher genannten Kriterien auch durch die Wahl des Rechners, des Betriebssystems, der Implementierungssprache (MS–PASCAL) und der Interndarstellung des tabellengesteuerten Topdown–Parsers beeinflußt.
Die experimentellen Ergebnisse lassen sich wie folgt verbal zusammenfassen:

Wenn der erhöhte Speicherbedarf nicht als nachteilig in die Effizienzberechnung einbezogen wird, so ergibt sich bezüglich des mittleren Zeitbedarfs eine Verringerung um etwa ein Drittel der Analysezeit beim Übergang von G_0 zu G_6, und zwar einigermaßen unabhängig von der Gewichtungsweise längerer Ausdrücke. Bei der geringen "Sophistication" des gewählten Ansatzes ist dies ein unerwartet gutes Ergebnis.

Für die Einbeziehung des Speicherbedarfs in ein kombiniertes Effizienzmaß wurden mehrere Möglichkeiten diskutiert. Wenn Speicherplatz als teuer gewertet wird, ist ein extremer Gebrauch von Fusions-Operationen zur Veränderung der Syntax geradezu schädlich. Sogar unter solchen Bedingungen führen einzelne, besonders vorteilhaft gewählte Fusionen jedoch zu besseren Bewertungszahlen für die veränderten Grammatiken.

5 Die Untersuchungen zu den Syntaxen von MODULA-2 und COBOL

Die Untersuchung zur MODULA-2-Syntax umfaßte unter anderem die Konstruktion der sLR(1)-Analysetafeln [1] für einige Dutzend äquivalente Varianten dieser Syntax und die Messung der zugehörigen Analysezeiten für bekannte Programme aus dem Lehrbuch [17] von N. Wirth.

Im Unterschied zur PASCAL-Syntax für Ausdrücke wurden hier alle syntaktischen Regeln des Sprachdokuments berücksichtigt. Dabei zeigte sich, daß die systematische Konstruktion von Programmen analog zum vorigen Abschnitt keine klare empirische Basis hatte. Dies führte zur Auswahl von zehn, als hinreichend typisch und unterschiedlich betrachteten Programmen aus [17].

Das Buch ist kein Einführungstext in die Programmierung, in dem möglicherweise aus didaktischen Gründen bestimmte Sprachmerkmale bevorzugt behandelt werden müßten. Stattdessen stehen Such- und Sortierverfahren und die Manipulation dynamischer Daten im Vordergrund.

Für die Ausführung der Tafelkonstruktion und die Messung der Analysezeiten wurde eine Micro-VAX II unter VMS verwendet.

Wir geben als Beispiele mehrere Regeln an, die sich durch eine oder mehrere Fusionen aus den vorgegebenen MODULA-2-Regeln gezielt konstruieren ließen, wobei zu beachten ist, daß jeweils viele weitere Regeln sozusagen als Seiteneffekte entstanden und ihrerseits wesentlich zur Änderung der Analysezeiten beitrugen. Weitere Beispiele finden sich in [6].

Zu jeder solchen Regel geben wir den minimalen und den maximalen Zeitgewinn in Prozenten bei der Analyse aller zehn Testprogramme bezogen auf die Analysezeiten derselben Programme mit der ursprünglichen Syntax an.

Konstruierte Regel(n):   Zeitgewinn (min, max):

```
type        → qualident
```
(0.0 %,   0.7 %)
```
IfStatement → IF expression THEN
              assignment END
```
(0.3 %,   3.1 %)

```
expression  →  integer | − integer        (10.6 %,   23.5 %)
expression  →  qualident + integer
             | qualident − integer         (18.9 %,   35.4 %)
```

Bei der Konstruktion mehrerer gezielt gewählter Regeln nacheinander läßt sich der Zeitgewinn unter diesen Bedingungen noch um einige Prozentpunkte erhöhen. Eine Syntax, die in [6] vollständig angegeben ist, liefert die Zeitgewinne (24.4 %, 39.2 %) bei gleichzeitiger Steigerung des Platzbedarfs um knapp 11 %.

Es sei auch erwähnt, daß wir in einer unveröffentlichten Zusatzstudie, die ausschließlich Ausdrücke behandelt, mit denselben Methoden Zeitgewinne von bis zu 63.8 % erhalten haben. Bei bestimmten Analysen wird also nur ein Drittel der ursprünglich erforderlichen Zeit aufgewendet. Ein Haken dabei ist der reale Platzzuwachs in den Parsertabellen von 617 % (!).

Um die Meßverfahren nicht auf die immerhin sehr ähnlichen Programmiersprachen PASCAL und MODULA–2 zu beschränken, wurde auch COBOL in allerdings geringerem Umfang herangezogen. In typischen COBOL–Programmen spielen die bei den zuvor betrachteten Sprachen ganz zentralen arithmetischen und sonstigen Ausdrücke eine verhältnismäßig weniger wichtige Rolle. Es wurde aus den Angaben der Sprachdokumente eine kontextfreie Grammatik aufgestellt. Wir machen auch hier einige Angaben über konstruierte Regeln und die dabei erzielten minimalen und maximalen Zeitgewinne. Gemessen wurden sechs Programme mit einfachen betriebswirtschaftlichen Funktionen.

(Das Symbol H bezeichnet eine Hilfsvariable, die bei der Erzeugung der Grammatik erforderlich war.)

Konstruierte Regel(n):   Zeitgewinn (min, max):

```
statement → MOVE ident TO ident        ( 3.1 %,   18.6 %)
H → ident , ident                      ( 1.5 %,   14.0 %)
Bed → ident > Zahl                     ( 0.0 %,   18.5 %)
```
Alle drei angegebenen Regeln zusammen   (20.3 %,   27.9 %)

Der Platzzuwachs in den LR(1)–Tafeln betrug bei keiner vorgenommenen Erzeugung über 35 %.

6   Frühere theoretische Aussagen zur strukturellen Ähnlichkeit von Grammatiken

Das Konzept der Fusion kontextfreier Regeln soll nun in einen allgemeineren Zusammenhang gestellt werden.

Ausgehend von der nicht entscheidbaren schwachen Äquivalenz von Grammatiken liegt es nahe, nach Einschränkungen zu suchen, die insbesondere die Eigenschaft der Entscheidbarkeit aufweisen. Eine eingeschränkte Form der Äquivalenz liegt dann vor, wenn neben der Sprachengleichheit eine surjektive Abbildung von der Menge der Ableitungen der einen Grammatik in diejenige der anderen Grammatik existiert und konstruktiv angegeben werden kann.

Dieser von G. Hotz begründete Ansatz [11, 12, 13, 16] führte zu einer Reihe von Entscheidbarkeitsaussagen, die jeweils das Erfülltsein zusätzlicher Anforderungen an die Gestalt der Grammatiken oder aber der Bilder einzelner Regeln voraussetzen. Zur Darstellung von Ableitungsmengen wurden in diesen Arbeiten monoidale Kategorien verwendet, die eine besonders knappe Schreibweise beim Rechnen mit baumwertigen Ausdrücken erlauben.

Beispielsweise konnte gezeigt werden:

- Für zwei lineare Grammatiken ist die Existenz einer surjektiven Abbildung entscheidbar [12].

- Für zwei kontextfreie Grammatiken ist die Existenz einer längenerhaltenden oder –reduzierenden surjektiven Abbildung entscheidbar [15].

- Für zwei kontextfreie Grammatiken ist die Existenz einer surjektiven Abbildung, die alle Bäume auf (möglicherweise längere) Bäume abbildet, entscheidbar [2].

Die allgemeinste Form der Abbildung, bei der einzelne Symbole auch zu Zeichenketten der Breite größer als 1 werden können, wurde nicht behandelt. Es erscheint jedoch plausibel, daß die Vorgehensweise der Urbildkonstruktion zu unterschiedlichen Positionen der Bildbäume, die einen zentralen Bestandteil der Beweisführung in [2] darstellt, auch auf Ableitungen mit konstant beschränkter Breite anwendbar ist. Das Zusammensetzen von Urbildern unterschiedlicher Breite ist technisch aufwendig, dürfte aber zu keinen schwerwiegenden Problemen führen. Als Korollare ergaben sich jeweils auch konstruktive Verfahren zur Gewinnung minimaler, im betrachteten Sinne strukturell verwandter Grammatiken.

In [3] wurde die konzeptionelle Nähe zwischen dem soeben vorgestellten Ansatz und dem Konzept der "Grammatikform" [8], einer Art Schablone zur zusammenfassenden Beschreibung ganzer Grammatikklassen, aufgezeigt.

Des weiteren war für längenerhaltende Abbildungen die Aussage möglich, daß zu jedem syntaktischen Analyseverfahren mithilfe der Bildgrammatik ein (bis auf konstante Faktoren) gleich schnelles oder schnelleres Analyseverfahren mithilfe der Urbildgrammatik existiert.

Der erwähnte Satz ist nun auch insofern interessant, als er einen Sachverhalt wiedergibt, der in ganz ähnlicher Weise auf die oben diskutierten Varianten von Programmiersprachensyntaxen zuzutreffen scheint. Durch Fusion erhält man erfahrungsgemäß deutlich schnellere Grammatiken. Eine Fusion läßt sich aber gerade als Konstruktion eines Urbildes für einen aus zwei Regeln zusammengesetzten Teilbaum der Ausgangsgrammatik auffassen, die dabei als Bildgrammatik dient. Die so gegebene Abbildung läßt sich in naheliegender Weise auf die Menge aller Ableitungsbäume der neuen Grammatik fortsetzen.

Hingegen ist es nicht möglich, die bei der Fusion entstandene Grammatik als Bildgrammatik im Sinne von [2] zu verstehen, da alle Bildbäume aus den Bildern einzelner Regeln der noch nicht fusionierten Grammatik zusammengesetzt werden müßten.

Ein Unterschied zu der in [3] betrachteten Situation besteht freilich darin, daß wir es bei den Messungen fusionierter Grammatiken durchweg mit einer Zeitkomplexität in Größenordnung $\mathcal{O}(n)$ zu tun haben. Die Verbesserungen betreffen also konstante multiplikative Faktoren. Eine eher plakative Darstellung der seinerzeit gewonnenen Erkenntnisse lautete, daß beim Übergang zu homomorphen Bildern von Ableitungen Information verlorengeht, deren Fehlen dann eben aufwendigere Fallunterscheidungen zur Laufzeit eines Parsers nach sich zieht. Diese Formulierung ist sicher viel zu vage, um die Vorgänge beim Fusionieren – beispielsweise das Ersetzen von Reduce-Aktionen durch Shift–Aktionen in LR(1)-Tabellen – als konkrete Spezialfälle ausweisen zu können.

Es dürfte deutlich geworden sein, daß Fusionen sehr spezielle Operationen sind und daß der Spielraum für praktisch nutzbare strukturelle Ähnlichkeiten zwischen Grammatiken durch den umfangreichen Einsatz des Fusionskonzepts keineswegs ausgeschöpft, sondern allenfalls angedeutet worden ist. Eine konkrete Implementierung der früher untersuchten Minimierungsverfahren wird im Rahmen der Diplomarbeit von Frau Sabine Zentarra [18] in Angriff genommen.

7 Automatisierte Techniken zur Effizienzsteigerung bei Grammatiken

Nachdem die Messungen auf der Micro–VAX ergeben hatten, daß einige durch interaktives Editieren erhaltene Syntaxvarianten zu erheblich verschnellerten Analysen führen, stellte sich die Frage, ob eine zweckdienliche Veränderung von Grammatiken auch ohne menschlichen Eingriff, also allein aufgrund von vorhandenen Teststrings und deren Ableitungsbäumen möglich wäre. Diese Frage bildete den Rahmen für die Diplomarbeit von E. Heckert [10].

Ausgangspunkt für die einschlägigen Heuristiken ist die Erwartung, daß besonders häufig vorkommende Teile der Grammatik bei Einbeziehung in Fusionen zu auffälligen Verbesserungen der durchschnittlichen Analysezeit führen müßten. Für die Häufigkeitsberechnung von Bestandteilen der konkret vorkommenden Ableitungsbäume bestehen mehrere alternative Möglichkeiten, die sich nach der Größe des jeweils als elementar betrachteten Kontextes gliedern lassen.

Zeichenbezogene Häufigkeiten lassen sich verfeinern durch regelbezogene Häufigkeiten und schließlich durch Häufigkeiten auf der Grundlage kleiner Teilbäume. Dabei ist wichtig und interessant, daß die Reihenfolge der einzelnen Aufbauschritte beim Transformieren von Teilbäumen Einfluß auf die Größe der entstehenden Regelmenge ausübt. Heckert gibt theoretische Abschätzungen der Regelanzahl an.

Als Ergebnis dieser Untersuchung ist festzuhalten, daß die auf automatischem Wege gewonnenen Syntaxvarianten bezüglich ihrer Zeiteffizienz weitestgehend besser sind als die zuvor manuell interaktiv erzeugten. Ganz generell wird von den Experimenten die intuitive Annahme bestätigt, daß mehr Speicherplatz zu entsprechenden Verringerungen der Analysezeit führt.

Erkenntnisse darüber, welche absoluten Häufigkeitsschwellen für die Auswahl einer zu fusionierenden Regel bei einer bestimmten angestrebten Verschnellerung (in Prozent) den geringsten Speicherzuwachs erfordern, waren nicht möglich. Dies überrascht aber kaum, da so komplizierte Parameter wie Grammatiken und MODULA-Programme kein flexibles Einstellen erlauben.

8 Einige allgemeine Überlegungen

Da der Fusionsbegriff in den erwähnten Arbeiten fast ausschließlich zum Zweck der zeitlichen Effizienzsteigerung eingesetzt wurde, könnte der Blick für andere Zusammenhänge etwas verstellt worden sein.

Eine Grammatik muß vielfältige Adäquatheitsforderungen zur Darstellung einer potentiell unendlichen Menge von Wörtern erfüllen.

Sowohl bei Programmiersprachen als auch bei natürlichen Sprachen wird meistens zwischen einer syntaktischen Verarbeitung im engeren Sinne und einer lexikalischen Analyse unterschieden. Die Möglichkeit, äquivalente Grammatiken als praktische Alternativen aufzufassen, erlaubt nun einen erheblichen Spielraum bei der Abgrenzung zwischen lexikalischer und syntaktischer Komponente im konkreten Einzelfall.

Wenn die lexikalische Analyse über reguläre Ausdrücke abgewickelt werden soll, so läßt sich die Optimierungsaufgabe formulieren, in möglichst sinnvoller Weise eine Zerlegung einer äquivalenten Grammatik in einseitig lineare (Chomsky–Typ–3) und andere (Chomsky–Typ–2) Regeln zu finden.

Dieses Beispielproblem mag Anlaß bieten, überhaupt den Bestand an möglichen Datenstrukturen und Verfahrensschritten beim Einsatz von Grammatiken zu sichten, um zu einem angemessenen und guten Entwurf bei der Implementierung einer vorgegebenen Sprache zu gelangen.

Neben dem Variieren von Grammatikregeln, das sicherlich zu den eher theoretisch fundierten Methoden zu rechnen ist, kommen dann ausgesprochene Ad–hoc–Verfahren beim Zugriff auf LR–Tabellen und der Auswertung von Einträgen in Frage. Auch die Berücksichtigung von Grammatikmodellen, die radikal vom Chomsky'schen Ansatz abweichen, sollte zulässig sein [9].

Auf halbem Wege zwischen Ad–hoc–Ideen (dem praktisch–bastlerischen Extrem) und der Auswahl als vorteilhaft erkannter Grammatiken (einem theoretisch begründeten Vorgang) stehen unseres Erachtens gezielte, systematische Anpassungen von Parser–Mechanismen für Einzelgrammatiken. Etwas allgemeiner läßt sich eine solche Vorgehensweise als Abtausch zwischen Prozeduranweisungen einerseits und passiven Daten andererseits auffassen.

Interessant in diesem Zusammenhang sind die "Idiome" der natürlichen Sprachen, zu deren korrekter Behandlung sowohl die zugehörigen Einträge im Lexikon als auch syntaktische Analyseschritte erforderlich sind. Wir beabsichtigen, in einer zukünftigen Untersuchung Analoga von Idiomen in gängigen Programmiersprachen aufzuzeigen und Vorschläge zu ihrer Behandlung im Rahmen von Compiler–Systemen zu unterbreiten.

Andererseits dürfte die Bereitstellung geeigneter Verfahren für die syntaktische Analyse menschlicher Sprachen eine an Bedeutung zunehmende Herausforderung für die Informatik sein. Da jegliche Benutzerschnittstellen immer weniger an den Rechnerstrukturen und immer mehr an den Denkweisen der betroffenen Personen orientiert werden (sollten), ist eine allmähliche Annäherung zwischen Programmiersprachen im weiteren Sinne und formal beschriebenen Teilmengen natürlicher Sprachen zu erwarten.

Dabei kann eine informatische Methodenlehre zur Prüfung äquivalenter Grammatiken wesentliche Beiträge leisten.

Literaturverzeichnis

[1] Aho, A.V., R. Sethi und J.D. Ullman. *Compilers: Principles, Techniques, and Tools.* Addison–Wesley 1986.

[2] Bertsch, E. Surjectivity of Functors on Grammars. *Mathematical Systems Theory* **9** (1975/76), 298–307.

[3] Bertsch, E. An Observation on Relative Parsing Time. *Journal of the ACM* **22** (1975), 493–498.

[4] Bertsch, E. Optimization of Expression Syntax – An Experimental Approach. *Software – Practice and Experience* **15** (1985), 269 – 276.

[5] Bertsch, E. Preservation of LL– and LR–properties in Flattened Grammars. *Journal of Information Processing and Cybernetics* **26** (1990), 445–452.

[6] Bertsch, E. Variation of Parsing Speed – A Case Study for MODULA–2. Erscheint im *Journal of Information Processing and Cybernetics* **28**(1992).

[7] Carter, L.R. *An Analysis of PASCAL programs.* UMI Research Press. Ann Arbor, 1982.

[8] Cremers, A.B. und S. Ginsburg. Context–free grammar forms. *J. Computer and Systems Sciences* **11** (1975), 86 – 117.

[9] Hausser, R. *Computation of Language.* Springer Series: Symbolic Computation – Artificial Intelligence. 1989.

[10] Heckert, E. Transformationsverfahren zur zeitoptimierten Syntaxanalyse mit kontextfreien Grammatiken. *Diplomarbeit. Ruhr–Universität Bochum*, 1990.

[11] Hotz, G. Eindeutigkeit und Mehrdeutigkeit formaler Sprachen. *Elektr. Inform. und Kybern.* **2** (1966), 235 – 246.

[12] Hotz, G. Reduktionssätze über eine Klasse formaler Sprachen mit endlich vielen Zuständen. *Math. Zeitschrift* **104** (1968), 205 – 221.

[13] Hotz, G. Übertragung automatentheoretischer Sätze auf Chomsky–Sprachen. *Computing* **4** (1969), 30 – 42.

[14] Knuth, D.E. On the Translation of Languages from Left to Right. *Inform. and Control* **8** (1965), 607 – 639.

[15] Schnorr, C. Transformational Classes of Grammars. *Inform. and Control* **14** (1969), 252 – 277.

[16] Schnorr, C. und H. Walter : Pullbackkonstruktionen bei Semi–Thue–Systemen. *Elektr. Inform. und Kybern.* **5** (1969), 27 – 36.

[17] Wirth, N. *Algorithmen und Datenstrukturen mit MODULA-2.* Teubner, 1986

[18] Zentarra, S. *Zur Implementierbarkeit grammatischer Optimierungstechniken (Arbeitstitel).* Diplomarbeit in Vorbereitung.

Verteilung der Nullstellen von Polynomen auf Jordanbögen

Hans-Peter Blatt

Lehrstuhl für Mathematik
Katholische Universität Eichstätt
8078 Eichstätt
Germany

Zusammenfassung

Die Diskrepanz eines Polynoms p_n mit reellen Nullstellen über dem Intervall $[-1, 1]$ beschreibt den Abstand der Nullstellenverteilung zur Arkussinusverteilung. Erdős und Turán bewiesen die im allgemeinen scharfe Abschätzung

$$c\,(\log A_n/\,n)^{1/2},$$

wobei $\max\limits_{-1 \leq x \leq 1} |p_n(x)| = A_n 2^{-n}$. Dagegen läßt sich die Schranke entscheidend verbessern, wenn man voraussetzt, daß das Polynom nur einfache Nullstellen besitzt. Grob gesagt, die Erdős-Turán-Schranke kann in diesem Fall quadriert werden. Gerade bei vielen Anwendungen, beispielsweise bei orthogonalen Polynomen, ist diese Eigenschaft gewährleistet. In dieser Arbeit werden Abschätzungen dieser Art dargestellt und auf ihre Schärfe eingegangen. Außerdem werden die Ergebnisse auf Jordanbögen verallgemeinert.

1   Erdős-Turán-Sätze

Im Jahre 1940 untersuchten erstmals Erdős und Turán in der berühmten Arbeit [5] die Verteilung der Nullstellen von Polynomen, allein aus der Kenntnis der Tschebyscheff-Norm der Polynome über dem Intervall $[-1, 1]$ und der zusätzlichen Bedingung, daß alle Nullstellen im Intervall $[-1, 1]$ liegen.

Ist nämlich $p_n(x) = x^n + \cdots$ ein Polynom vom Grad n mit höchstem Koeffizienten 1, dann ist aufgrund der Extremaleigenschaft der Tschebyscheff-Polynome

$$\|p_n\| := \max_{-1 \leq x \leq 1} |p_n(x)| \geq 2^{1-n}, \tag{1}$$

und das Gleichheitszeichen gilt nur für die Tschebyscheff-Polynome

$$T_n(x) = 2^{1-n} \cos(n \arccos x).$$

Außerdem ist bekannt, daß für eine Folge von Polynomen $\{p_n\}$, $p_n(x) = x^n + \cdots$, mit der Eigenschaft

$$\lim_{n \to \infty} \|p_n\|^{1/n} = 2^{-1} \tag{2}$$

die Verteilung der Nullstellen schon mit der Arcussinusverteilung vergleichbar ist. Um dieses Ergebnis präzis zu formulieren, sei

$$\nu_n(A) = \frac{\text{Anzahl der Nullstellen von } p_n \text{ in } A}{n} \tag{3}$$

das Nullstellenmaß von p_n, wobei A eine beliebige Menge in $\mathbb{C}$ ist. Weiterhin sei μ die **Arcussinusverteilung** (oder **Gleichgewichtsverteilung** von $[-1, 1]$), d.h.

$$\mu([a, b]) = \frac{1}{\pi} \int_a^b \frac{dx}{\sqrt{1 - x^2}} \tag{4}$$

für irgendein Intervall $[a, b]$ von $[-1, 1]$. Dann folgt aus (2), daß ν_n schwach gegen das Maß μ konvergiert (siehe [1, Satz 2.1]). Insbesondere gilt

$$\lim_{n \to \infty} |(\nu_n - \mu)([a, b])| = 0 \tag{5}$$

für jedes beliebige Intervall $[a, b] \subset [-1, 1]$, falls alle Nullstellen von p_n reell sind.

Liegen alle Nullstellen von p_n bereits im Intervall $[-1, 1]$, so gaben Erdős und Turán [5] folgende quantitative Abschätzung für die **Diskrepanz**

$$D[\nu_n - \mu] := \sup_{-1 \leq a \leq b \leq 1} |(\nu_n - \mu)([a, b])| \tag{6}$$

zwischen den Maßen ν_n und μ an, nämlich

$$D[\nu_n - \mu] \leq \frac{8}{\log 3} \sqrt{\frac{\log A_n}{n}}, \tag{7}$$

wobei die Konstante A_n durch

$$\|p_n\| = A_n \frac{1}{2^n} \tag{8}$$

festgelegt ist. Bis auf die Konstante $8/\log 3$ ist die Abschätzung (7) scharf, was wir durch das folgende Beispiel sofort einsehen: Sei

$$U_n(x) = c_n(x+1)^k P_{n-k}^{(0,2k)}(x), \tag{9}$$

wobei

$$k = [\sqrt{n \log n}] + 1,$$

$$c_n = 2^{n-k} \binom{2n}{n-k}^{-1},$$

und $P_m^{(\alpha,\beta)}(x)$ jeweils das Jacobi-Polynom vom Grad m zu den Parametern α und β bedeutet. Dann ist $U_n(x)$ ein Polynom mit höchstem Koeffizienten 1 und besitzt k Nullstellen in -1, also ist die Diskrepanz zwischen der Nullstellenverteilung ν_n von U_n und der Arcussinusverteilung

$$D[\nu_n - \mu] \geq \frac{k}{n} \geq \sqrt{\frac{\log n}{n}}.$$

Andererseits wird wegen der Monotonie der Gewichtsfunktion $w(x) = (x+1)^{2k}$ das Maximum von

$$|w(x)^{1/2} P_{n-k}^{(0,2k)}(x)| = |(x+1)^k P_{n-k}^{(0,2k)}(x)|$$

im Punkt 1 angenommen (Szegő [9, Satz 7.2, S.163]). Außerdem ist wegen der Normierung der Jacobi-Polynome

$$P_m^{(\alpha,\beta)}(1) = \binom{m+\alpha}{m},$$

also

$$P_{n-k}^{(0,2k)}(1) = 1.$$

Somit gilt insgesamt

$$\|U_n\| = 2^n \binom{2n}{n-k}^{-1}.$$

Da aus der Stirlingschen Formel

$$\binom{2n}{n-k}^{-1} \leq c_0 \frac{1}{2^{2n}} n^3$$

mit einer Konstanten $c_0 > 0$ folgt, ergibt sich

$$\|U_n\| \leq c_0 \frac{n^3}{2^n}$$

und damit aus (7)

$$D[\nu_n - \mu] \leq c_1 \sqrt{\frac{\log n}{n}}$$

mit einer geeigneten Konstanten c_1, die unabhängig von n ist.

Eine analoge Diskrepanzaussage wurde von Erdős und Turán [6] für den Kreis bewiesen. Verallgemeinerungen dieser Sätze auf Jordankurven und Jordanbögen finden sich in [2]. Außerdem wird dort auf die Zusatzforderung verzichtet, daß die Nullstellen auf der betreffenden Kurve (bzw. Bogen) liegen. Dabei sei aber betont, daß auch bei Erdős und Turán dies keine eigentliche Einschränkung bedeutet, da man mit dem Trick von Schur (siehe [6]) diese Forderung für den Kreis und das Intervall erfüllen kann. Ein solcher Kunstgriff ist jedoch für allgemeinere Kurven bzw. Bögen unbekannt.

2   Verteilung einfacher Nullstellen auf $[-1, 1]$

Obwohl wir im vorigen Abschnitt gesehen haben, daß die Erdős-Turán-Abschätzung scharf ist, lassen sich für eine wichtige Klasse von Polynomen die Ergebnisse verbessern. Wir setzen jetzt nämlich zusätzlich voraus, daß neben der alten Bedingung

(A) $$\|p_n\| = A_n \frac{1}{2^n}$$

noch alle Nullstellen x_i von $p_n(x)$, $1 \leq i \leq n$, im Intervall $[-1, 1]$ liegen und

(B) $$|p_n'(x_i)| \geq \frac{1}{B_n} \frac{1}{2^n}$$

erfüllt ist, d.h. alle Nullstellen von p_n sind einfach und die Ableitung ist durch (B) **nach unten** beschränkt. Unter diesen Voraussetzungen wurde in [3] folgender Satz bewiesen.

Satz 1. *p_n sei ein Polynom vom Grad n mit höchstem Koeffizienten 1 und nur einfachen Nullstellen, die alle in $[-1, 1]$ liegen. Außerdem seien die Bedingungen (A) und (B) erfüllt, $n \geq 2$. Dann gibt es eine Konstante $c > 0$, die unabhängig von n ist, mit der Eigenschaft*

$$D[\nu_n - \mu] \leq c \frac{\log C_n}{n} \log n, \tag{10}$$

wobei ν_n das Nullstellenmaß von p_n, μ die Arcussinusverteilung und

$$C_n = \max(A_n, B_n, n)$$

ist.

Kürzlich hat V. Totik [10] die Abschätzung (10) verschärft zu

$$D[\nu_n - \mu] \leq c \frac{\log C_n}{n} \log \frac{n}{\log C_n}. \tag{11}$$

Dabei ist selbstverständlich C_n einzuschränken, um überhaupt zu einer sinnvollen Diskrepanzaussage zu kommen. Wir setzen daher in (11) stets $\log C_n \leq n - 1$ voraus.

Außerdem ist es V. Totik in einer eleganten potentialtheoretischen Konstruktion gelungen zu zeigen, daß die Ungleichung bis auf die Konstante c scharf ist. Mit diesen neuen Ergebnissen lassen sich bisher bekannte Verteilungsaussagen für Nullstellen orthogonaler Polynome wesentlich verschärfen, ebenso auch Verteilungsaussagen für die Extremalpunkte bester Approximationen (siehe dazu [3]). Außerdem gilt analog zu Satz 1 ein entsprechender Satz für Polynome mit einfachen Nullstellen auf dem Einheitskreis [3]. Anwendungen auf die Verteilung von Fekete-Punkten (auch für allgemeinere Situationen) werden in [4] behandelt und zur Approximation konformer Abbildungen herangezogen.

3 Verteilung einfacher Nullstellen auf Jordanbögen

Wir wollen die Abschätzung (11) für Polynome mit einfachen Nullstellen auf Jordanbögen verallgemeinern. Dazu setzen wir im folgenden voraus, daß E ein Jordanbogen der Klasse C^{2+} ist, d.h. E ist rektifizierbar und die Koordinaten sind C^{2+}-Funktionen bezüglich der Bogenlänge. Hierbei gehört eine Funktion einer reellen Variablen zur Klasse C^{2+}, wenn die zweite Ableitung einer Hölder-Bedingung mit einem positiven Exponenten genügt.

Um die Ergebnisse zu formulieren und die Beweise durchzuführen, brauchen wir einige potentialtheoretische Hilfsmittel. Mit $G(z)$ bezeichnen wir die Greensche Funktion von $\overline{\mathbb{C}} \setminus E$ mit Pol bei ∞ und Randwerten 0. Ist $\mathrm{cap}(E)$ die logarithmische Kapazität von E, so ist $\mathrm{cap}(E) > 0$. Im Falle $E = [-1, 1]$ gilt $\mathrm{cap}(E) = 1/2$. Nun gibt es ein Wahrscheinlichkeitsmaß mit Träger in E, das das Energieintegral

$$I[\nu] := \iint \log \frac{1}{|z - \zeta|} d\nu(\zeta) d\nu(z) \tag{12}$$

minimiert. Dieses optimale Wahrscheinlichkeitsmaß $\mu = \mu_E$ ist eindeutig und heißt **Gleichgewichtsverteilung** von E. Zu jedem Wahrscheinlichkeitsmaß ν wird durch

$$U^\nu(z) := \int \log \frac{1}{|z - \zeta|} d\nu(\zeta) \tag{13}$$

das **logarithmische Potential** definiert. Das zu $\mu = \mu_E$ gehörige logarithmische Potential $U^\mu(z)$ heißt auch **Gleichgewichtspotential von** E und ist durch

$$U^\mu(z) = -G(z) - \log \mathrm{cap}(E), \quad z \in \overline{\mathbb{C}} \setminus E, \tag{14}$$

mit der Greenschen Funktion $G(z)$ verknüpft (siehe [11, Theorem III.37, S. 82]). Im Fall $E = [-1, 1]$ ist μ die obige Arcussinusverteilung.

60   *Hans-Peter Blatt*

Weiterhin seien

$$\Gamma_\sigma := \{z \in \mathbb{C} \colon G(z) = \log(1+\sigma)\}, \ \sigma \geq 0, \tag{15}$$

die Niveaulinien der Greenschen Funktion. Jede Niveaulinie $\Gamma_\sigma, \sigma > 0$, ist analytisch und stellt eine Jordankurve dar.

Sei nun $p(z) = z^n + \cdots$ ein Polynom vom Grad n mit höchstem Koeffizienten 1 und einfachen Nullstellen $z_i, 1 \leq i \leq n$, in E. $p_n(z)$ erfülle die Bedingungen

(A1)
$$\|p_n\|_E = \max_{z \in E} |p_n(z)| = A_n (\mathrm{cap}(E))^n,$$

(B1)
$$|p'_n(z_i)| \geq \frac{1}{B_n} (\mathrm{cap}(E))^n.$$

Für die **Diskrepanz zwischen** ν **und** μ, nämlich

$$D[\nu_n - \mu] := \sup_{J \subset E} |(\nu_n - \mu)(J)|, \tag{16}$$

wobei J alle Teilbögen von E durchläuft, gilt dann folgende Verallgemeinerung zu Satz 1, nämlich

Satz 2. *p_n sei ein Polynom vom Grad n mit höchstem Koeffizienten 1 ($n \geq 2$). Alle Nullstellen z_i, $1 \leq i \leq n$, von $p_n(z)$ seien einfach und liegen auf einem Jordanbogen E der Klasse C^{2+}. Außerdem seien (A1) und (B1) erfüllt, $C_n = \max(A_n, B_n, n)$, $\log C_n \leq n - 1$. Dann gibt es eine Konstante $c > 0$, die unabhängig von n ist, mit der Eigenschaft*

$$D[\nu_n - \mu] \leq c \frac{\log C_n}{n} \log \frac{n}{\log C_n}, \tag{17}$$

wobei ν_n das Nullstellenmaß von p_n, μ die Gleichgewichtsverteilung von E ist.

Ähnlich wie in [3], [10] beruht der Beweis dieses Satzes entscheidend auf einer Umformulierung der Bedingungen (A1) und (B1) in Bedingungen an die logarithmischen Potentiale $U^{\nu_n}(z)$ und $U^\mu(z)$: Aus (A1) ergibt sich durch Logarithmieren

$$\frac{1}{n} \log |p_n(z)| - \log \mathrm{cap}(E) \leq \frac{\log A_n}{n} \quad \text{für } z \in E. \tag{18}$$

Nun ist

$$U^{\nu_n}(z) = -\frac{1}{n} \log |p_n(z)|$$

und

$$U^\mu(z) = -\log \mathrm{cap}(E) \quad \text{für } z \in E.$$

Also läßt sich (18) schreiben als

$$U^\mu(z) - U^{\nu_n}(z) \leq \frac{\log A_n}{n} \quad \text{für } z \in E. \tag{19}$$

Da $U^\mu(z) - U^{\nu_n}(z)$ harmonisch ist in $\overline{\mathbb{C}} \setminus E$, so folgt damit aus dem Maximumprinzip, daß (19) für alle $z \in \mathbb{C}$ gilt. Aus (B1) folgt mit Lagrange-Interpolation

$$1 = \sum_{i=1}^{n} \frac{p_n(z)}{p_n'(z_i)(z - z_i)}$$

und damit

$$1 \leq \frac{n \, B_n \, |p_n(z)|}{(\mathrm{cap}(E))^n \mathrm{dist}(E, \Gamma_\sigma)} \tag{20}$$

für $z \in \Gamma_\sigma$. Nun existiert eine Konstante $c_1 > 0$ mit

$$\mathrm{dist}(E, \Gamma_\sigma) \geq c_1 \, \sigma^2$$

für alle $0 \leq \sigma \leq 1$ (Siciak [8, Lemma 1]). Somit folgt für $\sigma = 1/n^2$ und $z \in \Gamma_\sigma$ aus der Ungleichung (20)

$$0 \leq \frac{1}{n} \log |p_n(z)| - \log \mathrm{cap}(E) + \frac{\log B_n}{n} + c_2 \frac{\log n}{n} \tag{21}$$

mit einer geeigneten Konstanten $c_2 > 0$ für alle $n \geq 2$. Führt man in (21) wieder die potentialtheoretische Schreibweise ein, so erhält man

$$U^{\nu_n}(z) - U^\mu(z) \leq c_3 \frac{\log C_n}{n} \tag{22}$$

für $z \in \Gamma_\sigma$, $\sigma = 1/n^2$. Da die linke Seite harmonisch ist in $\overline{\mathbb{C}} \setminus E$, so gilt (22) für alle z mit $G(z) \geq \log(1 + 1/n^2)$.

Insgesamt haben wir somit für ein diskretes Wahrscheinlichkeitsmaß $\nu = \nu_n$ mit Träger in E folgende Eigenschaften vorliegen:

Es gibt eine Konstante $\epsilon > 0$, so daß

$$\text{(A2)} \qquad\qquad U^\mu(z) - U^\nu(z) \leq \epsilon \quad \text{für alle } z \in \mathbb{C},$$

$$\text{(B2)} \qquad U^\mu(z) - U^\nu(z) \geq -\epsilon \quad \text{für alle } z \text{ mit } G(z) \geq \log(1 + \epsilon^2).$$

Es zeigt sich nun, daß aus (A2) und (B2) bereits eine Diskrepanzaussage zwischen ν und μ gewonnen werden kann und somit Satz 1 und Satz 2 eine Folgerung des folgenden Satzes sind.

Satz 3. *Sei ν ein diskretes Wahrscheinlichkeitsmaß auf einem Jordanbogen E der Klasse C^{2+} und μ das Gleichgewichtsmaß von E, so daß die Bedingungen (A2) und (B2) erfüllt sind mit $\epsilon \leq 1/e$. Dann gibt es eine Konstante $c > 0$, unabhängig von ϵ, so daß*

$$D[\nu - \mu] \leq c \, \epsilon \log \frac{1}{\epsilon}, \tag{23}$$

wobei die Diskrepanz $D[\nu - \mu]$ analog zu (16) definiert ist.

Beim Beweis von Satz 3 werden wir auf Techniken zurückgreifen, die in [2] zum Beweis von Erdős-Turán-Sätzen entwickelt werden. Insbesondere sind Eigenschaften der ersten und zweiten Ableitung der Greenschen Funktion und ihres Stetigkeitsverhaltens auf E wesentlich, bei denen die charakteristischen Eigenschaften einer C^{2+}-Kurve eingehen. Außerdem wird folgender elementarer Hilfssatz benötigt, nämlich

Hilfssatz [2]: Es existiert eine Funktion $h \in C^2(\mathbb{R})$, $h : \mathbb{R} \to [0,1]$ mit

(i) $h(x) = 1$ für $x \leq 0$

(ii) $h(x) = 0$ für $x \geq 1$

(iii) $-3 \leq h'(x) \leq 0$ für $0 \leq x \leq 1$

(iv) $|h''(x)| \leq 5$ für $0 \leq x \leq 1$

Beweis von Satz 3: Im folgenden bezeichnen wir mit $c_1, c_2, \ldots$ jeweils absolute Konstanten, die wohl von E, aber nicht von ϵ abhängen. Es genügt, ähnlich wie in [2], die Ungleichung

$$(\mu - \nu)(J) \leq c\,\epsilon \log \frac{1}{\epsilon} \tag{24}$$

für alle Teilbögen $J \subset E$ zu beweisen. Wir setzen zur Abkürzung

$$\delta := 2\,\epsilon \log \frac{1}{\epsilon}.$$

Außerdem dürfen wir noch $\mu(J) > \delta$ voraussetzen und uns auf Teilbögen J beschränken, die einen Endpunkt von E enthalten.

Betrachten wir die orthogonalen Trajektorien zu den Niveaulinien Γ_σ, so sind diese Trajektorien paarweise zueinander disjunkt, und durch jeden Punkt $z \notin E$ verläuft genau eine Trajektorie, da grad $G(z) \neq 0$. Außerdem erkennt man mit dem Aufbiegetrick von Widom [12] (siehe auch [2]), daß jeder Punkt von E Häufungspunkt von genau einer Trajektorie ist. Für jeden Punkt $z \in E$ bezeichnen wir mit $L(z)$ diese eindeutig bestimmte Trajektorie. Weiter sei $\pi(z)$, $z \in \mathbb{C}$, die Projektion von z auf E entlang der abgeschlossenen Trajektorie, auf der z liegt.

Seien nun a und b die Endpunkte von E, $a \in J$. Durch den Schnitt $L(b)$ erhalten wir ein einfach zusammenhängendes Gebiet $F = \mathbb{C} \setminus (E \cup L(b))$. Auf F wählen wir eine zu $G(z)$ konjugierte Funktion $\varphi(z)$ aus. $\varphi(z)$ kann stetig auf E von jedem Ufer erweitert werden. Diese stetigen Erweiterungen nennen wir φ_+ und φ_-, die wir noch jeweils stetig in die Endpunkte von E fortsetzen. Damit ist auch $\varphi_+(a) = \varphi_-(a) = \varphi(a)$ und φ im Punkt a stetig. Außerdem können wir φ noch so normieren, daß $\varphi(a) = 0$ ist.

Wir zerlegen J in zwei disjunkte Teilbögen J_1 und J_2, so daß J_1 kompakt ist und a enthält und $\mu(J_2) = \delta$ ist. Da $2\pi\mu(J_2)$ gerade die Summe der beiden Intervalle $\varphi_+(J_2)$ und $\varphi_-(J_2)$ ist, dürfen wir o.B.d.A. $\varphi_+(J_2) \geq \pi\delta$ voraussetzen. Nun ist

$$\varphi_-(E) \cup \varphi_+(E) = [\alpha, \beta] \quad \text{mit } \beta = \alpha + 2\pi,$$

da die Abbildung

$$t = \Phi(z) = e^{G(z)+i\varphi(z)} \tag{25}$$

das Äußere von E konform auf das Äußere des Einheitskreises abbildet. Wegen $\varphi(a) = 0$ ist

$$\varphi_-(J_1) = [\alpha_1, 0], \quad \varphi_+(J_1) = [0, \beta_1],$$

und wir definieren mit vorigem Hilfsatz eine Funktion $g_1 : [0, \beta] \to [0, 1]$, so daß

$$
\begin{aligned}
g_1(\varphi) &= 1 \quad \text{für alle } \varphi \text{ in einer Umgebung von } [0, \beta_1], \\
g_1(\varphi) &= 0 \quad \text{für alle } \varphi \text{ in einer Umgebung von } [0, \beta] \setminus \varphi_+(J)
\end{aligned}
$$

und außerdem

$$\left| \frac{d^2}{d\varphi^2} g_1(\varphi) \right| \leq \frac{6}{(\pi\delta)^2}, \tag{26}$$

$$-\frac{4}{\pi\delta} \leq \frac{d}{d\varphi} g_1(\varphi) \leq 0 \tag{27}$$

für alle $\varphi \in [0, \beta]$. Schließlich setzen wir g_1 stetig auf $[\alpha, \beta]$ fort durch

$$g_1(\varphi) = g_1(\varphi_+(z)) \quad \text{für } \varphi = \varphi_-(z),\ z \in E.$$

Dann ist für $\varphi = \varphi_-(z) \in [\alpha, 0]$, $z \in E$,

$$\frac{d}{d\varphi} g_1(\varphi) = \frac{d}{d\varphi} g_1(\varphi_+(z)) = \frac{d}{d\varphi_+} g_1(\varphi_+) \cdot \frac{d\varphi_+}{d\varphi_-}.$$

Sind n_+ und n_- die beiden Normalen in einem Punkt z im Innern von E, die ins Gebiet $\mathbb{C} \setminus E$ zeigen, so gilt

$$\frac{d\varphi_+}{d\varphi_-} = -\frac{\partial G/\partial n_+}{\partial G/\partial n_-}$$

(beachte das Vorzeichen!) und nach Lemma 4 in [2] ist dieser Ausdruck stetig, fortsetzbar auf E und dann positiv und beschränkt auf ganz E. Somit existiert eine Konstante c_1, so daß

$$0 \leq \frac{d}{d\varphi} g_1(\varphi) \leq \frac{c_1}{\delta} \quad \text{für } \varphi \in [\alpha, 0]. \tag{28}$$

Die Funktion g_2 wird definiert durch

$$g_2(z) = h(e^{G(z)} - 1 - \delta)$$

mit h aus vorigem Hilfssatz. Schließlich setzen wir

$$g(z) = g_1(\varphi_+(\pi(z)))g_2(z), \quad z \in \mathbb{C},$$

und beachten, daß $g_1(\varphi_+(\pi(z)))$ konstant ist für alle Punkte z einer festen orthogonalen Trajektorie der Greenschen Funktion.

Analog zu der Vorgehensweise in [2, Beweis von Satz 1] können wir mittels (26) – (28) den Laplace-Operator, angewandt auf g im Gebiet $\mathbb{C} \setminus E$, abschätzen durch

$$|\Delta g(z)| \leq c_2 r^2 \left(\frac{\partial G}{\partial n}\right)^2 \frac{1}{\delta^2} \qquad (29)$$

wobei c_2 eine absolute Konstante ist und $r = r(z) = e^{G(z)}$, d.h. mit

$$t = \Phi(z) = e^{G(z) + i\varphi(z)} = r(z)e^{i\varphi(z)}$$

haben wir in der t-Ebene die Polarkoordinaten r und φ eingeführt (Greensche Koordinaten). Da $\Delta g = 0$ ist in einer Umgebung der Endpunkte von E, so ist $|\Delta g|$ beschränkt in $\mathbb{C} \setminus E$. Also gilt wie im Beweis vom Satz 1 in [2] die Darstellungsformel

$$g(\xi) = \frac{1}{2\pi} \iint \Delta g(z) \log|z - \xi| dx dy, \qquad (30)$$

($z = x + iy$), und wir erhalten mit dem Satz von Fubini

$$\int g(d\mu - d\nu) = \frac{1}{2\pi} \iint \Delta g(z) U^{\nu - \mu}(z) dx dy \qquad (31)$$

und durch Einführen der Greenschen Koordinaten $r = r(z)$, $\varphi = \varphi(z)$ schließlich

$$\int g(d\mu - d\nu) = \frac{1}{2\pi} \int_1^{2+\delta} \frac{1}{r} \int_0^{2\pi} \Delta g(z) U^{\nu - \mu}(z) \left(\frac{\partial G}{\partial n}\right)^{-2} d\varphi dr. \qquad (32)$$

Wir zerlegen nun das Integral rechts in zwei Anteile, nämlich

$$I_1 = \frac{1}{2\pi} \int_1^{1+\delta^2} \frac{1}{r} \int_0^{2\pi} \Delta g(z) U^{\nu - \mu}(z) \left(\frac{\partial G}{\partial n}\right)^{-2} d\varphi dr$$

und

$$\begin{aligned} I_2 &= \frac{1}{2\pi} \int_{1+\delta^2}^{2+\delta} \frac{1}{r} \int_0^{2\pi} \Delta g(z) U^{\nu - \mu}(z) \left(\frac{\partial G}{\partial n}\right)^{-2} d\varphi dr \\ &= \frac{1}{2\pi} \iint_V \Delta g(z) U^{\nu - \mu}(z) dx dy, \end{aligned}$$

wobei

$$V = \{z = x + iy : \log(1 + \delta^2) \leq G(z) \leq \log(2 + \delta)\}.$$

Nun ist wegen (A2)

$$|U^{\nu - \mu}(z)| \leq U^{\nu - \mu}(z) + 2\epsilon$$

für alle $z \in \mathbb{C}$. Also folgt für I_1 mit (29)

$$\begin{aligned} I_1 &\leq \frac{c_2}{2\pi\delta^2} \int_1^{1+\delta^2} r \int_0^{2\pi} (U^{\nu - \mu}(z) + 2\epsilon) d\varphi dr \\ &= \frac{c_2 \epsilon}{\pi\delta^2} \int_1^{1+\delta^2} r \int_0^{2\pi} d\varphi dr \leq c_3 \epsilon, \end{aligned} \qquad (33)$$

da nämlich wegen der Mittelwerteigenschaft für harmonische Funktion

$$\frac{1}{2\pi}\int_0^{2\pi} U^{\nu-\mu}(z)d\varphi = \frac{1}{2\pi}\int_{\Gamma_{r-1}} U^{\nu-\mu}(z)\frac{\partial G}{\partial n}ds = U^{\nu-\mu}(\infty) = 0.$$

Zur Abschätzung von I_2 verwenden wir die Greensche Formel für die Funktionen g und $U^{\nu-\mu}$ im abgeschlossenen Gebiet V und erhalten

$$\int_V \left(U^{\nu-\mu}\Delta g - g\Delta U^{\nu-\mu}\right) = -\int_{\partial V}\left(U^{\nu-\mu}\frac{\partial g}{\partial n} - g\frac{\partial U^{\nu-\mu}}{\partial n}\right),$$

wobei n die innere Normale auf dem Rand ∂V von V bedeutet. Da $U^{\nu-\mu}$ harmonisch in V und damit $\Delta U^{\nu-\mu} = 0$ ist, $(\partial g/\partial n) = 0$ auf dem Rand von V, und $g = 0$ auf der Niveaulinie $\Gamma_{2+\delta}$, so folgt

$$I_2 = \frac{1}{2\pi}\int_V U^{\nu-\mu}\Delta g = \frac{1}{2\pi}\int_{\Gamma_{\delta^2}} g\,\frac{\partial U^{\nu-\mu}}{\partial n}ds. \tag{34}$$

Sei nun $H(z)$ diejenige zu $U^{\nu-\mu}(z)$ in $\overline{\mathbb{C}}\setminus E$ konjugierte harmonische Funktion mit $H(\infty) = 0$. Aufgrund von (A2) und (B2) ist

$$\left|U^{\nu-\mu}(z)\right| \leq \epsilon \quad \text{für alle } z \text{ mit } G(z) \geq \log(1+\epsilon^2).$$

Damit ist nach Polya, Szegö [7, Problem 288, S. 140] aber auch für alle z mit $G(z) \geq \log(1+\epsilon^2)$

$$|H(z)| \leq c_4\,\epsilon\,\log\frac{1}{\epsilon} \tag{35}$$

mit $c_4 > 0$, wenn wir die Einschränkung $\epsilon \geq e^{-1}$ beachten. Sei $\Psi(t)$ die inverse Funktion zu $\Phi(z)$ aus (25). Dann parametrisieren wir die Niveaulinie Γ_{δ^2} durch

$$\Gamma_{\delta^2} = \{z = \Psi(re^{i\varphi}) : r = 1+\delta^2, \alpha \leq \varphi \leq \beta\}.$$

Identifiziert man die Punkte auf Γ_{δ^2} durch die Bogenlänge $s = s(\varphi)$, gemessen vom Punkt $z_0 = \Psi(re^{i\alpha})$ im mathematisch positiven Sinne, so gilt

$$\frac{d}{d\varphi}s(\varphi) = \left[\frac{\partial G}{\partial n}(z)\right]^{-1}, \quad z = \Psi(re^{i\varphi}),$$

wobei $\partial G/\partial n \neq 0$ für alle $z \in \Gamma_{\delta^2}$. Aus den Cauchy-Riemannschen Differentialgleichungen folgt

$$\frac{\partial U^{\nu-\mu}}{\partial n} = \frac{\partial H}{\partial s} = \frac{\partial H}{\partial\varphi}\frac{\partial G}{\partial n}$$

und somit aus (34)

$$I_2 = \frac{1}{2\pi}\int_\alpha^\beta g\,\frac{\partial H}{\partial\varphi}d\varphi,$$

woraus sich mit partieller Integration, unter Beachtung von $g(\alpha) = g(\beta) = 0$,

$$I_2 = -\frac{1}{2\pi} \int_\alpha^\beta H \, \frac{\partial g}{\partial \varphi} \left(\Psi(re^{i\varphi}) \right) d\varphi$$

ergibt. Da

$$g\left(\Psi(re^{i\varphi}) \right) = g_1\left(\Psi(e^{i\varphi}) \right) = g_1(\varphi),$$

erhält man mit (35)

$$|I_2| \leq \frac{c_4}{2\pi} \epsilon \log \frac{1}{\epsilon} \int_\alpha^\beta \left| \frac{dg_1}{d\varphi} \right| d\varphi.$$

Wegen der Monotonie der Ableitung von g_1 in den Teilintervallen $[\alpha, 0]$ und $[0, \beta]$ folgt

$$\int_\alpha^\beta \left| \frac{dg_1}{d\varphi} \right| d\varphi = 2.$$

Also ist insgesamt

$$\int g(d\mu - d\nu) \leq c_5 \, \epsilon \log \frac{1}{\epsilon},$$

und wegen

$$(\mu - \nu)(J) \leq \int g(d\mu - d\nu) + \mu(J_2)$$

folgt somit (24), wenn man noch $\mu(J_2) = 2 \, \epsilon \log(1/\epsilon)$ beachtet.

Literaturverzeichnis

[1] H.-P. Blatt, E.B. Saff und M. Simkani, *Jentzsch-Szegő type theorems for the zeros of best approximants*, J. London Math. Soc. 38 (1988), 307–316.

[2] H.-P. Blatt und R. Grothmann, *Erdős-Turán theorems on a system of Jordan curves and arcs*, Constr. Approx. 7 (1991), 19–47.

[3] H.-P. Blatt, *On the distribution of simple zeros of polynomials*, erscheint in J. Approx. Theory.

[4] H.-P. Blatt und H.N. Mhaskar, *A general discrepancy theorem*, Manuskript, 1991.

[5] P. Erdős und P. Turán, *On the uniformly dense distribution of certain sequences of points*, Ann. of Math. 41 (1940), 162–173.

[6] P. Erdős und P. Turán, *On the distribution of roots of polynomials*, Ann. of Math. 51 (1950), 105–119.

[7] F. Pólya und G. Szegő, "Aufgaben und Lehrsätze aus der Analysis I", Springer Verlag Berlin, 1925.

[8] J. Siciak, *Degree of convergence of some sequences in the conformal mapping theory*, Colloq. Math. 16 (1967), 49–59.

[9] G. Szegö, "Orthogonal polynomials", Amer. Math. Soc., Colloquium publications, Providence, Rhode Island, 1975.

[10] V. Totik, *Distribution of simple zeros of polynomials*, Manuskript, 1991.

[11] M. Tsuji, "Potential theory in modern function theory", Chelsea Publ., New York, 1950.

[12] H. Widom, *Extremal polynomials associated with a system of curves in the complex plane*, Adv. in Math. 3 (1969), 127–232.

Distributed Class Group Computation

Johannes Buchmann
Stephan Düllmann

FB 14-Informatik
Universität des Saarlandes
6600 Saarbrücken
Germany

Abstract

We present an improved sequential and a parallel version of the algorithm of Hafner and McCurley for the computation of the class group of imaginary quadratic fields. We describe the implementation of this algorithm on a network of UNIX-workstations using the system LIPS.

1   Introduction

In [7] Hafner and McCurley present a probabilistic algorithm which under the assumption of the generalized Riemann hypothesis (GRH) computes the class number and the structure of the class group of an imaginary quadratic field in expected running time $L(D)^{\sqrt{2}+o(1)}$ where D is the absolute discriminant of the field and where $L(D) = \exp(\sqrt{\log D \log \log D})$.

In [3] we describe a practical version of that algorithm and we give several numerical examples and timings for discriminants up to 40 decimal digits.

In this paper we present an improved sequential version and a parallel version of the class group algorithm and we describe its implementation using the system LIPS which is described in [2]. Numerical examples demonstrate how the new algorithm drastically reduces the effective running time of the algorithm.

In section 2 we briefly describe the idea of the algorithm. In section 3 we describe in detail the most time consuming part of the algorithm and we present some refinements for the sequential algorithm. In section 4 we show how to parallelize the algorithm. In section 5 we give computational results for a 55-digit discriminant.

2   The algorithm

2.1   The general idea

We briefly describe the main idea of the algorithm of Hafner and McCurley [7]:

Let $\mathcal{K}$ be an imaginary quadratic field of absolute discrimant D, Cl its class group and h its class number. Let

$$F = \{\wp_1, \ldots, \wp_k\}$$

be a set of prime ideals of $\mathcal{K}$ generating Cl. This set can easily be computed, since under GRH the class group Cl of $\mathcal{K}$ is generated by the prime ideals of norm below $6(\log D)^2$ (see [10] and [1]).

We define the mapping

$$\varphi : \quad \begin{aligned} \mathbb{Z}^k &\rightarrow Cl \\ (e_1, \ldots e_k) &\mapsto \prod_{i=1}^{k} \wp_i^{e_i} . \end{aligned}$$

Obviously φ is a homomorphism and, since F generates Cl, φ is surjective. The kernel of φ

$$L = \ker \varphi = \left\{ (e_1, \ldots e_k) \ : \ \prod_{i=1}^{k} \wp_i^{e_i} = 1 \right\}$$

is a k-dimensional lattice in $\mathbb{Z}^k$ and we have

$$\mathbb{Z}^k / L \cong Cl$$

and so

$$\det L = \#Cl = h.$$

The basic idea of the algorithm is to find vectors in L (so-called *relations*) by means of a probabilistic method until the full lattice is generated.

Assuming GRH and using the analytic class number formula an approximation h^* to the class number h with

$$h^* < h < 2h^* \tag{1}$$

can be computed in polynomial time (see [5]). Therefore, we have an easy criterion to decide whether the current set of relations generates the full lattice L: Each set of relations generates a sublattice L' of L. If L' has dimension k we know $h \mid h' = \det L'$. So if $h' < 2h^*$ we know $L' = L$ and $h' = h$.

After having computed h, the structure of the class group, i.e. integers m_i ($1 \le i \le k$) such that

$$m_i \mid m_{i+1} \qquad (1 \le i \le k-1),$$
$$h = \prod_{i=1}^{k} m_i,$$
$$Cl \cong \bigoplus_{i=1}^{k} \mathbb{Z}/m_i\mathbb{Z},$$

can be found easily by computing the Smith normal form (SNF) of a basis of L. For details see [7].

Moreover, if we keep track of the unimodular transformations occuring in the computation of the SNF, we obtain a minimal generating set for the class group in terms of the elements of F.

2.2 The implementation

The general idea described above and our practical experience lead to the following algorithm:

Algorithm 1
    Input: *Absolute discriminant D*
    Output: *Class number h and structure and minimal generating set of the class group Cl.*

(1) *Compute the set $F = \{\wp : N(\wp) \le 6(\log D)^2\} = \{\wp_1, \ldots, \wp_k\}$ of prime ideals generating Cl and the approximation h^* satisfying (1).*

(2) *Find at least k vectors of L generating a sublattice L' of L.*

(3) *Compute the dimension k' of L'. If $k' = k$ compute the determinant h' of L'.*

(4) *If $k' = k$ and $h' < 2h^*$ go to step (5) else find an additional vector of L to enlarge the sublattice L' and repeat step (3).*

(5) *Compute the structure and a minimal generating set of Cl.*

Here are some remarks concerning the details of the algorithm:

Step (1): The details are given in [3].

Step (2): The computation of the relations is the most time consuming part of the algorithm. The version presented in [3] has the following features:

- We generate a sparse relation matrix, since otherwise the computation of the determinant is infeasible.

- We generate $(k + 5)$ relations. Then the determinant, that is computed in step (3), is only a small multiple of h. In many cases, we already find the class number in this step; step (4) does not have to be executed.

Step (3): The computation of the determinant of L' is done by computing the Hermite normal form (HNF) of the relation matrix. There are two stages:

(3.1) Perform a slightly modified version of the structured Gaussian elimination algorithm [8] to decrease the size of the matrix. Here we use the sparsity of the relation matrix.

(3.2) By Gaussian elimination compute a multiple Δ of the lattice determinant and then perform modular Hermite reduction [6] modulo Δ. For details see [3].

In very few cases it turns out in step (3) that the lattice found in step (2) is not of dimension k.

Step (4): The additional relations are found by the same technique as in step (2). The computation of the new determinant is done by updating the HNF using standard techniques as described in [6].

Step (5): For the computation of the SNF of the basis of L we use standard techniques (see [7]). Since the size of the matrix to be considered is very small, no special treatment is necessary.

3   Improvements of the sequential algorithm

3.1   Finding one relation

First of all we describe the general idea how to find one random relation in L. This method is a modification of an algorithm of Seysen presented in [11].

Let $B \in \mathbb{Z}_{>0}$ be fixed. According to a heuristic strategy described below we pick a vector $\underline{v} = (v_1, \ldots v_k) \in \{0, \ldots B\}^k$ and compute the reduced ideal $\mathbf{a}'$ in the class of

$$\mathbf{a} = \prod_{i=1}^{k} \wp_i^{v_i}.$$

Then we try to factor $\mathbf{a}'$ over the set F. If there is a factorization

$$\mathbf{a}' = \prod_{i=1}^{k} \wp_i^{v_i'}$$

with $v_i' \in \mathbb{Z}$ $(1 \leq i \leq k)$ then the ideal

$$\mathbf{a}(\mathbf{a}')^{-1} = \prod_{i=1}^{k} \wp_i^{v_i - v_i'}$$

is principal and so the vector $\underline{e} = \underline{v} - \underline{v}'$ is a relation.

A practical algorithm for the computation of power products of ideals has been presented in [4], see also [12].

A way of factoring the ideal $\mathbf{a}'$ over F can be derived from [11], Theorem 3.1. The main work is to find a factorization of $N(\mathbf{a}')$ over the set $\hat{F}$ of primes built by the norms of the ideals in F.

3.2 Finding the relation matrix

We present a heuristic strategy for finding the relation matrix. This strategy is a modification of the method described in [3] and is very efficient in practice. We define the set

$$K = \{i \ : \ 1 \leq i \leq k, \ N(\wp_i) < \log D, \ \wp_i \text{ not ambigious}\}$$

and we let $k_0 = \#K$, $k_1 = k_0/2$, $B = 30$. The size of the parameters k_1 and B varies with the size of the discriminant. The values chosen here have experimentally turned out to be optimal for a 55-digit discriminant. The vectors $\underline{v}$ used to generate the relations are chosen from the set

$$V = \left\{ \underline{v} = (v_1, \ldots v_k) \in \mathbb{Z}^k \ : \ \sum_{i=1}^{k_0} |\text{sign } v_i| = k_1, \ \sum_{i=k_0+1}^{k} | v_i | \leq 1, 0 \leq v_i \leq B, \ 1 \leq i \leq k \right\}.$$

Those vectors have k_1 non zero entries in the first k_0 positions and at most one non zero entry on the last $k - k_0$ positions. Using such vectors has two advantages:

- The corresponding power products can be computed very quickly if the powers $\wp_i^{\,j}$, $1 \leq i \leq k_0$, $1 \leq j \leq B$ have been precomputed.

- The resulting relations matrix is very sparse.

The non zero entry on the last $k - k_0$ positions is used to make the relation matrix very likely to be non singular.

Here is the algorithm for finding the relation matrix:

Algorithm 2

(1) Determine K and k_0.

(2) Precompute the ideal powers $\mathbf{a}_{ij} = \wp_j^{\,i}$ for $j \in K$, $1 \leq i \leq B$.

(3) Set $n = 0$ (number of relations already found) and set $m_i = 0$ $(1 \leq i \leq k)$ (vector to mark the positions already represented in the relation matrix).

(4) If $n \geq k+5$ (enough relations are found) and $m_i = 1$ for $1 \leq i \leq k$ (all positions are marked) then the algorithm terminates.

(5) Choose at random pairwise different positions $\nu_1, \ldots \nu_{k_1} \in K$ and exponents $v_{\nu_1}, \ldots v_{\nu_{k_1}} \in \{1, \ldots B\}$.
Set $v_i = 0$ for $1 \leq i \leq k$, $i \notin \{\nu_1, \ldots \nu_{k_1}\}$.
If the set $J = \{i \ : \ 1 \leq i \leq k, \ m_i = 0, \ , \ i \neq \nu_j \text{ for } 1 \leq j \leq k_1\}$ is not empty, choose at random a position $\mu \in J$ and set $v_\mu = 1$.

(6) Compute the reduced ideal $\mathbf{a}'$ in the class of $\mathbf{a} = \prod_{i=1}^{k} \wp_i^{v_i}$.

(7) Try to find a factorization $\mathbf{a}' = \prod_{i=1}^{k} \wp_i^{v_i'}$.
If there is no such factorization go back to step (5).

(8) Compute the relation vector $\underline{e} = \underline{v} - \underline{v}'$.

(9) Find a position $i \in \{1, \ldots k\}$ with $m_i = 0$ and $e_i \neq 0$. If there is one, set $m_i = 1$. Set $n = n + 1$ and go to step (4).

The computation of the power product in step (6) is done by means of Shanks algorithms NUCOMP and NUDUPL [13] which yield a considerable speedup compared to the standard methods. The factorization process (step (7)) is speeded up by a modification of the early abort strategy (see e.g. [9]).

3.3   Large prime relations

Another speed up of the algorithm is achieved by using large prime relations. The idea of collecting large prime relations is well known from integer factoring algorithms (see e.g. [9]). It can be adopted for our purposes.

In step (7) of Algorithm 2 we have to test whether there is a complete factorization of $\mathbf{a}'$ over F. Now assume we find two incomplete factorizations with the same cofactor $\wp'$:

$$\mathbf{a}_j' = \prod_{i=1}^{k} \wp_i^{v_{ij}'} \cdot \wp'^{w_j'}$$

for $j = 1, 2$ where $\wp' \notin F$ is a prime ideal and $w_1', w_2' \in \{\pm 1\}$. Then we can construct a new relation vector from the equation

$$1 = (\mathbf{a}_1 \cdot (\mathbf{a}_1')^{-1})(\mathbf{a}_2 \cdot (\mathbf{a}_2')^{-1})^{-w_1' w_2'} = \prod_{i=1}^{k} \wp_i^{(v_{i1} - v_{i1}') - w_1' w_2'(v_{i2} - v_{i2}')}.$$

This idea can be used to modify Algorithm 2 as follows: Whenever we find an incomplete factorization with a prime cofactor $\wp'$ satisfying $N(\wp') \leq C$ for an appropriately chosen constant $C < (\log D)^4$ we store that factorization. If another incomplete factorization with the same cofactor is found, we have a new relation. We currently use $C = 10(\log D)^2$ and are experimenting to find out optimal values for C for different sizes of D.

This means that step (7) of the algorithm is replaced by two steps:

(7.1) Try to find a factorization of $\mathbf{a}'$ over F. If there is a complete factorization goto step (8). If there is a large prime factorization with $N(\wp') \leq C$ go to step (7.2). Else go back to step (5).

(7.2) If the the cofactor $\wp'$ also occures as a cofactor in a large prime factorization that was found previously compute the new relation and go to step (8) else store $\wp'$, $\mathbf{a}'$, $\underline{v}$ and $\underline{v}'$ in the list of large prime factorizations.

4 Parallelization

4.1 The general idea

In this section we describe a parallel version of the class group algorithm that can be implemented using the LIPS-system for distributed computation [2]. The LIPS-system works as follows:

There is one *host computer* which manages the distribution and puts the results together and an indefinite number of *client computers* which perform the actual computation.

The LIPS-system has two important features which influence the design of the parallel algorithm:

- The system only uses the idle time of the client computers. So one cannot be sure that a certain amount of CPU time is available on a certain machine.

- The LIPS system uses the standard computer network Ethernet. So there is no memory sharing and there should not be much communication between the clients.

Algorithm 2 for constructing the relation matrix consists of a big loop (steps (5) to (9)). In each iteration the algorithm checks whether a random vector yields a relation. These iterations are relatively independand of each other and therefore this part of the algorithm is well suited for distributed computation. So in our algorithm the clients find the relations and large prime factorizations.

Since large prime factorizations having the same cofactor may be found on different clients there is a need of communication to combine large prime factorizations to relations. So relations and large prime factorizations have to be sent from the clients to the host and handled there.

Moreover, the vector $(m_1, \ldots m_k)$, which carries information about the factor basis elements that are already represented in the relation matrix, has to be updated on the host. Since changes to this vector influence the choice of the random vectors $\underline{v}$, those changes have to be sent from the host to the clients.

So the communication complexity of this algorithm is higher than for the Multiple polynomial quadratic sieve factoring algorithm.

4.2 The algorithms

The ideas mentioned above lead to the following algorithms:

Algorithm 3 (Host)

(1) Start the clients, send D and the set F to them.

(2) Set $n = 0$ and $m_i = 0$ $(1 \leq i \leq k)$.

(3) If $n \geq k+5$ (enough relations are found) and $m_i = 1$ for $1 \leq i \leq k$ (all positions are marked) stop all clients and terminate the algorithm.

(4) Wait for relations or large prime factorizations sent from the clients. If a relation is received go to step (6). If a large prime factorization is received go to step (5).

(5) If the prime ideal $\wp'$ occures as a cofactor in the stored list of large prime factorizations compute the new relation and go to step (6) else store the large prime factorization in the list and go back to step (4).

(6) Find a position $i \in \{1, \ldots k\}$ with $m_i = 0$ and $e_i \neq 0$. If there is one set $m_i = 1$ and send i to all clients.

Algorithm 4 (Client)

(1) Determine K and k_0.

(2) Precompute the ideal powers $\mathbf{a}_{ij} = \wp_j{}^i$ for $j \in K$, $1 \leq i \leq B$.

(3) Set $m_i = 0$ for $1 \leq i \leq k$.

(4) If there is a message i from the host, set $m_i = 1$.

(5) Choose the random exponent vector $\underline{v}$ as described in step (5) of Algorithm 2.

(6) Compute the reduced ideal $\mathbf{a}'$ in the class of $\mathbf{a} = \prod\limits_{i=1}^{k} \wp_i^{v_i}$.

(7) Try to find a factorization of $\mathbf{a}'$ over F. If there is a complete factorization or a large prime factorization with $N(\wp) \leq C$ send the corresponding data to the host.

(8) Go back to step (4).

5   Computational results

As an example for the efficiency of our parallel algorithm we give computational results for a 55-digit discriminant. We computed

Table 1

absolut discriminant d	$4 \cdot 10^{54} + 4$
size k of the factor basis	4585
class number h	1056175002108254379317829632
structure of the class group $(m_i,\ m_i \neq 1,\ 1 \leq i \leq k)$	2, 2, 2, 2, 2, 3300546881588294935368 2176

In Table 2 below we list the data for the parallel computation of the relation matrix on 14 clients (Algorithm 4). The computation started on November 30th at 13.49 and was finished on December 12th at 7.58. So the real running time was 282.2 hours.

In the first column we give the name of the clients. In the second column we give the CPU-time used on the clients. In the third column we give the number of vectors $\underline{v}$ which were tested. In the fourth column we give the number of successes, i.e. relations or large prime factorizations sent from the clients to the host.

Table 2

client name	CPU time (in hours)	number of trials ($\cdot 10^6$)	number of successes
crypt1	149.4	3.18	665
crypt2	213.8	4.58	953
cauchy	281.9	6.91	1412
fermat	279.8	6.86	1390
fourier	85.3	2.17	423
gauss	281.1	6.89	1395
hilbert	281.2	6.86	1392
laplace	276.2	6.78	1363
riemann	281.5	6.86	1431
batman	234.3	4.83	983
robin	125.3	2.46	529
p1sun	212.8	5.37	1037
p2sun	125.2	3.07	635
cadsun	104.5	2.69	523

Algorithm 3 (host) was also performed on one of the workstations mentioned above (*crypt1*). Its CPU time was 5.7 hours.

Concluding this description we list the running times of all steps of Algorithm 1:

Table 3

	CPU time
step (1)	7 seconds
step (2)	see above
step (3)	27.3 hours
step (4)	not performed
step (5)	3 seconds

Bibliography

[1] E. Bach, *Explicit bounds for primality testing and related problems*, Math. Comp **55** (1990), 355-380.

[2] J. Buchmann, M. Diehl, R. Roth, *LIPS - a system for distributed applications*, preprint 1991.

[3] J. Buchmann, S. Düllmann, *A probabilistic class group and regulator algorithm and its implementation*, Computational Number Theory (A. Pethö, M. Pohst, H.C. Williams, H.G. Zimmer, eds.), Walter de Gruiter, Berlin, 1991, 53-72.

[4] J. Buchmann, S. Düllmann, H.C. Williams, *On the complexity and efficiency of a new key exchange system*, Advances in Cryptology - Proceedings EURO-CRYPT'89 (J. Quisquater, J. Vandewalle, eds.), Lecture Notes in Computer Science **434**, Springer, Berlin, 1990, 597-616.

[5] J. Buchmann, H.C. Williams, *On the computation of the class number of an algebraic number field*, Math. Comp. **53** (1989), 679-688.

[6] P.D. Domich, R. Kannan, L.E. Trotter, *Hermite normal form computation using modulo determinant arithmetic*, Math. Op. Res. **12** (1987), 50-59.

[7] J.L. Hafner, K.S. McCurley, *A rigorous subexponential algorithm for computation of class groups*, Journal AMS, to appear.

[8] B.A. LaMacchia, A.M. Odlyzko, *Computation of discrete Logarithms in Prime Fields*, to appear.

[9] C. Pomerance, *Analysis and Comparison of some Integer Factoring Algorithms*, Computational Methods in Number Theory (R. Tijdeman and H.W. Lenstra, Jr., eds.), Mathematisch Centrum Tract 154, Amsterdam, 1982, 89-139.

[10] R. Schoof, *Quadratic fields and factorization*, Computational Methods in Number Theory (R. Tijdeman and H.W. Lenstra, Jr., eds.), Mathematisch Centrum Tract 154, Amsterdam, 1982, 235-286.

[11] M. Seysen, *A Probabilistic Factorization algorithm with quadratic Forms of Negative Discriminant*, Math. Comp. **48** (1987), 737-780.

[12] D. Shanks, *Class number, a theory of factorization and genera*, Proc. Symp. Pure Math. **20** (1969 Institute on number theory), Amer. Math. Soc., Providence, 1971, 415-440.

[13] D. Shanks, *On Gauss and Composition*, Number Theory and Applications (R.A. Mollin, ed.), Kluwer Academic Publishers, 1989, 163-204.

Complexity Measures on Permutations

Volker Claus

Theoretische Informatik
Fachbereich Informatik
Universität Oldenburg
2900 Oldenburg
Germany

Abstract

Complexity measures on permutations are defined which can be used to measure pre-sortedness or to prove lower bounds for sorting. We present four functions C, D, E and F_s, but concentrate on the function C that was proposed by G. Hotz. The minimum and the average values can be determined, but the maximum value turns out to be difficult to compute. Given a sequence, consider the number of exchanges of two arbitrary elements that are necessary for ordering this sequence; then permutations with maximal C-values seem to describe sequences with a maximal number of such exchanges.

1 Introduction

When investigating lower bounds, a typical problem can be described as follows: Let
A be a set of elements and Σ be a set of operations on A, i.e. $\Sigma \subseteq \{f \mid f : A \to A\}$;
then, given two elements $y, z \in A$ we want to know how many operatins of Σ are
needed at least for transforming y into z. This problem can be solved easily if there
is a complexity-measure $\gamma : A \to \mathbb{R}^+$ (= the set of nonnegative real numbers) such
that for every $f \in \Sigma$ there exists a number $n_f \in \mathbb{R}^+$ satisfying

$$|\gamma(f(x)) - \gamma(x)| \leq n_f \qquad \text{for all } x \in A.$$

Consider a sequence for transforming y into z (with $f_j \in \Sigma$ and $u_j \in A$)

$$y = u_0 \xrightarrow{f_1} u_1 \xrightarrow{f_2} u_2 \to \ldots \xrightarrow{f_m} u_m = z$$

then we get:
$$
\begin{aligned}
|\gamma(z) - \gamma(y)| &\leq \sum_{j=0}^{m-1} |\gamma(u_{j+1}) - \gamma(u_j)| \\
&= \sum_{j=0}^{m-1} |\gamma(f_{j+1}(u_j)) - \gamma(u_j)| \\
&\leq \sum_{j=0}^{m-1} n_{f_{j+1}} \quad \leq \quad m \cdot N
\end{aligned}
$$

with $N = Max\{n_f \mid f \in \Sigma\}$. Hence, the number of operations m is bounded by

$$m \geq \tfrac{1}{N} \cdot |\gamma(z) - \gamma(y)|$$

Obviously, this method can be used for many-sorted operations, too. When applying
this idea to a given situation the complexity-measure γ should be chosen in such a
way that the operations of Σ are producing only small differences, i.e. for every $f \in \Sigma$
the number n_f should be as small as possible.

Many problems are based on sorting and searching. Often, sorting means to find a
suitable permutation. Therefore, we will investigate complexity-measures on permutations in this paper we consider, i.e. the set A to be the set $\mathcal{S}_n$ of all permutations
of the first n natural numbers

$$\mathcal{S}_n = \Big\{\pi \mid \pi : \{1, \ldots, n\} \to \{1, \ldots, n\} \text{ bijective}\Big\}.$$

A well known complexity-measure is the number of inversions $INV : \mathcal{S}_n \to \mathbb{R}^+$ defined
by

$$INV(\pi) = |\{(i, j) \mid 1 \leq i < j \leq n, \text{and } \pi(i) > \pi(j)\}|$$

for all $\pi \in \mathcal{S}_n$. Let Σ be the set of all transpositions, i.e.

$$\Sigma = \{\varrho_1, \varrho_2, \ldots, \varrho_{n-1}\},$$

and for all $\pi \in \mathcal{S}_n$ and $1 \leq i \leq n - 1$ the permutation $\pi' = \varrho_i(\pi)$ is defined by

$$\pi'(j) = \begin{cases} \pi(j) & \text{if } i \neq j \text{ and } i+1 \neq j \\ \pi(i) & \text{if } i+1 = j \\ \pi(i+1) & \text{if } i = j \end{cases}$$

(Note that Σ can be regarded as a subset of $\mathcal{S}_n$.) Let $id_n \in \mathcal{S}_n$ be the identity, and $\alpha \in \mathcal{S}_n$ be the reverse ordering, i.e.

$$\alpha(j) = n - j + 1 \qquad \text{(for all } 1 \leq j \leq n\text{)}.$$

Then, given a permutation $\pi \in \mathcal{S}_n$ we ask for the minimal number m of transpositions that are necessary to transform π into id_n.

Choosing the complexity-measure INV we get immediately:

$$|INV(\varrho_i(\pi)) - INV(\pi)| = 1 \qquad \text{for all } i = 1, \ldots, n-1$$

and therefore

$$m \geq \tfrac{1}{1} \cdot |INV(\pi) - INV(id_n)| = INV(\pi)$$

because of $INV(id_n) = 0$. From

$$0 \leq INV(\pi) \leq \tfrac{1}{2} \cdot n \cdot (n-1) \quad \text{and} \quad INV(\alpha) = \tfrac{1}{2} \cdot n \cdot (n-1)$$

we conclude that there exists a permutation π which cannot be transformed by Σ into id_n with less than $\tfrac{1}{2} \cdot n \cdot (n-1)$ steps.

If we replace Σ by another set of operations (e.g. $\Sigma' = \{\alpha, \varrho_1, shiftleft\}$) we can try to prove lower bounds by using other complexity-measures. There are some measures known in the literature. They are often introduced to make use of the presortedness of lists. Examples are *inv, runs, las, rem,* and *exc* ([1], [2]). But it is not the aim of this paper to restrict the set of sequences to a subset each element of which can be sorted in optimal time using a (special) algorithm. Here, we follow a suggestion of G. Hotz, and present a complexity measure C that is supposed to produce (sometimes) good lower bounds w.r.t. given sets of operations Σ.

2 Definitions and Basic Properties

In the following we assume n to be an arbitrary but fixed natural number (greater than 1) and $\mathcal{S}_n$ to be the set of all permutations on the first n natural numbers. A permutation $\pi \in \mathcal{S}_n$ is usually represented by the n-tupel $(\pi_1, \pi_2, \ldots, \pi_n)$ with $\pi_i = \pi(i)$ for $i = 1, \ldots, n$.

We have to deal with absolute values. Before starting with the proper definitions we prove two helpful lemmas.

Lemma 1 *Consider a natural number m and two ordered m-tupels over $\mathbb{R}^+$ $(a_1, a_2, \ldots \ldots, a_m)$ and $(b_1, b_2, \ldots, b_m)$ with $a_i \leq a_j$ and $b_i \leq b_j$ for $i \leq j$. Concatenate these two m-tupels yielding a $2m$-tupel and reorder this. The resulting $2m$-tupel is denoted by $(c_1, c_2, \ldots, c_m, d_1, d_2, \ldots, d_m)$ with $c_1 \leq c_2 \leq \ldots \leq c_m \leq d_1 \leq d_2 \leq \ldots \leq d_m$. Then, for all permutations $\pi, \eta \in \mathcal{S}_m$ the following inequality holds:*

$$\sum_{i=1}^{m} |a_{\pi(i)} - b_{\eta(i)}| \quad \leq \quad \sum_{i=1}^{m} d_i - \sum_{i=1}^{m} c_i$$

This lemma is evident because of

$$\sum_{i=1}^{m} |a_{\pi(i)} - b_{\eta(i)}| = \sum_{i=1}^{m} Max(a_{\pi(i)}, b_{\eta(i)}) - \sum_{i=1}^{m} Min(a_{\pi(i)}, b_{\eta(i)})$$
$$\leq \sum_{i=1}^{m} d_i - \sum_{i=1}^{m} c_i$$

Lemma 2 *Consider a natural number m and two ordered m-tupels over $\mathbb{R}^+$ $(a_1, a_2, \ldots \ldots, a_m)$ and $(b_1, b_2, \ldots, b_m)$ with $a_i \leq a_j$ and $b_i \leq b_j$ for $i \leq j$. For all permutations $\pi, \eta \in \mathcal{S}_m$ the following inequality holds:*

$$\sum_{i=1}^{m} |a_i - b_i| \leq \sum_{i=1}^{m} |a_{\pi(i)} - b_{\eta(i)}| \leq \sum_{i=1}^{m} |a_i - b_{m+1-i}|.$$

Proof: By induction. Consider $m = 2$. Only three of the six possibilities to order the four numbers $a_1, a_2, b_1,$ and b_2 are really different: $a_1 \leq a_2 \leq b_1 \leq b_2$, $a_1 \leq b_1 \leq a_2 \leq b_2$, and $a_1 \leq b_1 \leq b_2 \leq a_2$. The reader may verify the inequality for these three cases.

We are only considering the right part of the inequality. The left part can be proven in a similar way. Assume the right hand side of the inequality holds for some m and for all permutations of $\mathcal{S}_m$. Consider the next natural number $m + 1$ and two arbitrary permutations $\pi, \eta \in \mathcal{S}_{m+1}$. By rearranging the sum one can determine a permutation $\sigma \in \mathcal{S}_{m+1}$ such that

$$\sum_{i=1}^{m+1} |a_{\pi(i)} - b_{\eta(i)}| = \sum_{j=1}^{m+1} |a_j - b_{\sigma(j)}|$$
$$= \sum_{j=1}^{m} |a_j - b_{\sigma(j)}| + |a_{m+1} - b_{\sigma(m+1)}|.$$

If $\sigma(m + 1) = 1$ the inequality follows by induction. Let $\sigma(m + 1) \neq 1$. Fix the position q with $\sigma(q) = 1$ and define the permutation $\sigma' \in \mathcal{S}_{m+1}$ by

$$\sigma'(j) = \begin{cases} \sigma(j) & \text{if } q \neq j \neq m + 1 \\ \sigma(m + 1) & \text{if } j = q \\ \sigma(q) = 1 & \text{if } j = m + 1 \end{cases}$$

Thus, we get

$$\sum_{i=1}^{m+1} |a_i - b_{\sigma(i)}| = \sum_{i=1, i \neq q}^{m} |a_i - b_{\sigma(i)}| + |a_q - b_1| + |a_{m+1} - b_{\sigma(m+1)}|$$
$$\leq \sum_{i=1, i \neq q}^{m} |a_i - b_{\sigma(i)}| + |a_q - b_{\sigma(m+1)}| + |a_{m+1} - b_1|$$
$$\text{(because the lemma holds for sets of two elements)}$$
$$= \sum_{i=1}^{m} |a_i - b_{\sigma'(i)}| + |a_{m+1} - b_1|.$$

In this sum σ' can be interpreted as a permutation of m elements, and therefore the inequality follows by induction. $\blacksquare$

By H_k we denote the harmonic series

$$H_k := \sum_{j=1}^{k} \tfrac{1}{j}$$

especially: $H_0 := 0$. The values of H_k are growing like the natural logarithm $ln(k)$, expressed by the well-known relation

$$H_k = ln(k) + \hat{\gamma} + \tfrac{1}{2k} - \tfrac{1}{12k^2} + O(\tfrac{1}{k^4})$$

with $\hat{\gamma} = 0.57721566\ldots$ (Eulerian constant).

G. Hotz suggested [3] to define a measure in accordance with the differential quotient $\frac{(f(a)-f(b))}{(a-b)}$ of the analytic functions, but using the absolute values and adding all combinations.

Definition 1 $C : S_n \to \mathrm{I\!R}^+$ *is defined by*

$$
\begin{aligned}
C(\pi) &= \sum_{i,j=1;i\neq j}^{n} \left| \tfrac{\pi_j - \pi_i}{j-i} \right| \\
&= 2 \cdot \sum_{i=1}^{n-1} \sum_{j=i+1}^{n} \tfrac{|\pi_j - \pi_i|}{j-i}
\end{aligned}
$$

(C depends on n but we omit n for brevity.)

It will be shown that $C(\pi) \geq n \cdot (n-1)$ and $C(id_n) = n \cdot (n-1)$. Therefore the normalized C-function is denoted by

$$\hat{C}(\pi) := C(\pi) - n \cdot (n-1).$$

Definition 2 *For* $1 \leq i \leq n-1$ $\quad S_i : S_n \to \mathrm{I\!R}^+$ *is defined by* $S_i(\pi) = \sum_{j=1}^{n-i} |\pi_{j+i} - \pi_j|$
(The sum of the absolute differences of numbers of distance i in the representation of π. S_i depends on n but n is omitted, too.)

By collecting the terms with the same denominator we get:

$$
\begin{aligned}
C(\pi) &= 2 \cdot \sum_{i=1}^{n-1} \sum_{j=i+1}^{n} \tfrac{|\pi_j - \pi_i|}{j-i} = 2 \cdot \sum_{k=1}^{n-1} \tfrac{1}{k} \cdot \sum_{j=1}^{n-k} |\pi_{j+k} - \pi_j| \\
&= 2 \cdot \sum_{k=1}^{n-1} \tfrac{1}{k} \cdot S_k(\pi)
\end{aligned}
$$

Example 1 *Consider* $n = 4, \pi = (1\ 2\ 4\ 3)$ *and* $\pi' = (2\ 4\ 3\ 1)$, *then*

$$
\begin{array}{ll}
S_1(\pi) = 4 & S_1(\pi') = 5 \\
S_2(\pi) = 4 & S_2(\pi') = 4 \\
S_3(\pi) = 2 & S_3(\pi') = 1 \\
C(\pi) = 2 \cdot (4 + \tfrac{4}{2} + \tfrac{2}{3}) = \tfrac{40}{3} & C(\pi') = 2 \cdot (5 + \tfrac{4}{2} + \tfrac{1}{3}) = \tfrac{44}{3}
\end{array}
$$

The functions D, E and F_s are obtained from C by small alternations.

Definition 3 $D : \mathcal{S}_n \to \mathrm{I\!R}^+$ *is defined by*

$$D(\pi) = \sum_{i,j=1; i \neq j}^{n} \frac{\pi_j - \pi_i}{j - i} = 2 \cdot \sum_{i=1}^{n-1} \sum_{j=i+1}^{n} \frac{\pi_j - \pi_i}{j - i}$$

For $i = 1, \ldots, n-1$ the functions $T_i : \mathcal{S}_n \to \mathrm{I\!R}^+$ are given by

$$T_i(\pi) = \sum_{j=1}^{n-i} (\pi_{j+i} - \pi_j)$$

Again, it is easy to conclude

$$D(\pi) = 2 \cdot \sum_{k=1}^{n-1} \tfrac{1}{k} \cdot T_k(\pi)$$

Definition 4 *Let $E : \mathcal{S}_n \to \mathrm{I\!R}^+$ be the function*

$$E(\pi) = 2 \cdot \sum_{k=1}^{n-1} \tfrac{1}{k} \cdot |T_k(\pi)|$$

Definition 5 *Let $s \geq 1$ be a natural number, then $F_s : \mathcal{S}_n \to \mathrm{I\!R}^+$ is defined by*

$$F_s(\pi) = 2 \cdot \sum_{k=1}^{n-1} \left(\tfrac{1}{k}\right)^s \cdot S_k(\pi)$$

Especially: $F_1 = C$.

Lemma 3 *For every $\pi \in \mathcal{S}_n$:*

$$\sum_{k=1}^{n-1} S_k(\pi) = \tfrac{1}{6}(n-1) \cdot n \cdot (n+1) = \binom{n+1}{3}$$

Proof:

$$
\begin{aligned}
2 \cdot \sum_{k=1}^{n-1} S_k(\pi) &= 2 \cdot \sum_{k=1}^{n-1} \sum_{j=1}^{n-k} |\pi_{k+j} - \pi_j| \quad \text{(reorder again)} \\
&= \sum_{i=1}^{n} \sum_{j=1, j \neq i}^{n} |\pi_j - \pi_i| \quad \text{(Every difference } |j - i| \text{ occurs} \\
&\qquad\qquad\qquad\qquad\qquad \text{exactly once in this sum.)} \\
&= \sum_{i=1}^{n} \sum_{j=1, j \neq i}^{n} |j - i| \\
&= 2 \cdot \sum_{i=1}^{n-1} \sum_{j=i+1}^{n} (j - i) \\
&= 2 \cdot \sum_{i=1}^{n-1} \sum_{k=1}^{n-i} k = \ldots = \tfrac{1}{3}(n-1) \cdot n \cdot (n+1)
\end{aligned}
$$

∎

As a corollary we get $C(\pi) < \tfrac{1}{3}n^3$. We will improve this bound in the next chapter.
Denotation: $\lfloor x \rfloor$ denotes the greatest natural number less than or equal to x, for any $x \in \mathrm{I\!R}^+$.

Lemma 4 *For every permutation $\pi \in \mathcal{S}_n$ the functions S_i are satisfying the following inequalities:*

$$(i) \quad n - i \leq S_i(\pi) \qquad \text{for } i = 1, 2, \ldots, n - 1.$$
$$(ii) \quad S_i(\pi) \leq \lfloor \tfrac{1}{2} n^2 \rfloor - i^2 \quad \text{for } i = 1, 2, \ldots, \lfloor \tfrac{n-1}{2} \rfloor.$$
$$(iii) \quad S_i(\pi) \leq i \cdot (n - i) \quad \text{for } i = \lfloor \tfrac{n-1}{2} + 1 \rfloor, \ldots, n - 1.$$

Proof: The lower bound (i) is obvious because of

$$S_i(\pi) \geq \sum_{j=1}^{n-i} 1 = n - i.$$

Consider the definition of $S_i(\pi)$

$$S_i(\pi) = \sum_{j=1}^{n-i} |\pi_{j+i} - \pi_j|.$$

According to lemma 1 we assign the two $(n - i)$-tupels $t_1 := (\pi_1, \pi_2, \ldots, \pi_{n-i})$ and $t_2 := (\pi_{i+1}, \pi_{i+2}, \ldots, \pi_n)$ to this representation. t_1 and t_2 are concatenated and the resulting $2(n-i)$-tupel is reordered, yielding $t = (c_1, c_2, \ldots, c_{n-i}, d_1, d_2, \ldots, d_{n-i})$ with $1 \leq c_1 \leq c_2 \leq \ldots \leq c_{n-i} \leq d_1 \leq d_2 \leq \ldots \leq d_{n-i} \leq n$.

Case 1: $i > \lfloor \tfrac{n-1}{2} \rfloor$. Then $n - i < i + 1$, and all components of t_1 and t_2 are pairwise different. Therefore:
$1 \leq c_1 < c_2 < \ldots < c_{n-i} < d_1 < d_2 < \ldots < d_{n-i} \leq n$, and
$1 \leq c_1, \ 2 \leq c_2, \ldots, (n - i) \leq c_{n-i}, \ d_{n-i} \leq n, \ d_{n-i-1} \leq n - 1, \ldots$ etc.
Now we apply lemma 1:

$$
\begin{aligned}
S_i(\pi) &= \sum_{j=1}^{n-i} |\pi_{j+i} - \pi_j| \ \leq \ \sum_{j=1}^{n-i} d_j - \sum_{j=1}^{n-i} c_j \\
&\leq \ \sum_{k=i+1}^{n} k - \sum_{j=1}^{n-i} j \ = \ i \cdot (n - i)
\end{aligned}
$$

Case 2: $i \leq \lfloor \tfrac{n-1}{2} \rfloor$. Then there exist exactly $(n - 2 \cdot i)$ components occuring both in t_1 and t_2. Furthermore all natural numbers from 1 to n must be components of t. In the case that n is even we can assume

$$\{1, 2, \ldots, \tfrac{n}{2}\} = \{c_1, c_2, \ldots, c_{n-i}\}$$
and the first $(\tfrac{n}{2} - i)$ numbers are occuring twice in t,
$$\{\tfrac{n}{2} + 1, \ldots, n\} = \{d_1, d_2, \ldots, d_{n-i}\}$$
and the numbers $\tfrac{n}{2} + i + 1, \ldots, n$ are occuring twice in t.

Hence,

$$
\begin{aligned}
S_i(\pi) &\leq \ \sum_{j=1}^{n-i} d_j - \sum_{j=1}^{n-i} c_j \\
&\leq \ \sum_{k=\frac{n}{2}+1}^{n} k + \sum_{k=\frac{n}{2}+i+1}^{n} k - \sum_{j=1}^{\frac{n}{2}} j - \sum_{j=1}^{\frac{n}{2}-i} j \\
&= \ldots = \tfrac{1}{2} n^2 - i^2
\end{aligned}
$$

Analogously, if n is odd the upper bound is $\frac{1}{2}(n^2 - 1) - i^2$. This completes the proof of lemma 4. ∎

Lemma 5 *For all permutations $\pi \in \mathcal{S}_n$ and all $i = 1, 2, \ldots, n - 1$:*

$$S_i(\pi) + S_{n-i}(\pi) \leq \lfloor \tfrac{1}{2}n^2 \rfloor$$

Proof: (This proposition does not follow from lemma 4.) We consider a permutation π and extend the representation of π by the first i components, i.e. let $\pi_{n+j} := \pi_j$ for $j = 1, 2, \ldots, i$. Now it is easy to verify:

$$S_i(\pi) + S_{n-i}(\pi) = \sum_{j=1}^{n} |\pi_{j+i} - \pi_j|.$$

(S_{n-i} can be regarded as the cyclic completion of S_i, which becomes clear when representing π by a cycle instead of a vector.)

Using lemma 2 both n-tupels $(a_1, \ldots, a_n)$ and $(b_1, \ldots, b_n)$ are chosen as the n-tupel $(1, 2, \ldots, n)$, and from lemma 2 we get (assume n is even):

$$\begin{aligned}
S_i(\pi) + S_{n-i}(\pi) &\leq \sum_{j=1}^{n} |(n + 1 - j) - j| \\
&= 2 \cdot \sum_{j=1}^{\frac{n}{2}} (n + 1 - 2j) = \ldots = \tfrac{1}{2}n^2.
\end{aligned}$$

If n is odd, then the upper bound is $\frac{1}{2}(n^2 - 1)$. ∎

The functions T_i are simpler than the S_i's. From the definition we derive

$$T_i(\pi) = \sum_{j=i+1}^{n} \pi_j - \sum_{j=1}^{n-i} \pi_j, \quad \text{and so we get:}$$

Lemma 6 *For all $\pi \in \mathcal{S}_n$ and $1 \leq i \leq n - 1$:*
  (i) $T_i(\pi) = T_{n-i}(\pi)$
  (ii) $T_i(\pi) = T_{i-1}(\pi) + (\pi_{n-i+1} - \pi_i)$ *(Define: $T_0(\pi) := 0$)*
  (iii) $|T_i(\pi)| \leq S_i(\pi)$
  (iv) $|T_i(\pi)| \leq i \cdot (n - i)$ *(from (i), (iii), and lemma 4 (iii))*

Lemma 7 *For all $\pi \in \mathcal{S}_n$:*

$$\begin{aligned}
D(\pi) &= 2 \cdot \sum_{j=1}^{n} \pi_j \cdot (H_{j-1} - H_{n-j}), \text{ and} \\
D(\pi) &= 2 \cdot \sum_{i=1}^{n-1} H_i \cdot (\pi_{i+1} - \pi_{n-i}).
\end{aligned}$$

Proof:

$$
\begin{aligned}
D(\pi) &= 2 \cdot \sum_{i=1}^{n-1} \frac{T_i(\pi)}{i} \\
&= 2 \cdot \sum_{i=1}^{n-1} \frac{1}{i} \cdot \sum_{j=1}^{i} (\pi_{n-j+1} - \pi_j) \quad \text{because of lemma 6 (ii)} \\
&= 2 \cdot \sum_{j=1}^{n-1} (\pi_{n-j+1} - \pi_j) \cdot \sum_{k=j}^{n-1} \frac{1}{k} \\
&= 2 \cdot \sum_{j=1}^{n-1} (\pi_{n-j+1} - \pi_j) \cdot (H_{n-1} - H_{j-1})
\end{aligned}
$$

Lemma 7 follows from this equation by rearranging the terms of the sum. ∎

3 The function C

We are interested in the maximum value, the minimum value and the average value
of C for a given natural number n.

Definition 6

$$
\begin{aligned}
MaxC_n &:= Max\{C(\pi) \,|\, \pi \in \mathcal{S}_n\} \\
MinC_n &:= Min\{C(\pi) \,|\, \pi \in \mathcal{S}_n\} \\
AvC_n &:= \frac{1}{n!} \cdot \sum_{\pi \in \mathcal{S}_n} C(\pi)
\end{aligned}
$$

The calculation of AvC_n is simple.

Lemma 8 *For $n \geq 2$:*
$$
\begin{aligned}
(i) \qquad AvC_n &= \tfrac{2}{3}(n+1) \cdot (n \cdot H_{n-1} - n + 1) \\
&= \tfrac{2}{3}(n+1) \sum_{i=1}^{n-1} H_i \\
(ii) \quad AvC_{n+1} - AvC_n &= \tfrac{4}{3}(n+1) \cdot H_n - \tfrac{2}{3} \cdot n
\end{aligned}
$$

Proof: From definition 6 and the representation of C we get:

$$
\begin{aligned}
AvC_n &= \frac{2}{n!} \sum_{\pi \in \mathcal{S}_n} \sum_{i=1}^{n-1} \frac{1}{i} \sum_{j=1}^{n-1} |\pi_{j+i} - \pi_j| \\
&= \frac{2}{n!} \sum_{i=1}^{n-1} \frac{1}{i} \sum_{j=1}^{n-1} \left(\sum_{\pi \in \mathcal{S}_n} |\pi_{j+i} - \pi_j| \right)
\end{aligned}
$$

When summing up all permutations every difference $|k - l|$ must occur as $|\pi_{j+i} - \pi_j|$
exactly $(n-2)!$-times (note: $n \geq 2$); therefore:

$$
\begin{aligned}
\sum_{\pi \in \mathcal{S}_n} |\pi_{j+i} - \pi_j| &= (n-2)! \sum_{k=1}^{n} \sum_{l=1, l \neq k}^{n} |k - l| \\
&= 2 \cdot (n-2)! \sum_{k=1}^{n-1} \sum_{l=k+1}^{n} (l - k) = \ldots = \tfrac{1}{3}(n+1)!
\end{aligned}
$$

Thus

$$AvC_n \;=\; \tfrac{2}{n!} \cdot \tfrac{1}{3}(n+1)! \cdot \sum_{i=1}^{n-1} \tfrac{1}{i}\left(\sum_{j=1}^{n-i} 1\right) \;=\; \tfrac{2}{3}\cdot(n+1)\cdot \sum_{i=1}^{n-1}\tfrac{n-i}{i}$$
$$= \tfrac{2}{3}(n+1)\cdot(n\cdot H_{n-1} - n + 1)$$

Because of $n\cdot H_{n-1} - n + 1 = \sum_{i=1}^{n-1} H_i$ equation (i) follows. Proposition (ii) is a simple consequence of (i).   ∎

The minimum value of C is given by $C(id)$.

Lemma 9 *For $n \geq 2$:* $MinC_n = n\cdot(n-1)$.

Proof: For the identity id and the reverse ordering α we get

$$S_i(id) = S_i(\alpha) = i\cdot(n-i) \quad, \text{ for } i = 1, 2, \ldots, n-1,$$

and therefore $C(id) = C(\alpha) = n\cdot(n-1)$.

From the proof of lemma 3 we conclude that for any permutation π the sum

$$\sum_{i=1}^{n-1} S_i(\pi)$$

consists of all differences $j - m$ with $1 \leq j < m \leq n$, i.e. in this sum the number 1 occurs $(n-1)$-times, the number 2 $(n-1)$-times, $\ldots$, and the number $(n-1)$ once. Any sum $\sum_{i=1}^{k} S_i(\pi)$ (for $k \leq n-1$) must contain exactly $\tfrac{1}{2}\cdot k\cdot(2n - k - 1)$ of these numbers. Therefore, a lower bound of this sum is given by:

$$\sum_{i=1}^{k} S_i(\pi) \;\geq\; (n-1)\cdot 1 + (n-2)\cdot 2 + \ldots + (n-k)\cdot k.$$

For a fixed π and for all $0 \leq k \leq n-1$ let V_k be defined by

$$V_k := \sum_{i=1}^{k} S_i(\pi) - \sum_{i=1}^{k}(n-i)\cdot i$$

(especially: $V_0 = 0$). We have already shown: $V_k \geq 0$. Consider now

$$\tfrac{1}{2}(C(\pi) - n\cdot(n-1)) = \sum_{i=1}^{n-1}\tfrac{S_i(\pi)}{i} - \sum_{i=1}^{n-1}(n-i)$$
$$= \sum_{i=1}^{n-1}\tfrac{S_i(\pi)-(n-i)\cdot i}{i} \qquad \text{(Observe:}$$
$$\qquad\qquad\qquad\qquad S_i(\pi)-(n-i)\cdot i = V_i - V_{i-1})$$
$$= \sum_{i=1}^{n-1}\tfrac{1}{i}(V_i - V_{i-1}) = \sum_{i=1}^{n-2}\left(\tfrac{V_i}{i} - \tfrac{V_i}{i+1}\right) + \tfrac{V_{n-1}}{n-1}$$
$$= \sum_{i=1}^{n-2} V_i\cdot\tfrac{1}{i\cdot(i+1)} + \tfrac{V_{n-1}}{n-1} \geq 0 \quad, \text{ and the lemma follows.}$$

∎

From the proof of lemma 9 we get: $C(\pi) = n \cdot (n - 1)$ iff every $V_k = 0$. The reader may verify that this is equivalent to $\pi = id$ or $\pi = \alpha$.

The determination of $MaxC_n$ seems to be difficult. Using a computer we calculated the exact values for $n = 2, 3, \ldots, 11$. By stepwise exchanging (controlled by a trial-and-error-procedure) we got some high C-values for $n = 12, \ldots, 21$, which approximate the corresponding $MaxC_n$. The following table shows these results:

n	permutation $\pi \in \mathcal{S}_n$ with high C-value	$C(\pi)$
2	2 1	2,0000
3	2 3 1	7,0000
4	3 1 4 2	16,6667
5	3 1 5 2 4	29,1667
6	3 5 1 6 2 4	47,4000
7	3 5 1 7 2 6 4	68,4333
8	5 3 7 1 8 2 6 4	95,8857
9	4 7 2 9 5 1 8 3 6	126,9381
10	5 8 2 10 4 7 1 9 3 6	163,9365
11	7 4 9 2 11 6 1 10 3 8 5	205,4635
12	7 4 10 2 12 5 8 1 11 3 9 6	253,0659
13	8 5 11 2 7 13 1 10 4 12 3 9 6	305,1803
14	8 5 12 2 11 6 14 1 9 4 13 3 10 7	363,5144
15	7 11 4 14 2 10 6 15 1 8 13 3 12 5 9	427,2018
16	8 11 4 14 2 12 7 16 1 10 5 15 3 13 6 9	496,6970
17	8 12 4 15 9 2 14 5 17 1 11 7 16 3 13 6 10	571,9664
18	10 6 15 3 12 8 17 1 14 5 18 2 11 7 16 4 13 9	653,2838
19	9 14 4 17 7 12 2 19 5 15 10 1 18 8 13 3 16 6 11	740,4151
20	9 14 5 17 11 3 19 6 13 1 20 8 15 2 18 10 4 16 7 12	833,2263
21	12 7 16 4 19 9 13 2 21 5 17 11 1 20 8 15 3 18 6 14 10	933,0037

Remark: The value of C is invariant with respect to the composition of α, i.e. for all π:

$$C(\pi) = C(\pi\alpha) = C(\alpha\pi) = C(\alpha\pi\alpha).$$

Therefore, further permutations are corresponding to the C-values listed in the table. In the cases $n = 5$ and $n = 7$ there are essentially different permutations with the maximum value of C.

An upper bound for $MaxC_n$ can be derived from lemma 4. Let n be odd and $q = \frac{n-1}{2}$. Then, for any π:

$$\begin{aligned}
\tfrac{1}{2}C(\pi) &= \sum_{i=1}^{n-1} \frac{S_i(\pi)}{i} = \sum_{i=1}^{q} \frac{S_i(\pi)}{i} + \sum_{i=q+1}^{n-1} \frac{S_i(\pi)}{i} \\
&\leq \sum_{i=1}^{q} \tfrac{1}{2}(n^2 - 1) \cdot \tfrac{1}{i} - \sum_{i=1}^{q} i + \sum_{i=q+1}^{n-1} (n - i) = \tfrac{1}{2}(n^2 - 1) \cdot H_q
\end{aligned}$$

Let n be even and $q = \frac{n}{2} - 1$. Then:

$$\tfrac{1}{2}C(\pi) \;=\; \sum_{i=1}^{q}\frac{S_i(\pi)}{i} + \sum_{i=q+1}^{n-1}\frac{S_i(\pi)}{i} \;\leq\; \sum_{i=1}^{q}\tfrac{1}{2}n^2\cdot\tfrac{1}{i} - \sum_{i=1}^{q} i + \sum_{i=q+1}^{n-1}(n-i)$$

$$= \tfrac{1}{2}n^2 H_q + \tfrac{n}{2}$$

It follows:

Lemma 10

$$MaxC_n \;\leq\; \begin{cases} n^2 H_{\frac{n-2}{2}} + n & ,\ n\ \text{even} \\[2mm] (n^2-1)H_{\frac{n-1}{2}} & ,\ n\ \text{odd} \end{cases}$$

Using lemma 5 this upper bound can be sharpened but we were not able to decrease the factor 1.0 of the term $n^2 \cdot H_{\lfloor\frac{n-1}{2}\rfloor}$. Therefore, we omit those improvements.

Example: We construct a series of permutations $\varrho^{(2)}, \varrho^{(4)}, \varrho^{(6)}, \ldots$ with $\varrho^{(m)} \in \mathcal{S}_m$ such that $S_i(\varrho^{(m)})$ is maximal for odd i and $i \leq \frac{m}{2}$. Let $\varrho^{(2)} := (2\ 1)$ and define recursively

$$\varrho_1^{(m+2)} \;=\; \text{if } \varrho_1^{(m)} = \tfrac{m}{2} \text{ then } \tfrac{m}{2}+2 \text{ else } \tfrac{m}{2}+1$$

$$\varrho_{m+2}^{(m+2)} \;=\; \text{if } \varrho_1^{(m)} = \tfrac{m}{2} \text{ then } \tfrac{m}{2}+1 \text{ else } \tfrac{m}{2}+2$$

$$\varrho_{i+1}^{(m+2)} \;=\; \varrho_i^{(m)} + \big(\text{if } \varrho_i^{(m)} > \tfrac{m}{2} \text{ then } 2 \text{ else } 0\big)$$

This determines the series

$$\varrho^{(4)} = (2\ 4\ 1\ 3),\ \varrho^{(6)} = (4\ 2\ 6\ 1\ 5\ 3),\ \varrho^{(8)} = (4\ 6\ 2\ 8\ 1\ 7\ 3\ 5)\ \text{etc.}$$

A straightforward calculation yields

$$S_i(\varrho^{(m)}) = \begin{cases} \tfrac{1}{2}m^2 - i^2 & i\ \text{odd},\quad i \leq \tfrac{m}{2} \\[2mm] (m - \tfrac{3}{2}i)\cdot i & i\ \text{even},\quad i \leq \tfrac{m}{2} \\[2mm] (m-i)^2 & i\ \text{odd},\quad i > \tfrac{m}{2} \\[2mm] \tfrac{1}{2}(m-i)^2 & i\ \text{even},\quad i > \tfrac{m}{2} \end{cases}$$

and finally

$$C(\varrho^{(m)}) = 2\cdot m^2\cdot\big(H_{m-1} - \tfrac{3}{4}H_{\frac{m}{2}-1}\big) - \tfrac{3}{4}m(m+2).$$

In fact, this is a lower bound for $MaxC_n$ but a bad one. Though these values are maximal for $m = 2, 4, 6, 8$, they become smaller than AvC_n for $m > 400$. This behaviour results from the fact that for every even i the S_i-value is very small. By fixing the first and the last number in the vector-representation of $\varrho^{(m)}$ and by permuting first those numbers being less than $\frac{m}{2}$ and then (independently) those numbers being greater than $\frac{m}{2} + 1$, the S_i-values for some even i's (especially for $i = 2$) can be increased without decreasing the S_i-values for odd i's.

E.g. change $\varrho^{(14)} = (8\ 6\ 10\ 4\ 12\ 2\ 14\ 1\ 13\ 3\ 11\ 5\ 9\ 7)$ into $(8\ 4\ 10\ 2\ 12\ 6\ 14\ 1\ 13\ 5\ 11\ 3\ 9\ 7)$ and afterwards into $\overline{\varrho}^{(14)} = (8\ 4\ 11\ 2\ 13\ 6\ 9\ 1\ 14\ 5\ 10\ 3\ 12\ 7)$, then the C-value increases from $C(\varrho^{(14)}) = 358,312434$ to $C(\overline{\varrho}^{(14)}) = 363,128090$.

4 A hint for applications

We have to define a set of operations Σ on $\mathcal{S}_n$. We choose

$$\hat{\Sigma} = \{\alpha, \varrho_1, \ldots, \varrho_{n-1}, shiftleft, shiftright\}$$

where $\alpha, \varrho_1, \ldots, \varrho_{n-1}$ are described in the introduction, and *shiftleft* (resp. *shiftright*) is the operation that transforms $\pi = (\pi_1, \pi_2, \ldots, \pi_n)$ into $(\pi_2, \ldots, \pi_n, \pi_1)$, resp. into $(\pi_n, \pi_1, \ldots, \pi_{n-1})$.

Let $\varrho_{i,j}$ be the permutation that exchanges the positions i and j, i.e.

$$\varrho_{i,j}(\pi)(k) = \begin{cases} \pi(k) & \text{for} \quad i \neq k \neq j \\ \pi(i) & \text{if} \quad k = j \\ \pi(j) & \text{if} \quad k = i \end{cases}$$

Especially: $\varrho_{i,i+1} = \varrho_i$.

A straightforward calculation shows:

$$C(\varrho_{i,j}(\pi)) - C(\pi) = 2 \cdot \sum_{k=1; k \neq i, k \neq j}^{n} (|\pi_i - \pi_k| - |\pi_j - \pi_k|) \cdot \left(\frac{1}{|j-k|} - \frac{1}{|i-k|}\right)$$

From this, you may prove

$$|C(\varrho_{i,j}(\pi)) - C(\pi)| \leq 4 \cdot |\pi_i - \pi_j| \cdot H_{|j-i|} \leq 4 \cdot (n-1) \cdot H_{n-1},$$

but we are interested in the special case $j = i + 1$, and get

$$\begin{aligned}
C(\varrho_i(\pi)) - C(\pi) = \ & 2 \cdot \sum_{k=1}^{i-1} (|\pi_i - \pi_k| - |\pi_{i+1} - \pi_k|) \cdot \left(\frac{1}{i+1-k} - \frac{1}{i-k}\right) \\
& + 2 \cdot \sum_{k=i+2}^{n} (|\pi_i - \pi_k| - |\pi_{i+1} - \pi_k|) \cdot \left(\frac{1}{k-i-1} - \frac{1}{k-i}\right)
\end{aligned}$$

An upper bound for $|\pi_i - \pi_k| - |\pi_{i+1} - \pi_k|$ is $n - 2$, and the worst case can be bounded by

$$\begin{aligned}
C(\varrho_i(\pi)) - C(\pi) \ & \leq \ (n-2) \cdot 4 \cdot \sum_{k=1}^{q-1} \frac{1}{(q+1-k)(q-k)} \quad \text{with } q = \lfloor \tfrac{n}{2} \rfloor \\
& = \ (n-2) \cdot 4 \cdot \sum_{k=1}^{q-1} \frac{1}{k(k+1)} < 4 \cdot (n-2)
\end{aligned}$$

because of $\sum_{k=1}^{q-1} \frac{1}{k(k+1)} = \sum_{k=1}^{q-1} (\frac{1}{k} - \frac{1}{k+1}) = 1 - \frac{1}{q} < 1$. For the operation *shiftright* we get

$$C(\textit{shiftright}(\pi)) - C(\pi) = 2 \cdot \sum_{k=1}^{n-1} \frac{|\pi_n - \pi_k| - |\pi_{n-k} - \pi_n|}{k}$$

An upper bound is (with $q = \lfloor \frac{n}{2} \rfloor$)

$$\begin{aligned} |C(\textit{shiftright}(\pi)) - C(\pi)| &<& 2 \cdot (\tfrac{n-2}{1} + \tfrac{n-4}{2} + \tfrac{n-6}{3} + \ldots + \tfrac{n-2 \cdot q}{q}) \\ &\approx& 2 \cdot n \cdot (H_q - 1) \approx 2 \cdot n \cdot (H_n - 2). \end{aligned}$$

This bound is true for *shiftleft*, too. Because of $C(\alpha(\pi)) = C(\pi)$ we can conclude from lemma 8: Using the operation set $\hat{\Sigma}$ there are needed at least $\frac{n}{3}$ operations in the average for transforming an arbitrary permutation into the identity.

This is not a good bound. But a careful refinement of the method presented here should yield better results. In addition the functions D, E and F_s may be helpful and should be investigated.

The function C could turn out to be most interesting. Its values seem to be maximal for those sequences for which many operations $\varrho_{i,j}$ are needed when transforming them into an ordered sequence.

Bibliography

[1] Knuth, D.E., *The Art of Computer Programming*, Vol. III: Sorting and Searching, Addison Wesley

[2] Mannila, H., *Measures of Presortedness and Optimal Sorting Algorithms*, Technical Report 1984/14, Sarja C, University of Helsinki

[3] Hotz, G., *private communication*, Oberwolfach 1982

Dynamic Hashing in Real Time[1]

Martin Dietzfelbinger[2]
Friedhelm Meyer auf der Heide[2]
Fachbereich 17 · Mathematik – Informatik
und Heinz-Nixdorf-Institut
Universität-GH-Paderborn,
4790 Paderborn
Germany

Abstract

A dynamic hashing scheme that performs in real time is presented. It uses linear space and needs worst case constant time per instruction. Thus instructions can be given in constant length time intervals. Answers to queries given by the algorithm are always correct, the space bound is always satisfied. The algorithm is of the Monte Carlo type; it fails to meet the real time assumption only with probability $O(n^{-c})$, where n is the number of data items currently stored. The constant c can be chosen arbitrarily large.

Further, a parallel version of this dictionary for a p-processor CRCW PRAM is described, where in constant length time intervals p instructions are given via the p processors. For dictionaries of size $n \geq p^{1+\varepsilon}$, the same performance bounds as for the sequential case are obtained.

The construction is based on a new high performance universal class of hash functions.

---

[1]A preliminary version of parts of the material in this paper has been presented at ICALP 90
[2]Supported in part by DFG Grant ME 872/1-4

1   Introduction

A *dictionary* is a data structure that supports three kinds of instructions, namely "Insert x", "Delete x", and "Lookup x". Here the possible *keys* x are taken from some finite *universe* U. This is to be understood as follows: the instruction "Insert x" causes x together with some information associated with x to be stored in the data structure, the instruction "Delete x" causes x to be removed from the data structure; the instruction "Lookup x" returns the information associated with x (or a default message if x is currently not stored). We always assume that the user that provides the instructions waits for one instruction to be processed before entering the following one.

We will be interested in time requirements (for single instructions as well as amortized over a sequence of instructions) and space requirements for implementations of such a data structure.

Classical implementations of dictionaries are dynamic search trees such as AVL-trees, 2-3-trees, etc. (cf. [17]). They need $\Theta(\log n)$ time per instruction for a dictionary of size n.

A dynamic hashing strategy with *constant* time per instruction on the average and linear space was presented in [1]. Here the average is taken over all input sequences. A significant improvement was shown in [4]. Here the expected amortized time per instruction is constant, where the time bound holds for each sequence of instructions; the expected value is taken over all random choices of the randomized algorithm. A scheme with similar features was described in [3]. The main difference is that lookups take worst case constant time in [4] and expected constant time in [3].

Dynamic hashing in real time   The main result of the present paper is an implementation for an optimal dictionary that essentially has worst case constant time per instruction and needs linear space. The algorithm performs in real time and is probabilistic of the Monte Carlo type; that means, there is a certain (small) probability that the scheme fails. Note that the probability space underlying the analysis is induced by the random choices the algorithm makes; no assumptions are made concerning the distribution of the keys occurring in the instructions. We formulate the result in more detail.

Theorem 1 *There is a probabilistic (Monte Carlo type) algorithm for implementing a dictionary with the following features. Let $c \geq 1$ be constant.*

(i) *At any time, the space used is proportional to the number of keys currently stored in the dictionary.*

(ii) *Performing one lookup takes constant time in the worst case.*

(iii) *Insertions and deletions take constant time per operation.*

(iv) *The assertions (i) and (ii) always hold; the probability that (iii) is violated during a sequence of $\frac{1}{2}n$ instructions executed in a dictionary of size n is $O(n^{-c})$.*

(The constants in the time bounds and the space bound depend on c.)

High performance hash functions   The construction of the dictionary is based on a dynamic hashing strategy that uses a novel type of hash functions. We introduce a new class $\mathcal{R}$ of high performance hash functions that can be evaluated in constant time but share many properties of n-universal hash functions, i.e., hash functions that map n keys independently, uniformly into $\{0, \ldots, n-1\}$. The most striking properties of the new class are the following. For each $S \subseteq U$, $|S| \leq n$, there is $\mathcal{R}_S \subseteq \mathcal{R}$ such that:

(*)   Given S, a randomized construction of a random $h \in \mathcal{R}_S$ and the corresponding search table can be done in time $O(n)$ with high probability. The construction of h needs time $O(n^\varepsilon)$, where $0 < \varepsilon \leq 1$ may be arbitrary.

(**)   For a random $h \in \mathcal{R}_S$ and a fixed $j \in \{0, \ldots, n-1\}$, $\Pr(|\{x \in S \mid h(x) = j\}| \geq u)$ is exponentially small in u.

(***) $E\left(\max\{|\{x \in S \mid h(x) = j\}|, 0 \leq j < n\}\right) = O(\log n / \log \log n)$.

Previous work on high performance hash functions   Polynomials of degree up to $d-1$ as hash functions are used e.g. in [8], [4], [5] and many other papers. Such a random polynomial h is only d-wise independent, that means, the random variables $h(x_1), \ldots, h(x_d)$ are independent for any d distinct keys $x_1, \ldots, x_d \in U$, but not for larger sets of keys. If we demand constant evaluation time, thus take d to be constant, we get much weaker properties than mentioned above; for example, if polynomials of degree up to $d-1$ are used, one can only prove $E\left(\max\{|\{x \in S \mid h(x) = j\}|, 0 \leq j < n\}\right) = O(n^{1/d})$.

In [19], a construction of a class of n^η-wise independent hash functions is given that can be evaluated in constant time and constructed in time $O(n^\varepsilon)$, where $0 < \eta < \varepsilon < 1$. This class has similar properties as (**) and (***) from above. The constants in the time and space bounds for that class depend exponentially on r, where n^r is the size of the universe.

Parallel dynamic hashing in real time   We further present a parallel dictionary implemented on a CRCW PRAM with ARBITRARY write conflict resolution. We assume the following consistency rules: If several processors want to access key x, first the deletions, then the insertions, and finally the lookups are executed. If several processors try to insert the same key x in the same round, an arbitrary one of these instructions is performed.

Theorem 2 *There is a probabilistic (Monte Carlo type) algorithm for implementing a parallel dictionary on a CRCW PRAM with p processors with the following features. Let $c \geq 1$, $\varepsilon > 0$ be arbitrary constants.*

(i)   *If n keys are stored in the dictionary, then space $O(\max\{n, p^{1+\varepsilon}\})$ is occupied.*

(ii) *Each instruction needs worst case constant time, i. e., p instructions, one at each processor, can simultaneously be input in constant time intervals.*

(iii) *The algorithm fails to meet (ii) with probability at most n^{-c}, during the execution of $\frac{1}{2}n$ instructions in a dictionary of size n.*

Previously known parallel dictionaries are dynamic perfect hashing schemes that guarantee worst case constant lookup time but only amortized constant update time, see [5]. Very recently, a parallel dictionary with comparable features as ours was presented in [9]. It already works in optimal space for a dictionary of size p. On the other hand, it can only be executed with $p/\log^*(p)$ processors, executing p instructions in optimal time $O(\log^*(p))$, whereas our dictionary allows more parallelism: p processors execute p instructions in constant time. The construction in [9] generalizes static parallel hashing schemes as presented in [2] and [15].

Applications of the new dictionary   (a) *Distributed dictionaries.* Assume that each one of p processors maintains a dictionary, working independently. It is very useful to have a bound for the time the slowest processor needs to perform n instructions. The best estimate that could be obtained by using schemes available before was $O(n \log p)$ (expected). With our scheme, the probability that all dictionaries execute each instruction in constant time is at least $1 - O(p \cdot n^{-c})$. Sequential dictionaries that operate as evenly as that are indispensable when constructing efficient distributed dictionaries (cf. [6] for details of such techniques).

(b) *"Clocked adversaries".* It is shown in [14] that several schemes for implementing dynamic hashing strategies by use of chaining are susceptible to attacks by adversaries that are able to time the algorithm. Even though the adversary (who chooses the instructions) does not have access to the random numbers the algorithm chooses, he can draw conclusions about the structure of the hash function being used from the time the execution of certain instructions takes, and subsequently choose keys to be inserted that make the algorithm perform badly. Lipton and Naughton ask whether the algorithm of [4] is susceptible to such attacks. This question is still open; but the algorithm described in the present paper is immune to such attacks in a natural way: Normally, each instruction takes constant time, which gives no information to the adversary at all. From time to time the algorithm may crash; but the consequence is that the data structure is built anew, redoing all random choices. This results in a data structure that is independent of all previous events, in particular of the situation that caused the crash.

Structure of paper   In Section 2, some basic definitions are given and the technical background for our construction is reviewed. Section 3 contains several probability bounds useful for the analysis of the performance of our hash functions and dictionaries. Section 4 presents the new hash functions; Section 5 the new sequential dictionary, and Section 6 its parallelization.

A more complicated version of the sequential dictionary and a slightly different version of the new hash functions are described in [7].

2   Background: Polynomials as hash functions

Our constructions will be based on the following universal class of hash functions: Let p be prime, $U = \{0, 1, \ldots, p-1\}$ be the *universe*. Consider two parameters: $d \geq 2$, the *degree*, and $s \geq 1$, the *table size*. Define

$$\mathcal{H}_s^d := \{\, h_\alpha \mid \alpha = (\alpha_0, \ldots, \alpha_{d-1}) \in U^d \,\},$$

where for $\alpha = (\alpha_0, \ldots, \alpha_{d-1}) \in U^d$ we let

$$h_\alpha(x) := \left(\sum_{0 \leq i < d} \alpha_i x^i \bmod p \right) \bmod s, \qquad \text{for } x \in U.$$

We will have h chosen uniformly at random from $\mathcal{H}_s^d$. All probabilities are with respect to this probability space; no assumptions about the distribution of the input keys are made. We give some basic definitions and recall two useful facts. Assume for the following that some set $S \subseteq U$ with $|S| = n$ is given.

Definition 1 *Let* $h : U \to \{0, \ldots, s-1\}$.

(a) *The jth bucket is* $B_j^h := \{\, x \in S \mid h(x) = j \,\}$.

(b) *Its size is* $b_j^h := |B_j^h|$.

(c) *The set of keys colliding with* $x \in U$ *is* $B_x^{\mathrm{coll},h} := \{\, y \in S \mid h(y) = h(x) \,\}$.

(d) *The number of keys colliding with* $x \in U$ *is* $b_x^{\mathrm{coll},h} := |B_x^{\mathrm{coll},h}|$.

(e) h *is called l-perfect for* S *if* $b_i^h \leq l$ *for all* $i \in \{0, \ldots, s-1\}$.

(f) *If h is chosen at random from some class, we write B_j for the random set B_j^h and b_j for the random variable b_j^h, analogously for B_x^{coll} and b_x^{coll}.*

Fact 1 *([4]) Let $n \leq s$. For each d there is a constant c_d (which can be assumed to be smaller than $\frac{1}{2}$) so that for all $S \subseteq U$ with $|S| = n$ and all h randomly chosen from $\mathcal{H}_s^d$ we have*

$$\Pr\Big(h \text{ is } (d-1)\text{-perfect for } S\Big) \geq 1 - c_d \cdot n \cdot (n/s)^{d-1}.$$

We immediately apply this fact to obtain highly reliable real-time dictionaries in the case where much more space is available than the number of keys to be stored.

Fact 2 *Let $0 < \varepsilon < 1$, let $d \geq 2$, and let n be given. There is a dictionary that uses $d \cdot n$ space and that with probability exceeding $1 - O(n^{\varepsilon - (1-\varepsilon)(d-1)})$ executes a sequence of n^ε instructions in such a way that each one takes constant time (in the worst case).*

Proof: Use a hash table of size $s = n$, each table position comprising d slots for one key each. By Fact 1, a hash function h chosen randomly from $\mathcal{H}_n^d$ maps at most $d - 1$ of the n^ε keys to each of the table positions, with the claimed probability. Note that if a sequential list of length n^ε is maintained in which all keys are recorded that were ever entered in the dictionary, the memory space used can be cleared in $O(n^\varepsilon)$ steps after finishing the n^ε instructions. ∎

In a different situation, namely where the hash table size s is much smaller than the number n of keys occurring, the following result tells us that with high probability all buckets have size close to the average. (This fact will be used for the construction of the parallel dictionary.)

Fact 3 ([13, Section 4]) *Assume that $S \subseteq U$ with $|S| = n$ is given, and that $s = n^{1-\delta}$ for some fixed δ, $0 < \delta < 1$. Let $h \in \mathcal{H}_s^d$ be randomly chosen. Then*

$$\Pr(b_j \leq 2n^\delta \text{ for } 0 \leq j < n) \geq 1 - \alpha(d) \cdot n^{1-\delta-\delta d/2},$$

where $\alpha(d) = (d/2)^{d+1}$ is constant.

Finally, we sketch the famous construction of Fredman, Komlós, and Szemerédi [8] that provides a *static* dictionary with constant lookup time in the worst case. (The construction of the real-time dictionary given below is based on this prototype.) Assume a set $S \subseteq U$ of n keys is given. A level-1 hash function $h: U \to \{0, \ldots, s - 1\}$ splits S into the buckets $B_j^h = \{x \in S \mid h(x) = j\}$, for $0 \leq j < s$. For each of the buckets B_j^h there is a secondary hash table ST_j of size $2|B_j^h|^2$ and a level-2 hash function $h_j: U \to \{0, \ldots, 2|B_j^h|^2 - 1\}$ that is one-to-one on B_j^h. The element $x \in S$ is stored in position $h_j(x)$ of subtable ST_j, where $j = h(x)$. Access to the subtables and to (the programs for) the functions h_j is facilitated by a primary hash table HT with s entries. For $0 \leq j < s$, entry $HT[j]$ of this table contains a pointer to ST_j and a description of h_j. In order to construct (probabilistically) such a structure that takes space $O(n) = O(|S|)$, one chooses functions h and h_j at random from suitable classes $\mathcal{H}_s^2$. (See [8] for details.)

3   Probability bounds

In this section, we first quote a classical theorem that estimates the probability for a sum of independent *bounded* random variables to deviate far from the expected value. (For an overview over variants of this result, see [11].) Next, we prove a theorem in the same spirit dealing with sums of independent random variables that are approximately geometrically distributed, thus *unbounded*. Both estimates will be used in the next section for establishing the main properties of a new class of hash functions to be defined there. For the analysis of the real-time dictionary algorithm we need another type of estimate for the tails of the distribution of variables that depend on many independent random variables, each of which has little influence. Such an estimate is provided by the "martingale tail theorem" quoted at the end of this section.

Theorem 3 ([12]) *Let $X_1, \ldots, X_n$ be independent random variables with values in the interval $[0, z]$ for some $z > 0$. Let $Y := \sum_{1 \leq i \leq n} X_i$ and $m := E(Y)$. Then for all a, $0 < a < nz - m$, we have*

$$\Pr\big(Y \geq m + a\big) \leq \left[\left(\frac{m}{m+a}\right)^{m+a}\left(\frac{nz-m}{nz-m-a}\right)^{nz-m-a}\right]^{1/z} \leq \left(\frac{m}{m+a}\right)^{(m+a)/z} \cdot e^{a/z}.$$

(The second inequality holds since $(1 + \frac{x}{\alpha})^\alpha \leq e^x$ for $\alpha > 0$.)   □

We need another technical fact about sums of independent random variables with distribution bounded by a geometrical distribution.

Lemma 1 *Let $m \geq 1$, $w_1, \ldots, w_m > 0$ be arbitrary. Abbreviate $\sum_{i=1}^m w_i$ by W and $\max\{w_i \mid 1 \leq i \leq m\}$ by M. Assume that $X_1, \ldots, X_m$ are independent random variables so that for $1 \leq i \leq m$ holds*

$$\Pr(X_i > L \cdot w_i) \leq 2^{-L}, \quad \text{for } L = 1, 2, 3, \ldots.$$

Then

$$\Pr\left(\sum_{i=1}^m X_i \geq 3 \cdot l \cdot W\right) \leq e^{-(l-1)W/M}, \quad \text{for } l = 1, 2, 3, \ldots.$$

(Note that $E(\sum_{i=1}^m X_i) \leq 2W$.)

Proof: W. l. o. g. we may assume that X_i is as large as possible while still satisfying the assumption $\Pr(X_i > L \cdot w_i) \leq 2^{-L}$, that means, $\Pr(X_i \in \{Lw_i \mid L = 1, 2, 3, \ldots\}) = 1$ and

$$\Pr(X_i = L \cdot w_i) = 2^{-L} \quad \text{for } 1 \leq i \leq m, \ L = 1, 2, 3, \ldots. \tag{1}$$

Define $Y := \sum_{i=1}^m X_i$. It is a well-known trick to use the following inequality for proving estimates for sums of independent random variables. (See, e. g., the proof of the preceding theorem in [12].) For arbitrary $h > 0$ the following holds:

$$\Pr(Y \geq 3l \cdot W) \leq E\big(e^{h(Y - 3lW)}\big). \tag{2}$$

Thus, in order to prove the lemma, we need only estimate $E(e^{hY})$. By the independence of $X_1, \ldots, X_m$ we have

$$E(e^{hY}) = \prod_{i=1}^m E(e^{hX_i}). \tag{3}$$

Clearly, by (1),

$$E(e^{hX_i}) = \sum_{L=1}^\infty \Pr(X_i = Lw_i) \cdot e^{hLw_i} = \sum_{L=1}^\infty \left(\frac{e^{hw_i}}{2}\right)^L.$$

Hence, if h is so small that $e^{hw_i} < 2$, for which it is certainly sufficient if

$$h < M^{-1} \cdot \ln 2, \tag{4}$$

we obtain

$$E(e^{hX_i}) = \frac{e^{hw_i}}{2 - e^{hw_i}}. \tag{5}$$

It is an easy observation that $\frac{t}{2-t} \le t^3$ for $1 \le t \le \frac{1}{2}(1 + \sqrt{5})$. Hence, for h so small that $e^{hw_i} < \frac{1}{2}(1 + \sqrt{5})$, which is certainly satisfied if

$$h < M^{-1} \cdot \ln(\tfrac{1}{2}(1 + \sqrt{5})), \tag{6}$$

we get from (5) that

$$E(e^{hX_i}) < e^{3hw_i}. \tag{7}$$

By substituting (7) into (3) and by the definition of W, we conclude

$$E(e^{hY}) < e^{3hW}; \tag{8}$$

hence, in combination with (2) we get

$$\Pr(Y \ge 3l \cdot W) \le e^{-(l-1)\cdot 3h \cdot W}. \tag{9}$$

We let $h := (3M)^{-1}$. Then clearly (4) and (6) are satisfied, and (9) reads $\Pr(Y \ge 3lW) \le e^{-(l-1)W/M}$, which was to be shown. ∎

Remark 1 In [18], mean, variance, and higher moments for sums of independent geometrically distributed random variables are calculated. Starting from those formulas, estimates similar to that in Lemma 1 can be derived. By different methods, in [20] an analogous result is proved for sums of independent, unbounded, but not exactly geometrically distributed, random variables.

The following theorem (a proof of which can be found in [16]) is extremely helpful when we need to estimate the probability that a random variable T deviates far from its mean in case T is a function of many independent variables each of which has only small influence on T. The usefulness of this theorem for analyzing hash functions and parallel dictionary algorithms has been noted before, see, e. g., [2] and [9].

Theorem 4 ([16]) *Let $X_1, \ldots, X_n$ be independent random variables with ranges $\Omega_1, \ldots, \Omega_n$. Further, let $g: \prod_{1 \le i \le n} \Omega_i \to \mathbb{R}$ be an arbitrary function so that g changes by at most a constant $\vartheta > 0$ in response to a change in a single component, i. e.,*

$$\left| g(\omega_1 \ldots, \omega_{i_0}, \ldots, \omega_n) - g(\omega_1 \ldots, \omega'_{i_0}, \ldots, \omega_n) \right| \le \vartheta,$$

for $\omega_i \in \Omega_i$, $1 \le i \le n$, and $\omega'_{i_0} \in \Omega_{i_0}$ arbitrary. Let $T := g(X_1, \ldots, X_n)$. Then for all $t \ge 0$ we have

$$\Pr\left(T \ge E(T) + t\right) \le e^{-t^2/(2\vartheta^2 n)}.$$

□

The proof of this theorem is based on a "martingale tail estimate" ("Azuma's inequality"); this is why also arguments based on Theorem 4 will be referred to as "martingale estimates".

4   A new class of hash functions

In this section, we define a new class of hash functions and establish some of the properties of this class. The technical definition is as follows.

Definition 2 *Assume $r, s \geq 1$ and $d \geq 2$.*

(a) *For arbitrary $f: U \to \{0, 1, \ldots, rs - 1\}$ define $f_1(x) := f(x) \operatorname{div} s$ and $f_2(x) := f(x) \bmod s$, for $x \in U$.   ($a \operatorname{div} b$ means $\lfloor a/b \rfloor$.)*

(b) *For $f: U \to \{0, 1, \ldots, rs - 1\}$ and $a_0, \ldots, a_{r-1} \in \{0, 1, \ldots, s - 1\}$ let the function*
$$h =$$
$$h(f, a_0, \ldots, a_{r-1}) \text{ be defined by}$$

$$h(x) := \Big(f_2(x) + a_{f_1(x)}\Big) \bmod s.$$

(c) *The class $\mathcal{R}(r, s, d)$ consists of all functions $h(f, a_0, \ldots, a_{r-1})$ with $f \in \mathcal{H}_{rs}^d$ and $a_0, \ldots, a_{r-1} \in \{0, \ldots, s - 1\}$ arbitrary.*

Remark 2 (a) Note that f and the pair (f_1, f_2) are practically the same function. It will be convenient to regard the range of (f_1, f_2) as an $r \times s$-array.

(b) If $f \in \mathcal{H}_{rs}^d$ is given as $f(x) = F(x) \bmod rs$, where

$$F(x) = \Big(\sum_{0 \leq l < d} \alpha_l x^l \Big) \bmod p,$$

an efficient way for evaluating $h = h(f, a_0, \ldots, a_{r-1})$ is given by the formula

$$h(x) = \Big(F(x) + a_{(F(x) \operatorname{div} s) \bmod r}\Big) \bmod s.$$

(c) Obviously, a function $h \in \mathcal{R}(r, s, d)$ can be stored in $O(d + r)$ cells, can be generated in $O(d + r)$ steps, and can be evaluated in time $O(d)$.

Before embarking on the rigorous analysis of the class $\mathcal{R}(r, s, d)$, let us (informally) explain the effect of such a function on the keys $x \in U$. First, the function f maps the keys into an $r \times s$-array (x is mapped into the cell in row $f_1(x)$ and column $f_2(x)$ of the array). In a second step, row i is shifted cyclically by an offset a_i, i.e., key x is moved from column $f_2(x)$ to column $(f_2(x) + a_i) \bmod s$. These shifts are independent for the different rows. The hash value $h(x)$ is given by the number of the column to which x is moved by these two steps.

For the following, we assume that a set $S \subseteq U$ is given, with $|S| = n$, and that $rs > n$. The nice behaviour of the class $\mathcal{R}(r, s, d)$ rests essentially on the fact that if $n/(rs)$ is sufficiently small then with high probability f will be $(d - 1)$-perfect on S; this property implies that each row i of the array can contribute at most $d - 1$ keys to the quantity $b_j = |\{x \in S \mid h(x) = j\}|$.

Definition 3 *For $0 \leq i < r$, let $B_i^f := \{x \in S \mid f_1(x) = i\}$ (the elements of S mapped to the ith row of the array).*

Lemma 2 *We have $\Pr(f$ is $(d-1)$-perfect for $S) = 1 - O\left(n^d/(rs)^{d-1}\right)$, if f is chosen at random from $\mathcal{H}_{rs}^d$.*

Proof: This is immediate from Fact 1. ∎

Definition 4 *Let $f: U \to \{0, 1, \ldots, rs - 1\}$ be a function that is $(d-1)$-perfect on S. We fix f and consider only the random experiment of choosing $a_0, \ldots, a_{r-1}$. Formally, let*

$$\mathcal{R}_f(r, s) := \left\{ h(f, a_0, \ldots, a_{r-1}) \,\middle|\, a_0, \ldots, a_{r-1} \in \{0, \ldots, s - 1\} \right\}.$$

We write $\Pr_f(\mathcal{A})$ for $\Pr(\mathcal{A} \mid \mathcal{R}_f(r, s))$ and $E_f(X)$ for $E(X \mid \mathcal{R}_f(r, s))$, for arbitrary events $\mathcal{A}$ and random variables X.

The following theorem lists the most important properties of functions from $\mathcal{R}(r, s, d)$, for the case $s \geq n$. Note that these properties are close to what one would get if h mapped the keys from S uniformly and independently into $\{0, \ldots, s - 1\}$, i.e., in the case of uniform hashing (cf. [10]). We just remark that corresponding statements are also true for s smaller than n.

Theorem 5 *For $S \subseteq U$ fixed, $|S| \leq n = s$, and h randomly chosen from $\mathcal{R}_f(r, s)$ the following holds.*

(a) $\Pr_f(b_j \geq u) \leq \left(\dfrac{e^{u-1}}{u^u}\right)^{1/(d-1)}$, *for $0 \leq j < n$.*

(b) $\Pr_f\left(b_x^{\mathrm{coll}} \geq (d-1) + u\right) \leq \left(\dfrac{e^{u-1}}{u^u}\right)^{1/(d-1)}$, *for all $x \in U$.*

(c) *If $u \geq 4(d-1)\ln n / \ln\ln n$ then $\Pr_f\left(\max\{b_j \mid 0 \leq j < n\} \geq u\right) \leq u^{-u/(2(d-1))}$.*

(d) $E_f\left(\max\{b_j \mid 0 \leq j < n\}\right) = O(\log n / \log\log n)$.

Proof: Assume that $|S| = n = s$. (Otherwise add some dummy elements to S.) Fix some f that is $(d-1)$-perfect on S.

(a) Fix j, $0 \leq j < n$. We define random variables X_i that measure the contribution of B_i^f to the bucket B_j, as follows:

$$X_i := \left|\left\{x \in S \mid h(x) = j \text{ and } f_1(x) = i\right\}\right|, \quad \text{for } 0 \leq i < r.$$

Obviously, we have

$$b_j = \sum_{0 \leq i < r} X_i.$$

We observe that, by definition,

$$X_i = \left| \left\{ x \in S \mid f_1(x) = i \text{ and } f_2(x) = (j - a_i) \bmod n \right\} \right|, \quad \text{for } 0 \le i < r.$$

Now a_i is randomly chosen, hence $(j - a_i) \bmod n$ is a random element of $\{0, \dots, n - 1\}$. Thus, X_i is the number of elements of S in a randomly chosen cell of the ith row of the $r \times n$-array that forms the range of f. This implies the following:

(i) $0 \le X_i \le d - 1$, for $0 \le i < r$;

(ii) the X_i, $0 \le i < r$, are independent;

(iii) $E_f(X_i) = \frac{1}{n} \cdot |\{x \in S \mid f_1(x) = i\}| = \frac{1}{n} \cdot |B_i^f|$, for $0 \le i < r$.

From (iii) we get the (obvious) expected value of b_j:

$$E_f(b_j) = E_f\left(\sum_{0 \le i < r} X_i \right) = \frac{1}{n} \cdot \sum_{0 \le i < r} |B_i^f| = \frac{1}{n} \cdot |S| = 1.$$

Applying Hoeffding's Theorem (Theorem 3) immediately yields

$$\Pr_f(b_j \ge u) \le \left(\frac{1}{u} \right)^{u/(d-1)} \cdot e^{(u-1)/(d-1)},$$

which is (a).

(b) This is proved in exactly the same way as (a), excepting that in addition to f also $a_{f_1(x)}$ is considered fixed. At most $d - 1$ elements of $B_{f_1(x)}^f$ are mapped to the same cell as x by h; the contribution of the other B_i^f, $i \ne f_1(x)$, to b_x^{coll} is analyzed just as in (a).

(c) This follows easily from (a), as follows. Clearly,

$$\Pr_f\left(\max\{b_j \mid 0 \le j < n\} \ge u \right)$$

$$\le \min\left\{ 1, \sum_{0 \le j < n} \Pr_f(b_j \ge u) \right\} \le \min\left\{ 1, n \cdot \left(\frac{e^{u-1}}{u^u} \right)^{1/(d-1)} \right\}.$$

We must show that no u with $u \ge 4(d - 1) \ln n / \ln \ln n$ can satisfy the inequality $n \cdot (e^{u-1}/u^u)^{1/(d-1)} > u^{-u/(2(d-1))}$. But this is clear: the second inequality implies $2(d - 1) \ln n + 2(u - 1) > u \ln u$, while the first one implies $u \ln u \ge 2(u - 1) + \frac{4}{5} \cdot (4(d-1) \ln n / \ln \ln n) \cdot (\ln \ln n - \ln \ln \ln n) > 2(u-1) + \frac{16}{5} \cdot (d-1) \ln n \cdot (1 - \frac{\ln \ln \ln n}{\ln \ln n}) > 2(u - 1) + 2(d - 1) \ln n$. (We use that $\frac{\ln \ln \ln n}{\ln \ln n} < 1/e \approx 0.368$.)

(d) This is immediate from (c), by using that $E(Z) = \sum_{u \ge 1} \Pr(Z \ge u)$ for integral, nonnegative random variables Z. ∎

The basic properties given in the previous theorem are already very useful if the class $\mathcal{R}(r, s, d)$ is to be used in a simple hashing scheme, e. g., in chained hashing. For our purposes, we have to study the class a little more closely.

Below, we will only consider the case $|S| = r = s = n$.

Remark 3 Note that if it is desirable that the space needed to store h is smaller than n, also $r = n^\delta$ for some $\delta < 1$ is a suitable choice as long as $n/(rs)$ remains sufficiently small. Also, for $n/\log n \leq s \leq n$ the class $\mathcal{R}(r, s, d)$ exhibits a behaviour comparable to that of random functions.

It is our aim to obtain a sharper version of the estimate

$$E\left(\sum_{0 \leq j < n} (b_j)^2 \right) = O(n),$$

which $\mathcal{R}(n, n, d)$ shares with all other universal classes. We are interested in sums of the form $\sum_{j \in J} (b_j)^2$ for only a subset $J \subseteq \{0, \ldots, n-1\}$ (where J may depend on h), and we need to establish that this sum is close to its expectation with high probability. The essential tool for the proof will be the martingale tail estimate (Theorem 4). — We start with some technical preparations.

Definition 5 *Assume $r = s = n$. We abbreviate $\mathcal{R}(n, n, d)$ by $\mathcal{R}(d)$; if the value of d is inessential, we also write simply $\mathcal{R}$.*

Definition 6 *Let $S \subseteq U$ with $|S| = n$.*

(a) *We say that a function $f: U \to \{0, \ldots, n^2 - 1\}$ distributes S well if it is $(d-1)$-perfect on S and*

$$|B_i^f| = |\{x \in S \mid f_1(x) = i\}| \leq n^{1/4}, \quad \text{for } 0 \leq i < n.$$

(b) *The subclass $\mathcal{R}_S(d)$ (or $\mathcal{R}_S$) of $\mathcal{R}(d)$ consists of all functions $h = h(f, a_0, \ldots, a_{n-1})$ $\in \mathcal{R}(d)$ for which f distributes S well.*

If we regard the range of f as an $n \times n$-array, as discussed above, then the subclass $\mathcal{R}_S(d)$ consists of those functions $h = h(f, a_0, \ldots, a_{n-1})$ for which f maps at most $(d-1)$ keys from S into each cell of the array and at most $n^{1/4}$ into each row. We show that $\mathcal{R}_S(d)$ contains almost all functions from $\mathcal{R}(d)$.

Lemma 3 *For arbitrary $c \geq 1$, if the constant d is chosen sufficiently large, the following holds: For arbitrary $S \subseteq U$ with $|S| = n$ we have*

$$\frac{|\mathcal{R}_S(d)|}{|\mathcal{R}(d)|} = 1 - O(n^{-c}).$$

Proof: For $f: U \to \{0, \ldots, n^2 - 1\}$, consider the function $\hat{f}$ defined by

$$\hat{f}(x) := \left(f_1(x), f_2(x) \bmod \left\lfloor \frac{n^{1/4}}{d-1} \right\rfloor \right), \quad \text{for } x \in U.$$

This function has range $\{0, \ldots, n-1\} \times \{0, \ldots, \lfloor n^{1/4}/(d-1) \rfloor - 1\}$, which has $\hat{s} \approx n^{5/4}/(d-1)$ elements. It is not hard to check that the class $\{\hat{f} \mid f \in \mathcal{H}_{n^2}^d\}$ has essentially the same properties as the class $\mathcal{H}_s^d$ (cf. Fact 1); in particular, for f chosen at random from $\mathcal{H}_{n^2}^d$ we have

$$\Pr(\hat{f} \text{ is } (d-1)\text{-perfect on } S)$$

$$= 1 - O\left(n \cdot \left(\frac{n \cdot (d-1)}{n^{5/4}}\right)^{d-1}\right)$$

$$= 1 - O\left(n^{1-(d-1)/4}\right).$$

The last expression is $1 - O(n^{-c})$ for d sufficiently large. Finally, observe that clearly if $\hat{f}$ is $(d-1)$-perfect on S then f distributes S well. This finishes the proof of the lemma. $\blacksquare$

Now we turn to the study of certain sums $\sum_{j \in J}(b_j)^2$.

Definition 7 *For $S' \subseteq S$ arbitrary and $h \in \mathcal{R}$ random we define the random set $J(S')$ as the set of all indices j of buckets B_j that are hit by elements of S', i. e.,*

$$J(S') := \{j \mid 0 \le j < n \text{ and } \exists x \in S' : h(x) = j\}.$$

Further, we let

$$M_2(S') := \sum_{j \in J(S')} (b_j)^2.$$

Theorem 6 *Let $c \ge 1$ be fixed. Then there is a constant $C > 0$ such that for all $S \subseteq U$ with $|S| = n = s$ and all $S' \subseteq S$ with $|S'| \ge n^{7/8}$ the following holds:*

(a) $E\left(M_2(S')\right) \le (C-1) \cdot |S'|$

(b) $\Pr\left(M_2(S') \ge C \cdot |S'|\right) = O(n^{-c}).$

Proof: By Lemma 3, it is sufficient to show the following, for some fixed f that distributes S well (for the notation E_f and $\Pr_f$, see the proof of Theorem 5):

(a') $E_f(M_2(S')) \le (C-1) \cdot |S'|$;

(b') $\Pr_f\left(M_2(S') \ge E_f(M_2(S')) + |S'|\right) = O(n^{-c}).$

Thus, fix such an f. We first prove (a'). For this, recall from Theorem 5 that for arbitrary $x \in S$ the random variable $b_x^{\text{coll}} = |B_x^{\text{coll}}| = |\{y \in S \mid h(x) = h(y)\}|$ satisfies $\Pr_f(b_x^{\text{coll}} \ge u + (d-1)) \le (e/u)^{u/(d-1)}$, for $u \ge 1$. This implies that for all $v \ge 1$ and all $x \in S$ we have

$$\Pr_f\left((b_x^{\text{coll}})^2 \ge v\right) = \Pr_f\left(b_x^{\text{coll}} \ge \lfloor\sqrt{v}\rfloor\right) \le \left(\frac{e}{\lfloor\sqrt{v}\rfloor - (d-1)}\right)^{(\lfloor\sqrt{v}\rfloor - (d-1))/(d-1)}.$$

Thus,

$$E_f\left((b_x^{\mathrm{coll}})^2\right) = \sum_{v \geq 1} \sum_{w \geq 1} (2w + 1) \cdot \left(\frac{e}{w - (d-1)}\right)^{(w-(d-1))/(d-1)} = O(1).$$

Choose the constant C so that the last sum is bounded by $C - 1$. By the obvious inequality

$$M_2(S') \leq \sum_{x \in S'} (b_x^{\mathrm{coll}})^2,$$

we conclude (by linearity of expectation) that

$$E_f(M_2(S')) \leq \sum_{x \in S'} E_f\left((b_x^{\mathrm{coll}})^2\right) \leq (C - 1) \cdot |S'|.$$

Thus, (a') is proved. Now we turn to the proof of (b'). For technical reasons (which will become clear below), we need a truncated version of the random variable $M_2(S')$, which is defined as follows. Let

$$\hat{b}_j := \min\{b_j, \log n\}, \text{ for } 0 \leq j < n.$$

Note that by Theorem 5(c) we have

$$\mathrm{Pr}_f(b_j \leq \log n \text{ for } 0 \leq j < n) = O(n^{-c}).$$

Now, let

$$\hat{M}_2(S') := \sum_{j \in J(S')} (\hat{b}_j)^2.$$

By the above, we have

$$\mathrm{Pr}_f\left(M_2(S') \neq \hat{M}_2(S')\right) = O(n^{-c});$$

further it is clear, by definition, that

$$E_f(M_2(S')) \geq E_f(\hat{M}_2(S')).$$

Thus, in order to prove (b'), it suffices to establish the following:

(b") $\mathrm{Pr}_f\left(\hat{M}_2(S') \geq E_f(\hat{M}_2(S')) + |S'|\right) = O(n^{-c})$.

For fixed f, the hash function $h = h(f, a_0, \ldots, a_{n-1})$, and hence also $\hat{M}_2(S')$ is a function of the independent random variables $a_0, \ldots, a_{n-1}$. In order to apply Theorem 4 we must estimate the effect a change in a single one of the offsets a_i has on $\hat{M}_2(S')$. If a_i is changed, only the hash values $h(x)$ of the up to $n^{1/4}$ many keys x in B_i^f change. This can change the set $J(S')$ (if $h(x)$ changes for some $x \in S'$) and it can change $\hat{b}_j$ for some j that is in $J(S')$ before and after changing a_i. However, it is clear that the total number of $\hat{b}_j$'s that increase (or for which j enters $J(S')$) is bounded by $|B_i^f| \leq n^{1/4}$;

similarly for the total number of $\hat{b}_j$'s that decrease (or for which j leaves $J(S')$). Since, by definition, all $\hat{b}_j$ are in $[0, \log n]$, the total change of the sum $\hat{M}_2(S') = \sum_{j \in J(S')}(\hat{b}_j)^2$ in response to changing one a_i is bounded by $\vartheta := n^{1/4} \cdot (\log n)^2$.

Applying Theorem 4 now yields (using the notation $\exp(y)$ for e^y):

$$\Pr_f\Big(\hat{M}_2(S') \geq E_f(\hat{M}_2(S') + |S'|)\Big)$$
$$\leq \exp\Big(-|S'|^2/(2n \cdot (n^{1/4} \cdot (\log n)^2)^2)\Big)$$
$$= \exp\Big(-\Omega(n^{1/4}/(\log n)^4)\Big)$$
$$= O(n^{-c}).$$

(The second to the last estimate uses the assumption $|S'| \geq n^{7/8}$.) Thus, (b") (and hence also (b)) is proved. ∎

Remark 4 In [7], a slightly different type of hash functions with the same performance is used. The main advantage of the functions introduced in this section is the relatively simple proof of Lemma 3. The analogous argument in [7] is much more complicated.

5 Dynamic hashing in real time

In this section, we construct the dictionary with the features claimed in Theorem 1.

5.1 An auxiliary construction

As a first but crucial step, we present a probabilistic algorithm for a variant of a dictionary with the following properties: it is able to store n keys; lookups are performed right away; "batches" of n^ε update operations are performed in $O(n^\varepsilon)$ time; the algorithm works correctly with high probability. The algorithm differs from a standard dictionary in that the updates contained in a "batch" are performed in an arbitrary order, not in the order prescribed by the input. This auxiliary data structure constitutes the core of our real-time dictionary.

Definition 8 *Let $0 < \varepsilon \leq 1$, and let n be given. A type-A dictionary with respect to ε is a data structure with the following properties.*

(a) *The input is provided in two sequences of instructions:*

$(x_1, Op_1), \ldots, (x_n, Op_n)$ *are the insertion and deletion instructions;*

$(x'_1, Lookup), \ldots, (x'_n, Lookup)$ *are the lookup instructions.*

(b) *The operations are performed in phases $l = 1, 2, \ldots, n^{1-\varepsilon}$; in the lth phase the update instructions $\{(x_i, Op_i) \mid (l-1)n^\varepsilon < i \leq ln^\varepsilon\}$ and the lookup instructions $\{(x'_i, Lookup) \mid (l-1)n^\varepsilon < i \leq ln^\varepsilon\}$ are performed. All keys occurring in update instructions for phase l are assumed to be distinct.*

(c) *Each phase takes at most time Kn^ε in the worst case, for a constant K.*

(d) *The lookup instructions of phase l are performed at a constant rate, in the order given by the input sequence. For keys x_i' that do not occur in update instructions for phase l the answer is correct, that means, it reflects the state of the dictionary at the end of phase $l-1$.*

(e) *The algorithm uses space $O(n)$.*

Theorem 7 *For arbitrary $\frac{7}{8} < \varepsilon \leq 1$ and $c \geq 1$ there is a probabilistic algorithm for implementing a type-A dictionary with respect to ε. The probability that the algorithm fails to stay within the time bounds of Definition 8(c) is $O(n^{-c})$. The other properties are always satisfied.*

Proof: We describe the algorithm, which is a generalization of the perfect hashing schemes from [8] and [4] (cf. Section 1). Let $S = \{x_1, \ldots, x_n\}$ be the set of keys used in the n update instructions.

Algorithm 1 (Type-A dictionary)

Choose $h \in \mathcal{R} = \mathcal{R}(n, n, d)$ at random (cf. Definitions 2 and 5), for d large enough for the probability estimates below to hold. The function h splits S into buckets B_j, $0 \leq j < n$. The keys $x \in B_j$ are stored in a subtable ST_j. Initially, all ST_j are empty, as well as the header table HT, which is an array with n positions. Then the algorithm proceeds in phases $l = 1, 2, \ldots, n^{1-\varepsilon}$.

Phase l:

Part 1: Perform update instructions

Evaluate $h(x)$ for each $x \in S(l) := \{x_{(l-1)n^\varepsilon}, \ldots, x_{ln^\varepsilon}\}$. Let $J_l := \{j \mid S(l) \cap B_j \neq \emptyset\}$. For all $j \in J_l$, collect $x \in S(l) \cap B_j$ in a sequential list T_j'', construct a list T_j' of those keys from B_j already stored in ST_j, and construct $T_j :=$ concatenation of T_j' and T_j''.

Then, for each $j \in J_l$ in turn do the following: Randomly choose a function h_j from $\mathcal{H}_{2|T_j|^2}^2$ and test whether h_j is one-to-one on T_j. Repeat this until such a function is found. Set up a new subtable ST_j of size $2|T_j|^2$, and enter the keys $x \in T_j$ into this new table according to the hash function h_j, together with the information fields that belong to these keys according to the entry in the old subtable ST_j or to the input sequence. Deletions are effected by adding a tag "deleted" to the entry. When this is done, change the pointer in the header table $HT[j]$ to point to the new subtable ST_j.

Part 2: Perform lookup instructions

The n^ε lookup instructions are performed in an interleaved manner concurrently with the update procedure of Part 1, so that lookup instructions are accepted at constant time intervals. (Clearly, each lookup instruction takes constant time.)

Analysis: Let $c \geq 1$ be given. For d large enough, we have

$$\Pr(h \in \mathcal{R}_S) \geq 1 - O(n^{-c-1}),$$

by Lemma 3. Further, by Theorem 6 there is a constant C so that

$$\Pr\left(\sum_{j \in J_l} (b_j)^2 \geq C \cdot n^\varepsilon \,\Big|\, h \in \mathcal{R}_S \right) = O(n^{-c-1}),$$

where $b_j = |B_j|$, $0 \leq j < n$, and the set

$$J_l = J(S(l)) = \left\{ j \,\Big|\, 0 \leq j < n \text{ and } \exists x \in S(l) \colon h(x) = j \right\}$$

is as in Definition 7. Finally, we have noted already in the previous section that Theorem 5(c) implies that

$$\Pr\left(\max\{b_j \mid 0 \leq j < n\} \geq \log n \,\Big|\, h \in \mathcal{R}_S \right) = O(n^{-c}).$$

Thus, for $h \in \mathcal{R}$ randomly chosen, we have that with probability at least $1 - O(n^{-c})$ both the following two properties are satisfied:

$$\forall l \in \{1, \dots, n^{1-\varepsilon}\} \colon \sum_{j \in J_l} (b_j)^2 < C \cdot n^\varepsilon, \tag{1}$$

and

$$\forall j \in \{0, \dots, n-1\} \colon b_j \leq \log n. \tag{2}$$

As we allow the algorithm to fail to work properly with a probability bounded by $O(n^{-c})$, we may assume for the following that properties (1) and (2) are satisfied for the level-1 hash function h we have chosen. Under this assumption, we analyze the time needed for a single fixed phase l. The following obvious fact is crucial:

$$|T_j| \leq b_j, \text{ for all } j \in J_l. \tag{3}$$

(A subtle point to note here is that for this to be true, if T_j is regarded as a list, we need the condition that the keys occurring in update instructions in phase l are all distinct.) Evaluating $h(x)$ for $x \in S(l)$ takes total time $O(|S(l)|) = O(n^\varepsilon)$. Constructing the lists T_j', T_j'', and T_j, for $j \in J_l$, can easily be done in time $O(\sum_{j \in J_l} |T_j|^2)$, which is $O(n^\varepsilon)$, by inequalities (1) and (3). Transferring the data from the old ST_j to the new one and changing pointers can also clearly be done in time $O(\sum_{j \in J_l} |T_j|^2)$. The only point left to consider is the construction of the new level-2 hash functions h_j, $j \in J_l$, by the randomized procedure described above. Clearly, choosing one function h_j from $\mathcal{H}^2_{2|T_j|^2}$ and testing it for injectivity on T_j takes time $O(|T_j|^2)$. The probability of one such trial being successful is at least $\frac{1}{2}$, by Fact 1; hence the probability that more than L trials are needed until success occurs is at most 2^{-L}. If X_j, $j \in J_l$, denotes the time needed for constructing h_j, then X_j is bounded by a geometrically distributed random variable with mean $O(|T_j|^2)$ (which is $O((\log n)^2)$, by inequality (2)). Since

$$\sum_{j \in J_l} E(X_j) = O\left(\sum_{j \in J_l} |T_j|^2 \right) = O(n^\varepsilon)$$

and $\max\{E(X_j) \mid j \in J_l\} = O((\log n)^2)$, a straightforward application of Lemma 1 yields that, for some constant C', we have

$$\Pr\left(\sum_{j \in J_l} X_j \geq C' \cdot n^\varepsilon\right) = e^{-\Omega(n^\varepsilon/(\log n)^2)} = O(n^{-c-1}).$$

Thus, the update part of phase l is finished within $O(n^\varepsilon)$ steps with probability $1 - O(n^{-c-1})$, hence *all* phases are finished within this time bound with probability $1 - O(n^{-c})$.

As for the space requirements, it is obvious that the number of memory cells needed in phase l (for storing the new subtables) is bounded by $O(\sum_{j \in J_l} |T_j|^2) = O(\sum_{j \in J_l} (b_j)^2)$, which is $O(n^\varepsilon)$ as soon as (1) holds. This shows that in this case the total space needed is $O(n)$.

Performing the lookup operations in an interleaved manner with the update procedure causes no problems at all. It is obvious from the algorithm that the consistency condition stated in Definition 8(d) is satisfied.

This finishes the proof of Theorem 7. ∎

Remark 5 Starting with a clear memory, setting up an empty type-A dictionary takes constant time, if we assume that for choosing the level-1 function h from $\mathcal{R}$ only f is chosen at the beginning and the offset a_i is chosen only when the first key x with $f_1(x) = i$ appears, for each $i \in \{0, \ldots, n-1\}$.

5.2   The real-time dictionary

We describe two versions of the dynamic dictionary: one for the case that n, the number of instructions to be performed, is known in advance, and a fully dynamic one, where no such assumption is made.

First, assume that the number n of instructions is given in advance. The dictionary uses space $O(n)$; it is initially empty.

Definition 9 *A type-B dictionary is a data structure with the following properties. Let n arbitrary instructions to be performed. Then*

(a) *Space $O(n)$ is used;*

(b) *$O(1)$ steps are used for each instruction, in the worst case.*

Theorem 8 *Let $c \geq 1$ be arbitrary. There is a probabilistic algorithm for implementing a type-B dictionary. The probability that the algorithm fails to fulfill property (b) from Definition 9 is $O(n^{-c})$; property (a) is always satisfied.*

Proof: We first give an overall description of the parts of the data structure. Let $\frac{7}{8} < \varepsilon < 1$. The data structure consists of three layers:

(α) the *(current) temporary dictionary* D_{new} of capacity n^ε;

(β) the *closed temporary dictionary* D_{old} of capacity n^ε;

(γ) the *background dictionary* D_{back} of capacity n.

The algorithm proceeds in phases $l = 1, 2, \ldots, n^{1-\varepsilon}$. In each phase, n^ε arbitrary instructions are processed. The effect of insert- and delete-instructions during phase l is recorded in D_{new}. The closed dictionary D_{old} is the temporary dictionary of the previous phase $l-1$, in the state it had at the end of this phase. At the beginning of phase l, D_{back} is the complete dictionary in the state of the beginning of phase $l-1$. During phase l, concurrently with processing the n^ε new instructions, the information stored in D_{old} is used to update D_{back}.

Let $S(l)$ denote the keys occurring in phase l (i.e., those keys in instructions $(l-1)n^\varepsilon + 1, \ldots, l \cdot n^\varepsilon$). We now describe the algorithm in detail.

Algorithm 2 (Type-B dictionary)

Preparation: (During this period no instructions are processed.) Set up D_{back} as an empty type-A dictionary of capacity n. Set up empty dictionaries D_{new} and D_{old}, space $O(n)$, choose randomly one $h \in \mathcal{H}_n^d$ that serves as hash function for both D_{new} and D_{old} (cf. Fact 2).

Phase l:

The phase consists of n^ε rounds; in each round one instruction is processed. The actions in a single round are as follows.

1. Read next instruction (x, Op).

    (a) If $Op = Insert$, then enter key x in D_{new}. Overwrite the information associated with x, if x is already present in D_{new}.

    (b) If $Op = Delete$, then enter key x in D_{new} with tag "deleted", or add this tag, if x is already present in D_{new}.

    (c) If $Op = Lookup$, then look for x in $D_{\mathrm{new}}, D_{\mathrm{old}}$, and D_{back}, in this order; return the information associated with x in the first dictionary in which x is present. If x is found with tag "deleted" or not found at all, return "not found".

2. Perform a constant number of steps of the following procedure.
    Transfer D_{old} to D_{back}: Collect the keys in D_{old} in a sequential list. Insert the keys in the list into D_{back} (including tags "deleted", if applicable), as described in Algorithm 1. When finished, make D_{old} inaccessible for lookups, then erase all entries in D_{old}.

This finishes the description of a round. After all n^ε rounds that belong to phase l are finished, the roles of D_{new} and D_{old} are switched. (At the beginning of a phase, D_{new} is empty and D_{old} contains the result of the phase that just ended.)

Time analysis:

"Preparation": Setting up D_{new} and D_{old} clearly takes constant time, if the memory is clear initially. Setting up D_{back} also takes constant time, by Remark 5.

"Process Instructions": Executing an instruction (Part 1 of a round) takes constant time, if the hash function h of D_{new} is $(d-1)$-perfect for $S(l)$. By Fact 2, this is the case for *all* phases l with probability

$$1 - O\left(n^{1-\varepsilon} \cdot n^{\varepsilon} \cdot n^{-(\varepsilon-1)(d-1)}\right) = 1 - O(n^{-c}),$$

for d large enough. Clearing D_{old} takes time $O(n^{\varepsilon})$ by Fact 2. The transfer of elements from D_{old} to D_{back} takes $O(n^{\varepsilon})$ steps *in each phase l* with probability $1 - O(n^{-c})$ (Theorem 7). (Note that for each key stored in D_{old} there is only one entry; thus all keys to be inserted into D_{back} are distinct.) Thus, we must only set the constant large enough that directs how many steps of the "transfer" procedure are executed in each round to make sure that n^{ε} rounds are sufficient to finish the procedure.

Correctness:

We need to check that the answer for a task $(x, Lookup)$ given by the algorithm is correct, i. e., reflects the last change to the information associated with key x. There are three cases.

Case 1: x occurred last in a delete- or insert-instruction in phase l. Then D_{new} has the correct information on x, and this is found first.

Case 2: x occurred last in a delete- or insert-instruction in phase $l-1$. Then x is not found in D_{new}, but it is found together with the correct information in D_{old}, if the lookup takes place before D_{old} is made inaccessible. After that, the correct information concerning x is found in D_{back}.

Case 3: x occurred neither in phase l non in phase $l-1$. Then D_{back} contains the correct information on x, and it is found there.

This finishes the proof of Theorem 8. ∎

Remark 6 At the end of Algorithm 2, i. e., directly after the last instruction is processed, the data structure looks as follows. D_{back} reflects the state of the dictionary at the beginning of the last phase, D_{old} reflects the updates from the last phase, D_{new} is empty. It is easy to see that from this structure a list of the keys that eventually belong to the dictionary may be computed in $O(n)$ steps (worst case).

We still have to handle the case that the number of instructions to be performed by a dictionary may not be known in advance and/or exceed the number of keys stored in the dictionary at any one time. For these cases, a type-B dictionary is not flexible enough. Thus, we need a means for adjusting the size of the dictionary in use. This can easily be done if during these periods of readjustment no instructions are accepted, but the resulting behaviour of the dictionary would not have the desired real-time feature. We can do better by employing a similar strategy as in Algorithm 2: use a temporary dictionary for recording the changes made to the dictionary; simultaneously transfer the information contained in the previous temporary dictionary to a background dictionary. (In this approach we combine ideas from [4] and [3].) The construction finally yields the fully dynamic real-time dictionary claimed to exist in Theorem 1.

Algorithm 3 The algorithm again works in phases. The first phase starts when the first instruction is read. It ends as soon as a fixed constant number of keys are in the dictionary. If phase $l-1$ finishes with n keys in the dictionary, phase l consists of processing the next $n/2$ instructions. For each phase l we use a temporary dictionary D_{NEW} to keep track of the instructions of this phase. D_{NEW} is a Type-B dictionary. At the beginning of phase l, D_{NEW} is empty. A further dictionary, D_{OLD}, is the dictionary D_{NEW} of phase $l-1$ at the end of phase $l-1$ (cf. Remark 6). The complete dictionary in the state of the beginning of phase $l-1$ is stored in D_{OLDBACK}. D_{OLDBACK} is a Type-A dictionary with $\varepsilon = 1$. D_{NEWBACK} is of the same type; it is empty at the beginning of phase l. During phase l, the keys from D_{OLDBACK} and D_{OLD} will be inserted in the dictionary D_{NEWBACK}.

Assume that presently n keys are stored in the dictionary. We describe phase l.

Preparation: Set up an empty dictionary D_{NEW} of capacity $\frac{1}{2}n$ and an empty dictionary D_{NEWBACK} of capacity n.

Process Instructions:

This is done in $\frac{1}{2}n$ rounds. In each round one instruction is read and executed.

(a) Insertions are made into D_{NEW}.

(b) Deletions are executed by entering the key with tag "deleted" into D_{NEW}.

(c) Lookups are executed by searching for the key in $D_{\mathrm{NEW}}, D_{\mathrm{OLD}}, D_{\mathrm{OLDBACK}}$, in this order.

Also, in each round, a constant number of steps of the following procedure is performed:

Construct D_{NEWBACK}:

Collect the keys from D_{OLD} and D_{OLDBACK} into one sequential list. Eliminate keys marked "deleted" in D_{OLD}. (This leaves a list of those n keys present in the dictionary at the end of the previous phase.) Insert the keys from the list into D_{NEWBACK}. When the construction of D_{NEWBACK} is finished, rename D_{NEWBACK} into D_{OLDBACK} and mark D_{OLD} "empty". Then clear the two old dictionaries.

After all $n/2$ rounds that belong to phase l are finished, D_{NEW} is declared "closed" and the names of D_{NEW} and D_{OLD} are switched.

This finishes the description of phase l.

Analysis:

If we set the constant large enough that directs how many steps of the construction of D_{NEWBACK} are performed in each round, this construction will be finished within $n/2$ rounds, with probability exceeding $1 - O(n^{-c})$, by Theorem 7. Each operation in D_{NEW} will take constant time with probability $1 - O(n^{-c})$, by Theorem 8. Thus the overall probability that each round takes constant time and the construction of D_{NEWBACK} is finished after $n/2$ rounds is $1 - O(n^{-c})$. It is clear that $O(n)$ space is used. The correctness proof is similar to that in Theorem 8. $\square$

Remark 7 *Recovery from failure.*   If in a phase any one of the dictionaries or procedures fails to perform properly, i. e., exceeds its time bounds, we abort the phase. This happens with probability $O(n^{-c})$. The following recovery procedure is invoked in this case: All keys currently stored in the dictionary are collected and inserted into a new background dictionary D_{NEWBACK}. No instructions are accepted during this time. The expected time needed for this reconstruction is $O(n)$. Note that no information stored in the dictionary is lost if such a failure occurs.

6   Parallel dynamic hashing in real time

In this section we present a parallel version of the real-time dictionary, thus proving Theorem 2.

The basic idea of the algorithm is simple: Let M be a p-processor CRCW PRAM with processors $P_1, \ldots, P_p$ and ARBITRARY write-conflict resolution. Let $0 < \varepsilon < 1$ be fixed. A set S of $n \geq p^{1+\varepsilon}$ keys is stored in the shared memory as follows: S is split into $s = n^{1-\varepsilon/2}$ buckets $B_0^H, \ldots, B_{s-1}^H$ by a hash function $H \in \mathcal{H}_s^d$, for sufficiently large d. For each $i \in \{0, \ldots, s-1\}$, the keys from B_i^H are organized as a sequential real time dictionary D_i as presented in the previous section. The following algorithm shows how to perform n instructions, n/p per processor, in a real-time manner in space $O(n)$ if $n \geq p^{1+\varepsilon}$.

Algorithm 4 (Parallel dictionary) Let $(x_1^t, Op_1^t), \ldots, (x_p^t, Op_p^t)), t = 1, \ldots, n/p$ be the sequence of p-tuples of instructions to be executed, where (x_r^t, op_r^t) is known to processor P_r. Let $S = \{x_1, \ldots, x_n\}$ be the set of keys occurring in the n instructions. Let $s := n^{1-\varepsilon/2}$. Set up empty real-time dictionaries D_i, $0 \leq i < s$, each suitable for performing $2n^{\varepsilon/2}$ instructions, of the kind described in Algorithm 2.

 – For $t := 1$ to n/p do

         – For all $r \in \{1, \ldots, p\}$ in parallel do

            – P_r computes $i_r := H(x_r^t)$.

            – P_r executes (x_r^t, Op_r^t) in the sequential dictionary D_{i_r}. Different instructions arriving at the same D_i are executed sequentially. If several update instructions (x, Op_r^t), with the same key x arrive at the same D_i simultaneously, only one deletion (if any), then an arbitrary one of the insertions (if any) is executed. (Note that lookups for the same key do not interfere, since concurrent reading is permitted.)

Lemma 4 *For each $c \geq 1$ and $0 < \varepsilon < 1$, there is a $d \geq 2$ such that Algorithm 4 needs constant time for each pass of the outer loop, with probability at least $1 - n^{-c}$.*

Proof: By Facts 3 and 1 and by Theorem 8, each of the following properties hold with probability at least $1 - n^{-c'}$, for arbitrary $c' \geq 1$, if d is sufficiently large.

(i) H splits S into buckets of (roughly equal) size at most $2n^{\varepsilon/2}$.

(ii) For each $t = 1, \ldots, n/p$, the function H is $(d-1)$-perfect on $\{x_1^t, \ldots, x_p^t\}$; in particular, at most $d - 1$ different instructions arrive at each D_i during the tth pass through the outer loop.

(iii) For each $i \in \{0, \ldots, s-1\}$, the instructions arriving at D_i are executed in real time.

Thus, if c' is sufficiently large, (i), (ii) for all $t = 1, \ldots, n/p$, and (iii) for $i \in \{0, \ldots, s-1\}$ hold simultaneously with probability at least $1 - O(n^{-c})$. As each pass of the outer loop needs constant time if (i), (ii), (iii) hold, the lemma follows. ∎

In order to obtain the general parallel dictionary as described in Theorem 2, we have to show how to handle the situation where the size of the dictionary is not known in advance. This is done in a way very similar to the method used in the last section to get the general sequential dictionary from Type-A and Type-B dictionaries; the details are omitted here.

Bibliography

[1] H. V. Aho and D. Lee. Storing a dynamic sparse table. In *Proc. of the 27th IEEE Ann. Symp. on Foundations of Computer Science*, pages 55–60. IEEE, 1986.

[2] H. Bast and T. Hagerup. Fast and reliable parallel hashing. In *Proc. of the 3rd Ann. ACM Symp. on Parallel Algorithms and Architectures*, pages 50–61, 1991.

[3] G. Brassard and S. Kannan. The generation of random permutations on the fly. *Inform. Proc. Letters*, **28**:207–212, 1988.

[4] M. Dietzfelbinger, A. Karlin, K. Mehlhorn, F. Meyer auf der Heide, H. Rohnert, and R. E. Tarjan. Dynamic perfect hashing: Upper and lower bounds. Technical Report 77, Universität–GH–Paderborn, Fachbereich Mathematik/Informatik, Jan. 1991. *Revised Version* of the paper of the same title that appeared in *Proc. of the 29th IEEE Ann. Symp. on Foundations of Computer Science*, pages 524–531, 1988.

[5] M. Dietzfelbinger and F. Meyer auf der Heide. An optimal parallel dictionary. In *Proc. of the 1989 ACM Symp. on Parallel Algorithms and Architectures*, pages 360–368, 1989. (Revised version to appear in *Information and Computation*).

[6] M. Dietzfelbinger and F. Meyer auf der Heide. How to distribute a dictionary in a complete network. In *Proc. of the 22nd Ann. ACM Symp. on Theory of Computing*, pages 117–127, 1990.

[7] M. Dietzfelbinger and F. Meyer auf der Heide. A new universal class of hash functions and dynamic hashing in real time. In M. S. Paterson, editor, *Proceedings of 17th ICALP*, pages 6–19. Springer, 1990. Lecture Notes in Computer Science 443.

[8] M. L. Fredman, J. Komlós, and E. Szemerédi. Storing a sparse table with $O(1)$ worst case access time. *J. Assoc. Comput. Mach.*, **31**(3):538–544, July 1984.

[9] J. Gil, Y. Matias, and U. Vishkin. Towards a theory of nearly constant time parallel algorithms. In *Proc. of the 32nd IEEE Ann. Symp. on Foundations of Computer Science*, 1991.

[10] G. H. Gonnet. Expected length of the longest probe sequence in hash code searching. *J. Assoc. Comput. Mach.*, **28**(2):289–304, Apr. 1981.

[11] T. Hagerup and C. Rüb. A guided tour of Chernoff bounds. *Inform. Proc. Letters*, **33**:305–308, 1989/90.

[12] W. Hoeffding. Probability inequalites for sums of bounded random variables. *J. Am. Stat. Ass.*, **58**:13–30, 1963.

[13] C. P. Kruskal, L. Rudolph, and M. Snir. A complexity theory of efficient parallel algorithms. *Theoret. Comput. Sci.*, **71**:95–132, 1990.

[14] R. J. Lipton and J. G. Naughton. Clocked adversaries for hashing. Technical Report CS-TR-203-89, Princeton, 1989.

[15] Y. Matias and U. Vishkin. Converting high probability into nearly-constant time – with applications to parallel hashing. In *Proc. of the 23rd Ann. ACM Symp. on Theory of Computing*, pages 307–316, 1991.

[16] C. McDiarmid. On the method of bounded differences. In J. Siemons, editor, *Surveys in Combinatorics, 1989*, pages 148–188. Cambridge University Press, 1989. London Math. Soc. Lecture Note Series 141.

[17] K. Mehlhorn. *Data Structures and Algorithms 1: Sorting and Searching*. Springer-Verlag, Berlin, 1984.

[18] J. H. Reif and P. G. Spirakis. Random matroids. In *Proc. of the 12th Ann. ACM Symp. on Theory of Computing*, pages 385–397, 1980.

[19] A. Siegel. On universal classes of fast high performance hash functions, their time-space tradeoff, and their applications. In *Proc. of the 30th IEEE Ann. Symp. on Foundations of Computer Science*, pages 20–25, 1989. *Revised Version.*

[20] L. G. Valiant. General purpose parallel architectures. In J. van Leeuwen, editor, *Handbook of Theoretical Computer Science, Vol. A: Algorithms and Complexity*, chapter 18, pages 943–971. Elsevier, Amsterdam, 1990.

Baumautomaten zur Codeselektion

Christian Ferdinand   Helmut Seidl   Reinhard Wilhelm
Universität des Saarlandes
FB 14 – Informatik
Im Stadtwald 15
D-6600 Saarbrücken 11

Zusammenfassung

Die vorliegende Arbeit behandelt die Generierung von Codeselektoren. Die grundlegenden Methoden werden systematisch aus der Theorie der regulären Baumgrammatiken und endlichen Baumautomaten entwickelt. Es werden Algorithmen zur Generierung von Baumanalysatoren abgeleitet, die bestehende Ansätze verallgemeinern bzw. verbessern.

1 Einleitung

Bei der Codeerzeugung gehen wir von einer Zwischendarstellung IR des zu übersetzenden Programms aus, die in einem Übersetzer von den der Codeerzeugung vorangehenden Übersetzungsphasen aufgebaut wurde. Diese Zwischendarstellung kann als Code für eine **abstrakte Maschine** aufgefaßt werden. Die Codeerzeugung hat die Aufgabe, die Zwischendarstellung in eine möglichst effiziente Befehlsfolge für die konkrete Zielmaschine zu überführen.

Die Codeselektion, das heißt die Auswahl der Instruktionen, ist neben der Registerverteilung und der Instruktionsanordnung (bei Prozessoren mit Pipeline-Architekturen) eine wichtige Teilaufgabe der Codeerzeugung. Sie spielt eine besondere Rolle für die sogenannten CISC (Complex Instruction Set Computer) Maschinen, bei denen es meistens mehrere Möglichkeiten gibt, für ein Programmstück Code zu erzeugen.

Beispiel 1
Wir betrachten den einfachsten Vertreter der MC680x0 Reihe, den MC68000. Der Befehl

$$\text{MOVE.B } \textit{dist}(\text{A1, D1.W)}, \text{D5}$$

lädt ein Byte in das untere Viertel des Datenregisters D5. Die Adresse des Operanden ergibt sich folgendermaßen: Auf den Inhalt des Basisregisters A1 werden die untere Hälfte des Inhalts von D1 und die 16-Bit-Konstante *dist* addiert. Diese Adressierungsart hat den Namen: "Adressregister indirekt mit Index und 16-Bit-Distanz".

Mißt man die Kosten von Befehlen in Form von Ausführungszeit, also der Anzahl von Prozessorzyklen, so hat dieser Befehl die Kosten 14. Eine alternative Codesequenz ohne Indexadressierung und Adressdistanzwert wäre:

ADD.W	#*dist*, A1	Kosten: 16
ADD.W	D1, A1	Kosten: 8
MOVE.B	(A1), D5	Kosten: 8
	mit Gesamtkosten 32	

Eine weitere ist:

ADD.W	D1, A1	Kosten: 8
MOVE.B	*dist*(A1), D5	Kosten: 12
	mit Gesamtkosten 20	

Dieses Beispiel zeigt die Notwendigkeit einer geschickten Codeselektion bei CISC–Architekturen.

Die beiden vorgestellten alternativen Codesequenzen verhalten sich nur auf dem Speicher und auf dem Ergebnisregister D5 äquivalent. Der Endzustand bzgl. des Bedingungscodes und des Registers A1 ist unterschiedlich! □

Durch die Verwendung von Codeselektorgeneratoren kann der Aufwand zum Erstellen eines Codeerzeugers für eine neue Maschine erheblich reduziert werden.

Generatoren von Codeselektoren benötigen als Eingabe eine Maschinenbeschreibung. Eine Möglichkeit ist, Maschineninstruktionen durch die Regeln einer regulären Baumgrammatik zu beschreiben. Dabei beschreibt die rechte Seite einer Regel die "Bedeutung" einer Instruktion (siehe Abb. 1). Die Terminalsymbole stehen für (Teil-) Operationen, welche die Instruktion ausführt. Nichtterminale bezeichnen Locationen bzw. Resourcenklassen, wie zum Beispiel die verschiedenen Registertypen. Das Nichtterminal auf der linken Seite einer Regel gibt an, wo (bzw. in welcher Resourcenklasse) das Ergebnis der Instruktion abgelegt wird.

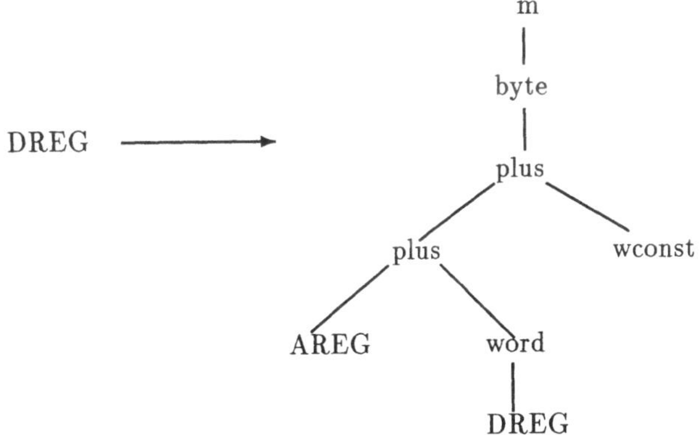

$$DREG \rightarrow m(byte(plus(plus(AREG, word(DREG)), wconst)))$$

Abbildung 1: Instruktion aus Beispiel 1

Aus einer solchen **Maschinengrammatik** lassen sich IR-Bäume für Ausdrücke ableiten. Der Ableitungsbaum für einen IR-Baum beschreibt eine Möglichkeit, für den zugehörigen Ausdruck Code zu erzeugen. In der Regel (und im besonderem Maße bei CISC-Prozessoren) ist die Maschinengrammatik mehrdeutig. In diesem Falle kann es mehrere Ableitungsbäume für den gleichen Ausdruck geben, die für verschiedene Instruktionsfolgen stehen. Zur Auswahl einer besonders günstigen Instruktionsfolge annotiert man die Regeln mit **Kosten**, die beispielsweise die Anzahl der Maschinenzyklen einer Instruktion angeben. Damit kann ein Ableitungsbaum, bzw. eine **lokal optimale** Instruktionsfolge mit geringsten Kosten ausgewählt werden. Allerdings gibt es Prozessoren (z.B. Motorola 68020), bei denen die Anzahl der benötigten Maschinenzyklen wegen ihrer Abhängigkeit vom Ausführungskontext nicht genau vorhergesagt werden kann. Hier muß mit Näherungen gearbeitet werden.

Die Anzahl der Kombinationen von Instruktionen, Adressierungsmodi und Wortbreiten der gebräuchlichen CISC-Prozessoren ist in der Regel sehr groß. Um zu vermeiden, für alle möglichen Kombinationen eine Regel mit Kosten angeben zu müssen, werden Adressierungsmodi und Operanden- bzw. Ergebnisbreiten häufig ausfaktorisiert und mit eigenen Kosten versehen.

Um den Einsatzbereich der Codeselektion zu erweitern, kann man als Wurzelknoten eines Baumes auch Sprung- bzw. Zuweisungsoperatoren zulassen. Sprünge und Zuweisungen liefern in dem oben beschriebenen Sinne kein "Resultat", so daß in Regeln, die solche Instruktionen beschreiben, auf der linken Seite ein *dummy*-Nichtterminal verwendet wird. Dieses Nichtterminal kommt in den rechten Seiten der Regeln nicht mehr vor und stellt sozusagen ein Startnichtterminal für Sprünge und Zuweisungen dar.

2　Einordnung der vorliegenden Arbeit

Ripken [25] war der erste, der eine Maschinenbeschreibungssprache für realistische Maschinen vorschlug, aus der Codeerzeuger generiert werden sollten. Da die dazu benötigten Baumanalyseverfahren noch nicht verbreitet waren, schlugen Glanville und Graham [11, 12] eine Linearisierung der Zwischensprache und Codeselektion durch einen modifizierten LALR–Parser vor. Dieser Ansatz wurde von Henry bis zur Anwendungsnähe weiterverfolgt [14]. Andere Ansätze kombinieren effiziente Mustererkennungsalgorithmen [18, 17, 3, 22] mit dynamischem Programmieren zur Bestimmung lokal optimaler Befehlssequenzen [1, 15, 26] bzw. erweitern die Mustererkenner zur direkten Auswahl lokal optimaler Befehlssequenzen [23, 16].

In einer Reihe von Veröffentlichungen bemühen sich Giegerich und Schmal [10, 9], die verschiedenen Ansätze der Mustererkennung und Baumanalyse in einem formalen Rahmen zu behandeln. Das Problem der Codeselektion wird auf die Invertierung eines "Derivors", d.h. eines Baum-Homomorphismus, zurückgeführt.

Pelegri-Llopart [23] und Emmelmann [6] beschreiben Maschinen durch Termersetzungssysteme und leiten daraus "Automaten" ab, die eine effiziente Codeselektion erlauben.

Die meisten dieser Arbeiten ignorieren allerdings die theoretisch seit langem wohluntersuchten Zusammenhänge zwischen Methoden zur Baummustererkennung beziehungsweise Baumanalyse und endlichen Baumautomaten. Einen guten Überblick über die klassischen Ergebnisse der Baumautomatentheorie findet man etwa in [8]. Ohne explizit auf Codeselektion einzugehen, nimmt Kron in [18] eine Reihe von gegenwärtigen Techniken zur Baummustererkennung vorweg.

Die gegenwärtige Arbeit entwickelt systematisch die grundlegende Theorie, die bei der Codeselektion Anwendung findet. Der Einfachheit halber beschränken wir uns auf Maschinenbeschreibungen mithilfe von regulären Baumgrammatiken. Eine Übersetzung von etwa variablenfreien Termersetzungssystemen oder anderen eingeschränkten Klassen von Termersetzungssystemen in Baumautomaten ist ebenfalls möglich [23, 6, 4].

Der Zusammenhang zu Baumautomaten wird deutlich gemacht. Die beiden Ansätze der Baummustererkennung und der Baumanalyse werden auf die selben Grundlagen gestellt. Es werden generelle Methoden zur Implementierung von Baumautomaten angegeben. Insbesondere verallgemeinern wir den Kron'schen Ansatz auf beliebige Baumautomaten. Dies gestattet es, effiziente Generatoren effizienter Codeselektoren abzuleiten.

Die vorgestellten Resultate sind auch in anderen Bereichen nützlich. Sowohl Baummustererkennung als auch Baumanalyse finden Verwendung zur Überprüfung syntaktischer Anwendbarkeitsbedingungen von Transformationsregeln in Programmtransformationssystemen [7, 21]. Baummustererkennung wird beispielsweise auch bei der Implementierung von Termersetzungssystemen benutzt [13].

3 Das Mustererkennungsproblem

Ein **Alphabet mit Stelligkeit** ist eine endliche Menge Σ von Operatoren[1] zusammen mit einer Funktion $\varrho : \Sigma \to \mathbb{N}_0$, der **Stelligkeit**. Wir schreiben Σ_k für $\{a \in \Sigma \mid \varrho(a) = k\}$. Die **homogene Baumsprache** über Σ ist die folgende induktiv definierte Menge $B(\Sigma)$:

- $a \in B(\Sigma)$ für alle $a \in \Sigma_0$;

- Sind $b_1, \ldots, b_k$ in $B(\Sigma)$ und ist $f \in \Sigma_k$, so ist $f(b_1, \ldots, b_k) \in B(\Sigma)$.

Wie üblich bezeichnen wir Knoten oder Stellen in einem Baum mit Adressen aus $\mathbb{N}^*$. Die Menge $S(t)$ der Knoten in $t = a(t_1, \ldots, t_k)$ ist dabei induktiv definiert durch $S(t) := \{\epsilon\} \cup \bigcup_{j=1}^{k} j.S(t_j)$. Der **Teilbaum** t/n von t am Knoten n ist definiert durch $t/\epsilon := t$ und $t/n := t_j/n'$ falls $t = a(t_1, \ldots, t_k)$ und $n = j.n'$.

Sei zusätzlich eine unendliche Menge V von Variablen gegeben. Wir ordnen ihnen die Stelligkeit 0 zu. Ein Element aus $B(\Sigma \cup V)$ heißt ein **Muster** über Σ. Ein Muster heißt **linear**, wenn keine Variable mehr als einmal in ihm vorkommt.

Beispiel 2 Sei $\Sigma = \{a, cons, nil\}$ mit $\varrho(a) = \varrho(nil) = 0$, $\varrho(cons) = 2$. Bäume über Σ, d.h. Elemente der Baumsprache über Σ sind z.B. a, $cons(nil, nil)$, $cons(cons(a, nil), nil)$.
Sei $V = \{X\}$. Dann sind X, $cons(nil, X)$, $cons(X, nil)$ Muster über Σ. □

Eine **Substitution** Θ ist eine Abbildung $\Theta : V \to B(\Sigma \cup V)$. Θ wird zu einer Abbildung $\Theta : B(\Sigma \cup V) \to B(\Sigma \cup V)$ fortgesetzt durch $t\Theta = x\Theta$, falls $t = x \in V$ und $t\Theta = a(t_1\Theta, \ldots, t_k\Theta)$, falls $t = a(t_1, \ldots, t_k)$.[2] Wir schreiben auch $t\Theta = t[t_1/x_1, \ldots, t_k/x_k]$, falls die in t vorkommenden Variablen eine Teilmenge von $\{x_1, \ldots, x_k\}$ sind und $x_j\Theta = t_j$ für alle j.

Ein Muster $\tau \in B(\Sigma \cup V)$ mit Variablen $x_1, \ldots, x_k$ **trifft** einen Baum t, wenn es Bäume $t_1, \ldots, t_k$ in $B(\Sigma)$ gibt, so daß $t = \tau[t_1/x_1, \ldots \ldots, t_k/x_k]$.

---

[1]in funktionalen Sprachen auch Konstruktoren genannt.

[2]Traditionell wird die Anwendung einer Substitution Θ auf ein Muster t als $t\Theta$ geschrieben.

Definition 1 (Mustererkennungsproblem)
Eine Instanz des **Mustererkennungsproblems** besteht aus einer endlichen Menge
von Mustern $T = \tau_1, \ldots, \tau_k \in B(\Sigma \cup V)$ und einem Eingabebaum $t \in B(\Sigma)$. Die
Lösung des Mustererkennungsproblems für diese Instanz ist die Menge aller Paare
(n, i), so daß das Muster τ_i den Teilbaum t/n trifft. Als **Baummustererkenner** für T
bezeichnen wir einen Mechanismus, der bei Eingabe eines Baums $t \in B(\Sigma)$ die Lösung
des Mustererkennungsproblems für (T, t) liefert. Ein **Mustererkennungsgenerator**
wiederum konstruiert zu jeder Menge T von Mustern einen Mustererkenner für T.

□

Beispiel 3
Seien die Muster $\tau_1 = cons(X, nil)$ und $\tau_2 = cons(a, X)$ und der Eingabebaum
$t = cons(cons(a, nil), nil)$ gegeben. Die Lösung der Instanz des Mustererkennungsproblems $(\{\tau_1, \tau_2\}, t)$ ist die Menge $\{(\varepsilon, 1), (1, 1), (1, 2)\}$. □

Die Algorithmen zur Generierung von Mustererkennern, die wir hier vorstellen, beschränken sich auf die Bearbeitung **linearer** Muster. Zur Lösung des Mustererkennungsproblems mit nichtlinearen Mustern gibt es zwei Möglichkeiten. Entweder man
verwendet bei Bedarf Gleichheitstests auf Teilbäumen. Dies kann allerdings sehr teuer
sein, da man eventuell gezwungen ist, dieselben Teilbäume mehrmals zu besuchen. Die
andere Möglichkeit besteht darin, gewissermaßen sämtliche möglichen Gleichheitstests
vorher auszuführen. Dazu repräsentiert man den Eingabebaum als gerichteten azykli-
schen Graphen, der genau einen Knoten für jeden möglichen Teilbaum enthält. Dieser
sogenannte **Teilbaumgraph** läßt sich in linearer Zeit berechnen [5].

4   Das Baumanalyseproblem

Eine **reguläre Baumgrammatik** G ist ein Tripel (N, Σ, P). Dabei ist

- N eine endliche Menge von **Nichtterminalen**,

- Σ ein endliches Alphabet (mit Stelligkeit) von **Terminalen** [3],

- P eine endliche Menge von **Regeln** der Form $X \rightarrow s$ mit $X \in N$ und $s \in B(\Sigma \cup N)$.

Sei $p : X \rightarrow s$ eine Regel aus P. p heißt **Kettenregel**, falls $s \in N$; andernfalls
heißt p **Nicht-Kettenregel**. p hat den **Typ** $(X_1, \ldots, X_k) \rightarrow X$, falls das j-te
Vorkommen eines Nichtterminals in s das Nichtterminal X_j ist. Für eine rechte Seite
s definieren wir das Muster $\tilde{s}$ als das Muster in $B(\Sigma \cup \{x_1, \ldots, x_k\})$, das man aus
s erhält, indem man für alle j das j-te Vorkommen eines Nichtterminals durch die
Variable x_j ersetzt.

---

[3]Der Begriff Terminal wurde wegen der Analogie zum kontextfreien Fall gewählt. Oben heißen
die Terminale Operatoren.

Ein X-**Ableitungsbaum** für einen Baum $t \in B(\Sigma \cup N)$ ist ein Baum $\psi \in B(P \cup N)$, der den folgenden Bedingungen genügt.

- Ist $\psi \in N$, dann ist $\psi = X = t$.

- Ist $\psi \notin N$, dann ist $\psi = p(\psi_1, \ldots, \psi_k)$ für eine Regel $p : X \to s \in P$ vom Typ $(X_1, \ldots, X_k) \to X$, so daß $t = \tilde{s}[t_1/x_1, \ldots, t_k/x_k]$ und ψ_j X_j-Ableitungsbäume sind für die Bäume t_j.

Schließlich benötigen wir den Begriff des **Ableitungskopfes** $head(\psi)$ eines X-Ableitungsbaums ψ.

- Ist $\psi = X$, dann ist $head(\psi) := X$.

- Ist $\psi = p(\psi_1, \ldots, \psi_k)$ und p keine Kettenregel, dann ist $head(\psi) := p$.

- Ist $\psi = p(\psi_1)$ und p eine Kettenregel, dann ist $head(\psi) := p \cdot head(\psi_1)$.

Den Ableitungskopf eines X-Ableitungsbaums nennen wir auch X-Ableitungskopf.

Der Ableitungskopf von ψ beschreibt den "oberen" Abschnitt von ψ, der aus Anwendungen von Kettenregeln besteht, gefolgt von der ersten Anwendung einer Nicht-Kettenregel bzw. einem Nichtterminalsymbol.

Für $X \in N$ definieren wir die gemäß G bezüglich X erzeugte **Sprache** als $L(G, X) := \{ t \in B(\Sigma) \mid \exists \psi \in B(P \cup N) : \psi$ ist X-Ableitungsbaum für $t \}$.

Beispiel 4 Sei G_1 die reguläre Baumgrammatik (N_1, Σ, P_1);
$\Sigma = \{ a, cons, nil \}$ mit $\varrho(a) = \varrho(nil) = 0, \varrho(cons) = 2, N_1 = \{ E, S \}$ und
$P_1 = \{\ S\ \to\ nil,$
$\phantom{P_1 = \{\ }S\ \to\ cons(E, S),$
$\phantom{P_1 = \{\ }E\ \to\ a \}$
$L(G_1, S)$ ist gleich der Menge der linearen Listen von a's einschließlich der leeren Liste, d.h. $L(G_1, S) = \{ nil, cons(a, nil), cons(a, cons(a, nil)), \ldots \}$.

$\square$

Beispiel 5 Sei G_m die reguläre Baumgrammatik (N_m, Σ, P_m);
$\Sigma = \{ const, m, plus, REG \}$ mit $\varrho(const) = 0; \varrho(m) = 1, \varrho(plus) = 2,$
$N_m = \{ REG \}$ und
$P_m = \{\ addmc :\quad REG\ \to\quad plus(m(const), REG),$
$\phantom{P_m = \{\ }addm :\quad REG\ \to\quad plus(m(REG), REG),$
$\phantom{P_m = \{\ }add :\quad REG\ \to\quad plus(REG, REG),$
$\phantom{P_m = \{\ }ldmc :\quad REG\ \to\quad m(const),$
$\phantom{P_m = \{\ }ldc :\quad REG\ \to\quad const,$
$\phantom{P_m = \{\ }ld :\quad REG\ \to\quad REG \}$
P_m beschreibt einen Ausschnitt eines Instruktionssatzes eines einfachen Prozessors, wobei die Regeln mit dem Namen der beschriebenen Instruktion markiert wurden. Die ersten drei Regeln stehen für Additionsinstruktionen, die

- den Inhalt einer Speicherzelle, deren Adresse durch eine Konstante gegeben ist,

- den Inhalt einer Speicherzelle, deren Adresse in einem Register liegt,

- den Inhalt eines Registers

zu dem Inhalt eines Registers addieren und das Ergebnis in einem Register ablegen. Die letzten drei Regeln beschreiben Ladeinstruktionen, die

- den Inhalt einer Speicherzelle, deren Adresse durch eine Konstante gegeben ist,

- eine Konstante bzw.

- den Inhalt eines Registers

in ein Register laden.   □

Definition 2 (Baumanalyseproblem)
Eine Instanz des **Baumanalyseproblems** besteht aus einer regulären Baumgrammatik G und einem Baum t. Die Lösung des Baumanalyseproblems ist die Menge aller Ableitungsbäume von G für t (bzw. eine Repräsentation davon). Als **Baumanalysator** für G bezeichnen wir einen Mechanismus, der für jeden Baum $t \in B(\Sigma)$ die Lösung für die Instanz (G, t) des Baumanalyseproblems liefert. Ein **Baumanalysegenerator** schließlich konstruiert zu jeder regulären Baumgrammatik einen Baumanalysator für G.   □

Wie reguläre Wortsprachen sind reguläre Baumsprachen (d.h. die Sprachen, die von regulären Baumgrammatiken erzeugt werden) abgeschlossen unter Durchschnitt, Vereinigung und Komplementbildung. Da Leerheit entschieden werden kann, folgt damit etwa auch, daß es ebenfalls entscheidbar ist, ob eine reguläre Baumsprache in einer anderen enthalten ist.

Wir werden nun die Generierung von Mustererkennern und Baumanalysatoren auf die Konstruktion endlicher Baumautomaten zurückführen.

5 Endliche Baumautomaten

Ein **endlicher Baumautomat** A ist ein 4-Tupel $A = (Q, \Sigma, \delta, Q_F)$. Dabei ist

- Q eine endliche Menge von Zuständen,

- $Q_F \subseteq Q$ die Menge der Endzustände,

- Σ das Eingabealphabet (mit Stelligkeit) und

- $\delta \subseteq \bigcup_{j \geq 0} Q \times \Sigma_j \times Q^j$ die Menge der Übergänge.

Der Automat A heißt **deterministisch**, falls es für jedes $a \in \Sigma_k$ und jede Folge $q_1, \ldots, q_k$ von Zuständen höchstens einen Übergang $(q, a, q_1 \ldots q_k) \in \delta$ gibt. In diesem Falle kann δ auch als partielle Funktion geschrieben werden.

- $\delta : \bigcup_{j \geq 0} \Sigma_j \times Q^j \to Q$

Bei der Bearbeitung eines Eingabebaums t durchmustert A (man kann sich vorstellen in einem DFS-Durchlauf) den Baum t und nimmt in jedem Knoten von t einen bestimmten Zustand an, wobei der an jedem Knoten gewählte Übergang aus δ sein muß. Ist A deterministisch, gibt es in jedem Knoten höchstens eine mögliche Wahl des Übergangs, andernfalls eventuell mehrere. Technisch beschreiben wir eine solche Bearbeitung als *Annotation* des Eingabebaums t. Dafür führen wir ein erweitertes Alphabet $\Sigma \times Q$ ein, dessen Operatoren aus *Paaren* von Operatoren aus Σ und Zuständen bestehen.

Sei $\Sigma \times Q$ das Alphabet $\{\langle a, q \rangle \mid a \in \Sigma, q \in Q\}$, wobei $\langle a, q \rangle$ den gleichen Rang hat wie a. Eine q–**Berechnung** φ des endlichen Automaten A auf einem Eingabebaum $t = a(t_1, \ldots, t_m)$ definieren wir induktiv über die Struktur von t als einen Baum $\langle a, q \rangle(\varphi_1, \ldots, \varphi_m) \in B(\Sigma \times Q)$, wobei die φ_j q_j–Berechnungen für die Teilbäume t_j, $j = 1, \ldots, m$, sind und $(q, a, q_1 \ldots q_m)$ ein Übergang aus δ ist. Ist $q \in Q_F$, dann heißt φ **akzeptierend**. Die Sprache $L(A)$ der von A akzeptierten Bäume besteht aus allen Bäumen, für die eine akzeptierende Berechnung existiert. Ein Übergang $\tau \in \delta$ heißt **überflüssig**, falls er in keiner Berechnung von A vorkommt. Überflüssige Übergänge können offenbar weggelassen werden, ohne das "Verhalten" des Automaten zu beeinträchtigen.

Falls A deterministisch ist, gibt es für jeden Eingabebaum höchstens eine Berechnung. Dann können wir die partielle Funktion δ zu einer partiellen Funktion $\delta^* : B(\Sigma) \to Q$ fortsetzen durch: $\delta^*(t) = \delta(a, \delta^*(t_1) \ldots \delta^*(t_k))$, falls $t = a(t_1, \ldots, t_k)$. Der Einfachheit halber nennen wir δ^* ebenfalls wieder δ. Durch Induktion über die Struktur von t zeigt man, daß $\delta^*(t) = q$ genau dann gilt, wenn es eine q–Berechnung für t gibt.

Wenden wir uns als erstes der Generierung von Mustererkennern zu. Sei τ ein lineares Muster in $B(\Sigma \cup V)$. Wir wollen einen (evt. nichtdeterministischen) endlichen Baumautomaten A_τ bauen, der erkennt, ob das Muster τ einen gegebenen Eingabebaum trifft. Intuitiv arbeitet A_τ wie folgt. Außerhalb des Musters nimmt A_τ einen unspezifischen Zustand $\bot$ an; innerhalb des Musters τ bezeichne der Zustand gerade das bereits gelesene Teilmuster. Da es uns (hier) auf die genaue Numerierung der Variablen nicht ankommt, ersetzen wir in τ sämtliche Variablen durch $\bot$ ("eine Variable trifft alles"). Nehmen wir darum an, daß $\tau \in B(\Sigma \cup \{\bot\})$. Dann definieren wir $A_\tau := (Q_\tau, \Sigma, \delta_\tau, Q_{\tau,F})$, wobei $Q_\tau := \{s \mid s \text{ Teilbaum von } \tau\} \cup \{\bot\}$, $Q_{\tau,F} := \{\tau\}$ und δ wie folgt definiert ist:

- $(\bot, a, \bot \ldots \bot) \in \delta$;

- ist $s \in Q_\tau$ und $s = a(s_1, \ldots, s_k)$, dann ist $(s, a, s_1 \ldots s_k) \in \delta$.

Offenbar gilt:

1. für jeden Baum t gibt es eine $\bot$–Berechnung;

2. es gibt für einen Baum t genau dann eine τ–Berechnung, wenn τ t trifft.

Das Beispiel läßt sich leicht auf den Fall einer Menge linearer Muster $T = \{\tau_1, \ldots, \tau_n\}$ verallgemeinern, von denen wir wieder o.B.d.A. annehmen, daß sämtliche vorkommenden Variablen durch das Symbol $\perp$ ersetzt wurden. Als Menge der Zustände für unseren Automaten A_T wählen wir $Q_T := \bigcup_{j=1}^{n} Q_{\tau_j}$ mit $Q_{T,F} := T$, während sich die Definition von δ textuell nicht ändert (aber die resultierende Menge der Übergänge natürlich).

Wollen wir herausfinden, welche Muster einen Eingabebaum t an der Wurzel treffen, müssen wir uns eine Übersicht über alle möglichen Berechnungen von A_T auf t verschaffen. Dies gelingt mithilfe der **Teilmengenkonstruktion** für Baumautomaten.

Definition 3 (Teilmengenkonstruktion I)
Sei $A = (Q, \Sigma, \delta, Q_F)$ ein endlicher Baumautomat. Der zugehörige Teilmengenautomat ist der deterministische endliche Baumautomat $P(A) = (Q_1, \Sigma, \delta_1, Q_{1,F})$ mit

- $Q_1 := 2^Q$ ist die Potenzmenge von Q;
- $Q_{1,F} := \{B \subseteq Q \mid B \cap Q_F \neq \emptyset\}$;
- δ_1 ist die Funktion mit $\delta_1(a, B_1 \ldots B_k) = \{q \in Q \mid \exists q_1 \in B_1, \ldots, q_k \in B_k : (q, a, q_1 \ldots q_k) \in \delta\}$.   □

Mittels Induktion über die Größe des Eingabebaums zeigt man:

Lemma 1 *Sei $t \in B(\Sigma)$. Dann ist $\delta_1(t)$ die Menge aller Zustände $q \in Q$, für die eine q–Berechnung auf t existiert. Insbesondere gilt: $L(A) = L(P(A))$.*   □

Die Teilmengenkonstruktion erlaubt es uns, Baummustererkenner zu generieren. Zu einer Mustermenge T konstruieren wir den deterministischen Baumautomaten $P(A_T) = (Q, \Sigma, \delta, Q_F)$. Die Menge $\delta(t) \cap T$ enthält dann genau die Muster aus T, die den Eingabebaum t treffen.

Beispiel 6 *Sei etwa $T = \{\tau_1, \tau_2\}$ mit*

$$\tau_1 = b(a(a(X_1, X_2), X_3), X_4) \text{ und } \tau_2 = b(X_1, c(X_2, c(X_3, X_4))).$$

Dann ist $A_T = (Q_T, \Sigma, \delta_T, Q_{T,F})$ mit
$$Q_T = \{\perp, \quad a(\perp, \perp), \quad a(a(\perp, \perp), \perp), \quad b(a(a(\perp, \perp), \perp), \perp),$$
$$c(\perp, \perp), \quad c(\perp, c(\perp, \perp)), \quad b(\perp, c(\perp, c(\perp, \perp))) \, \}.$$
  □

Wir benötigen 7 Zustände. Zu unserem Entsetzen stellen wir fest, daß unsere Konstruktion I im Beispiel (wie in den meisten praktischen Fällen) hoffnungslos ineffizient ist: der generierte Automat für die zwei angegebenen kleinen (!) Muster hätte bereits $2^7 = 128$ Zustände. Es läßt sich zeigen, daß im schlimmsten Fall exponentiell viele Zustände benötigt werden. Oft werden aber ein Großteil der durch Konstruktion I eingeführten neuen Zustände nicht gebraucht. In unserem Beispiel würde etwa auch die Menge $\{a(\perp, \perp), b(\perp, c(\perp, c(\perp, \perp)))\}$ generiert werden, die "widersprüchliche" Muster enthält, also solche, die gar nicht denselben Baum treffen können. Darum geben wir eine sparsamere Konstruktion II an, die von vornherein nur solche Zustandsmengen generiert, die tatsächlich in Berechnungen vorkommen können.

Definition 4 (Teilmengenkonstruktion II)

Sei $A = (Q, \Sigma, \delta, Q_F)$ ein endlicher Baumautomat. Der zugehörige (reduzierte) Teilmengenautomat ist der deterministische endliche Baumautomat $P_r(A) = (Q_r, \Sigma, \delta_r, Q_{r,F})$ mit $Q_{r,F} := \{B \in Q_r \mid B \cap Q_F \neq \emptyset\}$, dessen Zustandsmenge und Übergänge iterativ berechnet werden durch $Q_r := \bigcup_{n \geq 0} Q_r^{(n)}$ und $\delta_r := \bigcup_{n \geq 0} \delta_r^{(n)}$, wobei:

- $Q_r^{(0)} = \emptyset$;

- sei $n > 0$. Für $a \in \Sigma_k$ und $B_1, \ldots, B_k \in Q_r^{(n-1)}$ sei $B := \{q \in Q \mid \exists q_1 \in B_1, \ldots, q_k \in B_k : (q, a, q_1 \ldots q_k) \in \delta\}$. Ist $B \neq \emptyset$, dann ist $B \in Q_r^{(n)}$ und $(B, a, B_1 \ldots B_k) \in \delta_r^{(n)}$. $\qquad\square$

Da für alle n gilt, daß $Q_r^{(n)} \subseteq Q_r^{(n+1)}$ und $\delta_r^{(n)} \subseteq \delta_r^{(n+1)}$, können wir die Iteration abbrechen, sobald keine neuen Zustände mehr erzeugt werden, d.h. $Q_r = Q_r^{(n)}$ und $\delta_r = \delta_r^{(n)}$ für das erste n mit $Q_r^{(n)} = Q_r^{(n+1)}$. Folglich bricht das Verfahren nach spätestens $2^{|Q|}$ Iterationen ab.

Mittels Induktion über die Größe des Eingabebaums zeigt man:

Lemma 2

1. *Für jedes $t \in B(\Sigma)$ gilt:*

   - *Ist $\delta_r(t)$ nicht definiert, dann gibt es für kein $q \in Q$ eine q–Berechnung von A für t.*

   - *Ist $\delta_r(t)$ definiert, dann ist $\delta_r(t)$ die Menge aller Zustände q, für die eine q–Berechnung auf t existiert.*

2. $L(A) = L(P_r(A))$.

3. *Für jeden Zustand $B \in Q_r$ gibt es einen Baum t, so daß $\delta_r(t) = B$.* $\qquad\square$

Betrachten wir den Automaten A_T. Dann finden wir, daß nun nicht mehr *alle* Mengen von Teilmustern als Zustände generiert werden, sondern nur solche, die **maximal kompatibel** sind. Dabei heißt eine Menge $S \subseteq T$ von Mustern **kompatibel**, falls es einen Baum t gibt, den jedes Muster aus S trifft. S heißt **maximal** kompatibel, falls es einen Baum gibt, den alle Muster aus S treffen und alle Muster aus $T \backslash S$ nicht treffen.

Die Zustandsmenge Q_r des reduzierten Teilmengenautomaten für A_T besteht genau aus den maximal kompatiblen Mengen von Teilmustern. Folglich ergibt sich in unserem Beispiel:

$$Q_r = \{ \quad \{\bot\}, \qquad \{\bot, b(a(a(\bot, \bot), \bot), \bot), b(\bot, c(\bot, c(\bot, \bot)))\},$$
$$\{\bot, a(\bot, \bot)\}, \quad \{\bot, a(\bot, \bot), a(a(\bot, \bot), \bot)\}, \quad \{\bot, b(a(a(\bot, \bot), \bot), \bot)\},$$
$$\{\bot, c(\bot, \bot)\}, \quad \{\bot, c(\bot, \bot), c(\bot, c(\bot, \bot))\}, \quad \{\bot, b(\bot, c(\bot, c(\bot, \bot)))\}\}.$$

Der reduzierte Teilmengenautomat besitzt nur acht Zustände! Gegenüber den 128 Zuständen der Konstruktion I eine beträchtliche Ersparnis.

6 Die Generierung von Baumanalysatoren

Sei $G = (N, \Sigma, P)$ eine Grammatik und $X \in N$. Um alle möglichen X–Ableitungs-bäume zu gegebenen Bäumen zu berechnen, gehen wir ähnlich vor wie im Falle des Mustererkennungsproblems. Wir konstruieren zuerst einen nichtdeterministischen Automaten $A_{G,X}$, dessen Berechnungen den Ableitungsbäumen bzgl. G entsprechen. Auf $A_{G,X}$ wenden wir in einem zweiten Schritt die Teilmengen-Konstruktion an. Der resultierende Teilmengenautomat ist die Grundlage unseres Baumanalysators.

Intuitiv arbeitet der Automat $A_{G,X}$ auf einem Eingabebaum $t \notin N$ wie folgt. Er rät an der Wurzel einen X–Ableitungskopf $p_1 \ldots p_k p$ eines X–Ableitungsbaums für t mit $p : X' \to s$. Dann verifiziert $A_{G,X}$, daß s tatsächlich "paßt", d.h. daß das aus s gewonnene Muster den Baum t trifft. Stößt $A_{G,X}$ bei der Verifikation dabei auf eine Stelle, der die rechte Seite s erneut ein Nichtterminal, etwa X_j, zuordnet, rät $A_{G,X}$ an dieser Stelle einen X_j–Ableitungskopf und so fort. Formal definieren wir darum $A_{G,X} = (Q_G, \Sigma, \delta_G, \{X\})$, wobei $Q_G = N \cup \{s' \mid \exists X \to s \in P$, mit s' ist Teilmuster von $s\}$.

δ_G besteht aus zwei Bestandteilen; dem ersten, der für die Verifikation einer ausgewählten rechten Seite verantwortlich ist, und dem zweiten, der Ableitungsköpfe rät. Wir definieren:

$$\delta_G := \quad \{(X, a, \epsilon) \mid a \in \Sigma \text{ und } \exists X\text{–Ableitungsbaum für } a\} \ \cup$$
$$\{(s, a, s_1 \ldots s_k) \mid s = a(s_1, \ldots, s_k) \in Q_G\} \ \cup$$
$$\{(X, a, s_1 \ldots s_k) \mid \exists X' \to s \in P : \exists X\text{–Ableitungsbaum für } X'$$
$$\text{und } s = a(s_1, \ldots, s_k)\}$$

Es läßt sich zeigen, daß sich aus dem so definierten Automaten $A_{G,X}$ mithilfe etwa der reduzierten Teilmengenkonstruktion II in der Literatur beschriebene Baumanaly-satoren ableiten lassen. Betrachtet man die Definition von δ_G genauer, findet man, daß δ_G eine Reihe von überflüssigen Übergängen enthält! Der Automat gestattet auch Übergänge $(s, a, s_1 \ldots s_k)$, bei denen s die rechte Seite einer Regel, aber selbst nicht *echtes* Teilmuster einer weiteren rechten Seite ist. Es ist leicht zu sehen, daß solche Übergänge in keiner X–Berechnung verwendet werden. Darum wählen wir stattdessen:

$$\delta_G := \quad \{(X, a, \epsilon) \mid a \in \Sigma \text{ und } \exists X\text{–Ableitungsbaum für } a\} \ \cup$$
$$\{(s, a, s_1 \ldots s_k) \mid s = a(s_1, \ldots, s_k) \text{ echtes Teilmuster einer rechten}$$
$$\text{Seite}\} \ \cup$$
$$\{(X, a, s_1 \ldots s_k) \mid \exists X' \to s \in P : \exists X\text{–Ableitungsbaum für } X'$$
$$\text{und } s = a(s_1, \ldots, s_k)\}$$

Lemma 3 *Sei G eine reguläre Baumgrammatik und t ein Eingabebaum.*

- *Es existiert ein X–Ableitungsbaum für t bzgl. G genau dann, wenn es eine X–Berechnung für t bzgl. $A_{G,X}$ gibt.*
  Insbesondere gilt: $L(G, X) = L(A_{G,X})$.

- *Sei $A = (Q, \Sigma, \delta, Q_F)$ der (reduzierte) Teilmengenautomat zu $A_{G,X}$. Dann ist*
  $\delta(t) \cap N = \{X' \in N \mid \exists X'\text{–Ableitungsbaum für } t\}$. $\qquad\square$

Aus den X–Berechnungen des nichtdeterministischen Automaten $A_{G,X}$ für einen Eingabebaum t lassen sich leicht die X–Ableitungsbäume für t rekonstruieren.[4] Damit können wir das Baumanalyse-Problem auf das Problem reduzieren, aus der Berechnung eines (reduzierten) Teilmengenautomaten für t sämtliche akzeptierenden Berechnungen des zugrunde liegenden nichtdeterministischen Automaten zu rekonstruieren.

Sei $A = (Q, \Sigma, \delta, Q_F)$ ein nichtdeterministischer Automat und $A_r = (Q_r, \Sigma, \delta_r, Q_{r,F})$ der reduzierte Teilmengenautomat zu A. Wir stellen einen einfachen Algorithmus vor, der bei Eingabe einer B–Berechnung von A_r für einen Baum t und $q \in B$ die q–Berechnungen für t konstruieren kann.

Für einen Übergang $\tau = (B, a, B_1 \ldots \ldots B_k) \in \delta_r$ und $q \in B$ sei $\Theta(\tau)_q := \{(q, a, q_1 \ldots q_k) \in \delta \mid q_1 \in B_1, \ldots, q_k \in B_k\}$ die Menge der Übergänge des nichtdeterministischen Automaten A, die zu τ gehören und q als Nachfolgezustand haben.

Sei $\varphi = \langle a, B \rangle (\varphi_1, \ldots, \varphi_k)$ die B–Berechnung von A_r. Der Algorithmus durchmustert den Baum φ in Pre-Order. Sei τ der Übergang an der Wurzel von φ. Der Algorithmus wählt einen Übergang $(q, a, q_1 \ldots q_k) \in \Theta(\tau)_q$ aus. Dann werden rekursiv q_j–Berechnungen ψ_j, $j = 1, \ldots, k$, zu den Berechnungen φ_j bestimmt. Als Resultat wird $\langle a, q \rangle (\psi_1, \ldots, \psi_k)$ ausgegeben. Verschiedene Wahlen eines Überganges entsprechen verschiedenen Ableitungsbäumen. Somit können mit diesem Algorithmus alle Ableitungsbäume aufgezählt werden.

7 Anwendung auf die Codeselektion

Wir wollen unsere Methode zur Erzeugung von Baumanalysatoren anwenden, um Codeselektoren zu generieren. Dies konfrontiert uns mit dem Problem, aus der Fülle der möglichen Ableitungsbäume einen günstigsten auszuwählen. Wie bei der Generierung eines Baumanalysators gehen wir in drei Schritten vor. Wir nehmen an, daß die Regeln der Grammatik mit Kostenfunktionen annotiert sind, die die Kosten des durch diese Regel modellierten Befehls beschreiben. Wir übersetzen diese in Kostenfunktionen für die Übergänge des nichtdeterministischen Automaten. Aus der Berechnung des zugehörigen Teilmengenautomaten läßt sich dann eine preiswerteste akzeptierende Berechnung des nichtdeterministischen Automaten bestimmen.

Sei also jeder Regel p des Typs $(X_1, \ldots, X_k) \to X$ eine k-stellige Funktion $C(p) : \mathbb{N}_0^{\ k} \to \mathbb{N}_0$ zugeordnet.

Ein Kostenmaß C läßt sich zu einer Funktion fortsetzen, die jedem Ableitungsbaum ψ Kosten $C(\psi) \in \mathbb{N}_0$ zuordnet. Ist $\psi = X \in N$, dann ist $C(\psi) := 0$. Ist $\psi = p(\psi_1, \ldots, \psi_k)$, dann ist $C(\psi) := C(p)C(\psi_1) \ldots C(\psi_k)$, d.h. wir wenden die Funktion $C(p)$ auf die rekursiv bereits berechneten Werte $C(\psi_1), \ldots, C(\psi_k)$ an.

---

[4]Man beachte, daß die Anzahl der X–Berechnungen für t stets endlich ist, auch wenn die Anzahl der X–Ableitungsbäume eventuell *unendlich* ist. Dies entspricht der Tatsache, daß die Menge der Ableitungsköpfe, die zu einem ratenden Übergang gehören, unendlich sein kann.

Analog können wir die Übergänge eines nichtdeterministischen Automaten mit Kostenfunktionen annotieren und zu Kosten von Berechnungen fortsetzen.

Das Kostenmaß C heißt **monoton** bzw. **additiv**, falls $C(p)$ für alle $p \in P$ monoton ist bzw. die Form $C(p) = c_p + x_1 + \cdots + x_k, c_p \in \mathbb{N}_0$, hat.

Kostenmaße, die in der Praxis verwendet werden, sind i.a. monoton. Übliche Kostenmaße sind z.B. die zur Ausführung benötigten Prozessorzyklen, die Anzahl der referierten Speicherzellen oder die Anzahl der Operanden einer Instruktion. Ein Beispiel für ein nicht-additives Kostenmaß ist C_R, das die zur Berechnung eines Ausdrucks benötigte Registeranzahl ermittelt. Häufig werden (selbst wenn sie die "Realität" bei sehr komplizierten Prozessorarchitekturen nur noch grob approximieren) additive Kostenmaße benutzt, da sie leicht zu verwalten sind. Ein einfaches Maß ist $C_\sharp$, das jeder Kettenregel p die Kosten $C_\sharp(p) = x_1$ und jeder Nicht-Kettenregel p vom Typ $(X_1, \ldots, X_k) \to X$ die Kosten $C_\sharp(p) = 1 + x_1 + \cdots + x_k$ zuordnet. Der Wert $C_\sharp(\psi)$ etwa liefert die Anzahl der Nicht-Kettenregeln in ψ.

Wir übersetzen die Kostenannotation C der Grammatik G in eine Kostenannotation C^* des zugehörigen Automaten $A_{G,X}$ für ein Nichtterminal X. Das hier vorgestellte Verfahren setzt ein additives Kostenmaß C voraus. In diesem Fall können die Kosten jeder Regel durch eine Konstante beschrieben werden, d.h. wir fassen C als eine Funktion $P \to \mathbb{N}_0$ auf. Dann definieren wir C^* wie folgt.

- Ist $\tau = (X, a, \epsilon)$ für $X \in N$, dann sind $C^*(\tau)$ die minimalen Kosten eines X–Ableitungsbaums für a.

- Ist $\tau = (s, a, s_1 \ldots s_k)$ mit $s = a(s_1, \ldots, s_k)$, dann ist $C^*(\tau) := 0$.

- Ist $\tau = (X, a, s_1 \ldots s_k)$, dann ist $C^*(\tau)$ das Minimum der Werte $(\gamma + C(p))$ für Regeln $p : X' \to a(s_1, \ldots, s_k)$ und minimale Kosten γ eines X–Ableitungsbaums für X'.

Die Kosten $C^*(\varphi)$ einer X–Berechnung φ sind gerade die minimalen Kosten eines X–Ableitungsbaums, der durch φ repräsentiert wird.

Sei allgemein $A = (Q, \Sigma, \delta, Q_F)$ ein endlicher Baumautomat und $C : \delta \to \mathbb{N}_0$ eine (additive) Kostenfunktion für die Übergänge von A. Sei $A_r = (Q_r, \Sigma, \delta_r, Q_{r,F})$ der zu A gehörige (reduzierte) Teilmengenautomat.

Wir geben eine Modifikation des oben beschriebenen Algorithmus zur Konstruktion von Berechnungen von A an, so daß eine günstigste Berechnung ausgegeben wird. Die Idee besteht darin, dem Algorithmus an jedem Auswahlpunkt Informationen für die Auswahl zur Verfügung zu stellen. Dazu ordnen wir jeder B–Berechnung φ des Teilmengenautomaten A_r für einen Eingabebaum t zwei Tupel $C(\varphi) = \langle C(\varphi)_q \rangle_{q \in B}$ und $D(\varphi) = \langle D(\varphi)_q \rangle_{q \in B}$ zu, wobei $C(\varphi)_q$ die Kosten einer billigsten q–Berechnung für t und $D(\varphi)_q$ den an der Wurzel gewählten Übergang einer q–Berechnung für t mit Kosten $C(\varphi)_q$ enthalten. Die Tupel $C(\varphi/n)$ und $D(\varphi/n)$ für alle Knoten n von φ können während eines Post-Order-Durchlaufs durch φ berechnet werden.

Mithilfe der Tupel $D(\varphi/n)$ für alle Knoten n von φ kann der obige Algorithmus so modifiziert werden, daß er in einer Durchmusterung der Berechnung φ von A_r für

t nun in Pre-Order eine q–Berechnung von A für t mit minimalen Kosten ausgibt, indem immer die entsprechende Komponente von $D(\varphi/n)$ gewählt wird.

Die Verwaltung der Kostentupel kann sehr teuer sein. Darum wird man sich bemühen, die Kostenberechnung so weit wie möglich in die Zustandsübergänge des Teilmengenautomaten selbst zu integrieren. Ein Versuch in diese Richtung wurde von Pelegri-Llopart in [23], [24] unternommen. Pelegri-Llopart beobachtete, daß bei den üblichen Maschinengrammatiken $G_m = (N_m, \Sigma, P_m)$ die Differenzen der Kosten von X-Ableitungsbäumen mit minimalen Kosten für die verschiedenen $X \in N_m$ in der Regel durch eine Konstante beschränkt sind. Dies erklärt sich aus dem Umstand, daß es in Maschinengrammatiken üblicherweise ein zentrales Nichtterminal gibt, das "Register" in der Maschine beschreibt, aus dem sich (fast) alle anderen Nichtterminale durch Anwendung von Kettenregeln ableiten lassen [24]. Dadurch kann bei der Auswahl einer günstigsten Berechnung des nichtdeterministischen Automaten $A = (Q, \Sigma, \delta, Q_F)$ aus der Berechnung des entsprechenden Teilmengenautomaten ohne Einschränkung mit beschränkten Kostendifferenzen anstelle der realen Kosten gearbeitet werden.

Die endlich vielen Kostendifferenzen können direkt bei der Teilmengenkonstruktion mit in den Zustand des Teilmengenautomaten $A_c = (Q_c, \Sigma, \delta_c, Q_{c,F})$ integriert werden. In jedem Zustand B von A_c ordnen wir dazu jedem erreichbaren Zustand $q \in Q$ des nichtdeterministische Automaten eine Kostendifferenz d zu, d.h. $B \subseteq \{\langle q, d \rangle \mid q \in Q \text{ und } d \in \mathbb{N}_0\}$. Für $\langle q, d \rangle \in B$ beschreibt d die Kostendifferenz einer q–Berechnung von A zu einer billigsten Berechnung.

Definition 5 (Teilmengenkonstruktion III)
Sei $A = (Q, \Sigma, \delta, Q_F)$ ein endlicher Baumautomat und $C : \delta \to \mathbb{N}_0$ eine Kostenfunktion, die jedem Übergang aus δ Kosten aus $\mathbb{N}_0$ zuordnet. Der zugehörige (**reduzierte**) **Teilmengenautomat mit integrierten Kosten** ist der deterministische endliche Baumautomat $P_c(A) = (Q_c, \Sigma, \delta_c, Q_{c,F})$ mit $Q_{c,F} := \{B \in Q_c \mid \langle q, d \rangle \in B \text{ und } q \in Q_F\}$, dessen Zustandsmenge und Übergänge iterativ berechnet werden durch $Q_c := \bigcup_{n \geq 0} Q_c^{(n)}$ und $\delta_c := \bigcup_{n \geq 0} \delta_c^{(n)}$, wobei:

- $Q_c^{(0)} = \emptyset$;

- sei $n > 0$. Für $a \in \Sigma_k$ und $B_1, \ldots, B_k \in Q_c^{(n-1)}$ sei
  $B := \{\langle q, d \rangle \mid \exists \langle q_1, d_1 \rangle \in B_1, \ldots, \langle q_k, d_k \rangle \in B_k \text{ und } \tau = (q, a, q_1 \ldots q_k) \in \delta \text{ so daß } d = C(\tau) + d_1 + \ldots + d_k \text{ minimal }\}$.
  Ist $B \neq \emptyset$, dann ist $norm(B) \in Q_c^{(n)}$ und $(norm(B), a, B_1 \ldots B_k) \in \delta_c^{(n)}$ mit
  $norm(B) = \{\langle q, (d - \epsilon) \rangle \mid \langle q, d \rangle \in B \text{ und } \epsilon = min_{i=1}^{|B|}(d_i) \text{ mit } \langle q_i, d_i \rangle \in B\}$ □

Der Algorithmus zur Konstruktion einer q–Berechnung eines nichtdeterministischen Baumautomaten aus der Berechnung des zugehörigen (reduzierten) Teilmengenautomaten kann fast unverändert übernommen werden, um aus einer Berechnung eines (reduzierten) Teilmengenautomaten mit integrierten Kosten eine preiswerteste q–Berechnung des nichtdeterministischen Baumautomaten zu konstruieren. Sei $A = (Q, \Sigma, \delta, Q_F)$ ein nichtdeterministischer Baumautomat, $C : \delta \to \mathbb{N}_0$ eine

Kostenfunktion und $A_c = (Q_c, \Sigma, \delta_c, Q_{c,F})$ der zugehörige (reduzierte) Teilmengenautomat mit integrierten Kosten. Für einen Übergang $\tau = (B, a, B_1 \ldots B_k) \in \delta_c$ und $\langle q, d \rangle \in B$ sei $\Theta_c(\tau)_q := \{\eta = (q, a, q_1 \ldots q_k) \in \delta \mid \langle q_1, d_1 \rangle \in B_1, \ldots, \langle q_k, d_k \rangle \in B_k$ so daß $C(\eta) + d_1 + \ldots + d_k$ minimal$\}$ die Menge der billigsten Übergänge des nichtdeterministischen Automaten A, die zu τ gehören und q als Nachfolgezustand haben.

Sei $\varphi = \langle a, B \rangle(\varphi_1, \ldots, \varphi_k)$ die B-Berechnung von A_c. Analog zu oben durchmustert der Algorithmus den Baum φ in Pre-Order. Sei τ der Übergang an der Wurzel von φ. Der Algorithmus wählt einen Übergang $(q, a, q_1 \ldots q_k) \in \Theta_c(\tau)_q$ aus. Dann werden rekursiv q_j-Berechnungen ψ_j, $j = 1, \ldots, k$, zu den Berechnungen φ_j bestimmt. Als Resultat wird $\langle a, q \rangle(\psi_1, \ldots, \psi_k)$ ausgegeben. Alle möglichen Resultate sind billigste q-Berechnungen von A.

Die Teilmengenautomaten der Konstruktion III sind in der Regel größer als die entsprechenden Automaten der Konstruktion II ohne integrierten Kosten. Weiterhin ist die Konstruktion III nur für Baumautomaten möglich, bei denen die Kostenunterschiede von billigsten Berechnungen für Bäume t durch eine Konstante beschränkt sind. Der Vorteil von Teilmengenautomaten mit integrierten Kosten liegt darin, daß sie eine erheblich schnellere Konstruktion einer billigsten Berechnung des zugehörigen nichtdeterministischen Baumautomaten erlauben.

Der Einfachheit halber klammern wir bei den folgenden Teilmengenkonstruktionen die Integration der Kosten aus.

8 Implementierung deterministischer Baumautomaten

In diesem Abschnitt wollen wir uns mit geschickten Implementierungen deterministischer Baumautomaten beschäftigen. Am einfachsten stellt man die Menge δ_a der Übergänge für einen Operator a der Stelligkeit k als k-dimensionale Matrix M_a dar. Dabei ist $M_a[q_1, \ldots, q_k] = \delta(a, q_1 \ldots q_k)$, falls δ für diese Argumente definiert ist, andernfalls $\perp$, ein spezielles Fehlersymbol.

Nehmen wir an, der Eingabebaum t sei gegeben als knotenmarkierter geordneter Wurzelbaum im Sinne von [20]. Der Zustand an einem Knoten n von t mit Markierung $a \in \Sigma_k$ ist $M_a[q_1, \ldots, q_k]$, wobei $q_1, \ldots, q_k$ die Zustände an den Söhnen des Knotens n sind. Dies erfolgt etwa bei einer Post-Order-Durchmusterung des Baums t. Die Kosten für einen "Lauf" eines Baumautomaten über t bestehen somit neben dem Aufwand für die Post-Order-Durchmusterung, die in Zeit proportional zur Größe des Baumes durchgeführt werden kann, aus einem indizierten Matrizen- bzw. Feldzugriff für jeden Knoten von t.

Die Zeit für einen Feldzugriff $M[i_i, \ldots, i_n]$ ist auf den meisten realen Rechner linear abhängig von der Anzahl n der Indizes. Da aber jeder Unterbaum eines Baumes t nur einmal zu einer Indizierung beiträgt, ist die Gesamtlaufzeit linear zu der Anzahl der Knoten von t (unabhängig von den auftretenden Stelligkeiten).

Beispiel 7 Fortführung von Beispiel 5

Sei G_m die Grammatik aus Beispiel 5. Der nichtdeterministische Automat $A = (Q, \Sigma, \delta, Q_F)$ zu G_m hat die Zustände:

$$Q = \{const, REG, m(const), m(REG)\}$$

und die Übergänge:

$$\delta = \{ \ (const, const, \epsilon)$$
$$(REG, const, \epsilon)$$
$$(REG, REG, \epsilon)$$
$$(m(const), m, const)$$
$$(REG, m, const)$$
$$(m(REG), m, REG)$$
$$(REG, plus, m(const) \ REG)$$
$$(REG, plus, m(REG) \ REG)$$
$$(REG, plus, REG \ REG)\}$$

Der reduzierte Teilmengenautomat $A_r = (Q_r, \Sigma, \delta_r, Q_{F,r})$ zu G_m hat die Zustände:

$$Q_r = \{ \ q_1 \ = \ \{REG\}$$
$$q_2 \ = \ \{const, REG\}$$
$$q_3 \ = \ \{m(REG)\}$$
$$q_4 \ = \ \{m(const), REG, m(REG)\}$$

δ_r dargestellt in Tabellenform:

$$\delta_{r,const} = \qquad q_2$$

$$\delta_{r,REG} = \qquad q_1$$

$\delta_{r,m} =$

	Sohn		
q_1	q_2	q_3	q_4
q_3	q_4	$\perp$	q_3

$\delta_{r,plus} =$

linker Sohn	rechter Sohn			
	q_1	q_2	q_3	q_4
q_1	q_1	q_1	$\perp$	q_1
q_2	q_1	q_1	$\perp$	q_1
q_3	q_1	q_1	$\perp$	q_1
q_4	q_1	q_1	$\perp$	q_1

Die Darstellung von δ als eine Menge von Matrizen ist in der Regel sehr speicherplatzintensiv, zumal die Größe einer Matrix M_a für einen Operator $a \in \Sigma_k$ proportional zu $|Q|^k$ ist, d.h. exponentiell in der Stelligkeit von a unabhängig davon, wieviele (wie wenige) definierte Übergänge der Automat A für a besitzt. Mit üblichen Tabellenkompaktierungsmethoden kann der Platzbedarf zur Speicherung der Matrizen in den meisten in der Praxis auftretenden Fällen erheblich reduziert werden.

Eine andere Methode zur Darstellung der Übergangsfunktion δ_a für einen Operator a ist die Verwendung von Entscheidungsbäumen.

Seien Q und D endliche Mengen und $H : Q^k \to D$ eine partielle Abbildung. Ein **Entscheidungsbaum** für H ist ein blattmarkierter Baum der Höhe k, dessen Knotenmenge V gegeben ist als die Menge

$$V := \bigcup_{j=0}^{k} \{q_1 \ldots q_j \mid \exists q_{j+1}, \ldots, q_k \in Q : H(q_1 \ldots q_j q_{j+1} \ldots q_k) \text{ ist definiert } \}$$

Dabei sind die Knoten $q_1 \ldots q_j$ und $q_1 \ldots q_j q'$ durch eine Kante mit Beschriftung q' verbunden. Weiterhin sind die Blätter $b = q_1 \ldots q_k$ mit $H(q_1 \ldots q_k)$ markiert.

Für einen deterministischen Baumautomaten $A = (Q, \Sigma, \delta, Q_F)$ und einen Operator a mit Stelligkeit k können wir δ_a repräsentieren durch den Entscheidungsbaum zur Funktion H_a, die gegeben ist durch $H_a(q_1 \ldots q_k) := \delta(a, q_1 \ldots q_k)$. Die Knoten des Entscheidungsbaums repräsentieren genau die Präfixe der Zustandsfolgen, die in δ_a vorkommen. Der Zustand an einem Knoten n des Eingabebaums mit Markierung $a \in \Sigma_k$, für dessen Söhne die Zustände $q_1, \ldots, q_k$ bereits berechnet wurden, ergibt sich durch Verfolgen des Pfades im Entscheidungsbaum, dessen Kanten nacheinander mit $q_1, \ldots, q_k$ markiert sind. Die Markierung des Blattes am Ende des Pfades liefert den für n zu berechnenden Zustand.

Im Falle des oben konstruierten Baumanalysators zu einer regulären Baumgrammatik G sind wir aber nicht nur an dem Zustand für den Knoten n interessiert, sondern auch an der Menge der möglichen Übergänge des nichtdeterministischen Automaten $A_{G,X}$ an n. Hier wählen wir H_a entsprechend "informativer", d.h. als

$$H_a(B_1 \ldots B_k) := \langle \delta(a, B_1 \ldots B_k), \Theta \rangle$$

wobei $\Theta := \{(q, a, q_1 \ldots q_k) \in \delta_G \mid q_j \in B_j\}$ ist.

Bei der Verwendung von Entscheidungsbäumen ist der Platzbedarf des Automaten nun proportional zur Größe von δ. Ist die Übergangsfunktion δ des Automaten A total (wie z.B. bei den Teilmengenautomaten gemäß Konstruktion I), haben die Entscheidungsbäume die gleiche Größe wie die Matrizen.

Entscheidungsbäume können wir (eventuell) sparsamer repräsentieren, indem wir einige isomorphe Teilbäume identifizieren. Einen so erhaltenen Graphen nennen wir **komprimierten Entscheidungsbaum** oder **Entscheidungsgraphen**. Insbesondere können wir Entscheidungsbäume natürlich durch ihren Teilbaumgraphen darstellen. Einen Entscheidungsgraphen T_a wiederum repräsentieren wir durch eine zweidimensionale Matrix N_a, deren erste Komponente mit den Knoten v des Entscheidungsgraphen, die nicht Blätter sind, und deren zweite mit Markierungen q von Kanten indiziert wird. Der Eintrag $N_a[v, q]$ liefert dann den Nachfolgeknoten von v in T_a entlang der mit q beschrifteten Kante, falls ein solcher existiert, und andernfalls $\perp$.

Im schlimmsten Fall ist die Matrix N_a bis auf einen linearen Faktor genauso platzaufwendig wie die Matrix M_a. In der Praxis erweisen sich die Matrizen N_a allerdings

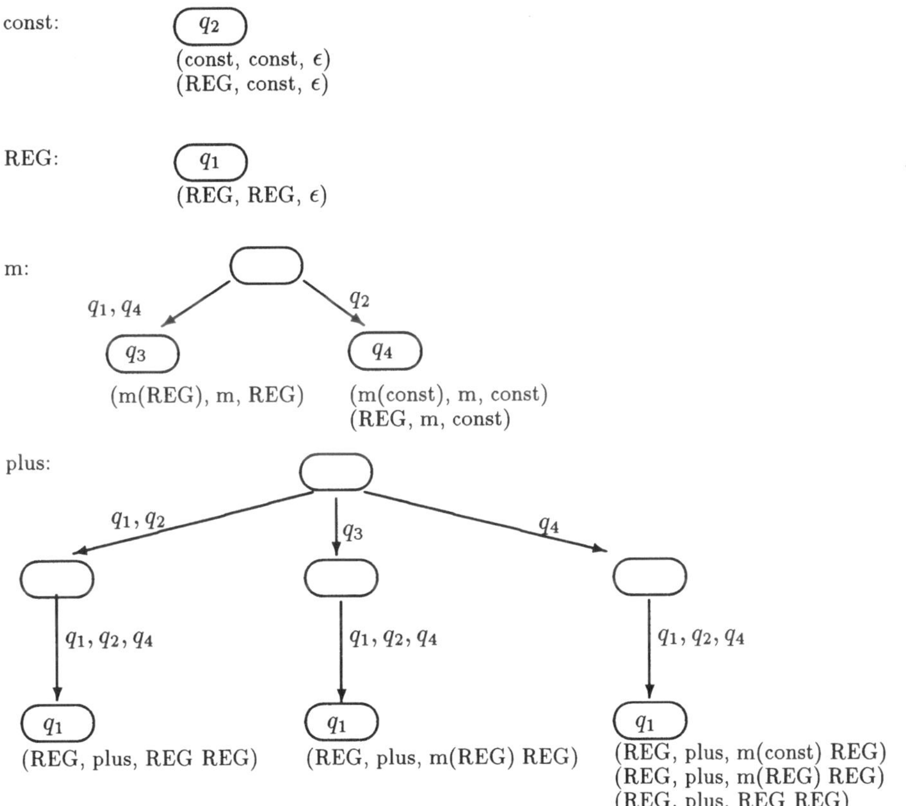

Abbildung 2: Komprimierte Entscheidungsbäume für die Grammatik aus Beispiel 5

als erheblich günstiger. Die Matrizen N_a können wiederum mit Tabellenkomprimierungsverfahren kompakt dargestellt werden. [2] stellt verschiedene Entscheidungsbaum- und Tabellenkomprimierungsverfahren vor.

Da die unkomprimierten Tabellen oder Entscheidungsbäume in vielen Fällen nicht in den Speicher passen, ist man in der Generierungsphase daran interessiert, bei der Anwendung der Teilmengenkonstruktion *direkt* eine komprimierte Darstellung der Entscheidungsbäume des Teilmengenautomaten zu erzeugen. Dies gelingt mithilfe der Konstruktion IV.

Sei $A = (Q, \Sigma, \delta, Q_F)$ ein (nichtdeterministischer) Baumautomat und $a \in \Sigma_k$. Die Idee der vierten Teilmengenkonstruktion besteht darin, einen Entscheidungsgraphen für H_a zu generieren, dessen Knoten aus Mengen von Übergängen von A bestehen. Zur Unterscheidung fügen wir die Nummer der jeweiligen Stufe als Markierung hinzu. Die Wurzel enthält ganz δ_a, d.h. ist $(\delta_a, 0)$. Die Kanten sind mit Zustandsmengen von A markiert. Eine Kante mit Beschriftung B führt dabei von einem Knoten $(v, j - 1)$ zu einem Knoten (v', j), genau dann wenn v' gerade aus allen Übergängen aus v

besteht, die als j-tes Argument ein $q \in B$ aufweisen und v' nicht leer ist. Ein Blatt b erhält schließlich zusätzlich als Markierung die Menge aller der Zustände, die als linke Seiten von Übergängen in b vorkommen. Dieses Vorgehen läßt sich mit dem iterativen Verfahren gemäß Teilmengenkonstruktion II verschränken.

Definition 6 (Teilmengenkonstruktion IV)

Sei $A = (Q, \Sigma, \delta, Q_F)$ ein Baumautomat. Für $n \geq 0$ definieren wir eine Menge von Zuständen $Q_s^{(n)}$ und Graphen $T_a^{(n)}$, $a \in \Sigma$, wie folgt.

$Q_s^{(0)} := \emptyset$ während $T_a^{(0)}$ leere Graphen sind.

Sei $n > 0$ und $a \in \Sigma_k$. Dann besitzt $T_a^{(n)}$ die Knotenmenge $V = V_0 \cup \ldots \cup V_k$ und die Kantenmenge E, die wie folgt definiert sind.

- $V_0 := \{(\delta_a, 0)\}$.

- Sei $j > 0$ und V_{j-1} bereits definiert. Dann betrachten wir für jedes $(v, j-1) \in V_{j-1}$ und $B \in Q_s^{(n-1)}$ die Menge $v' := \{(q, a, q_1 \ldots q_k) \in v \mid q_j \in B\}$. Ist $v' \neq \emptyset$, dann fügen wir (v', j) in die Menge V_j ein sowie eine Kante von $(v, j-1)$ nach (v', j) mit Beschriftung B in die Menge E.

- Jeden Knoten $(v, k) \in V_k$ markieren wir mit der Menge $\{q \in Q \mid \exists q_1, \ldots q_k \in Q : (q, a, q_1 \ldots q_k) \in v\}$.

Die Menge $Q_s^{(n)}$ ist die Menge der Blattbeschriftungen aller $T_a^{(n)}$, $a \in \Sigma$. $\square$

$T_a^{(n)}$ ist ein Teilgraph von $T_a^{(n+1)}$ und $T_a^{(n)} = T_a^{(n+1)}$ falls $Q_s^{(n)} = Q_s^{(n+1)}$. Mittels Induktion über n zeigt man:

Lemma 4 *Sei $A = (Q, \Sigma, \delta, Q_F)$ ein endlicher Baumautomat. Seien $Q_r^{(n)}$ und $\delta_r^{(n)}$ die n-te Approximation an die Zustandsmenge bzw. Übergangsrelation des reduzierten Teilmengenautomaten gemäß Teilmengenkonstruktion II. Dann gilt für alle $n \geq 0$:*

- *$Q_s^{(n)} = Q_r^{(n)}$;*

- *Entfernt man in $T_a^{(n)}$ alle Knoten, von denen kein beschriftetes Blatt erreicht werden kann, dann erhält man einen Entscheidungsgraphen für $(\delta_r^{(n)})_a$.* $\square$

Sei darum $T_a := T_a^{(n)}$ für das kleinste n mit $Q_s^{(n)} = Q_s^{(n+1)}$. Nehmen wir an, daß A selbst keine überflüssigen Übergänge enthielt. Dann läßt sich zeigen, daß T_a ebenfalls keine unnötigen Knoten enthält, d.h. ein komprimierter Entscheidungsbaum für $(\delta_r)_a$ ist.

Für unsere Lösung des Baumanalyse-Problems benötigen wir an einem Blatt b neben der Zustandscodierung ebenfalls die Menge der Übergänge des Automaten A, die dem Pfad zu b entsprechen, d.h. gerade b selber! Für diese erweiterte Beschriftung H_a ist T_a sogar **minimal**.

Lemma 5 *Sei $A = (Q, \Sigma, \delta, Q_F)$ ein endlicher Baumautomat ohne überflüssige Übergänge. Dann gilt für alle $a \in \Sigma$:*

- T_a *ist ein Entscheidungsgraph für δ_a;*

- T_a *ist der Teilbaumgraph des Entscheidungsbaums für H_a.*

Beweis: Wir beweisen nur den zweiten Punkt. Nehmen wir an, T_a wäre nicht minimal. Dann enthält T_a Knoten $(v, j) \neq (v', j)$, von denen die gleiche Menge von Blättern erreichbar ist. Kommt allerdings jeder Übergang von A in einer Berechnung vor, dann ist die Menge von Übergängen in jedem Knoten von T_a gleich der Vereinigung der Übergangsmengen in den Blättern, die von ihm aus erreichbar sind. Folglich ist $v = v'$ – im Widerspruch zu unserer Annahme. ∎

Unser Verfahren IV verallgemeinert das (nicht explizit beschriebene) Verfahren in [18], das für den Spezialfall von Baummustererkennern unmittelbar komprimierte Entscheidungsbäume erzeugt. [26] benutzen eine ähnliches Verfahren für Baumanalysatoren.

Die angegebene Konstruktion IV ist optimal in der Hinsicht, daß sie Entscheidungsgraphen mit minimaler Anzahl von Knoten liefert. Allerdings bemerkte Chase in [Ch87], daß sich viele Zustände eines generierten Teilmengenautomaten bei Übergängen gleich verhalten. Sei $A = (Q, \Sigma, \delta, Q_F)$ wiederum ein endlicher Baumautomat und $a \in \Sigma_k$. Für $j = 1, \ldots, k$ definieren wir die Menge $Q_{a,j} := \{q_j \mid (q, a, q_1 \ldots q_k) \in \delta\}$. Für eine Menge $B \subseteq Q$ sei der (a, j)-*relevante Anteil* die Menge $B \cap Q_{a,j}$. Im Entscheidungsgraphen T_a der Konstruktion IV führen Mengen von Zuständen, deren (a, j)-relevante Anteile gleich sind, von einem Knoten $(v, j - 1)$ der Stufe $j - 1$ jeweils zum selben Knoten der Stufe j.

Die fünfte Teilmengenkonstruktion liefert darum Entscheidungsgraphen T'_a, deren Knotenmengen jeweils mit denen der T_a übereinstimmen, deren Kanten der Stufe j aber nur noch mit (a, j)-relevanten Anteilen beschriftet sind. Die Entscheidungsgraphen selbst lassen sich dadurch erheblich kompakter darstellen. Dafür muß man in gesonderten Tabellen zu jeder Menge B jeweils die (a, j)-relevanten Anteile für alle $a \in \Sigma_k$ und $j \in \{1, \ldots, k\}$ verwalten. Zur Berechnung des Nachfolgezustands an einem Knoten n eines Eingabebaums mit Beschriftung $a \in \Sigma_k$ geht man nun zweistufig vor. Der Reihe nach werden für die Zustände $B_1, \ldots, B_k$ an den Söhnen jeweils die (a, j)-relevanten Anteile B'_j nachgeschlagen. Der Pfad mit Kantenbeschriftung $B'_1, \ldots, B'_k$ liefert dann den Zustand des Teilmengenautomaten für n.

Wie in der Teilmengenkonstruktion IV bauen wir die modifizierten Entscheidungsgraphen nach "Bedarf" auf.

Definition 7 (Teilmengenkonstruktion V)
Sei $A = (Q, \Sigma, \delta, Q_F)$ ein endlicher Baumautomat. Für $n \geq 0$ definieren wir eine Menge von Zuständen $Q_s^{(n)}$ und Graphen $T_a'^{(n)}$, $a \in \Sigma_k$, sowie Mengen $R_{a,j}^{(n)} = \{B \cap Q_{a,j} \mid B \in Q_s^{(n)}\} \setminus \{\emptyset\}$ Für $1 \leq j \leq k$ wie folgt.
$Q_s^{(0)} := \emptyset$, während $T_a'^{(0)}$ leere Graphen sind.
Sei $n > 0$ und $a \in \Sigma_k$.

Dann besitzt $T_a'^{(n)}$ die Knotenmenge $V = V_0 \cup \ldots \cup V_k$ und die Kantenmenge E, die wie folgt definiert sind.

- $V_0 := \{(\delta_a, 0)\}$.

- Sei $j > 0$ und V_{j-1} bereits definiert. Dann betrachten wir für jedes $(v, j-1) \in V_{j-1}$ und $B \in R_{a,j}^{(n-1)}$ die Menge $v' := \{(q, a, q_1 \ldots q_k) \in v \mid q_j \in B\}$. Ist $v' \neq \emptyset$, dann fügen wir (v', j) in die Menge V_j ein sowie eine Kante von $(v, j-1)$ nach (v', j) mit Beschriftung B in die Menge E.

- Jeden Knoten $(v, k) \in V_k$ markieren wir mit der Menge $\{q \in Q \mid \exists q_1, \ldots q_k \in Q : (q, a, q_1 \ldots q_k) \in v\}$.

Die Menge $Q_s^{(n)}$ ist die Menge der Blattbeschriftungen aller $T_a'^{(n)}$, $a \in \Sigma$. $\square$

Chase benutzt die Idee der Äquivalenzklasseneinteilung, um komprimierte Tabellen zur Mustererkennung zu erzeugen [3]. Unser Verfahren V verallgemeinert sowohl dieses Verfahren wie das Verfahren von Kron auf beliebige Baumautomaten.

9 Praktische Erfahrungen

An der Universität des Saarlandes wurde, basierend auf [26], ein Codeselektorgenerator [19] entwickelt, der als Eingabe eine annotierte Baumgrammik erhält, und als Ausgabe Tabellen und ein Treiberprogramm liefert, welches für einen IR-Baum Code selektiert. Als Beispieleingabe wurde eine Beschreibung des NSC32000 erstellt. Diese Grammatik umfaßt 763 Regeln mit 54 Nichtterminalen und 168 Terminalen. Der generierte Baumanalysator hat 970 Zustände. Die Tabellen M_a für den Baumalalysator würden etwa 180 Megabyte benötigen. Die Darstellung des Baumanalysators im Rechner als komprimierter Entscheidungsbaum, d.h. als Tabellen N_a, erfordert etwa 1 Megabyte. Durch die Anwendung von üblichen Tabellenkomprimierungen, wie row-displacement [2] kann der Platzbedarf auf 14 Kilobyte verringert werden.

Danksagung

Wir möchten Reinhold Heckmann für das sorgfältige Korrekturlesen des Manuskripts danken.

Die Arbeit von Christian Ferdinand wurde teilweise unterstützt von dem ESPRIT Project #5399 (COMPARE). Die Arbeit von Helmut Seidl wurde teilweise unterstützt von der Deutschen Forschungsgemeinschaft SFB #124 VLSI-Entwurf und Parallelität.

Literaturverzeichnis

[1] A.V. Aho, M. Ganapathi: *Efficient Tree Pattern Matching: An Aid to Code Generation.* Proc. of the 12th ACM Symp. on Principles of Programming Languages, pp. 334-340, 1985

[2] J. Börstler, U. Möncke, R. Wilhelm: *Table Compression for Tree Automata*. ACM Transactions on Programming Languages and Systems, Vol. 13, No. 3, July 1991, pp. 295-314

[3] D. R. Chase: *An improvement to bottom-up tree pattern matching*. Proc. of 14th ACM Symposium on Principles of Programming Languages, pp. 168-177, 1987

[4] M. Dauchet, A. Deruyver: *Compilation of Ground Term Rewriting Systems and Applications*. In Dershowitz (Ed.): Proceedings of the Conference: Rewriting Techniques and Applications, LNCS 355, pp. 556-558, Springer 1989

[5] P.J. Downey, R. Sethi, E.R. Tarjan: *Variations on the common subexpression problem*. JACM 27 (1980), pp. 758-771

[6] H. Emmelmann: *Code Selection by Regularly Controlled Term Rewriting*. In Proceedings of the Workshop: CODE'91 in Dagstuhl, 1991, *(to appear)*

[7] C. Ferdinand: *Pattern Matching in a Functional Transformation Language using Treeparsing*. Deransart, Małuszyński (Eds.): Proceedings of the Workshop: Programming Language Implementation and Logic Programming 90, LNCS 456, pp. 358-371, Springer 1990

[8] F. Gecseg, M. Steinby: *Tree Automata*. Akademiai Kiado, Budapest 1984

[9] R. Giegerich: *Code Selection by Inversion of Order-sorted Derivors*. Theoretical Computer Science 73, pp. 177-211, 1990

[10] R. Giegerich, K. Schmal: *Code Selection Techniques: Pattern Matching, Tree Parsing, and Inversion of Derivors*. H. Ganzinger (Ed.): Proc. ESOP 88, LNCS 300, pp. 247-268, Springer 1988

[11] R.S. Glanville: *A Machine Independent Algorithm for Code Generation and its Use in Retargetable Compilers*. Ph.D. Thesis, Univ. of California, Berkeley, 1977

[12] R.S. Glanville and S. L. Graham: *A new Method for Compiler Code Generation*. Proc. of the 5th ACM Symp. on Principles of Programming Languages, pp. 231-240, 1978

[13] A. Gräf: *Left-to-Right Tree Pattern Matching*. In Book (Ed.): Proceedings of the Conference: Rewriting Techniques and Applications, LNCS 488, pp. 323-334, Springer 1991

[14] R.R. Henry: *Graham–Glanville Code Generators*. Ph.D. Thesis, Univ. of California, Berkeley, 1984

[15] R.R. Henry, P.C. Damron: *Algorithms for Table-Driven Code Generators Using Tree-Pattern Matching*. University of Washington, Seattle, Technical Report # 89-02-03, 1989

[16] R.R. Henry, P.C. Damron: *Encoding Optimal Pattern Selection in a Table-Driven Bottom-Up Tree-Pattern Matcher.* University of Washington, Seattle, Technical Report # 89-02-04, 1989

[17] D.M. Hoffmann, M.J. O'Donnell: *Pattern Matching in Trees.* JACM 29,1, pp. 68-95, 1982

[18] H. Kron: *Tree Templates and Subtree Transformational Grammars.* Ph.D. Thesis, Univ. of California, Santa Cruz 1975

[19] N. Mathis: *Weiterentwicklung eines Codeselektorgenerators und Anwendung auf den NSC32000 .* Universität des Saarlandes, Diplomarbeit 1990

[20] K. Mehlhorn: *Datenstrukturen und Algorithmen.* Teubner 1986

[21] U. Möncke, B. Weisgerber, R. Wilhelm: *Generative support for transformational programming.* ESPRIT: Status Report of Continuing Work, Elsevier Sc., 1986, Brussels

[22] U. Möncke: *Simulating Automata for Weighted Tree Reductions.* Universität des Saarlandes, Technischer Bericht Nr. A10/87, 1987

[23] E. Pelegri-Llopart: *Rewrite Systems, Pattern Matching, and Code Selection.* Ph.D. Thesis, Univ. of California, Berkeley, 1988

[24] E. Pelegri-Llopart, S.L. Graham: *Optimal Code Generation for Expression Trees: An Application of BURS Theory.* Proc. of the 15th ACM Symposium on Principles of Programming Languages. San Diego, CA, Jan. 1988, 294-308

[25] K. Ripken: *Formale Beschreibungen von Maschinen, Implementierungen und optimierender Maschinencode–Erzeugung aus attributierten Programmgraphen.* Dissertation, TU München, 1977

[26] B. Weisgerber, R. Wilhelm: *Two tree pattern matchers for code selection.* In Hammer (Hrsg.): Proceedings of the Workshop: Compiler Compilers and High Speed Compilation, LNCS 371, pp. 215-229, Springer 1988

Decision Making in the Presence of Noise[1]

Michael J. Fischer   Sophia A. Paleologou

Department of Computer Science
Yale University
USA

Abstract

We consider problems of decision making based on imperfect information. We derive Bayesian optimal decision procedures for some simple one-person games on trees in which the player is given redundant but noisy information about the true configuration of the game. Our procedures are computationally efficient, and the decision rules which they implement are describable by simple formulas. Not surprisingly, the presence of noise greatly affects the decision procedure, and decisions procedures that are optimal for the corresponding noiseless games may be far from optimal in the presence of noise. In many cases, the optimal decision depends not only on the given noisy data but also on knowledge of the expected amount of noise present in the data. For arbitrary $m \in \mathcal{N}$, we present examples in which the optimal decision changes m times as the probability of error in an individual datum increases from 0 to 1/2. Thus, no decision procedure that is insensitive to (or does not know) the amount of uncertainty in the data can perform as well as one that is aware of the unreliability of its data.

---

[1]This research was supported in part by National Science Foundation grant IRI–9015570.

1 Introduction

Many complex real-life situations require that people make decisions based on imperfect information. Widely used algorithms for decision making in such complex environments often overlook the fact that the information with which they are provided is unreliable and use this information as if it were accurate, hoping that this will nevertheless lead to a good decision.

A typical example of such an algorithm is the Shannon chess playing algorithm, which looks k levels ahead in the game tree, evaluates the strength of each resulting board position, and then uses standard "min-max" techniques to choose the most promising next move. If the evaluations were 100% accurate, this would lead to optimal play, but it is unclear how good a move this produces in real chess programs. In a recent conference for Learning, Rationality, and Games at the Santa Fe Institute, John Geanakoplos and Larry Gray [2] gave examples of simple one-person games in which the Shannon algorithm was provably non-optimal and in which its performance actually deteriorated as the amount of permitted look-ahead (and hence the amount of data upon which to base one's decision) increased.[1]

In this paper, we investigate the structure of the Bayesian optimal decision in the simple games of Geanakoplos and Gray. Rather surprisingly, the optimal decision can be expressed by compact, closed-form formulas of low computational complexity. From these formulas, we gain qualitative insights into the Bayesian optimal decision. We observe, for example, that no algorithm that bases its decision solely on the information contained at the level-k nodes of the tree (as the Shannon algorithm does) is Bayesian optimal. We also give an example to show that the optimal decision sometimes depends not only on the information contained at the nodes, but also on knowledge of the expected amount of noise present in the data. Thus, an algorithm that is aware of the imperfection of its data can do better than one that is not.

It is tempting to apply these insights to chess in order to obtain improved algorithms, and we are hopeful that workers on chess will find our results enlightening. Nevertheless, we should point out a number of important differences between our games and the problem of playing chess. In our games, uncertainty arises from two underlying sources of randomness. The instance of the game to be played is chosen at random (as in card games such as bridge or poker), and the information about the chosen game that is given to the player is also chosen at random. Thus, a player is presented with probabilistic, partial information about the true underlying game. The difficulty the player faces is in knowing which game is being played, not how to play the game once it is known. In chess, on the other hand, the underlying game tree is fixed, and the player's information is computed by a deterministic procedure of low computational complexity. The barrier to optimal play is not the lack of accurate information but the apparent intractability of the computational problem of making good use of that information. An important research problem is to clarify the rela-

---

[1] Judea Pearl has also noted such phenomena in chess and given probabilistic game models in which reaching deeper consistently degrades the quality of the Shannon algorithm's decision [3].

tionship between probabilistic and computational sources of uncertainty in decision problems.

2   A Basis of Tree Games

In this section, we provide the notation and definitions we will be using throughout the paper.

For any complete binary tree T, $nodes(T)$ and $leaves(T)$ are the sets of nodes and leaves of T respectively. For notational convenience, we often identify T with $nodes(T)$ and use $x \in T$ to mean $x \in nodes(T)$. If $x \in nodes(T) - leaves(T)$, then $Lchild(x)$ and $Rchild(x)$ denote the left and right children of x respectively. Whenever we consider subtrees of T, we restrict ourselves to subtrees that satisfy the following property: if $x \in nodes(T)$ is the root of the subtree X, then all the descendants of x are also in $nodes(X)$. Similarly, whenever we consider paths in T, we restrict ourselves to paths from the root of T to a leaf; $paths(T)$ is the set of all such paths in T. Finally, if $\pi \in paths(T)$, $nodes(\pi)$ is the set of all nodes of T on path π.

Definition 1 *Let T be a complete binary tree. A function $\lambda : nodes(T) \to \{0,1\}$ is called a* labelling *of T, and the tuple (T, λ) is called a* labelled tree. *For $x \in nodes(T)$, $\lambda(x)$ is then called the* label *of node x under λ.*

If λ is a labelling of T, then λ_X is the restriction of λ to the nodes of X; i.e., $\lambda_X = \lambda|_{nodes(X)}$. Obviously, $\lambda_T = \lambda$. Also, if r is a single node of T, we sometimes write λ_r to denote $\lambda|\{r\}$.

Definition 2 *Let T be a complete binary tree. A labelling λ of T is called* proper *iff it satisfies the* max-property; *that is, for all $x \in nodes(T) - leaves(T)$,*

$$\lambda(x) = \max\{\lambda(Lchild(x)), \lambda(Rchild(x))\}.$$

The tuple (T, λ) is then called a MaxTree.

Definition 3 *Let T be a complete binary tree. A proper labelling λ of T is called* winning *iff there exists $\alpha \in leaves(T)$ such that $\lambda(\alpha) = 1$. A proper labelling that is not winning is called* losing.

Definition 4 *Let (T, λ) be a MaxTree. A path $\pi \in paths(T)$ is called* winning *iff $\lambda(x) = 1$ for all $x \in nodes(\pi)$. Similarly, a leaf $\alpha \in leaves(T)$ is called* winning *iff $\lambda(\alpha) = 1$.*

As a result of the max-property, any proper labelling λ can be uniquely specified by the labels it assigns to the leaves of T; the labels of all the internal nodes of T can then be recursively computed as the maximum of the labels of their children. This

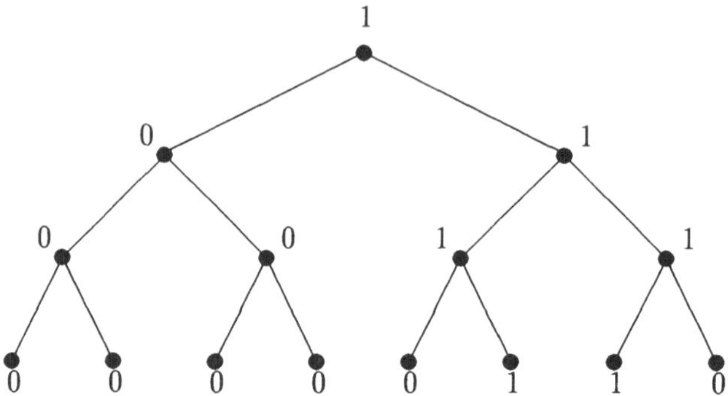

Figure 1: A MaxTree with two winning leaves.

recursive node-labelling process is widely known in game theory as *backward induction* (see [5]). An example of a MaxTree T of uniform depth $k = 3$ is given in Figure 1.

In this paper, we generally consider MaxTrees with *exactly one* leaf labelled 1 and all other leaves labelled 0. For every $\alpha \in leaves(T)$, λ^α denotes the proper labelling of T which assigns the label 1 to α and the label 0 to all other leaves of T; i.e., $\lambda^\alpha(\alpha) = 1$, and $\lambda^\alpha(\beta) = 0$ for all $\beta \in leaves(T) - \{\alpha\}$.

MaxTrees can be thought of as game trees for natural one-person games which model multistage decision processes. Each node of the tree models a state of the game. At every internal node, the unique player of the game is faced with two alternatives, going Left or going Right, while the leaves of the tree represent final states of the game. They can be used as a basis for defining a variety of related simple (single-stage) one-person games. An example of such a simple game follows:

Example 1 *Let (T, λ) be a MaxTree. Given (T, λ), the player is asked to choose a path from the root of T to a leaf. If the path she chooses leads to a leaf labelled 1 under λ, the player wins; otherwise, she loses.*

An obvious way for the player to play the game of example 1 is to choose Left and Right according to which of the two children of the current node is labelled 1. This strategy will cause the player to "walk" down a path labelled with 1's and to reach a winning leaf after k moves, where k is the depth of T. In the special case where all nodes of the given MaxTree are labelled 0, the player always loses.

The game of example 1 is not interesting, since it provides the player with complete and accurate information, thereby turning her decision-making into a trivial process. In example 2 below, we present a variant of that game where the information visible to the player has been corrupted by noise.

Before we proceed with the necessary definitions, we introduce the following notational conventions: if Z is a random variable, we identify Z with the corresponding random experiment and use z to denote the outcome of this experiment, sometimes also referred to as the realization of the random variable Z. Furthermore, whenever there are no grounds for confusion, we use $\mathbf{prob}[z]$ to denote $\mathbf{prob}[Z = z]$, the probability that $Z = z$.

Definition 5 *Let* $p \in (0, 1/2)$ *be a constant.*[2] *A 0/1 random variable* Z *is called a* random coin with bias p *iff*

$$\mathbf{prob}[Z = 0] = 1 - p \quad and \quad \mathbf{prob}[Z = 1] = p.$$

Definition 6 *Let* $p \in (0, 1/2)$ *be a constant. Let* $\{Z_x : x \in nodes(T)\}$ *be a collection of independent random coins with the same bias* p. *Let* (T, λ) *be a MaxTree and let* $x \in nodes(T)$. *A 0/1 random variable* V_x *is called a* random corruption *of the label* $\lambda(x)$ *with error probability* p *iff*

$$V_x = \lambda(x) \oplus Z_x.$$

According to definition 6, the random corruptions of the labels of the nodes in T satisfy the following two properties:

- *locality*: for all $x \in nodes(T)$, the corruption V_x depends only on the label $\lambda(x)$ and is independent of the labels $\lambda(y)$ of all $y \in nodes(T) - \{x\}$;

- *independence*: for all $x, y \in nodes(T)$ with $x \neq y$, the corruptions V_x and V_y of the labels $\lambda(x)$ and $\lambda(y)$ are independent random variables.

Definition 7 *Let* (T, λ) *be a MaxTree. If, for every node* x *in* T, v_x *is the outcome of a random corruption* V_x *of its label* $\lambda(x)$ *with error probability* p, *then the labelling* $\vartheta : nodes(T) \rightarrow \{0, 1\}$, *such that* $\vartheta(x) = v_x$, *is called a* corrupted view *of the proper labelling* λ *with error probability* p.

We refer to the labels of the nodes of T under λ as *actual labels*, while we refer to the labels of the nodes of T under ϑ as *corrupted* or *observed labels*. Unlike λ, ϑ does not necessarily satisfy the max-property of proper labellings.

Example 2 *Let* (T, Λ) *be a MaxTree with* Λ *a random labelling of* T *following a probability distribution* $\mathcal{P}$. *Let* λ *be the outcome of* Λ *and let* ϑ *be a corrupted view of* λ *with error probability* p. *Given* (T, ϑ, p) *and the probability distribution* $\mathcal{P}$, *the player is asked to choose a path from the root of* T *to a leaf. If the path she chooses ends in a leaf labelled 1 under* λ, *the player wins; otherwise, she loses.*

---

[2]If $p = 1/2$, the corruption V_x is independent of the labelling $\lambda(x)$. Also, if $p > 1/2$, the player can use the corrupted view ϑ to construct a new labelling ϑ' by taking $\vartheta'(x) = 1 \oplus \vartheta(x)$. We can think of ϑ' as another corrupted view of λ with error probability $p' = 1 - p < 1/2$. Thus, it is reasonable for us to focus on the case $p \in (0, 1/2)$, since all other cases are either not interesting or can be reduced to the case $p \in (0, 1/2)$.

In general, no algorithm for the game of example 2 can guarantee the player a win, since she has access only to the corrupted labels of the nodes in T. In the absence of an algorithm that guarantees success, the player might alternatively look for an algorithm that maximizes her chances of winning, thereby exploiting the probabilistic structure of the game under consideration. She can use the view ϑ and the error probability p to update her prior knowledge of the distribution $\mathcal{P}$ and choose the leaf that is most likely to be winning, given the data visible to her (see section 4). An algorithm that computes the decision with the maximum probability of winning, given all available information, is *Bayesian optimal*.

In this paper, we present and analyze two simple games defined on MaxTrees, the second of which is the game of example 2 above. We show that, in many cases, seemingly intractable computations can be reduced to efficient algoritms for computing Bayesian optimal decisions. (For a general introduction to probability theory, see [4].)

3   Game I: Choose a Subtree

We first consider a one-player, one-move game played on a MaxTree. The current state of the game is modelled by the root node, and the player is asked to choose one move—Left or Right, that takes her closer to a winning leaf. More formally, we have:

Game I: Let (T, Λ) be a MaxTree with Λ a random labelling following a probability distribution $\mathcal{P}$. Let λ be the outcome of Λ and let ϑ be a corrupted view of λ with error probability p. Given (T, ϑ, p) and the probability distribution $\mathcal{P}$, the player is asked to choose a subtree $Y \in \{L, R\}$, where L and R are the left and right subtrees of T. If Y contains a leaf labelled 1 under λ, the player wins; otherwise, she loses.

For the purposes of our probabilistic analysis, we fix the following probability distribution $\mathcal{P}$: for all proper labellings λ of T,

$$\mathbf{prob}[\lambda] = \begin{cases} 1/2^k & \text{if } \lambda = \lambda^\alpha \text{ for some } \alpha \in \mathit{leaves}(T) \\ 0 & \text{otherwise} \end{cases} \tag{1}$$

where k is the depth of T. Thus, we assume exactly one leaf is labelled 1. However, the techniques we use in this paper to analyze Games I and II can be extended so as to handle arbitrary probability distributions (see [1]).

Let X be a subtree of T and let L and R be the left and right subtrees of X. In general, the event (λ_X winning) can be viewed as the disjoint union of three events: (λ_L losing) & (λ_R winning), (λ_L winning) & (λ_R losing), and (λ_L winning) & (λ_R winning). However, in the special case of the distribution $\mathcal{P}$ that we fixed in equation 1 above, the first two of those events are equiprobable, while the third event is impossible; that is,

- $\mathbf{prob}[(\lambda_L \text{ losing}) \ \& \ (\lambda_R \text{ winning}) \mid \lambda_X \text{ winning}] = \frac{1}{2}$

- $\mathbf{prob}[(\lambda_L \text{ winning}) \ \& \ (\lambda_R \text{ losing}) \mid \lambda_X \text{ winning}] = \frac{1}{2}$

- $\mathbf{prob}[(\lambda_L \text{ winning}) \ \& \ (\lambda_R \text{ winning}) \mid \lambda_X \text{ winning}] = 0$

3.1 Probabilistic Analysis of Game I

In this section, we provide an exact probabilistic analysis of Game I. Given the corrupted view of the game tree and the a priori information about the underlying distribution of labellings, we compute the conditional probability of winning the game for both choices Left and Right. Although this analysis might look intractable at first glance, we show how the combinatorics nicely collapse to yield compact recursive formulas that are easy to compute.

Let X be a subtree of T and define the following two quantities:

- l_X is the probability that a random corruption of a losing labelling yields in ϑ_X; i.e.,

$$l_X = \mathbf{prob}[\vartheta_X \mid \lambda_X \text{ losing}],$$

- w_X is the probability that a random corruption of a winning labelling yields in ϑ_X; i.e.,

$$w_X = \mathbf{prob}[\vartheta_X \mid \lambda_X \text{ winning}].$$

In Lemmas 1 and 2, we derive recursive formulas that allow us to compute l_X and w_X for any subtree X.

Lemma 1 *Let X be a subtree of T. Then,*

$$l_X = \begin{cases} (1-p)^{1-v}p^v & \text{if } X \text{ is a single node} \\ l_L \cdot l_R \cdot (1-p)^{1-v}p^v & \text{otherwise} \end{cases} \tag{2}$$

where L and R are the left and right subtrees of X, and $v = \vartheta_X(r)$ is the observed label of the root r.

Proof: Let tree X consist of a single node r with observed label $v = \vartheta_X(r)$, and assume λ_X is losing; i.e., $\lambda_X(r) = 0$. Then, $v = 0$ with probability $(1-p)$, while $v = 1$ with probability p. The two possibilities can be expressed in one formula as follows:

$$l_X = \mathbf{prob}[\vartheta_X \mid \lambda_X \text{ losing}] = (1-p)^{1-v}p^v \tag{3}$$

In the case where X is not a single node, l_X can be computed in terms of l_L and l_R.

$$\begin{aligned} l_X &= \mathbf{prob}[\vartheta_X \mid \lambda_X \text{ losing}] \\ &= \mathbf{prob}[\vartheta_L \ \& \ \vartheta_R \ \& \ \vartheta_r \mid \lambda_X \text{ losing}] \end{aligned} \tag{4}$$

The independence of the corruptions allows us to express l_X in equation 4 as the product of the conditional probabilities of the independent events ϑ_L, ϑ_R, and ϑ_r.

$$l_X = \mathbf{prob}[\vartheta_L \mid \lambda_X \text{ losing}] \cdot \mathbf{prob}[\vartheta_R \mid \lambda_X \text{ losing}] \cdot \mathbf{prob}[\vartheta_r \mid \lambda_X \text{ losing}] \qquad (5)$$

Because of the locality of corruptions, the observed label of any node in X depends only on its own actual label. Furthermore, λ_X is losing if and only if all λ_L, λ_R, and λ_r are losing. Combining these observations with equation 3, we obtain:

$$\begin{aligned} l_X &= \mathbf{prob}[\vartheta_L \mid \lambda_L \text{ losing}] \cdot \mathbf{prob}[\vartheta_R \mid \lambda_R \text{ losing}] \cdot \mathbf{prob}[\vartheta_r \mid \lambda_r \text{ losing}] \\ &= l_L \cdot l_R \cdot (1-p)^{1-v} p^v \end{aligned}$$

$\blacksquare$

Lemma 2 *Let X be a subtree of T. Then,*

$$w_X = \begin{cases} (1-p)^v p^{1-v} & \text{if } X \text{ is a single node} \\ \frac{1}{2}(l_L \cdot w_R + w_L \cdot l_R) \cdot (1-p)^v p^{1-v} & \text{otherwise} \end{cases} \qquad (6)$$

where L and R are the left and right subtrees of X, and $v = \vartheta_X(r)$ is the observed label of the root r.

Proof: Let tree X consist of a single node r with observed label $v = \vartheta_X(r)$, and assume λ_X is winning; i.e., $\lambda_X(r) = 1$. Then, $v = 0$ with probability p, while $v = 1$ with probability $(1-p)$. The two possibilities can be expressed in one formula as follows:

$$w_X = \mathbf{prob}[\vartheta_X \mid \lambda_X \text{ winning}] = (1-p)^v p^{1-v} \qquad (7)$$

In the case where X is not a single node, w_X can be computed in terms of l_L, l_R, w_L, and w_R.

$$\begin{aligned} w_X &= \mathbf{prob}[\vartheta_X \mid \lambda_X \text{ winning}] \\ &= \mathbf{prob}[\vartheta_L \ \& \ \vartheta_R \ \& \ \vartheta_r \mid \lambda_X \text{ winning}] \end{aligned} \qquad (8)$$

The independence of the corruptions allows us to express w_X in equation 8 as the product of the conditional probabilities of the independent events $(\vartheta_L \ \& \ \vartheta_R)$ and ϑ_r.

$$w_X = \mathbf{prob}[\vartheta_L \ \& \ \vartheta_R \mid \lambda_X \text{ winning}] \cdot \mathbf{prob}[\vartheta_r \mid \lambda_X \text{ winning}] \qquad (9)$$

However, under $\mathcal{P}$, the event $(\lambda_X \text{ winning})$ is the disjoint union of the equiprobable events $(\lambda_L \text{ losing}) \ \& \ (\lambda_R \text{ winning})$ and $(\lambda_L \text{ winning}) \ \& \ (\lambda_R \text{ losing})$. Equation 9 then becomes:

$$\begin{aligned} w_X = \frac{1}{2}\Big[\ &\mathbf{prob}[\vartheta_L \ \& \ \vartheta_R \mid (\lambda_L \text{ losing}) \ \& \ (\lambda_R \text{ winning})] \\ &+ \mathbf{prob}[\vartheta_L \ \& \ \vartheta_R \mid (\lambda_L \text{ winning}) \ \& \ (\lambda_R \text{ losing})] \ \Big] \cdot \mathbf{prob}[\vartheta_r \mid \lambda_X \text{ winning}] \end{aligned} \qquad (10)$$

Finally, we use the independence and locality of the corruptions to further simplify equation 10:

$$
\begin{aligned}
w_X &= \frac{1}{2}\Big[\ \mathbf{prob}[\vartheta_L \mid \lambda_L \text{ losing}] \cdot \mathbf{prob}[\vartheta_R \mid \lambda_R \text{ winning}] \\
&\quad + \mathbf{prob}[\vartheta_L \mid \lambda_L \text{ winning}] \cdot \mathbf{prob}[\vartheta_R \mid \lambda_R \text{ losing}]\ \Big] \cdot \mathbf{prob}[\vartheta_r \mid \lambda_r \text{ winning}] \\
&= \frac{1}{2}(l_L \cdot w_R + w_L \cdot l_R) \cdot (1-p)^v p^{1-v}
\end{aligned}
$$

where $\mathbf{prob}[\vartheta_r \mid \lambda_r \text{ winning}]$ was substituted from equation 7. ∎

In Theorem 1 below, we derive formulas that allow us to compute exactly the conditional probabilities of winning for the two choices of the player—Left and Right—based on the corrupted labelling of the game tree visible to her.

Lemma 3 *Let (T, λ) be a MaxTree and let ϑ be a corrupted view of λ with error probability p. Then,*

$$\mathbf{prob}[\vartheta \mid \lambda_L \text{ winning}] = w_L \cdot l_R \cdot (1-p)^v p^{1-v} \tag{11}$$

$$\mathbf{prob}[\vartheta \mid \lambda_R \text{ winning}] = l_L \cdot w_R \cdot (1-p)^v p^{1-v} \tag{12}$$

where L and R are the left and right subtrees of T, and $v = \vartheta(r)$ is the observed label of the root r.

Proof: We show the derivation of the formula for $\mathbf{prob}[\vartheta \mid \lambda_L \text{ winning}]$; the formula for $\mathbf{prob}[\vartheta \mid \lambda_R \text{ winning}]$ is derived similarly.

From the independence of the actual label corruptions, we have:

$$
\begin{aligned}
\mathbf{prob}[\vartheta \mid \lambda_L \text{ winning}] &= \mathbf{prob}[\vartheta_L \ \& \ \vartheta_R \ \& \ \vartheta_r \mid \lambda_L \text{ winning}] \\
&= \mathbf{prob}[\vartheta_L \mid \lambda_L \text{ winning}] \cdot \mathbf{prob}[\vartheta_R \mid \lambda_L \text{ winning}] \cdot \mathbf{prob}[\vartheta_r \mid \lambda_L \text{ winning}] \tag{13}
\end{aligned}
$$

However, since λ was chosen from the probability distribution $\mathcal{P}$, λ is always winning, and λ_L is winning if and only if λ_R is losing. Combining this observation and the locality of the label corruptions, equation 13 yields:

$$
\begin{aligned}
\mathbf{prob}[\vartheta \mid \lambda_L \text{ winning}] &= \\
&= \mathbf{prob}[\vartheta_L \mid \lambda_L \text{ winning}] \cdot \mathbf{prob}[\vartheta_R \mid \lambda_R \text{ losing}] \cdot \mathbf{prob}[\vartheta_r \mid \lambda_r \text{ winning}] \\
&= w_L \cdot l_R \cdot (1-p)^v p^{1-v}
\end{aligned}
$$

∎

Theorem 1 *Let (T, λ) be a MaxTree and let ϑ be a corrupted view of λ with error probability p. Then,*

$$\mathbf{prob}[\lambda_L \text{ winning} \mid \vartheta] = \frac{w_L \cdot l_R}{w_L \cdot l_R + l_L \cdot w_R} \tag{14}$$

$$\mathbf{prob}[\lambda_R \text{ winning} \mid \vartheta] = \frac{w_R \cdot l_L}{w_L \cdot l_R + l_L \cdot w_R} \tag{15}$$

where L and R are the left and right subtrees of T.

Proof: We show the derivation of the formula for $\mathbf{prob}[\lambda_L$ winning $\mid \vartheta]$; the formula for $\mathbf{prob}[\lambda_R$ winning $\mid \vartheta]$ is derived similarly.

Since λ was chosen from $\mathcal{P}$, λ is always winning, and it is equiprobable that either one of λ_L and λ_R is also winning. Thus,

$$\mathbf{prob}[\vartheta] = \frac{1}{2}\left(\ \mathbf{prob}[\vartheta \mid \lambda_L \text{ winning}] + \mathbf{prob}[\vartheta \mid \lambda_R \text{ winning}]\ \right) \tag{16}$$

Using formulas 11 and 12, equation 16 becomes:

$$\mathbf{prob}[\vartheta] = \frac{1}{2}(w_L \cdot l_R + l_L \cdot w_R)(1-p)^v p^{1-v} \tag{17}$$

where $v = \vartheta(r)$ is the observed label of the root r of T. Finally, we use *Bayes' Theorem* to combine equation 17 with formulas 11 and 12:

$$
\begin{aligned}
\mathbf{prob}[\lambda_L \text{ winning} \mid \vartheta] &= \frac{\mathbf{prob}[\vartheta \mid \lambda_L \text{ winning}] \cdot \mathbf{prob}[\lambda_L \text{ winning}]}{\mathbf{prob}[\vartheta]} \\[2mm]
&= \frac{\frac{1}{2}w_L \cdot l_R \cdot (1-p)^v p^{1-v}}{\frac{1}{2}(w_L \cdot l_R + l_L \cdot w_R)(1-p)^v p^{1-v}} \\[2mm]
&= \frac{w_L \cdot l_R}{w_L \cdot l_R + l_L \cdot w_R}
\end{aligned}
$$

$\blacksquare$

In trying to compute $\mathbf{prob}[\lambda_L$ winning $\mid \vartheta]$ and $\mathbf{prob}[\lambda_R$ winning $\mid \vartheta]$ using equations 14 and 15, we run into computational difficulties. The quantities l_L, l_R, w_L, and w_R very quickly approach zero, so any arithmetic based on those values (using an ordinary floating-point representation) becomes impossible. However, we can rewrite those formulas in terms of the ratios w_L/l_L and w_R/l_R. It is convenient to also pull out some constants. Let $a = 2(1-p)/p$ and $c = ((1-p)/p)^2$. Let X be a k-depth subtree of T and define:

$$\Phi_X = \frac{a^{k+1}}{2} \cdot \frac{w_X}{l_X} \tag{18}$$

Lemma 4 gives a recursive formula for computing Φ_X for any subtree X.

Lemma 4 *Let X be a k-depth subtree of T. Then,*

$$\Phi_X = \begin{cases} c^v & \text{if } X \text{ is a single node} \\ (\Phi_L + \Phi_R) \cdot c^v & \text{otherwise} \end{cases} \tag{19}$$

where L and R are the left and right subtrees of X, and $v = \vartheta_X(r)$ is the observed label of the root r.

Proof: If X is a single node, X has depth $k = 0$, and formulas 2, 6, and 18 yield:

$$\Phi_X = \frac{a}{2} \cdot \frac{w_X}{l_X} = \frac{1-p}{p} \cdot \frac{(1-p)^v p^{1-v}}{(1-p)^{1-v} p^v} = \left(\frac{1-p}{p}\right)^{2v} = c^v$$

If X is not a single node, we can use formulas 2 and 6 to express Φ_X in terms of Φ_L and Φ_R:

$$
\begin{aligned}
\Phi_X &= \frac{a^{k+1}}{2} \cdot \frac{w_X}{l_X} \\
&= \frac{a^{k+1}}{2} \cdot \frac{\frac{1}{2}(l_L \cdot w_R + w_L \cdot l_R)(1-p)^v p^{1-v}}{l_L \cdot l_R \cdot (1-p)^{1-v} p^v}
\end{aligned}
\tag{20}
$$

Taking into account that both L and R have depth $(k-1)$, equation 20 can be rewritten as:

$$
\begin{aligned}
\Phi_X &= \frac{a}{2} \cdot \frac{p}{1-p} \cdot \left(\frac{a^k}{2} \cdot \frac{w_L}{l_L} + \frac{a^k}{2} \cdot \frac{w_R}{l_R} \right) \cdot \left(\frac{1-p}{p} \right)^{2v} \\
&= (\Phi_L + \Phi_R) \cdot c^v
\end{aligned}
$$

■

The following is a restatement of Theorem 1 in terms of Φ_L and Φ_R.

Theorem 2 *Let (T, λ) be a MaxTree. Then,*

$$
\mathbf{prob}[\lambda_L \text{ winning} \mid \vartheta] = \frac{\Phi_L}{\Phi_L + \Phi_R}
\tag{21}
$$

$$
\mathbf{prob}[\lambda_R \text{ winning} \mid \vartheta] = \frac{\Phi_R}{\Phi_L + \Phi_R}
\tag{22}
$$

where L and R are the left and right subtrees of T.

Corollary 1 *Let (T, λ) be a MaxTree and let ϑ be a corrupted view of λ with error probability p. Then*

$$
\mathbf{prob}[\lambda_L \text{ winning} \mid \vartheta] \geq \mathbf{prob}[\lambda_R \text{ winning} \mid \vartheta] \;\; \text{iff} \;\; \Phi_L \geq \Phi_R
\tag{23}
$$

Finally, we provide the solution to equation 19, the recursive definition of Φ_X. Define the function $f: paths(T) \rightarrow \mathcal{N}$ such that, for every path π,

$$
f(\pi) = \sum_{x \in nodes(\pi)} \vartheta(x)
\tag{24}
$$

Thus, $f(\pi)$ is the number of 1's among the corrupted labels of nodes on π.

Theorem 3 *Let X be a subtree of T. Then*

$$
\Phi_X = \sum_{\pi \in paths(X)} c^{f(\pi)}
\tag{25}
$$

Proof: By induction on the depth k of X. ■

We sometimes write $\Phi_X(c)$ to emphasize the fact that Φ_X depends on c as well as on X.

3.2 Bayesian Optimal Algorithms for Game I

In this paragraph, we turn our attention to the problem of *computing* the Bayesian optimal decision for Game I. The first algorithm, A_1, shown in Figure 2, is a direct application of Corollary 1 and Lemma 4.

ALGORITHM A_1

Input:
    T, a complete binary tree
    ϑ, a corrupted view of T
    p, the error probability of the corruption

Output:
    Left or Right

Description:
    Step 0:
        Compute $c = ((1 - p)/p)^2$.
    Step 1:
        Compute Φ_L recursively using equation 19.
    Step 2:
        Compute Φ_R recursively using equation 19.
    Step 3:
        Compare Φ_L and Φ_R and choose the direction—Left or Right—
        that corresponds to the maximum. Break ties arbitrarily.

Figure 2: A Bayesian optimal algorithm for Game I using $O(n)$ arithmetic operations.

Theorem 4 *Let A_1 be the algorithm shown in Figure 2. Let (T, λ) be a MaxTree of n nodes and let ϑ be a corrupted view of λ with error probability p. On input (T, ϑ, p), algorithm A_1 requires $O(n)$ additions and $O(n)$ multiplications/divisions to compute the Bayesian optimal decision for Game I.*

Proof: It can be easily seen that computing Φ_L recursively from equation 19 in step 1 of algorithm A_1 requires a number of additions (multiplications) equal to the number of internal nodes in L. Similarly, computing Φ_R recursively in step 2 requires a number of additions (multiplications) equal to the number of internal nodes in R. Thus, steps 1 and 2 of algorithm A_1 require a total of $O(n)$ additions and multiplications. Step 0 of the algorithm requires only a constant number of additional operations. ∎

By restructuring the computations involved in algorithm A_1 and using table lookup, we can reduce the number of multiplications/divisions to $O(\log(n))$. The resulting algorithm, A_2, shown in Figure 3, makes use of the results of Corollary 1 and Theorem 3.

ALGORITHM A_2

Input:
   T, a complete binary tree
   ϑ, a corrupted view of T
   p, the error probability of the corruption

Output:
   Left or Right

Description:
   Step 0:
      Compute $c = ((1 - p)/p)^2$
   Step 1:
      Compute and store in a look-up table all powers c^i,
      $i = 0, 1, \ldots, (\lceil \log(n) \rceil - 1)$.
   Step 2:
      Compute the values of $f(\pi)$, for all $\pi \in paths(L)$, as follows: Working down from the root, compute for each node α in L the number of nodes labelled 1 by ϑ on the path from α to the root.
   Step 3:
      Add up the values of $c^{f(\pi)}$ for all paths $\pi \in paths(L)$, to compute Φ_L.
   Step 4:
      Repeat steps 2 and 3 substituting R for L, to compute Φ_R.
   Step 5:
      Compare Φ_L and Φ_R and choose the direction—Left or Right— that corresponds to the maximum. Break ties arbitrarily.

Figure 3: A Bayesian optimal algorithm for Game I using $O(\log(n))$ multiplications.

Theorem 5 *Let A_2 be the algorithm shown in Figure 3. Let (T, λ) be a MaxTree of n nodes and let ϑ be a corrupted view of λ with error probability p. On input (T, ϑ, p), algorithm A_2 requires $O(n)$ additions and $O(\log(n))$ multiplications/divisions to compute the Bayesian optimal decision for Game I.*

Proof: Step 0 of algorithm A_2 takes a constant number of arithmetic operations. Step 1 requires $(\lceil \log(n) \rceil - 2) = O(\log(n))$ multiplications to compute all the powers $c^i, i = 0, 1, 2, \ldots, \lceil \log(n) \rceil - 1$. Step 2 requires at most one addition for each node of L for a total of $O(n)$ additions. Step 3 requires at most one addition for each paths

$\pi \in paths(L)$ for a total of $O(n)$ additions. Step 4 takes another $O(n)$ additions. The result follows. ∎

3.3 Dependence on p

In section 3.1, we showed that the Bayesian optimal decision for Game I is a function of the corrupted view ϑ and the error probability p. The dependence of the optimal decision on the view ϑ is obvious. For example, a subtree whose corrupted view assigns the label 1 to all nodes is always preferable to a subtree whose corrupted view assigns the label 0 to all nodes.

In this section, we analyze the dependence of the optimal decision on the error probability p for a fixed corrupted view of the game tree. For the purposes of this analysis, we treat c as the independent variable. As a result, the error probability p becomes a function of c:

$$p = p(c) = \frac{1}{1 + \sqrt{c}} \tag{26}$$

It is not difficult to see that $p(c)$ is a continuous and strictly decreasing function of c over $\mathcal{R}^+$. Furthermore, $c \in (1, +\infty)$ if and only if $p(c) \in (0, 1/2)$.

An example where knowledge of p matters

Definition 8 *Let (T, λ) be a MaxTree and let ϑ be a corrupted view of λ with error probability $p(c)$, $c \in (1, +\infty)$. Then the polynomial $q_T(c)$ is defined as follows:*

$$q_T(c) = \Phi_L(c) - \Phi_R(c) \tag{27}$$

where L and R are the left and right subtrees of T.

It follows from Corollary 1 that the sign of $q_T(c)$ at any point c determines the Bayesian optimal decision for the corresponding instance of Game I.

We show that there exist instances of Game I where the Bayesian optimal decision depends on p.

Theorem 6 *There exists a labelled tree (T, ϑ), such that the Bayesian optimal decision for Game I on input (T, ϑ, p) depends on p.*

Proof: Consider the labelled tree (T, ϑ) of Figure 4 and let L and R be the left and right subtrees of T. By inspection, L has 1 path with exactly 4 nodes labelled 1, 1 path with exactly 3 nodes labelled 1, 2 paths with exactly 2 nodes labelled 1, and 4 paths with exactly 1 node labelled 1. From equation 25, we obtain

$$\Phi_L(c) = \sum_{\pi \in paths(L)} c^{f(\pi)} = c^4 + c^3 + 2c^2 + 4c \tag{28}$$

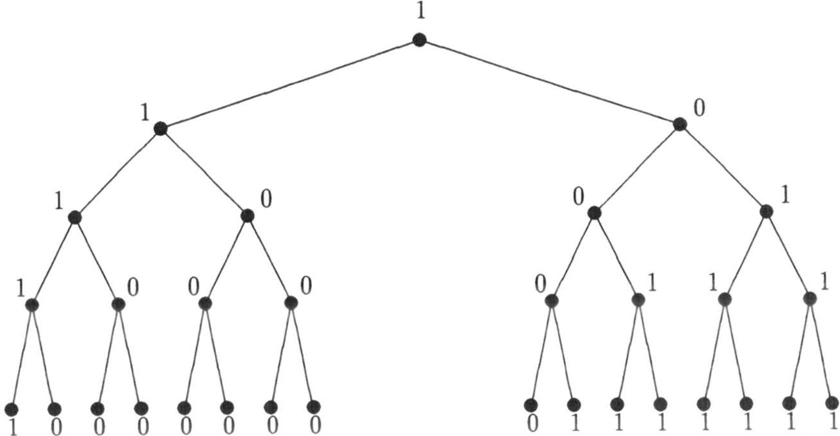

Figure 4: A view requiring knowledge of p.

By similar reasoning applied to R, we obtain:

$$\Phi_R(c) = \sum_{\pi \in paths(R)} c^{f(\pi)} = 4c^3 + 2c^2 + c + 1 \tag{29}$$

Subtracting equation 29 from 28 gives:

$$\begin{aligned} q_T(c) &= \Phi_L(c) - \Phi_R(c) \\ &= c^4 - 3c^3 + 3c - 1 \\ &= (c^2 - 1)(c - \varrho_1)(c - \varrho_2) \end{aligned} \tag{30}$$

where

$$\varrho_1 = \frac{3 - \sqrt{5}}{2} \simeq 0.382 \quad \text{and} \quad \varrho_2 = \frac{3 + \sqrt{5}}{2} \simeq 2.618$$

From equation 30, we see that $q_T(c)$ is strictly negative over the open interval $(1, \varrho_2)$ and strictly positive over the open interval $(\varrho_2, +\infty)$. Thus, the optimal decision depends on the value of c (which in turn is a function of the error probability p):

- If $c \in (1, \varrho_2)$, then the Bayesian optimal decision for (T, ϑ) is Right.

- If $c = \varrho_2$, then both Left and Right are Bayesian optimal decisions for (T, ϑ).

- If $c \in (\varrho_2, +\infty)$, then the Bayesian optimal decision for (T, ϑ) is Left.

Arbitrarily many flips as p varies

In section 3.3, we gave an example in which the optimal decision flips once as p varies from 0 to 1/2. In this section, we show that the decision may flip an arbitrary number of times. In particular, for each $m > 0$, we construct a tree T for which the optimal decision flips $m - 1$ times as p varies from 0 to 1/2.

The tree T will be built by embedding copies of trees with paths of various numbers of 1's. By adjusting the numbers of copies of each such tree placed in the left and right subtree of T, we will be able to control rather precisely both the degree of $q_T(c)$ and the placement of its roots. The claimed result follows by causing $m - 1$ simple roots to fall in the open interval $(1, +\infty)$.

Let $\mathcal{T} = \{(T_i, \vartheta_i) : i = 1, 2, \ldots, m\}$ be a collection of labelled trees such that for each i, the root of T_i is labelled 1, and the maximum number of nodes labelled 1 in any path from the root to a leaf in T_i is equal to i. The trees (T_i, ϑ_i) form the basis of our construction.

Let $\mathbf{s} = (s_1, s_2, \ldots, s_m) \in \mathcal{N}^m$ be an m-dimensional vector of non-negative integers. $\mathcal{T}$ and $\mathbf{s}$ define a forest $\mathcal{F}$ of trees which contains s_i copies of labelled tree (T_i, ϑ_i) for each i. In the following, we show how, given such a forest $\mathcal{F}$, we can construct a single labelled tree (X, ϑ_X) satisfying the following:

Path Property: The number of paths in (X, ϑ_X) with exactly j nodes labelled 1 is equal to the number of paths in the forest $\mathcal{F}$ with exactly j nodes labelled 1, for all $j = 1, 2, \ldots, m$.

The eventual construction of T will then be performed as follows. First, we construct two forests $\mathcal{F}_s$ and $\mathcal{F}_t$ for appropriate m-dimensional vectors $\mathbf{s}$ and $\mathbf{t}$. Next we construct trees (L, ϑ_L) and (R, ϑ_R) having the path property with respect to $\mathcal{F}_s$ and $\mathcal{F}_t$, respectively. Finally, we construct T by choosing a root node and making L and R the left and right subtrees, respectively. With appropriate choices of $\mathbf{s}$ and $\mathbf{t}$, the polynomial $q_T(c)$ can be made to coincide with an arbitrary polynomial $q(c)$ with integer coefficients, up to a constant factor. The details follow.

Let X be a complete binary tree, such that the number of leaves in X is greater or equal to the total number of leaves in all trees of the forest $\mathcal{F}$. Let $roots(\mathcal{F})$ be the set of all roots in the forest $\mathcal{F}$. We define an embedding $e \colon roots(\mathcal{F}) \to nodes(X)$ which maps each root of a tree in the forest $\mathcal{F}$ to a node of X. This embedding will satisfy the following two properties:

- If r is the root of a tree of height[3] k in the forest $\mathcal{F}$, then its image $e(x)$ has height k in X.

- If $r, r' \in roots(\mathcal{F})$ and $r \neq r'$, then $e(r) \neq e(r')$, and neither is a descendant of the other in X.

---

[3]The *height* of a node in a tree is its maximum distance to a leaf.

It is obvious that e can be naturally extended to an embedding $\hat{e}$ of all nodes of trees in $\mathcal{F}$ into the nodes of X. We define the labelling ϑ_X as follows: for all $x \in nodes(X)$,

$$\vartheta_X(x) = \begin{cases} \vartheta_i(\hat{e}^{-1}(x)) & \text{if } x \in \hat{e}(nodes(T_i)) \\ 0 & \text{otherwise} \end{cases} \tag{31}$$

The labelled tree (X, ϑ_X) satisfies the desired property by construction.

Let α_{ij} be the number of paths with exactly j nodes labelled 1 in (T_i, ϑ_i) and let β_j be the total number of paths with exactly j nodes labelled 1 in $\mathcal{F}$. Then,

$$\beta_j = \sum_{i=j}^{m} s_i \alpha_{ij} \tag{32}$$

for all $j = 1, 2, \ldots, m$. Since β_j is also the total number of paths in (X, ϑ_X) with exactly j nodes labelled 1, $\Phi_X(c)$ can be written in the form:

$$\Phi_X(c) = \sum_{j=0}^{m} \beta_j c^j \tag{33}$$

where β_0 is the number of leaves in X that are not images of any node x in the forest $\mathcal{F}$ under $\hat{e}$. We note that $\Phi_X(1)$ is equal to the total number of leaves in X.

Consider now a pair of vectors $\mathbf{s} = (s_1, s_2, \ldots, s_m)$, $\mathbf{t} = (t_1, t_2, \ldots, t_m) \in \mathcal{N}^m$. The collection of trees $\mathcal{T}$ of the previous construction and the two vectors $\mathbf{s}$ and $\mathbf{t}$ define two forests, $\mathcal{F}_\mathbf{s}$ and $\mathcal{F}_\mathbf{t}$. We use $\mathcal{F}_\mathbf{s}$ and $\mathcal{F}_\mathbf{t}$ to construct the left subtree (L, ϑ_L) and right subtree (R, ϑ_R) of a bigger labelled tree (T, ϑ). We choose L and R to have the same size, such that the number of leaves in L or R is greater than or equal to the maximum of the total number of leaves in the forests $\mathcal{F}_\mathbf{s}$ and $\mathcal{F}_\mathbf{t}$. For all $x \in nodes(T)$, define:

$$\vartheta(x) = \begin{cases} \vartheta_L(x) & \text{if } x \in nodes(L) \\ \vartheta_R(x) & \text{if } x \in nodes(R) \\ 0 & \text{if } x \text{ is the root of } T \end{cases} \tag{34}$$

We have thus defined a labelled tree (T, ϑ).

Lemma 5 *Let* $\mathbf{s}, \mathbf{t} \in \mathcal{N}^m$, *and let* $q_T(c) = \sum_{j=0}^{m} \gamma_j c^j$ *be the polynomial of the labelled tree* (T, ϑ) *constructed as above. Then:*

$$\gamma_j = \sum_{i=j}^{m} (s_i - t_i) \alpha_{ij} \tag{35}$$

for all $j = 1, 2, \ldots, m$, *and*

$$\gamma_0 = -\sum_{j=1}^{m} \gamma_j \tag{36}$$

Proof: From equations 27, 32, and 33, we have:

$$\gamma_j = \sum_{i=j}^{m} s_i \alpha_{ij} - \sum_{i=j}^{m} t_i \alpha_{ij}$$

$$= \sum_{i=j}^{m} (s_i - t_i) \alpha_{ij}$$

for $j = 1, 2, \ldots, m$, establishing equation 35. $q_T(1)$ is equal to the number of leaves in L minus the number of leaves in R. Since L and R were chosen to have the same size, $q_T(1) = 0$, and equation 36 follows. ∎

Theorem 7 *Let $q(c)$ be a polynomial of degree m with integer coefficients, such that $q(1) = 0$. Then there exists a labelled tree (T, ϑ) and a positive integer d such that $q_T(c) = d \cdot q(c)$ for all c, i.e., $q_T = d \cdot q$ as polynomials. In particular, q_T and q have the same roots.*

Proof: Let $q(c)$ be as in the theorem. Write $q(c) = \sum_{i=0}^{m} \delta_i c^i$, where $\delta_i \in \mathcal{Z}$. We compute a pair of vectors $\mathbf{s}$, $\mathbf{t}$, and a positive integer d so that the labelled tree (T, ϑ) defined by $\mathbf{s}$ and $\mathbf{t}$ as sketched above has polynomial $q_T(c) = d \cdot q(c)$.

Consider the following system of linear equations:

$$\sum_{i=j}^{m} \alpha_{ij} x_i = \delta_j \tag{37}$$

for all $j = 1, 2, \ldots, m$. This system always has a rational solution $\mathbf{x} = (x_1, x_2, \ldots, x_m)$ because it is upper triangular and all α_{ij} are positive integers. Let d be the least common multiple of the denominators of the x_i's. Then $d \cdot x_i \in \mathcal{Z}$ for all i, and $\mathbf{x}' = (d \cdot x_1, d \cdot x_2, \ldots, d \cdot x_m)$ solves the system of linear equations:

$$\sum_{i=j}^{m} \alpha_{ij} x_i' = d \cdot \delta_j \tag{38}$$

for all $j = 1, 2, \ldots, m$.

Now, define vectors $\mathbf{s}$ and $\mathbf{t}$ as follows. For all $i = 1, 2, \ldots, m$,

- if $x_i \geq 0$, set $s_i = d \cdot x_i$ and $t_i = 0$;

- if $x_i < 0$, set $s_i = 0$ and $t_i = -d \cdot x_i$.

It is obvious that $\mathbf{s}, \mathbf{t} \in \mathcal{N}^m$ and $d \cdot x_i = s_i - t_i$, for all $i = 1, 2, \ldots, m$. Using equations 35 and 38, we have

$$\gamma_j = \sum_{i=j}^{m} \alpha_{ij} (d \cdot x_i) = d \cdot \delta_j$$

for all $j = 1, 2, \ldots, m$. Because both $q_T(c)$ and $q(c)$ have a root at 1, $\gamma_0 = d \cdot \delta_0$. Hence, $q_T(c) = d \cdot q(c)$ as desired. ∎

As a result of Theorem 7, we can select a polynomial $q(c)$ of degree m with one root at 1 and $m - 1$ simple roots in the interval $(1, +\infty)$, so that the Bayesian optimal decision for the corresponding labelled tree (T, ϑ) flips (from Left to Right or from Right to Left) m times as c increases continuously in the interval $(1, +\infty)$. This observation is stated formally in the following corollary of Theorem 7.

Corollary 2 *For any $m \in \mathcal{N}$, there exist a labelled tree (T, ϑ) and $m - 1$ thresholds $0 < \tau_1 < \tau_2 < \ldots < \tau_{m-1} < 1/2$ defining intervals $I_0 = (0, \tau_1), I_1 = (\tau_1, \tau_2), \ldots, I_{m-1} = (\tau_{m-1}, 1/2)$ such that:*

1. *For all $j = 0, 1, \ldots, m$, the Bayesian optimal decision for (T, ϑ) is the same for all error probabilities $p \in I_j$.*

2. *For all $j = 1, \ldots, m - 1$, if the Bayesian optimal decision for error probability in the interval I_{j-1} is Left (Right), then the Bayesian optimal decision for error probability in the interval I_j is Right (Left).*

Proof: Consider the polynomial $q(c) = (c - 1)(c - 2) \ldots (c - (m - 1))(c - m)$. Obviously, all the coefficients of $q(c)$ are integers, and the m roots of $q(c)$ are the numbers $1, 2, \ldots, m$. By Theorem 7, we can find a tree (T, ϑ) and a positive integer d such that $q_T(c) = d \cdot q(c)$. Then $q_T(c)$ has also has simple roots $1, 2, \ldots, m$, all but one of which falls in the interval $(1, +\infty)$. Let $\tau_i = p(m + 1 - i)$ for $i = 1, 2, \ldots, m - 1$. Since $p(c)$ is a continuous and monotonic decreasing function, $\tau_1, \tau_2, \ldots, \tau_{m-1}$ satisfy properties (1) and (2) above.

$\blacksquare$

4 Game II: Choosing a Leaf

In this section, we consider a variant of Game I where the player is asked to choose a whole path from the root to a leaf of the game tree.

Game II: Let (T, Λ) be a MaxTree with Λ a random labelling following the probability distribution $\mathcal{P}$. Let λ be the outcome of Λ and let ϑ be a corrupted view of λ with error probability p. Given (T, ϑ, p), the player is asked to choose a leaf $\alpha \in leaves(T)$. If α is labelled 1 under λ, the player wins; otherwise, she loses.

We fix the underlying probability distribution $\mathcal{P}$ for Λ to be the one defined by equation 1 in section 3, that is, Λ is uniformly distributed among proper labellings in which a single leaf is labelled 1.

4.1   Probabilistic Analysis of Game II

We provide an exact probabilistic analysis of Game II. Given the corrupted view of the game tree and the a priori information about the underlying distribution of labellings, we compute the conditional probability of winning the game for all choices of leaves in T.

Let m be the number of nodes with observed label 1 and n the total number of nodes in (T, ϑ). Define:

$$b = (1 - p)^{n-(m+k+1)} p^{m+k+1}$$

where k is the depth of T. Note that b is a constant which depends only on the value of p and the corrupted labelling ϑ of the game tree.

Lemma 6 *Let (T, λ) be a MaxTree and let ϑ be a corrupted view of λ with error probability p. If $\alpha \in leaves(T)$ and π_α is the path from the root of T to the leaf α, then:*

$$\mathbf{prob}[\vartheta \mid \lambda^\alpha] = b \cdot c^{f(\pi_\alpha)} \tag{39}$$

where $c = ((1 - p)/p)^2$.

Proof: Let λ^α be the actual labelling of T. Then, for every $x \in nodes(T)$, the probability that the observed label $\vartheta(x)$ agrees with the actual label $\lambda^\alpha(x)$ is $(1 - p)$, and the probability that the two labels disagree is p. Since the label corruptions happen independently of each other, we can compute the probability of any corrupted view ϑ by taking the product of the probabilities of the individual label corruptions.

$$\mathbf{prob}[\vartheta \mid \lambda^\alpha] = \left(\prod_{x \in \pi_\alpha} (1 - p)^{\vartheta(x)} p^{1-\vartheta(x)} \right) \cdot \left(\prod_{x \in (T-\pi_\alpha)} (1 - p)^{1-\vartheta(x)} p^{\vartheta(x)} \right) \tag{40}$$

where the first product ranges over all nodes x on the path π_α, and the second product ranges over all nodes x in any other path of T. However, the second product can be rewritten as follows:

$$\prod_{x \in (T-\pi_\alpha)} (1 - p)^{1-\vartheta(x)} p^{\vartheta(x)} = \frac{\prod_{x \in T} (1 - p)^{1-\vartheta(x)} p^{\vartheta(x)}}{\prod_{x \in \pi_\alpha} (1 - p)^{1-\vartheta(x)} p^{\vartheta(x)}} \tag{41}$$

Substituting 41 in equation 40, we obtain:

$$\begin{aligned}
\mathbf{prob}[\vartheta \mid \lambda^\alpha] &= \left(\prod_{x \in T} (1 - p)^{1-\vartheta(x)} p^{\vartheta(x)} \right) \cdot \left(\prod_{x \in \pi_\alpha} \left(\frac{1 - p}{p} \right)^{\vartheta(x)} \cdot \left(\frac{p}{1 - p} \right)^{1-\vartheta(x)} \right) \\
&= (1 - p)^{n-m} p^m \cdot \left(\frac{p}{1 - p} \right)^{(k+1)} \cdot \prod_{x \in \pi_\alpha} \left(\frac{1 - p}{p} \right)^{2\vartheta(x)} \\
&= b \cdot c^{f(\pi_\alpha)} \tag{42}
\end{aligned}$$

■

Theorem 8 *Let (T, λ) be a MaxTree and let ϑ be a corrupted view of λ with error probability p. Then, for all $\alpha \in leaves(T)$,*

$$\mathbf{prob}[\lambda^\alpha \mid \vartheta] = \frac{c^{f(\pi_\alpha)}}{\sum_\beta c^{f(\pi_\beta)}} \tag{43}$$

where the sum in the denominator ranges over all $\beta \in leaves(T)$.

Proof: Since λ was chosen from the probability distribution $\mathcal{P}$, λ is always winning. Furthermore, it is equally probable that any one of $\beta \in leaves(T)$ is labelled 1 under λ; that is,

$$\mathbf{prob}[\vartheta] = \frac{1}{2^k} \sum_\beta \mathbf{prob}[\vartheta \mid \lambda^\beta] \tag{44}$$

Substituting $\mathbf{prob}[\vartheta \mid \lambda^\beta]$ from equation 42, we obtain:

$$\mathbf{prob}[\vartheta] = \frac{b}{2^k} \sum_\beta c^{f(\pi_\beta)} \tag{45}$$

Finally, we use *Bayes' Theorem* to combine formulas 42 and 45, in order to compute $\mathbf{prob}[\lambda^\alpha \mid \vartheta]$:

$$\begin{aligned} \mathbf{prob}[\lambda^\alpha \mid \vartheta] &= \frac{\mathbf{prob}[\vartheta \mid \lambda^\alpha] \cdot \mathbf{prob}[\lambda^\alpha]}{\mathbf{prob}[\vartheta]} \\ &= \frac{c^{f(\pi_\alpha)}}{\sum_\beta c^{f(\pi_\beta)}} \end{aligned}$$

$\blacksquare$

Corollary 3 *Let (T, λ) be a MaxTree and let ϑ be a corrupted view of λ with error probability p. If $\alpha, \beta \in leaves(T)$, then:*

$$\mathbf{prob}[\lambda^\alpha \mid \vartheta] \geq \mathbf{prob}[\lambda^\beta \mid \vartheta] \quad \textit{iff} \quad f(\pi_\alpha) \geq f(\pi_\beta) \tag{46}$$

Corollary 4 *For all instances of Game II, the Bayesian optimal decision does not depend on the value of the error probability p.*

4.2 A Bayesian Optimal Algorithm for Game II

We use Corollary 3 to design algorithm B_1, shown in Figure 5, which is Bayesian optimal for Game II. The player computes the number of nodes with observed label 1 on every path from the root of the game tree to a leaf and chooses the path that corresponds to the maximum. Algorithm B_1 is much easier than the Bayesian optimal algorithms A_1 and A_2 for Game I, in that it involves no arithmetic other than counting.

ALGORITHM B_1

Input:
    T, a complete binary tree
    ϑ, a corrupted view of T
    p, the error probability of the corruption

Output:
    α, a leaf of T

Description:
    Step 1:
        Compute the values of $f(\pi_\beta)$, for all $\beta \in leaves(T)$, as follows: Working
        down from the root, compute for each node β in T the number of nodes
        labelled 1 by ϑ on the path from β to the root. For $\beta \in leaves(T)$, this
        number is $f(\pi_\beta)$.
    Step 2:
        Compute $\max\{f(\pi_\beta), \beta \in leaves(T)\}$.
        Choose any leaf α, such that $f(\pi_\alpha)$ is maximum.

Figure 5: A Bayesian optimal algorithm for Game II.

Theorem 9 *Let B_1 be the algorithm shown in Figure 5. Let (T, λ) be a MaxTree of n nodes and let ϑ be a corrupted view of λ with error probability p. On input (T, ϑ, p), algorithm B_1 requires $O(n)$ additions and no multiplications to compute the Bayesian optimal decision for Game II.*

Proof: Step 1 of algorithm B_1 is the same as step 2 of algorithm A_2, which was previously shown to require only $O(n)$ additions. ∎

4.3 Similar Games—Different Strategies

In section 3.3, we showed that the Bayesian optimal strategy for Game I—choosing a subtree—depends on the error probability p. In other words, in order for the player to compute an optimal decision, it is necessary for her to have some knowledge of the accuracy of the data at hand. On the other hand, in section 4.2, we showed that the optimal strategy for Game II—choosing a leaf—does not depend on p. This observation leads us to an interesting, but somewhat counterintuitive, result: repeated application of a Bayesian optimal decision rule for Game I does not give a Bayesian optimal decision rule for Game II.

Let (T, ϑ, p) be an instance of Game II where T is of depth k. Consider the following decision algorithm B_2 for playing Game II:

- Set $X = T$;

- Repeat k times: compute the Bayesian optimal decision Y for the instance (X, ϑ_X, p) of Game I and set $X = Y$;

- Output X.

Theorem 10 B_2 *is not a Bayesian optimal decision algorithm for Game II.*

Proof: Let (T, ϑ) be the labelled tree of Figure 4 and let $c \in (1, \varrho_2)$.

The Bayesian optimal decision for Game II on input (T, ϑ, p) is the leftmost leaf, since the path from the root of T to the leftmost leaf has the maximum number of nodes with observed label 1 over all such paths.

However, in Theorem 6, we showed that the Bayesian optimal decision for Game I on input (T, ϑ, p) is Right. As a result, on input (T, ϑ, p), B_2 chooses a leaf in the right subtree of T. Hence, decision algorithm B_2 is non-optimal. $\blacksquare$

5 Conclusions and Future Work

In this paper, we look at the problem of game-playing/decision-making based on information that is inaccurate because of the presence of random noise. In order for the player in our games to compute a rational decision, she needs to somehow make use of all of all available information. A *Bayesian optimal* strategy accomplishes this since it maximizes the expected value of the player's choice. We show that the Bayesian optimal decision for these games is both easy to compute and expressible by simple, easily-understood formulas which allow for some interesting observations:

- In answering the question "Which subtree has the winning leaf?", the player can make a better decision if she also has some knowledge of the accuracy of the data. Consequently, any algorithm that bases its decision simply on the corrupted view of the game tree will be inferior to one that also makes use of knowledge of the error probabilities.

- In answering the question "Which is the winning leaf?", consistency of the observed data is most important, and the error probability does not affect the optimal decision.

Certain simplifying assumptions in our model seem to make our results very restrictive. For example, we assume that our trees have uniform depth and constant branching factor of two. We also assume a uniform error probability p at each node of the tree and binary node labels in the corrupted view. Geanakoplos and Gray show that these assumptions can be removed and similar results can still be obtained [1, 2].

On the other hand, certain other assumptions in our probabilistic setting are difficult to eliminate. For example, the independence of errors among the nodes can be easily characterized as unrealistic, yet removing it entirely makes the problem seem intractable. Perhaps it might still be possible to handle simple patterns of correlation between parent-child nodes or sibling nodes—this is one possible direction for further research. Other future directions include generalizing these results to two-person games and to iterated (multi-move) games, where new information becomes available after each move.

Finally, we would like to apply Bayesian optimal reasoning to other problems of a similar probabilistic flavor, where redundancy of the data can be used to offset the effect of data corruption. One practical example, pointed out to us by Linda Shapiro, is in computer vision, where noise in the visual data is an everpresent problem in the recognition of the underlying physical objects or patterns.

Bibliography

[1] John Geanakoplos and Larry Gray, June 1991. Personal communication.

[2] John Geanakoplos and Larry Gray. When seeing further is not seeing better. Manuscript, July 1991.

[3] Judea Pearl. On the nature of pathology in game searching. *Artificial Intelligence*, 20:427–453, 1983.

[4] Sheldon Ross. *A first course in Probability*. Macmillan, New York, NY, 1976.

[5] Martin Shubik. *Game Theory in the social sciences*. MIT Press, Cambridge, MA, 1982.

Über den Nutzen von Orakelfragen bei nichtdeterministischen Kommunikationsprotokollen

Bernd Halstenberg
Rüdiger Reischuk

Institut für Theoretische Informatik
Technische Hochschule Darmstadt
6100 Darmstadt
Germany

Zusammenfassung

Für das 2-Prozessor-Kommunikationsmodell untersuchen wir nichtdeterministische Orakelprotokolle. Die beiden Prozessoren dürfen in diesem Fall gemeinsam ein Orakel für eine vorgegebene Sprache oder Sprachklasse zu Hilfe nehmen. Mit diesem Modell läßt sich eine alternative Definition für die polynomielle Kommunikationshierarchie geben.

Es wird der Begriff der disjunktiven und konjunktiven Rechteckreduktion zwischen Sprachen eingeführt. Unter gewissen Vollständigkeits- und Abschlußbedingungen bezüglich dieser Reduktionen für die Orakelsprache zeigen wir, daß für derartige Orakelprotokolle eine einzige Orakelfrage ausreichend ist. Dies gilt beispielsweise für die Orakelmengen $C-\mathcal{PP}$ und $C-\mathcal{BPP}$, d.h. die Sprachen, die durch probabilistische Protokolle mit mäßig beschränktem bzw. beschränktem Fehler mit polylogarithmischem Kommunikationsaufwand akzeptiert werden können. Ähnliche Ergebnisse lassen sich auch für Berechnungen von Orakel-Turing-Maschinen zeigen.

1   Einleitung

Viele Arbeiten haben sich in den letzten Jahren damit befaßt, den erforderlichen Informationsfluß zwischen den einzelnen Teilen eines Berechnungssystems als untere Schranke für die Komplexität eines Problems zu nutzen. Eine kleine Auswahl dieser Arbeiten zeigt die weitreichende Bedeutung der Kommunikationskomplexität: ABELSON [1] untersuchte beispielsweise die Berechnung „glatter" reeller Funktionen auf einem Netz von Prozessoren, in dem reelle Zahlen zwischen den Prozessoren ausgetauscht werden. YAO [9] betrachtete die Berechnung boolescher Funktionen für verschiedene Zwei-Prozessor-Modelle. LIPTON und SEDGEWICK [7] und AHO, ULLMAN und YANNAKAKIS [2] untersuchten die Bedeutung der Kommunikationskomplexität für VLSI-Chips. KARCHMER und WIGDERSON [6] zeigten mit Hilfe eines Kommunikationsproblems untere Schranken für die Tiefe monotoner Schaltkreise.

Neben den verschiedenen Anwendungen solcher Informationsfluß-Argumente haben sich auch viele Arbeiten mit den verschiedenen Modellen der Kommunikationskomplexität als Technik befaßt. Analog zu den herkömmlichen Berechnungsmodellen wurden Simulationstechniken zwischen verschiedenen Modellen entwickelt, Tradeoff-Effekte zwischen unterschiedlichen Resourcen beobachtet und die Probleme durch Komplexitätsklassen strukturiert, wobei auch hier durch Orakel relativierte Klassen betrachtet wurden. Die vorliegende Arbeit zeigt einige hinreichende Bedingungen auf, unter denen nichtdeterministische Orakelprotokolle nur eine Frage benötigen.

2   Die Kommunikationskomplexität boolescher Funktionen

Betrachten wir zunächst das in Abb. 1 dargestellte Grundmodell: Gegeben sind zwei endliche Mengen X_0 und X_1. Zwei Prozessoren, P_0 und P_1, sollen eine Funktion $f : X_0 \times X_1 \to Y$ berechnen, d.h. Prozessor P_0 erhält eine Eingabe $x_0 \in X_0$, und Prozessor P_1 erhält eine Eingabe $x_1 \in X_1$; schließlich soll mindestens einer der beiden Prozessoren das Ergebnis $f(x_0, x_1)$ kennen. Dazu dürfen die Prozessoren beliebig komplexe lokale Berechnungen durchführen und untereinander – entsprechend einem vorher festgelegten Protokoll – Informationen austauschen.

Definition 1 Ein *deterministisches Protokoll* $\mathcal{A}$ zur Berechnung einer Funktion $f : X_0 \times X_1 \to Y$ wird formal definiert durch zwei partielle *Übertragungsfunktionen* $\varphi_i : X_i \times \{0,1\}^* \to \{0,1\}^*$ und zwei partielle *Ausgabefunktionen* $a_i : X_i \times \{0,1\}^* \to Y$ für $i \in \{0,1\}$. In Runde j sendet Prozessor P_i, $i = (j+1)_2$, eine Nachricht $w_j = \varphi_i(x_i, w_1 \ldots w_{j-1})$ in Abhängigkeit von seiner lokalen Eingabe x_i und den bereits ausgetauschten Nachrichten $w_1, \ldots, w_{j-1}$. Dabei nehmen wir einschränkend an, daß der empfangende Prozessor das Ende einer jeden Nachricht aus den vorher ausgetauschten Nachrichten und der Nachricht selbst erkennen kann. Schließlich ist für ein $k \in \mathbb{N}$ und ein $i \in \{0,1\}$ der Wert $a_i(x_i, w_1 \ldots w_k)$ definiert. Dieser ist dann der vom Protokoll $\mathcal{A}$ auf Eingabe (x_0, x_1) berechnete Wert $\mathcal{A}(x_0, x_1)$.

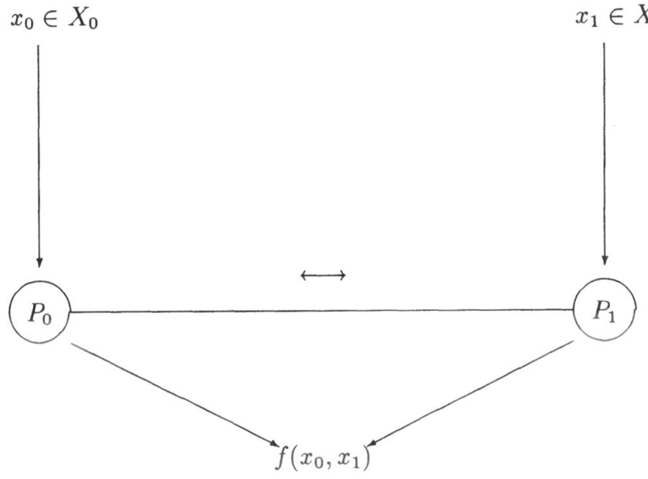

Abbildung 1: Das grundlegende Berechnungsmodell

Wir nennen $\mathcal{A}$ ein *k-Runden-Protokoll*, wenn für jede Eingabe $(x_0, x_1) \in X_0 \times X_1$ höchstens k Nachrichten ausgetauscht werden. 1-Runden-Protokolle heißen auch *Einweg-Protokolle*. Das Protokoll $\mathcal{A}$ *berechnet* die Funktion f, wenn für alle Eingaben $(x_0, x_1) \in X_0 \times X_1$ als Ergebnis $\mathcal{A}(x_0, x_1) = f(x_0, x_1)$ berechnet wird. Die *Länge* der Berechnung von $\mathcal{A}$ auf Eingabe (x_0, x_1) bezeichnen wir mit $\ell(\mathcal{A}, x_0, x_1)$; sie ist die Gesamtzahl der ausgetauschten Bits in der Berechnung. Mit $cost(\mathcal{A})$ bezeichnen wir die *Kosten* des Protokolls $\mathcal{A}$, d.h. die Länge einer Berechnung im schlechtesten Fall. Damit gilt:

$$cost(\mathcal{A}) := \max_{(x_0, x_1) \in X_0 \times X_1} \ell(\mathcal{A}, x_0, x_1)$$

Die *deterministische Kommunikationskomplexität* der Funktion f ist gleich den Kosten eines optimalen Protokolls, das f berechnet, also:

$$C_{\mathrm{det}}(f) := \min\{cost(\mathcal{A}) \mid \mathcal{A} \text{ berechnet } f\}$$

Wie bei Turingmaschinen kann man auch für Kommunikationsprotokolle nichtdeterministische und probabilistische Verfahren untersuchen. Diese können wie folgt formal definiert werden.

Definition 2 Ein *nichtdeterministisches Protokoll* $\mathcal{A}$ für eine Funktion $f : X_0 \times X_1 \to \{0, 1\}$ wird formal definiert durch zwei Mengen $Z_0 = \{0, 1\}^r$ und $Z_1 = \{0, 1\}^s$ sowie ein deterministisches Protokoll $\mathcal{A}'$ für eine Funktion $f' : (X_0 \times Z_0) \times (X_1 \times Z_1) \to \{0, 1\}$. Ein Element $(z_0, z_1) \in Z_0 \times Z_1$ stellt dann eine der möglichen Berechnungen von $\mathcal{A}$ auf Eingabe (x_0, x_1) dar. Diese Berechnung heißt *akzeptierend*, falls $f'((x_0, z_0), (x_1, z_1)) = 1$ ist; andernfalls heißt sie *verwerfend*. $\mathcal{A}$ *berechnet* die durch f definierte Sprache

$L(f) = f^{-1}(\{1\})$, falls für alle Eingaben (x_0, x_1) gilt:

$$f(x_0, x_1) = 1 \iff \exists\, (z_0, z_1) \in Z_0 \times Z_1 : f'((x_0, z_0), (x_1, z_1)) = 1.$$

Die *Kosten cost($\mathcal{A}$)* solch eines Protokolls $\mathcal{A}$ sind gleich den Kosten des zugehörigen deterministischen Protokolls $\mathcal{A}'$. Die *nichtdeterministische Kommunikationskomplexität* der Sprache $L(f)$ ist dann

$$C_{\mathrm{ndet}}(L(f)) := \min\{cost(\mathcal{A}) \mid \mathcal{A} \text{ berechnet } L(f)\}.$$

Gelegentlich interessiert man sich auch für die *Anzahl* der akzeptierenden oder der verwerfenden Berechnungen für eine Eingabe (x_0, x_1). Wir definieren daher

$$acc(\mathcal{A}, (x_0, x_1)) := \quad |\{(z_0, z_1) \in Z_0 \times Z_1 \mid f'((x_0, z_0), (x_1, z_1)) = 1\}|$$
$$rej(\mathcal{A}, (x_0, x_1)) := \quad |\{(z_0, z_1) \in Z_0 \times Z_1 \mid f'((x_0, z_0), (x_1, z_1)) = 0\}|$$

Definition 3 Ein *probabilistisches Protokoll* $\mathcal{A}$ für eine Funktion $f : X_0 \times X_1 \to \{0, 1\}$ wird formal definiert durch zwei endliche Mengen $Z_0 = \{0, 1\}^r$ und $Z_1 = \{0, 1\}^s$, zwei Wahrscheinlichkeitsverteilungen μ_0 und μ_1 über Z_0 bzw. Z_1 sowie ein deterministisches Protokoll $\mathcal{A}'$ für eine Funktion $f' : (X_0 \times Z_0) \times (X_1 \times Z_1) \to \{0, 1\}$. Das Protokoll $\mathcal{A}$ *berechnet* die Funktion f mit *Fehlerwahrscheinlichkeit* höchstens $\epsilon(\mathcal{A}, f)$, falls für die Produktwahrscheinlichkeitsverteilung $\mu = \mu_0 \times \mu_1$ über $Z_0 \times Z_1$ gilt:

$$\mu(f'((x_0, z_0), (x_1, z_1)) \neq f(x_0, x_1)) \leq \epsilon(\mathcal{A}, f)$$

für alle $(x_0, x_1) \in X_0 \times X_1$. Die *Kosten cost($\mathcal{A}$)* des Protokolls $\mathcal{A}$ sind gleich den Kosten des zugehörigen deterministischen Protokolls $\mathcal{A}'$. Die *probabilistische Kommunikationskomplexität* der Funktion f für Fehlerwahrscheinlichkeit $\epsilon < 1/2$ ist dann

$$C_\epsilon(f) := \min\{cost(\mathcal{A}) \mid \epsilon(\mathcal{A}, f) \leq \epsilon\}.$$

Während die Bestimmung der deterministischen Kommunikationskomplexität einer Funktion im allgemeinen sehr schwierig ist, konnten AHO, ULLMAN und YANNAKAKIS in [2] die folgende Charakterisierung der nichtdeterministischen Kommunikationskomplexität der Sprache $L(f)$ angeben.

Definition 4 Ein *Rechteck* einer Menge $X_0 \times X_1$ ist eine Teilmenge der Form $A \times B$. Dieses heißt *monochromatisch* bzgl. einer Funktion $f : X_0 \times X_1 \to Y$, wenn f auf $A \times B$ konstant ist. Ist $f(A \times B) = \{y\}$, so heißt $A \times B$ auch ein *y-Rechteck*. Mit $\delta_y(f)$ bezeichnen wir die *y-Überdeckungszahl* der Funktion f, d.h. die minimale Anzahl von y-Rechtecken deren Vereinigung $f^{-1}(\{y\})$ ergibt.

Satz 1 $\lceil \log \delta_1(f) \rceil \leq C_{\mathrm{ndet}}(L(f)) \leq \lceil \log(1 + \delta_1(f)) \rceil$.

Definition 5 Um die Notation etwas zu vereinfachen, definieren wir für natürliche Zahlen a und b die Intervallschreibweise

$$[a : b] := \{n \in \mathbb{N} \mid a \leq n \leq b\}.$$

Im folgenden beschränken wir uns, soweit nichts anderes gesagt wird, auf die Untersuchung der Kommunikationskomplexität boolescher Funktionen auf Argumentpaaren gleicher Länge. D.h. wir nehmen an, daß $Y = \{0,1\}$ und $X_0 = X_1 = \{0,1\}^n$ für ein $n \in \mathbb{N}$ ist. Daher lassen sich die hier vorgestellten Berechnungsprobleme auch als Entscheidungsprobleme auffassen: Ist die Eingabe (x_0, x_1) ein Element der Sprache $L(f) := f^{-1}(\{1\})$? Wir identifizieren im folgenden Sprachen $L \subseteq \{0,1\}^n \times \{0,1\}^n$ mit ihrer jeweiligen charakteristischen Funktion $\chi_L : \{0,1\}^n \times \{0,1\}^n \to \{0,1\}$.

Da wir uns hier für das asymptotische Wachstum der Kommunikationskomplexität interessieren, betrachten wir nicht nur endliche Sprachen $L \subseteq \{0,1\}^n \times \{0,1\}^n$, sondern auch unendliche Sprachen

$$L \subseteq \bigcup_{n \in \mathbb{N}} \left(\{0,1\}^n \times \{0,1\}^n\right) =: \{0,1\}^{**}$$

von Eingabepaaren gleicher Länge. Diese betrachten wir als Vereinigung endlicher Sprachen $L = \bigcup_{n \in \mathbb{N}} L^{=n}$ mit $L^{=n} \subseteq \{0,1\}^n \times \{0,1\}^n$. Ein *Protokoll* für solch eine unendliche Sprache ist dann eine Folge $\mathcal{A} = (\mathcal{A}_n)_{n \in \mathbb{N}}$ von Protokollen für die endlichen Sprachen $L^{=n}$. In natürlicher Weise ist also die Kommunikationskomplexität für L als Funktion der Eingabelänge n definiert. Diese Funktionen sind jedoch im allgemeinen nicht berechenbar.

3   Kommunikationskomplexitätsklassen

Anstatt die Komplexitäten einzelner Probleme für unterschiedliche Berechnungsmodelle zu untersuchen, definieren wir *Kommunikationskomplexitätsklassen* und untersuchen die Beziehungen zwischen diesen Klassen. Da in unserem Fall die Komplexitäten stets durch die Eingabelänge nach oben beschränkt sind, würde es z.B. wenig Sinn machen, die Klasse aller Probleme zu betrachten, deren deterministische Kommunikationskomplexität polynomiell in der Länge der Eingabe beschränkt ist. Andererseits sehen wir polynomielle Unterschiede zwischen Komplexitäten als nicht so gravierend an. Daher folgen wir den Definitionen von BABAI, FRANKL und SIMON und betrachten solche Klassen von Problemen, deren Kommunikationskomplexität polynomiell im *Logarithmus* der Eingabelänge beschränkt ist. Wir betrachten als Schranken für die Kommunikationskomplexität also Funktionen aus der Klasse

$$\mathrm{PLOG} := \{p : \mathbb{N} \to \mathbb{N} \mid \exists k \in \mathbb{N} : p(n) \leq O(\log^k n)\}$$

Diese Funktionen nennen wir dann *polylogarithmisch* beschränkt. Funktionen der Form 2^p mit $p \in \mathrm{PLOG}$ heißen auch *quasipolynomiell*. Analog zu [3] definieren wir die folgenden Klassen:

$$\mathrm{C}\text{-}\mathcal{P} \quad := \quad \{L \in \{0,1\}^{**} \mid C_{\mathrm{det}}(L) \in \mathrm{PLOG}\}$$

$$\begin{aligned}
\text{C}-\mathcal{NP} &:= \{L \in \{0,1\}^{**} | C_{\text{ndet}}(L) \in \text{PLOG}\} \\
\text{C}-\text{co}\mathcal{NP} &:= \{L \in \{0,1\}^{**} | C_{\text{ndet}}(\bar{L}) \in \text{PLOG}\} \\
\text{C}-\mathcal{BPP} &:= \{L \in \{0,1\}^{**} | C_{\epsilon}(L) \in \text{PLOG}\}
\end{aligned}$$

Diese Definition suggeriert durch die verwendete Notation zwar eine Bedeutung, die ihr so nicht zukommt, insbesondere da es sich ja hier um ein nichtuniformes Modell handelt. Dennoch wollen wir diese Notation verwenden, da sie Assoziationen an die entsprechend bezeichneten Zeitkomplexitätsklassen für Turingmaschinen weckt, die durchaus hilfreich sein können. Zudem gibt es eine Reihe recht bemerkenswerter Analogien zwischen den so bezeichneten Kommunikationskomplexitätsklassen und den entsprechenden Zeitkomplexitätsklassen für Turingmaschinen. Daß diese Analogien jedoch nicht dergestalt sind, daß sie ohne weiteres vom einen Modell auf das andere übertragbar wären, zeigen die beiden folgenden Sätze, die in dieser Formulierung erstmals in [3] zu lesen waren.

Satz 2 $\text{C}-\mathcal{P} \neq \text{C}-\mathcal{NP}$.

Satz 3 $\text{C}-\mathcal{P} = \text{C}-\mathcal{NP} \cap \text{C}-\text{co}\mathcal{NP}$.

Eine wichtige Rolle in der Komplexitätstheorie spielt der Begriff der *Reduktion* einer Sprache L auf eine andere L'. Für die Kommunikationskomplexität ist der folgende Reduktionsbegriff am wichtigsten.

Definition 6 Eine (polylogarithmische) *Rechteckreduktion* einer Sprache L auf eine andere L' ist eine Folge $(f_n, g_n)_{n \in \mathbb{N}}$ von Paaren von Abbildungen $f_n, g_n : \{0,1\}^n \to \{0,1\}^{\ell(n)}$ mit $\log \ell \in \text{PLOG}$, so daß für alle $(x_0, x_1) \in \{0,1\}^n \times \{0,1\}^n$ gilt:

$$(x_0, x_1) \in L \iff (f_n(x_0), g_n(x_1)) \in L'$$

L heißt *reduzierbar* auf L', wenn es eine polylogarithmische Rechteckreduktion von L auf L' gibt. L' ist *$\mathcal{C}$-hart* für eine Kommunikationskomplexitätsklasse $\mathcal{C}$, wenn jede Sprache $L \in \mathcal{C}$ auf L' reduzierbar ist. L' ist *$\mathcal{C}$-vollständig*, falls L' $\mathcal{C}$-hart und $L' \in \mathcal{C}$ ist.

Beispiel 1 *Sei SI das Schnittmengenproblem, d.h. für $x, y \in \{0,1\}^n$ gilt:*

$$(x, y) \in SI \iff \exists i \in [1:n] : x_i \wedge y_i$$

Es ist leicht zu verifizieren, daß SI eine $\text{C}-\mathcal{NP}$-vollständige Sprache ist. Das Komplement, $\overline{SI}$, ist damit $\text{C}-\text{co}\mathcal{NP}$-vollständig. Das Schnittmengenproblem ist daher ein guter Kandidat, um $\text{C}-\mathcal{NP}$ von anderen Komplexitätsklassen zu separieren.

Für diese Arbeit spielen auch andere Reduktionsbegriffe eine wichtige Rolle, nämlich die nachfolgend definierten disjunktiven und konjunktiven Rechteckreduktionen.

Definition 7 Eine (polylogarithmisch beschränkte) *disjunktive* bzw. *konjunktive Rechteckreduktion* einer Sprache L auf eine Sprache L' besteht aus einer Funktion $p \in \text{PLOG}$ und einer Folge von Abbildungspaaren $f_i, g_i : \{0,1\}^* \to \{0,1\}^*$ mit $|f_i(x)| = |g_i(x)| = 2^{\ell(|x|)}$ für $\ell \in \text{PLOG}$, so daß für alle $(x_0, x_1) \in \{0,1\}^n \times \{0,1\}^n$ gilt:

$$
\begin{aligned}
(x_0, x_1) \in L &\iff \exists\, i \in [1 : p(n)] : (f_i(x_0), g_i(x_1)) \in L' \qquad \text{bzw.} \\
(x_0, x_1) \in L &\iff \forall\, i \in [1 : p(n)] : (f_i(x_0), g_i(x_1)) \in L' \ .
\end{aligned}
$$

Eine Klasse $\mathcal{C}$ ist *abgeschlossen* unter disjunktiven bzw. konjunktiven Rechteckreduktionen, wenn aus der Existenz einer disjunktiven bzw. konjunktiven Rechteckreduktion einer Sprache L auf eine Sprache $L' \in \mathcal{C}$ bereits $L \in \mathcal{C}$ folgt.

Alle hier betrachteten Kommunikationskomplexitätsklassen sind abgeschlossen unter den vorher definierten simplen Rechteckreduktionen, da solche Reduktionen nur lokale Berechnungen erfordern. Der Abschluß unter disjunktiven Rechteckreduktionen aber impliziert beispielsweise auch die nichttriviale Eigenschaft der Abgeschlossenheit unter endlicher Vereinigung von Sprachen.

Definition 8 Um die Notation weiter zu vereinfachen, definieren wir die folgende, verkürzte Schreibweise für Paare von Bitfolgen gleicher Länge: Für $x \in \{0,1\}^{2n}$ bezeichne $\langle x \rangle$ das Paar $(x_0, x_1) \in \{0,1\}^n \times \{0,1\}^n$, das aus den ersten bzw. letzten n Bits von x besteht, für das also die Konkatenation $x_0 x_1 = x$ ist. Sind $y \in \{0,1\}^{2p}$, $z \in \{0,1\}^{2q}$ usw. weitere Bitfolgen, so ist $\langle y \rangle = (y_0, y_1)$, $\langle z \rangle = (z_0, z_1)$ usw. Wir definieren dann

$$
\begin{aligned}
\langle x, y \rangle &:= (x_0 y_0, x_1 y_1) \\
\langle x, y, z \rangle &:= (x_0 y_0 z_0, x_1 y_1 z_1) \\
&\ \ \vdots
\end{aligned}
$$

Als Erweiterung der Klassen $\text{C}-\mathcal{NP}$ und $\text{C}-\text{co}\mathcal{NP}$ wurde in [3] die *polynomielle Kommunikationshierarchie* durch alternierende Existenz- und Allquantoren über Folgen polylogarithmischer Länge definiert.

Definition 9 Sei $k \in \mathbb{N}$. Eine Sprache L ist in $\text{C}-\Sigma_k$ genau dann, wenn es Funktionen $\ell_1, \ldots, \ell_k \in \text{PLOG}$ gibt und, für $\ell := \ell_1 + \cdots + \ell_k$, zwei Funktionsfolgen $\varphi_n, \psi_n : \{0,1\}^{n+\ell(n)} \to \{0,1\}$, so daß für alle $n \in \mathbb{N}$ und $\langle x \rangle \in \{0,1\}^{2n}$ gilt:

$$
\langle x \rangle \in L \iff
$$
$$
\exists\, u_1 \in \{0,1\}^{\ell_1(n)} \,\forall\, u_2 \in \{0,1\}^{\ell_2(n)} \,\exists\, u_3 \in \{0,1\}^{\ell_3(n)} \ldots Q\, u_k \in \{0,1\}^{\ell_k(n)} :
$$
$$
\varphi_n(x_0, u_1 \ldots u_k) \diamond \psi_n(x_1, u_1 \ldots u_k)
$$

Für gerade k bezeichnet dabei Q den Allquantor und $\diamond$ die logische „oder"-Verknüpfung, für ungerade k ist Q der Existenzquantor und $\diamond$ das logische „und". Eine Sprache ist in $C-\Pi_k$, wenn ihr Komplement in $C-\Sigma_k$ ist. Die *polynomielle Kommunikations-hierarchie*, $C-\mathcal{PH}$, ist die Vereinigung aller Stufen $C-\Sigma_k$ und $C-\Pi_k$:

$$C-\mathcal{PH} := \bigcup_{k \in \mathbb{N}} C-\Sigma_k = \bigcup_{k \in \mathbb{N}} C-\Pi_k$$

Zur Vereinfachung der Notation definieren wir noch $C-\Sigma_0 := C-\Pi_0 := C-\mathcal{P}$.

Eine weitere wichtige Kommunikationskomplexitätsklasse, die in [3] definiert wur-de, die Klasse $C-\mathcal{PP}$, steht im Zusammenhang mit der Anzahl akzeptierender Be-rechnungen nichtdeterministischer Protokolle. Solche Klassen werden deshalb auch *Zählklassen* genannt.

Definition 10 Eine Sprache L liegt in $C-\mathcal{PP}$, wenn es eine Folge nichtdeterminisiti-scher Protokolle $\mathcal{A}_n$ für Eingaben aus $\{0,1\}^{2n}$ gibt, so daß die Kommunikationskosten polynomiell im Logarithmus der Eingabelänge beschränkt sind und für die Anzahl der akzeptierenden bzw. verwerfenden Berechnungen gilt:

$$\langle x \rangle \in L \implies acc(\mathcal{A}_n, \langle x \rangle) > rej(\mathcal{A}_n, \langle x \rangle)$$
$$\langle x \rangle \notin L \implies rej(\mathcal{A}_n, \langle x \rangle) > acc(\mathcal{A}_n, \langle x \rangle)$$

4   Orakelprotokolle

Das für Turingmaschinen schon länger bekannte Konzept des *Orakels* wurde von BA-BAI, FRANKL und SIMON in [3] auf deterministische Kommunikationsprotokolle über-tragen. Im folgenden wollen wir ihre Definitionen kurz vorstellen und diese auch auf nichtdeterministische Protokolle erweitern.

In einem *Orakelprotokoll* mit einer Sprache Y bzw. einer Funktion f als Ora-kel können die beiden kommunizierenden Prozessoren zusätzlich Fragen der Form „$(q_0, q_1) \in Y$?" bzw. „Was ist $f(q_0, q_1)$?" an das Orakel stellen. Dabei hängt der Wert von q_i nur vom Wissensstand des Prozessors P_i ab, d.h. vom Wert seiner Ein-gabe, den bereits ausgetauschten Informationen und den Antworten des Orakels auf vorangegangene Fragen. Solche Protokolle, bei denen die Orakelfragen auch von den Antworten auf vorangegangene Fragen abhängen, heißen auch *adaptiv* oder *sequentiell*. Im Gegensatz dazu nennen wir Orakelprotokolle *nichtadaptiv* oder *parallel*, wenn die Orakelfragen als Funktion der lokalen Eingabe und der zu Beginn, vor irgendwelchen Orakelbefragungen ausgetauschten Informationen darstellbar sind. Die *Kosten* einer Berechnung eines Orakelprotokolls sind definiert als die Summe aus der Anzahl der ausgetauschten Bits und den einzelnen Kosten der Orakelbefragungen. Dabei sind die Kosten einer Orakelbefragung als die Summe der Länge der Orakelantwort und des Logarithmus der Länge der Frage, $\log |q_i|$, definiert. Diese Definition wird dadurch motiviert, daß sinnvollerweise die Länge der Orakelfragen mit in die Kosten eingehen

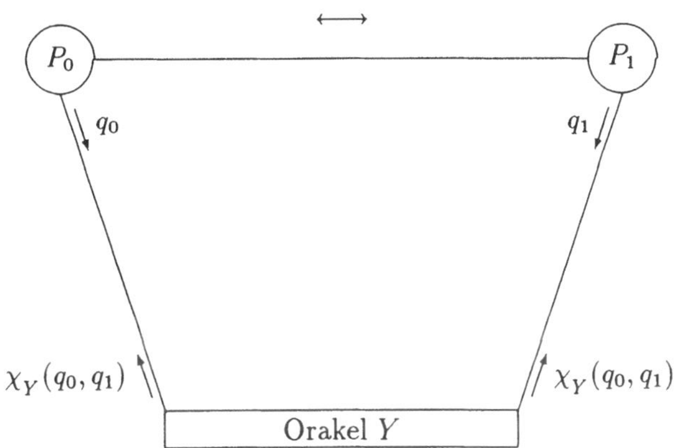

Abbildung 2: Kommunikation mit Orakel

muß; andererseits sollten die Fragen nicht mit ihrer vollen Länge zu Buche schlagen, da dann keine Verringerung der Kosten durch Orakel möglich wäre.

Mit C$-\mathcal{P}(Y)$ bezeichnen wir die Menge aller Sprachen L, die von einem deterministischen Orakelprotokoll mit Orakel Y mit höchstens polylogarithmischen Kosten erkannt werden können. Für eine Menge $\mathcal{C}$ von Orakeln definieren wir entsprechend C$-\mathcal{P}(\mathcal{C}) := \bigcup_{Y \in \mathcal{C}}$ C$-\mathcal{P}(Y)$. Betrachten wir nur Protokolle, die höchstens k Orakelfragen stellen, so bezeichnen wir die daraus resultierende Klasse mit C$-\mathcal{P}(\mathcal{C}[k])$. Für nichtadaptive Protokolle kennzeichnen wir die resultierenden Klassen durch das Zeichen „$\|$", also beispielsweise C$-\mathcal{P}(\mathcal{C} \| [k])$. Schränken wir uns auf Protokolle ein, die eine Eingabe genau dann akzeptieren, wenn die Orakelantworten alle positiv bzw. alle negativ sind, so kennzeichnen wir die entsprechenden Klassen mit C$-\mathcal{P}(\mathcal{C}+)$ bzw. C$-\mathcal{P}(\mathcal{C}-)$.

Die Definition von Orakelprotokollen kann auch auf nichtdeterministische Protokolle ausgedehnt werden. Dabei ist jedoch darauf zu achten, daß Orakelfragen wiederum nur vom aktuellen Wissensstand der Prozessoren abhängen dürfen. Würde man nämlich erlauben, daß die Prozessoren eine Frage nichtdeterministisch raten dürfen, ohne weitere Informationen auszutauschen, so könnten die Prozessoren mit der Menge $ID = \{(x_0, x_1) \mid x_0 = x_1\}$ als Orakel für ein nichtdeterministisches Protokoll jede Sprache L mit Kosten $1 + \log n$ akzeptieren. Dann könnte nämlich P_0 ein x mit $(x_0, x) \in L$ nichtdeterministisch raten und die Eingabe akzeptieren, falls das Orakel ihm bestätigt, daß er richtig geraten hat, daß also $x = x_1$ gilt.

Wir fordern also für nichtdeterministische Orakelprotokolle, daß die Orakelfragen jeweils als Funktion des lokalen Teils der Eingabe, der bereits ausgetauschten Nachrichten und der Orakelantworten auf vorangegangene Fragen darstellbar sind. Eine *Berechnung* α eines nichtdeterministischen Orakelprotokolls $\mathcal{A}(Y)$ mit Orakel Y auf

Eingabe $\langle x \rangle$ ist also eine Folge der Form

$$\alpha = w_{0,1} \ldots w_{0,m_0} a_1 w_{1,1} \ldots w_{1,m_1} a_2 \ldots a_k w_{k,1} \ldots w_{k,m_k}$$

Dabei ist $w_{i,\nu}$ eine von Prozessor P_j, $j = \nu + 1_2$, gesandte Nachricht, die nichtdeterministisch aus einer präfixfreien Menge von *zulässigen* Nachrichten gewählt werden kann:

$$w_{i,\nu} \in R_j(x_j, w_{0,1} \ldots w_{0,m_0} a_1 \ldots a_i w_{i,1} \ldots w_{i,\nu-1}) \subset \{0,1\}^*$$

$a_i = \chi_Y(q_0^i, q_1^i)$ ist die Antwort des Orakels auf die i-te Frage (q_0^i, q_1^i); dabei ist die Komponente q_j^i als Funktion der Eingabe x_j von Prozessor P_j und des Berechnungsanfangs vor der i-ten Frage darstellbar, also

$$q_j^i = f_j(x_j, w_{0,1} \ldots w_{0,m_0} a_1 \ldots a_{i-1} w_{i-1,1} \ldots w_{i-1,m_{i-1}}).$$

Die Berechnung α ist *akzeptierend*, falls für $j = j(\alpha)$ die *Entscheidungsfunktion* y_j des Prozessors P_j den Wert $y_j(x_j, \alpha) = 1$ liefert.

Analog zum deterministischen Fall bezeichnen wir mit $\mathrm{C}{-}\mathcal{NP}(Y)$ die Menge aller Sprachen L, für die es ein nichtdeterministisches Orakelprotokoll $\mathcal{A}$ mit polylogarithmischen Kosten gibt, so daß $\mathcal{A}(Y)$ die Sprache L akzeptiert. Entsprechend sind für Familien $\mathcal{C}$ von Orakelmengen und $k \in \mathbb{N}$ die Klassen $\mathrm{C}{-}\mathcal{NP}(\mathcal{C})$, $\mathrm{C}{-}\mathcal{NP}(\mathcal{C}[k])$ usw. definiert. Für die Zählklasse $\mathrm{C}{-}\mathcal{PP}$ definieren wir die relativierten Klassen $\mathrm{C}{-}\mathcal{PP}(Y)$ durch Orakel für die zugehörigen nichtdeterministischen Zählprotokolle.

Lemma 1 *Sei $\mathcal{A}(Y)$ ein nichtdeterministisches Protokoll mit Orakel Y. Dann gibt es ein nichtdeterministisches Orakelprotokoll $\mathcal{A}'(Y)$, das die gleiche Sprache wie $\mathcal{A}(Y)$ akzeptiert, höchstens doppelt so hohe Kosten hat und bei dem nur Prozessor P_0 eine Nachricht sendet, anschließend das Orakel befragt wird und schließlich P_1 akzeptiert bzw. verwirft.*

Beweis: Prozessor P_0 kann die gesamte Berechnung einschließlich der Antworten des Orakels nichtdeterministisch raten, wobei nur solche Folgen geraten werden, die für die Eingabe x_0 von P_0 zulässig sind und für die es eine Eingabe x_1' für P_1 gibt, so daß die Folge für x_1' zulässig ist. (Der lokale Teil der Fragen an das Orakel kann von P_0 aus dem Anfang der Berechnung und der Eingabe x_0 bestimmt werden.) Dann wird diese Berechnung an Prozessor P_1 gesandt. Ist diese Berechnung nicht für die Eingabe x_1 von P_1 zulässig, dann verhält sich P_1 zunächst so, als hätte er eine Eingabe x_1', auf der die geratene Folge eine zulässige Berechnung darstellt; schließlich verwirft er aber die Eingabe auf jeden Fall. Beide Prozessoren stellen nun entsprechend der geratenen Berechnung ihre Fragen an das Orakel Y und erhalten von diesem die Antwort. Stimmt die erhaltene Antwort nicht mit der zuvor geratenen überein, so wird die Berechnung abgebrochen und die Eingabe verworfen. Andernfalls wird mit der nächsten Orakelfrage fortgefahren bzw. – nach der letzten verifizierten Antwort – die Eingabe akzeptiert.   ■

Eine weitere interessante Eigenschaft nichtdeterministischer Orakelprotokolle, die wieder eine Analogie zu den Zeitkomplexitätsklassen für Turingmaschinen darstellt, ist die folgende in [5] gegebene Charakterisierung der polynomiellen Kommunikationshierarchie durch nichtdeterministische Orakelprotokolle.

Satz 4 $\mathrm{C}-\mathcal{NP}(\mathrm{C}-\Sigma_k) = \mathrm{C}-\mathcal{NP}(\mathrm{C}-\Sigma_k[1]-) = \mathrm{C}-\Sigma_{k+1}$.

Im Beweis dieses Satzes geht die im nachfolgenden Lemma genannte Abgeschlossenheitseigenschaft wesentlich ein.

Lemma 2 *Für alle* $k \in \mathbb{N}$ *ist* $\mathrm{C}-\Sigma_k$ *abgeschlossen unter disjunktiven und konjunktiven Rechteckreduktionen.*

Beweis: Sei $L' \in \mathrm{C}-\Sigma_k$ und seien $\ell'_1, \ldots, \ell'_k \in \mathrm{PLOG}$. Für $\langle x \rangle \in \{0,1\}^{2n}$ sei

$$\langle x \rangle \in L' \iff \exists\, u_1 \in \{0,1\}^{\ell'_1(n)} \forall\, u_2 \in \{0,1\}^{\ell'_2(n)} \ldots \mathcal{Q}\, u_k \in \{0,1\}^{\ell'_k(n)} :$$
$$\varphi_n(x_0, u_1 \ldots u_k) \diamond \psi_n(x_1, u_1 \ldots u_k)$$

Für gerade k bezeichnet dabei wieder $\mathcal{Q}$ den Allquantor und $\diamond$ die logische „oder"-Verknüpfung, für ungerade k ist $\mathcal{Q}$ der Existenzquantor und $\diamond$ das logische „und". Seien $p, \ell \in \mathrm{PLOG}$ und $f_i, g_i : \{0,1\}^* \to \{0,1\}^*$ mit $|f_i(x)| = |g_i(x)| = 2^{\ell(|x|)}$. Gilt (im Falle einer disjunktiven Rechteckreduktion) für alle $\langle x \rangle \in \{0,1\}^{2n}$

$$\langle x \rangle \in L \iff \exists\, i \in [1 : p(n)] : (f_i(x_0), g_i(x_1)) \in L',$$

so hat L die folgende $\mathrm{C}-\Sigma_k$-Darstellung:

$$\langle x \rangle \in L \iff \exists\, (u_1, i) \in \{0,1\}^{\ell_1(n) + \lceil \log p(n) \rceil} \forall\, u_2 \in \{0,1\}^{\ell_2(n)} \ldots \mathcal{Q}\, u_k \in \{0,1\}^{\ell_k(n)} :$$
$$\varphi_{2^{\ell(n)}}(f_i(x_0), u_1 \ldots u_k) \diamond \psi_{2^{\ell(n)}}(g_i(x_1), u_1 \ldots u_k)$$

Dabei ist $\ell_j(n) := \ell'_j(2^{\ell(n)})$. Liegt dagegen eine konjunktive Rechteckreduktion vor, gilt also für alle $\langle x \rangle \in \{0,1\}^{2n}$

$$\langle x \rangle \in L \iff \forall\, i \in [1 : p(n)] : (f_i(x_0), g_i(x_1)) \in L',$$

so hat L die $\mathrm{C}-\Sigma_k$-Darstellung

$$\langle x \rangle \in L \iff$$
$$\exists\, (u_1^1, \ldots, u_1^{p(n)}) \in \{0,1\}^{\ell_1(n) \cdot p(n)} \forall\, (u_2, i) \in \{0,1\}^{\ell_2(n) + \lceil \log p(n) \rceil} \ldots \mathcal{Q}\, u_k \in \{0,1\}^{\ell_k(n)} :$$
$$\varphi_{2^{\ell(n)}}(f_i(x_0), u_1^i \ldots u_k) \diamond \psi_{2^{\ell(n)}}(g_i(x_1), u_1^i \ldots u_k)$$

Man beachte, daß dies auch für $k = 1$ eine $\mathrm{C}-\Sigma_1$-Darstellung ist, da der Allquantor dann nur über einen Bereich polylogarithmischer (statt quasipolynomieller) Größe quantifiziert. ∎

Mit Hilfe des vorangegangenen Satzes läßt sich nun auch leicht die folgende Abschlußeigenschaft unter deterministischen Orakelreduktionen zeigen, die nach einer Bemerkung von TODA in [8] auch im klassischen Fall gilt.

Satz 5 $\mathrm{C}-\mathcal{P}(\mathrm{C}-\Sigma_k \cap \mathrm{C}-\Pi_k) = \mathrm{C}-\Sigma_k \cap \mathrm{C}-\Pi_k$.

Beweis: Die Inklusion in der Richtung „$\supseteq$" ist offensichtlich. Für die andere Richtung genügt es, die Inklusion in $\mathrm{C}-\Sigma_k$ zu zeigen. Sei $L_1 \in \mathrm{C}-\Sigma_k \cap \mathrm{C}-\Pi_k$, $L \in \mathrm{C}-\mathcal{P}(L_1)$ und $(\mathcal{A}_n^1(L_1))_{n\in\mathbb{N}}$ eine entsprechende Folge deterministischer Orakelprotokolle für L. Wir konstruieren eine Folge von $\mathrm{C}-\mathcal{NP}(\mathrm{C}-\Sigma_{k-1})$-Protokollen $(\mathcal{A}_n^2(L_2))_{n\in\mathbb{N}}$ für L wie folgt: Als L_2 wählen wir eine $\mathrm{C}-\Sigma_{k-1}$-vollständige Sprache. Für das nichtdeterministische Protokoll $\mathcal{A}_n^2(L_2)$ rät Prozessor P_0 eine für seinen Teil der Eingabe zulässige Berechnung des deterministischen Protokolls $\mathcal{A}_n^1(L_1)$ inklusive der Orakelantworten. Der lokale Teil Orakelfragen ergibt sich deterministisch aus der Eingabe und der Berechnung mit den vorangegangenen Antworten. Diese geratene Berechnung wird dann an Prozessor P_1 geschickt. Die geratenen Antworten des Orakels müssen nun verifiziert werden. Für positive Antworten nutzt man, daß $L_1 \in \mathrm{C}-\Sigma_k$ ist: Eine solche Antwort kann mit einem $\mathrm{C}-\mathcal{NP}(\mathrm{C}-\Sigma_{k-1})$-Protokoll verifiziert werden. Für die negativen Antworten wird genutzt, daß auch $L_1 \in \mathrm{C}-\Pi_k$ ist. Statt eine negative Antwort für das Orakel L_1 zu verifizieren kann man nun eine positive Antwort für das Komplement $\overline{L_1}$ verifizieren. Da $\overline{L_1} \in \mathrm{C}-\Sigma_k$ ist, ist dies wieder mit einem $\mathrm{C}-\mathcal{NP}(\mathrm{C}-\Sigma_{k-1})$-Protokoll möglich. Schließlich akzeptiert Prozessor P_1, wenn die geratene Berechnung auch für seine Eingabe zulässig ist, die Orakelantworten richtig geraten wurden und die so gefundene, eindeutige Berechnung des deterministischen Protokolls $\mathcal{A}_n^1(L_1)$ akzeptierend ist.   ■

Wie bei der Charakterisierung der polynomiellen Kommunikationshierarchie läßt sich auch für andere Klassen von Orakelmengen die Anzahl der Fragen eines $\mathrm{C}-\mathcal{NP}$-Orakelprotokolls auf eine einzige Frage begrenzen.

Satz 6 *Sei $\mathcal{C}$ eine Kommunikationskomplexitätsklasse, die abgeschlossen ist unter Komplementbildung und konjunktiven Rechteckreduktionen. Ferner gebe es eine $\mathcal{C}$-vollständige Sprache V. Dann ist $\mathrm{C}-\mathcal{NP}(\mathcal{C}) = \mathrm{C}-\mathcal{NP}(V[1])$.*

Beweis: Sei $L_1 \in \mathcal{C}$, $(\mathcal{A}_n(L_1))_{n\in\mathbb{N}}$ eine Folge von $\mathrm{C}-\mathcal{NP}$-Orakelprotokollen und L die von $(\mathcal{A}_n(L_1))_{n\in\mathbb{N}}$ akzeptierte Sprache. Wir zeigen, wie das Protokoll $\mathcal{A}_n$ so modifiziert werden kann, daß *eine* Frage an das Orakel V genügt. Da V $\mathcal{C}$-vollständig ist, können wir das Orakel L_1 durch V ersetzen, ohne die Kosten wesentlich zu erhöhen. Aus diesem Protokoll konstruiert man nun ein Einweg-Protokoll für L. Dazu rät Prozessor P_0 eine für seine Eingabe zulässige Berechnung inklusive der Orakelantworten und sendet diese an P_1. Prozessor P_1 prüft nun, ob diese Berechnung für seine Eingabe zulässig und akzeptierend ist. Ist das nicht der Fall, so sendet er eine 0 an Prozessor P_0; die Berechnung wird dann abgebrochen und die Eingabe wird verworfen. Andernfalls sendet P_1 eine 1 an P_0 und die geratenen Antworten des Orakels werden verifiziert. Dazu werden die Fragen, für die negative Antworten geraten wurden, in Fragen an das Orakel $\overline{V}$ geändert, die nun für akzeptierende Berechnungen positiv beantwortet werden sollten. Diese veränderten Fragen können mit Hilfe des Orakels V beantwortet werden, da $\mathcal{C}$ unter Komplementbildung abgeschlossen ist und V $\mathcal{C}$-vollständig ist. Um herauszufinden, ob alle Orakelfragen positiv beantwortet werden,

werden diese Fragen nun als konjunktive Rechteckreduktion betrachtet. Die dadurch definierte Sprache ist nach Voraussetzung wieder in C enthalten und kann daher auf V reduziert werden. Es muß also nur eine Frage an das Orakel V gestellt werden, und die Eingabe kann akzeptiert werden, wenn diese Frage positiv beantwortet wird. ∎

Um den vorangegangenen Satz auf die Klasse $C-\mathcal{PP}$ anwenden zu können, benötigen wir die Abgeschlossenheit dieser Klasse unter konjunktiven (oder disjunktiven) Rechteckreduktionen. Diese Eigenschaft wurde für das klassische Vorbild von BEIGEL, REINGOLD und SPIELMAN in [4] bewiesen und ist mit den entsprechenden Anpassungen auch für unser Modell gültig.

Satz 7 $C-\mathcal{PP}$ *ist abgeschlossen unter disjunktiven und konjunktiven Rechteckreduktionen.*

$C-\mathcal{PP}$ ist eine Komplexitätsklasse, die unter Komplementbildung und disjunktiven Rechteckreduktionen abgeschlossen ist. Das Majoritätsproblem

$$MAJ := \{(x,y) \mid \sum_{i=1}^{n} x_i \cdot y_i \geq \frac{n}{2}\}$$

ist $C-\mathcal{PP}$-vollständig. Die Voraussetzungen von Satz 6 sind also erfüllt. Damit folgt:

Korollar 1 $C-\mathcal{NP}(C-\mathcal{PP}) = C-\mathcal{NP}(MAJ[1])$.

Anzumerken ist noch, daß dieses Ergebnis auch für den klassischen Fall gilt, daß also $\mathcal{NP}(\mathcal{PP}) = \mathcal{NP}(\mathcal{PP}[1])$ ist. Dieses Ergebnis war bisher anscheinend nicht bekannt.

Wie der folgende Satz zeigt, ist die oben vorausgesetzte Existenz von C-vollständigen Sprachen keine notwendige Bedingung, um nichtdeterministische Orakelprotokolle auf eine einzige Frage an ein C-Orakel einschränken zu können. Vielmehr genügt auch der Abschluß der Klasse C unter deterministischen Orakelreduktionen.

Satz 8 *Für alle Orakelmengen Q gilt:*

$$C-\mathcal{NP}(Q) \subseteq C-\mathcal{NP}(C-\mathcal{P}(Q)\,[1])$$

Beweis: Sei $\mathcal{A}$ ein $C-\mathcal{NP}$-Orakelprotokoll und L die von $\mathcal{A}(Q)$ akzeptierte Sprache. Für Eingaben $\langle x \rangle \in \{0,1\}^{2n}$ sei $C_n(\langle x \rangle)$ die Menge der möglichen Berechnungen von $\mathcal{A}$ auf Eingabe $\langle x \rangle$, wobei wir die Orakelantworten (jedoch nicht die Fragen) wieder als Teil der Berechnung betrachten. Dabei fixieren wir noch kein Orakel, sondern betrachten die möglichen Berechnungen von $\mathcal{A}(X)$ für variable Orakelmengen X. Da $\mathcal{A}$ nur polylogarithmische Kosten hat, haben die Berechnungen c höchstens polylogarithmische Länge, und es ist möglich, mit höchstens polylogarithmischem Kommunikationsaufwand deterministisch zu überprüfen, ob ein gegebenes c in $C(\langle x \rangle)$ liegt. Die Orakelantworten in c müssen dazu nicht verifiziert werden, da ja kein Orakel fixiert wurde, sie können aber auf Konsistenz geprüft werden. Wir nehmen ohne Einschränkung an, daß als letztes Bit jeder Berechnung c ihr Ergebnis gesandt wird; sei $acc(c)$ das entsprechende Prädikat, das genau dann wahr ist, wenn das letzte Bit von c eine 1 ist.

Für eine Eingabe $\langle x \rangle$ und eine Berechnung $c \in C_n(\langle x \rangle)$ seien $\langle q^i(\langle x \rangle, c) \rangle$ für $i \in [1 : s(n)]$ die Orakelfragen für die Berechnung c auf $\langle x \rangle$; weiter sei $a^i(c)$ die i-te Orakelantwort in c. Dabei nehmen wir der Einfachheit halber an, daß für jede Eingabe der Länge n die gleiche Anzahl $s(n)$ von Orakelfragen gestellt wird. Wir definieren nun eine Menge $Y = \bigcup_{n \in \mathbb{N}} Y_n$ wie folgt:

$$\langle x, (c, c), r^1, \ldots, r^k \rangle \in Y_n : \iff \quad |x| = 2n \wedge k = s(n) \wedge c \in C_n(\langle x \rangle) \wedge acc(c) \wedge$$
$$\forall i \in [1 : k] : (\langle r^i \rangle = \langle q^i(\langle x \rangle, c) \rangle \wedge \chi_Q(\langle r^i \rangle) = a^i(c))$$

Dann ist $Y \in$ C$-\mathcal{P}(Q)$: In einem deterministisches Orakelprotokoll kann auf Eingabe $\langle x, (c, c'), r^1, \ldots, r^k \rangle$ mit $|x| = 2n$ zunächst Prozessor P_0 seinen Teil c an P_1 senden, der dann überprüft ob $c = c'$ ist und auch $acc(c)$ gilt. Dann prüfen beide deterministisch, ob $c \in C(\langle x \rangle)$ liegt. Anschließend prüfen die beiden Prozessoren P_ν lokal, ob $r^i_\nu = q^i_\nu(x_\nu, c)$ für alle $i \in [1 : k]$ gilt. Schließlich werden dem Orakel Q die Fragen r^i für $i \in [1 : k]$ gestellt, und als Ergebnis wird 1 berechnet, falls die Orakelantworten mit den erwarteten Antworten $a^i(c)$ übereinstimmen.

Ein nichtdeterministisches Protokoll $\mathcal{A}'$, dem als Orakel die Menge Y zur Verfügung steht, kann dies wie folgt verwenden, um die Sprache L zu erkennen: Auf Eingabe $\langle x \rangle$ rät Prozessor P_0 eine Berechnung c des Protokolls $\mathcal{A}$ und überträgt diese an P_1. Aus dieser Berechnung und der lokalen Eingabe berechnen dann beide Prozessoren P_ν die Komponenten $q^i_\nu(x_\nu, c)$ der Orakelfrage

$$\langle x, (c, c), \langle q^1(\langle x \rangle, c) \rangle, \ldots, \langle q^{s(n)}(\langle x \rangle, c) \rangle \rangle$$

Die Eingabe wird genau dann akzeptiert, wenn das Orakel die Frage positiv beantwortet. Da zwischen den Prozessoren nur polylogarithmisch viele Bits übertragen werden und die Orakelfrage nur quasipolynomielle Länge hat, folgt $L \in$ C$-\mathcal{NP}(Y[1])$. ∎

Mit Hilfe dieses Satzes erhält man das folgende Ergebnis, dessen Entsprechung nach [10] auch für den klassischen Fall der relativierten Zeitkomplexitätsklassen gilt.

Korollar 2 C$-\mathcal{NP}($C$-\mathcal{BPP}) =$ C$-\mathcal{NP}($C$-\mathcal{BPP}[1])$.

Beweis: Offenbar ist C$-\mathcal{P}($C$-\mathcal{BPP}) =$ C$-\mathcal{BPP}$. Nach dem vorangegangenen Satz gilt daher C$-\mathcal{NP}($C$-\mathcal{BPP}) \subseteq$ C$-\mathcal{NP}($C$-\mathcal{BPP}[1])$. Die Umkehrung der Inklusion ist offensichtlich. ∎

Literaturverzeichnis

[1] H. Abelson, Lower bounds on information transfer in distributed computations *Proceedings of the 19th Annual IEEE Symposium on Foundations of Computer Science*, 151–158, 1978

[2] A. V. Aho, J. D. Ullman, M. Yannakakis, On notions of information transfer in VLSI circuits *Proceedings of the 15th Annual ACM Symposium on Theory of Computing*, 133–139, 1983

[3] L. Babai, P. Frankl, J. Simon, Complexity classes in communication complexity *Proceedings of the 27th Annual IEEE Symposium on Foundations of Computer Science*, 337–347, 1986

[4] R. Beigel, N. Reingold, D. Spielman, *PP is closed under intersection*, Technischer Bericht YALEU/DCS/ TR-803, Yale University, 1990

[5] B. Halstenberg, R. Reischuk, Relations between communication complexity classes *Journal of Computer and System Sciences* **41** (1990), 402–429, Academic Press

[6] M. Karchmer, A. Wigderson, Monotone circuits for connectivity require super-logarithmic depth *Proceedings of the 20th Annual ACM Symposium on Theory of Computing*, 539–550, 1988

[7] R. J. Lipton, R. Sedgewick, Lower bounds for VLSI *Proceedings of the 13th Annual ACM Symposium on Theory of Computing*, 300–307, 1981

[8] S. Toda, *PP is $\leq_T^p$-hard for the polynomial-time hierarchy*, Technischer Bericht, Tokyo University of Electro-communications, Tokio, 1989

[9] A. C. Yao, Some complexity questions related to distributed computing *Proceedings of the 11th Annual ACM Symposium on Theory of Computing*, 209–213, 1979

[10] S. Zachos, Probabilistic quantifiers, adversaries, and complexity classes: an overview *Proceedings Structure in Complexity Theory 1st Annual Conference*, 383–400, 1986

Performance Optimization of Combinational Circuits

Uwe Hinsberger, Reiner Kolla

Institut für Informatik – Abteilung VI
Rheinische Friedrich-Wilhelms-Universität
5300 Bonn
Germany

Abstract

Performance optimization, i.e. the problem of finding an optimal investment of transistor area which meets given delay constraints, is considered from an abstract, cell based point of view which allows only solutions within a discrete solution space of coarse granularity. The main advantages of this problem formulation are the independence of the methods from concrete delay modelling (and thus from technology) and the applicability to even very restrictive design styles (as for example gate arrays or sea of gates). Our approach can be considered as a discrete version of the transistor sizing problem on one hand and generalizes to the library mapping problem on the other hand. The paper presents optimal dynamic programming algorithms for trees and heuristics together with first experimental results for general combinational circuits.

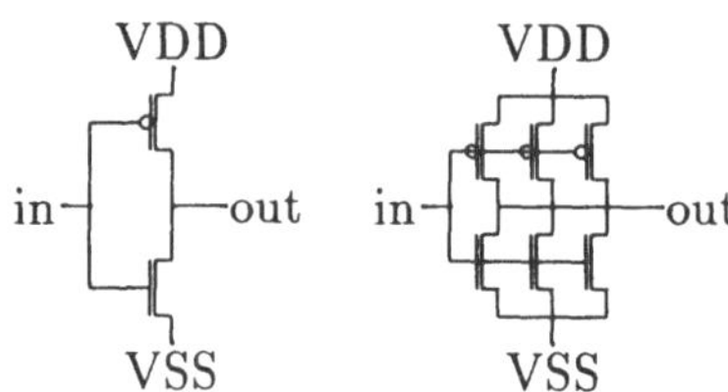

Figure 1: Different implementations of an inverter with fixed transistors

1   Introduction

One problem on the way from the logic design of a digital VLSI circuit to its physical
implementation is to make an optimal choice of transistor sizes in order to meet area,
power dissipation and delay constraints. How and whether this problem is solved
depends on the design style. If a full custom layout is developed, the designer is
responsible for the right choice of transistor sizes. He may use tools for simulation
[21] or timing analysis [23] in order to find the critical paths of his design, to verify
delay constraints and to control the effect of changing transistor sizes. For complex
VLSI circuits this manual approach is not only tedious and time consuming but will
probably not produce optimal sizes. More recent developments of algorithms and
tools offer automatic optimal sizing capabilities for critical paths or graphs. They are
based on MOS specific analytical delay formulas and translate the sizing problem into
a nonlinear program [6, 7, 16, 20, 22], a linear program [4] or use heuristic approaches
[22, 27]. As a consequence these tools change transistor widths continuously or at least
by very small amounts, because they compute optima or approximations of optima
within a continuous solution space. Therefore, this approach is only useful for full
custom design styles together with powerful layout generation and verification tools.

For semi custom design styles as gate arrays or standard cells, performance optimization by transistor sizing is impossible. Here the set of cells is fixed and if the
implementation of a circuit over a set of cells does not meet certain timing conditions,
its structure has to be changed. For standard cells more flexible libraries with different implementations (with respect to transistor sizes and electrical circuit structure)
of the same boolean function are possible. For gate arrays the transistor sizes are
determined by the master. Nevertheless, even in this case different implementations
of the same function with respect to pull-up and pull-down impedance and capacitive
load are possible e.g. by parallel connection of transistors (see figure 1).

In [9, 10] we proposed a cell based approach to the sizing problem which can also
be applied in such cases. We presented an abstract model which considers several
(but only a finite number of) possibilities to realize each gate of a circuit; the different
implementations of a gate may differ in delay and area. Such a discrete approach
to sizing (also investigated in [15]) is not only suitable for different design styles but
also (nearly) independent of delay modelling. In opposition to [15] who proposed

a global heuristic we tackled the discrete sizing problem by means of an optimum dynamic programming algorithm for trees and a heuristic for combinational circuits which uses this algorithm to speed up iteratively critical paths or trees. It has been shown independently in [13, 24, 26] that dynamic programming on trees also works well for the purpose of library mapping; i.e. for finding an optimum realization of a network which is specified over and/or/not primitives by means of more complex cells. Therefore library mapping and sizing of fanout-free regions can be done by the same methods. However it is still an open question whether this pays and how the heuristics for combinational circuits harmonize. In this paper we want to consider those problems.

Section 2 briefly repeats our sizing model. In section 3 we summarize our previous results on computational complexity of optimum sizing, our original algorithms for optimum sizing of trees and the most profitable heuristic we found so far for general combinational circuits. Then we extend our model and algorithms in such a way that they can also perform library mapping (section 4). We conclude in section 5 with a comparision of different methods to combine library mapping and sizing.

2   Model and problem definition

We consider combinational circuits build up of gates with k inputs and one output. It is well known that such a structure can be formally described by a directed acyclic graph. The following section contains a short summary of this formal representation together with the notation which we will use in the rest of this paper:

Definition 1 *A combinational circuit C is a directed acyclic graph $C = (V, E)$. The set of nodes V is given by*

$$V := PI \cup PO \cup G$$

where G is the set of gates, PI is the set of primary inputs and PO is the set of primary outputs.
There is an edge $(u, v) \in E$ if and only if the output of gate u (the primary input u) is connected with an input of gate v (the primary output v). For each $v \in V$ $succ(v)$ is the set of nodes w with $(v, w) \in E$, $pred(v)$ denotes the set of nodes u with $(u, v) \in E$ and $indeg(v) := \sharp pred(v), outdeg(v) := \sharp succ(v)$ are the number of incoming resp. outgoing edges of v.

Let us suppose that the set of sink nodes, i.e. nodes u with $outdeg(u) = 0$, is exactly the set of primary outputs and the set of source nodes $(indeg(u) = 0)$ is exactly the set of primary inputs. Figure 2 shows the circuit diagram and the graph representation which corresponds to a "nand"-implementation of a fulladder. The cell based sizing problem as mentioned above consists in choosing an implementation $\sigma(u)$ out of a finite set of possible implementations for each gate $u \in G$.

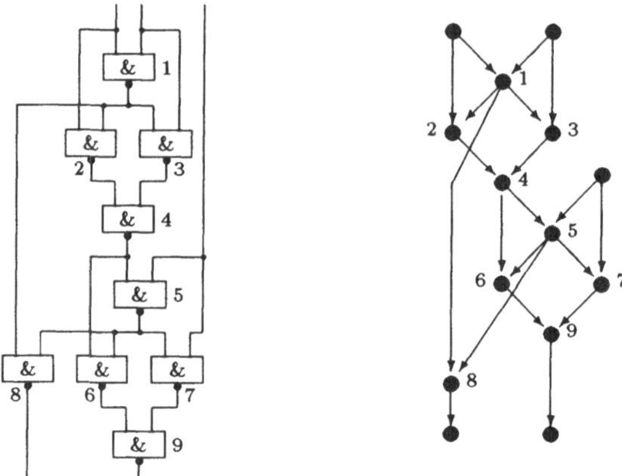

Figure 2: Circuit diagram and graph representation of a fulladder circuit

Definition 2 *Let C be a combinational circuit. A sizing of C is an assignment*

$$\sigma : G \longrightarrow \{0, \ldots, s-1\}$$

of implementation numbers to the gates of C. The area of a sizing σ is

$$area(\sigma) := \sum_{u \in G} area_u\big(\sigma(u)\big).$$

The area in this context means the cell area, not the area of the complete layout. We use this simple bound for the layout area, because it is not easy to analyse the interaction between the choice of a sizing and the final layout. The second measure of performance will be the delay of the circuit. The delay depends on the choice of a sizing as well as on physical properties of the primary inputs and primary outputs. Furthermore, it is difficult to determine the worst case delay of a circuit, because there are possibly different critical paths for different input patterns. A simple way to overcome this problem is to analyse the delay pattern independently by taking into account each potential path. This always yields an upper bound for the delay but it can be too pessimistic considering so called false paths [5, 18, 17]. For optimization problems, however, it is convenient to use this simple pattern independent approach, because it makes optimization algorithms feasible and we may hope that a good solution for a pessimistic delay model is also a good solution for a realistic model.

Definition 3 *Let Σ be the set of all possible sizings of a circuit $C = (V, E)$. The delay of an edge (u, v) is an assignment*

$$\delta_{u,v} : \Sigma \longrightarrow \mathcal{R}_0^+$$

of nonnegative real numbers, where $\delta_{u,v}(\sigma)$ is the delay between a transition of the output of gate u (the primary input u) and a transition of the output of gate v. This assignment depends on the physical properties of the gates and thus on the choice of a sizing σ. Since for a primary output v and $(u,v) \in E$ a transition of the output of u is already a transition of v, we set $\delta_{u,v}(\sigma) := 0$ for each $v \in PO$ and each sizing σ. The transition time of a primary input depends also on the sizing of the gates. In order to take this into account we consider an arrival time assignment

$$t_u : \Sigma \longrightarrow \mathcal{R}_0^+$$

which assigns an arrival time $t_u(\sigma)$ to each primary input $u \in PI$ subject to the sizing σ.

From this delay and arrival time assignments, the timing of a circuit can be modeled independent of input patterns by "longest" paths (if $\delta_{u,v}(\sigma)$ is considered as "length" of (u,v)).

Definition 4 *A sequence $p = u_1, \ldots, u_k$ of nodes $u_i \in V$ is a path with $source(p) = u_1$ and $sink(p) = u_k$ if $(u_i, u_{i+1}) \in E$ for all $1 \leq i < k$. The delay of a path p is*

$$\delta_p(\sigma) = \sum_{i=1}^{k-1} \delta_{u_i,u_{i+1}}(\sigma)$$

The worst case delay $d_{u,v}(\sigma)$ between node u and node v is either 0, if there is no path with source u and sink v, or the maximum of $\delta_p(\sigma)$ over all paths p with source u and sink v. The worst case delay of a circuit C for a sizing σ is

$$\Delta_C(\sigma) := \max\left\{\delta_p(\sigma) + t_{source(p)}(\sigma) \mid source(p) \in PI, sink(p) \in PO\right\}$$

In this paper we do not want to discuss how delay assignment can be done, because the results are more or less independent of any concrete delay model. Notice that each edge has an individual delay assignment. Therefore, it can take into account parasitic effects of the wires which realize the connections as well as individual physical properties of the sink gate of the edge and its load. Different delays between different inputs and the output of a gate can also be included if we keep track of which input corresponds to which edge.

In general, the problem of performance optimization consists in choosing a sizing in such a way that the delay is minimized for a given bound of the transistor area (or energy consumption) or the area is minimized for a given worst case delay constraint. We will present a dynamic programming approach to these problems in section 3.1. This approach requires that the delay of an edge may not depend on a sizing in an arbitrary way. What we need is to be able to compute the delay locally. In other words, if we know for an edge (u,v) the sizing of the driving gate (this is gate v for the delay between an output transition of u and an output transition of v) and the load at the output of the driving gate (the load can be computed from the sizing $\sigma(w)$ for each $w \in succ(v)$), then we know the delay. This condition is formally given by the following

Definition 5 *A delay assignment $\delta_{u,v}(\sigma), t_u(\sigma)$ is locally determined if for all sizings σ_1 and σ_2 and for all edges $(u,v) \in E$ the condition*

$$\text{if } \sigma_1(x) = \sigma_2(x) \text{ for all } x \in \{v\} \cup succ(v) \text{ then } \delta_{u,v}(\sigma_1) = \delta_{u,v}(\sigma_2)$$

and for all primary inputs $u \in PI$ the condition

$$\text{if } \sigma_1(x) = \sigma_2(x) \text{ for all } x \in succ(u) \text{ then } t_u(\sigma_1) = t_u(\sigma_2)$$

holds.

Within this abstract but simple model we consider the following performance optimization problems:

Problems:

- *Unconstrained optimal sizing:* Find a sizing σ with optimal time performance.

- *Constrained optimal sizing:* Consider a circuit C and a bound X.
  Find a "fastest" sizing σ with $area(\sigma) \leq X$ (minimize delay under area constraint).
  Find a "smallest" sizing σ with $\Delta_C(\sigma) \leq X$ (minimize area under delay constraint).

There are several papers in the recent literature which also consider sizing problems from a discrete, cell based point of view. The delay models in [13, 19] are restrictions of our delay model because they do not take different capacitive loads for different sizings into account. In [19] this is not necessary because they consider sizing problems for CMOS domino logic with different output driver sizes. In [13] the main objective is the mapping of a boolean network to complex gates. An extension of [13] to our delay model together with optimal sizing algorithms for trees has been introduced independently in [24].

3 Sizing algorithms

In [9, 8] we proved that constrained optimal sizing of general combinational circuits is strongly NP-complete. Furthermore we could show that even for simple chains the simpler unconstrained problem still remains weakly NP-complete. From these results we can conclude that

1. Optimal sizing algorithms for general combinational circuits are not feasible in practice.

2. For circuits with even a special structure (e.g. fanout-free circuits) optimal algorithms have at least pseudopolynomial time complexity.

The next section will demonstrate in fact that for tree-like circuits optimal sizings can be computed by dynamic programming if the delay assignment is locally determined.

3.1 Optimal algorithms for trees

Suppose that the circuit C is a tree. Then the delay of an edge (u, v) only depends on the sizing of two nodes, namely v and the successor of v (which is uniquely determined since C is a tree). We can rewrite the delay assignment for an edge (u, v) in this case as a function

$$\delta_{u,v}(\sigma) := \delta_{u,v}\big(\sigma(v), \sigma(succ(v))\big)$$

in two variables and the arrival time of a primary input u as a function

$$t_u(\sigma) := t_u\big(\sigma(succ(u))\big)$$

in one variable. As a consequence, we can consider the sizing problem for a subtree with root v independently from the rest of the tree, if we know the load which is imposed by the sizing of $succ(v)$. In other words, if we are looking for an optimal sizing $\overline{\sigma}$ of the whole tree and v is an inner node, then we can assume that $\overline{\sigma}$ is an optimal sizing of the subtree rooted at v with respect to the condition, that the successor of v has size $\overline{\sigma}(succ(v))$. So it is near at hand to compute bottom up from the leaves (primary inputs) to the root (the primary output) for each possible load imposed by a sizing of $succ(v)$ an optimal sizing of the tree rooted at v. Let $\tau(v, y)$ be the optimal arrival time of a signal at node v subject to $\sigma(succ(v)) = y$. Then, what we have to do in order to compute $\tau(v, y)$ is to try out each possible sizing x of v, to compute the delay of the output of v subject to "load" y and to time optimal sizings of the subtrees, and to take the best x:

$$(*) \qquad \tau(v, y) = \begin{cases} t_v(y) & v \in PI \\ \min_{x} \max_{u \,\in\, pred(v)} \big(\tau(u, x) + \delta_{u,v}(x, y)\big) & v \notin PI \end{cases}$$

The computation of this optimal arrival time for all nodes can obviously be done in time $O(n \cdot s^2)$, where n is the number of nodes and s is the maximum number of possible sizings per gate. A recovery of an optimal sizing is also straightforward if we keep for each $\tau(v, y)$ the sizing $x = S(v, y)$ of v which yields the minimum in $(*)$. (This computation of $\tau(v, y)$ simplifies to $\tau(v)$ if $succ(v)$ is a primary output, because the output load is usually fixed. In the rest of this section, we will not distinguish this special case. It is also possible to compute a sizing profile for different load situations at the primary output as is done for all other inner nodes.)

The problem becomes a little bit more difficult if we have constraints to resources like time or area. If we want to use again dynamic programming we have to discretize the constrained parameter. This directly yields an algorithm for delay constraints, because the delay as a resource must not be distributed to the subtrees: If we want to compute the optimal area $a(v, t, y)$ of a sizing of a subtree rooted at v subject to the delay constraint t and load y, we can do it again bottom up by trying out each possible sizing x of v and taking the best.

$$(**) \quad a(v, t, y) = \begin{cases} \infty & \text{if } v \in PI,\ t_v(y) > t \\ 0 & \text{if } v \in PI,\ t_v(y) \le t \\ \min_{x} \Big\{ area_v(x) + \sum_{u \in pred(v)} a\big(u, t - \delta_{u,v}(x, y), x\big) \Big\} & v \notin PI \end{cases}$$

Herein "∞" is a sufficiently large number indicating that no solution is possible. The time complexity to compute $a(v, t, y)$ for each node u of the tree and each of the at most s different sizings of the successor (the "load" y) is $O(n \cdot \Delta T \cdot s^2)$ where ΔT is the maximum number of different discrete times which can occur for a node u. This algorithm can be considered as a simplification of the *minimum_area_under_delay_constraint* procedure of [24]. In [24] the area is minimized for coverings of the tree with subtrees from a set of tree patterns which represent complex gates. In section 4 we will discuss a generalization to this case.

In opposition to delay constraints, for area constraints there is the problem that the resource "area" must be distributed optimally to the predecessor trees, i.e. if we discretize the area and consider the optimum delay $\tau(v, a, y)$ of a subtree rooted at node v which is loaded by y for a sizing which needs at most area a within this subtree, a naive adaptation of the case $v \notin PI$ in (**) would yield

$$\tau(v, a, y) = \min_{x} \quad \min_{a_1, \dots, a_{indeg(v)}} \quad \max_{u\, \in\, pred(v)} \Big(\tau(u, a_i, x) + \delta_{u,v}(x, y) \Big)$$

where we consider the minimum over all distributions $a_1, \dots, a_{indeg(v)}$ with

$$area_v(x) + \sum_{i=1}^{indeg(v)} a_i \le a$$

because the area which is still available has to be distributed optimally to the predecessor trees. If we look at all possible distributions this would result in a time complexity of $O(n \cdot s^2 \cdot \Delta A^m)$ where m is the maximum indegree of a node and ΔA is the difference between the maximum area and minimum area of a possible sizing of the tree, i.e. we may consider a as the area overhead instead of the area itself. The following lemma shows that it is not necessary to try out all possible distributions of the area to the subtrees rooted at the predecessors:

Lemma 1 *Let $a_1, \dots, a_k$ be an optimal distribution of area $a = area_v(x) + \sum_{i=1}^{k} a_i$ into the k subtrees rooted at the predecessors $u_1, \dots, u_k$ of a node v for $\sigma(v) = x$ and load y. Let u_s be a "slowest" (with respect to the delay of v) predecessor of v, i.e.*

$$\tau(u_s, a_s, x) + \delta_{u_s,v}(x, y) = \max_{1 \le i \le k} \Big(\tau(u_i, a_i, x) + \delta_{u_i,v}(x, y) \Big).$$

If each a_i satisfies

$$(I) \quad \tau(u_i, a_i - 1, x) + \delta_{u_i,v}(x, y) \ge \tau(u_s, a_s, x) + \delta_{u_s,v}(x, y)$$

then $a'_1, \ldots, a'_n$ with $a'_i = \begin{cases} a_i & i \neq s \\ a_s + 1 & i = s \end{cases}$ is an optimum distribution of area $a + 1$
for $\sigma(v) = x$ and load y and satisfies (I).

Proof: We first argue that $a'_1, \ldots, a'_k$ is an optimum distribution of area $a + 1$: The only way to achieve a smaller arrival time is to invest more area into a slowest subtree, i.e. it only becomes smaller if $a'_s > a_s$. Since $a_1, \ldots, a_k$ satisfy (I) it is impossible to take away any amount of area from any a_i without making the arrival time of v greater or equal to $\tau(u_s, a_s, x) + \delta_{u_s,v}(x, y)$. Therefore $a'_1, \ldots, a'_k$ is an optimum distribution of area $a + 1$.

In order to see that $a'_1, \ldots, a'_k$ again satisfy (I) we only have to check (I) for a'_s. But

$$\begin{aligned} \tau(u_s, a'_s - 1, x) + \delta_{u_s,v}(x, y) &= \tau(u_s, a_s, x) + \delta_{u_s,v}(x, y) \\ &\geq \max_{1 \leq i \leq k} \tau(u_i, a'_i, x) + \delta_{u_i,v}(x, y) \end{aligned}$$

since u_s was the "slowest" predecessor.   ■

By this lemma we know that if we invest the area unit by unit always into the "slowest" predecessor subtree, this strategy fulfills the invariant (I) and yields an optimal distribution. During the computation of $\tau(v, a, y)$ we only have to store for each x, y the distribution of area and the slowest predecessor.

Algorithm I. shows the details of this computation. A few comments are in order. Line (1) initializes the optimal arrival time $\tau(v, a, y)$ by a sufficiently large number for all nodes v, all possible loads y corressponding to sizings $y = \sigma(succ(v))$ and all area constraints $1 \leq a \leq \Delta A$ for the subtree with root v. The initialization in line (2) presets $\tau(v, 0, y)$ for all $v \in V$ by the arrival time at node v for a minimum area sizing of the subtree with root v and load $y = \sigma(succ(v))$. A minimum area sizing is not necessarily uniquely determined if there are different sizings of a gate with smallest area. In this case we determine a fastest sizing by solving an unconstraint problem for set of cells with minimum area. Line (3) stores the area imposed by this sizing to each subtree into the variables $a_{min}(v)$. The variable $p\_area(u, x, y)$ which is initialized in line (4) will contain the optimal area distribution of the previous step, where x is the sizing of the driving gate, u is the predecessor node and y is the sizing of the successor node. This is done for all tuples (u, x, y). The loops starting in line (6),(8),(9) and (10) perform the computation of minimum delay under area constraint, where the inner loop tries out all sizings of the driving gate. Loop (9) considers the different loads. Loop (8) performs the bottom up pass through the tree and finally loop (6) investigates all possible area investments. The computation of an optimal sizing can be easily established if we store the sizing $S(v, a, y)$ (line 20) and the area distribution $AD(u, a, y)$ (line 21) which yield a minimum for $\tau(v, a, y)$, and afterwards compute an optimal sizing backwards from the root to the leaves.

The time complexity of algorithm I. is $O(\Delta A \cdot \sum_{u \in V} s^2 \cdot indeg(u)) = O(\Delta A \cdot s^2 \cdot n)$. The space complexity is bounded by the space for the optimum arrival times $\tau(u, a, y)$,

Algorithm I.: *Minimum_delay_under_area_constraint*

1.   Initialize all $\tau(v, a, y)$ by a sufficiently large number $(a = 1, \ldots, \Delta A)$;
2.   Initialize $\tau(v, 0, y)$ by the arrival time of a minimum area sizing;
3.   Initialize $a_{min}(v)$ by the area of subtree rooted at v under a minimum area sizing;
4.   Initialize $p\_area(u, x, y) \leftarrow 0$;
5. $a \leftarrow 0$;
6.   **while** $a < \Delta A$
7. $a \leftarrow a + 1$;
8.   **for all** $v \in V$ from the leaves to the root
9.   **for all** possible sizings y of $succ(v)$
10.   **for all** possible sizings x of v with $area_v(x) + \sum\limits_{u \in pred(v)} a_{min}(u) \leq a_{min}(v) + a$
11.   /* area overhead a is sufficient for sizing x of v */
12.   **if** $area_v(x) + \sum\limits_{u \in pred(v)} a_{min}(u) < a_{min}(v) + a$
13.   **then** /* There even remains some area overhead for the predecessors, */
14.   /* namely one area unit more than during the previous execu- */
15.   /* tion of loop (6). Hence invest this unit according to lemma 1 */
16.   Find a predecessor u of max. $\tau(u, p\_area(u, x, y), x) + \delta_{u,v}(x, y)$;
17. $p\_area(u, x, y) \leftarrow p\_area(u, x, y) + 1$;
18.   **if** $\tau(v, a, y) > \max\limits_{u \in pred(v)} \tau(u, p\_area(u, x, y), x) + \delta_{u,v}(x, y)$;
19.   **then** $\tau(v, a, y) \leftarrow \max\limits_{u \in pred(v)} \tau(u, p\_area(u, x, y), x) + \delta_{u,v}(x, y)$;
20. $S(v, a, y) \leftarrow x$;
21.   **for all** $u \in pred(v)$ $AD(u, a, y) \leftarrow p\_area(u, x, y)$;

$S(v, a, y)$ and $AD(u, a, y)$ which is $O(n \cdot s \cdot \Delta A)$ and the space for $p\_area(u, x, y)$ which is $O(n \cdot s^2)$. Thus, we have an algorithm for minimum delay under area constraint with comparable complexity to area under delay constraint if ΔA is comparable with ΔT. Let us, therefore, conclude with

Theorem 1 *There is an $O(n \cdot s^2 \cdot \Delta A)$ time and $O(n \cdot s \cdot (\Delta A + s))$ space algorithm to compute optimal delays under area constraint for trees.*

The condition for the sizing x of v (lines 10 and 12) could be simplified by introducing a notion for the area overhead of gate v under sizing x. But this would not improve the overall complexity of the algorithm. Moreover this simplification is only possible as long as we are interested in the pure sizing problem. For the extension of the algorithm to library mapping the more generell condition will be necessary.

3.2 Heuristics for combinational circuits

Because there is no hope of finding tractable provable optimal sizing algorithms for combinational circuits in general, we have to look for heuristics. A popular method to solve combinatorial optimization problems is the use of iterative improvement, i.e. to take a part of a feasible solution, compute locally a best substitution of this part, and substitute it. In a first trial we will use our optimal algorithm for trees in order to perform an iterative improvement of so-called "critical paths". This has the advantage that the change of the set of gates which we consider in one step is always done nearly optimal (we will turn to this problem later). Of course we could improve even "critical trees" by means of algorithm I. but previous experiments have shown that this does not pay in general [9, 10]. Let us first give a definition of the notion "critical path". Notice, that since the delay depends on the sizing, also the definition of criticality is only possible with respect to a sizing.

Definition 6 *Let $C = (V, E)$ be a combinational circuit C, σ be the actual sizing, $\delta_{u,v}(\sigma)$, $t_v(\sigma)$ the corresponding delay assignment. Then a critical path for the sizing σ is a path from a primary input to a primary output which yields the maximum delay of C.*

Algorithm II. presents the heuristic which has turned out to be the most profitable. The main feature of this algorithm is the iterative "optimal investment" of area resources into critical paths of the circuit. The algorithm starts with a minimum area sizing. During an iteration step is invested unit by unit into the actual critcal path. The iteration step ends as soon as another path becomes more critical (also see line (8)); i.e. step width is dynamically adapted.

However, since the circuit itself is in general not a path, the locality condition of the delay assignment does not yield such an easy local correspondence between sizing and delay assignment for the critical path as is the case for tree-like circuits: Figure 3 shows a typical situation. The delay of node v is also affected by changing the sizing of w' via the edge (v, w') which does not belong to the path. In order to be able to

Algorithm II.: *Optimization of combinational circuits*

1.   **Foreach** gate u set $\sigma(u)$ to the smallest sizing;
2. $a \leftarrow 0$;
3.   **while** $a < A_{max}$
4.   choose a critical path p;
5. $c \leftarrow 0$;
6.   **repeat** $c \leftarrow c + 1$;
7.   Use Algorithm I. to compute an optimal sizing σ' of p
           with $area_u(\sigma'(u)) \geq area_u(\sigma(u))$ and area overhead constraint c.
8.   **until** $a + c = A_{max}$ or p is uncritical for σ'
9. $\sigma \leftarrow \sigma'$; $a \leftarrow a + c$;

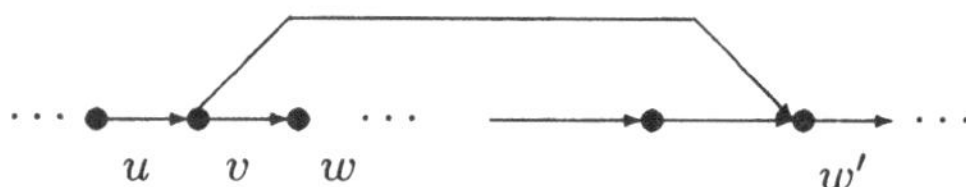

Figure 3: Side effects to the delay assignment

use algorithm I we have made the following oversimplification: During computation of optimal sizings of critical paths we have to compute the delay of an edge (u, v) as $\delta_{u,v}(x, y)$ where x is the sizing v and y is the sizing of the successor w of v in the critical path. Therefore the assignment $\delta_{u,v}()$ is done in such a way, that we suppose for all other successors $w' \neq w$ of v a load as it is imposed by the actual sizing $\sigma(w')$. This simplification does not matter if no other successor of a node v belongs to the critical path, because we do not change sizings of gates outside this path. If more than one successor belongs to the critical path as it is shown in figure 3 the delay computation during the optimization of the critical path is too optimistic. However, the termination criterion of the repeat loop can be tested exactly by a simple longest path analysis for the "best sizing" found in the actual iteration step.

4   The library mapping problem

So far we assumed the structure of the circuit to be determined in a previous design step and presented algorithms to perform sizing. At the same time as we found this result for sizing [9], a very similar approach for library mapping was proposed by [24, 26]. Library mapping is the task of finding an optimum realization of a boolean function by means of a cell library. Traditionally this problem is tackled by an approach that originates from Keutzer [13]: one starts with a network specified over some primitive gate types and substitutes these primitives by the (more complex) cells of the given library. The optimum substitution of primitives by more complex gates does not necessarily result in an optimum realization of the corresponding boolean function, because the input network determines the rough structure of the realizations which are taken into account. But assuming that the input network is not too bad this approach seems to be a very clever method to manage the library mapping process. Whilest the delay model in [13] was still very inexact and the possibilities to trade delay for area were quite restricted, the system of [26] models delay in the same way as we do and includes our sizing capabilities. But it is still an open question whether one should do library mapping and sizing in a single synthesis step (which can be very expensive) or whether one should separate these tasks. In order to examine this problem we have added a library mapping module to our sizing algorithms and made some experiments. In the following, we will describe at first the library mapping problem in more detail by extending our model and terminology in an appropriate manner. Then we will show how we have integrated this subject into our algorithms.

When considering different realizations of a boolean network over a library of complex gates one has to use a model that is general enough to express actions like replacement of several primitive gates by one complex gate, insertion of buffers or application of de Morgan's law. We want to continue to describe the input network, i.e. the rough structure of the circuit, as a directed acyclic graph $C = (V, E)$ with edges E representing the connections and nodes $V = PI \cup PO \cup V_{pr}$ representing primary inputs, primary outputs and (the output signals of) primitive gates. In the rest of the paper we will call this graph *subject graph* and assume that all its gates are either inverters or 2-input-nands. In addition we need some notion for complex gates embracing several primitives. That is why gates will be described as connected subgraphs of C. For each of these subgraphs there has to be a corresponding cell in the underlying library; i.e. a cell which computes exactly the same boolean function as the subgraph. The logical behaviour of a library cell will be described by a set of logically equivalent networks over the same gate primitives as used in the subject graph. These networks specifying which cell of the library is adequate to realize some part of the subject graph will be denoted as *pattern graphs*.

Figure 4 shows two different pattern graphs for a 4-input-nand. Remember that we want to consider only *not-* and *nand*-primitives; therefore the nodes with indegree 1 can be interpreted as inverters and the nodes with indegree 2 as 2-input-nands. As in this example we will confine ourselves in this paper to the case of all pattern graphs being trees; this includes all usual cell types except exclusive or's, multiplexers and cells with more than one output. The reason for this restriction is a considerable difference in problem complexity between the optimization with tree patterns and optimization with patterns representing acyclic graphs [13, 3]: even the simpler problem of finding a minimum area realization over a set of graph patterns is already NP-complete, whereas the same problem can easily be solved in linear time for tree patterns. Let us summarize these considerations in

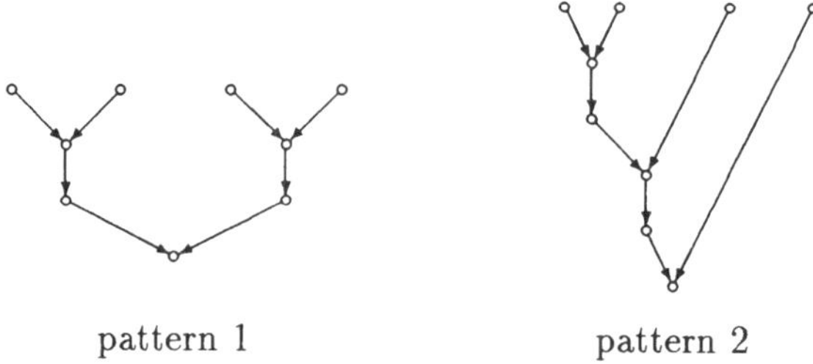

pattern 1   pattern 2

Figure 4: Two patterns for a 4-input-nand

Definition 7 *Let $C = (V, E)$ be a subject graph and P a set of pattern trees corresponding to the cells available in a given library. A connected subgraph $G = (V_G, E_G)$ of C represents a gate with inputs $inp(G)$ and an output $out(G)$ iff*

(1)   There is a pattern tree $T \in P$ which is isomorphic to G
(2) $inp(G)$ is the set of nodes of G corresponding to the leaves of T
(3) $out(G)$ is the node of G corresponding to the root of T
(4) $\forall v \in V_G - inp(G) : (u, v) \in E \implies (u, v) \in E_G$
(5) $\forall v \in V_G - inp(G) - \{out(G)\} : (v, u) \in E \implies (v, u) \in E_G$

The nodes $v \in V_G - inp(G) - \{outp(G)\}$ will be called internal nodes *of gate G. Further we will say that pattern T matches at node $out(G)$; $\pi(v)$ will denote the set of patterns matching at v, $\gamma(v)$ the set of the corresponding gates. The gates coming into question as* loads *of v are given by $\lambda(v) = \{G \mid v \in inp(G)\}$.*

The conditions (1)–(3) in the foregoing definition are evident; (4) is to guarantee that the indegree of internal nodes of the subject graph and the indegree of the corresponding pattern nodes are equal. Remember that the indegree of a node determines whether it represents a not- or a nand-primitive. By (5) we claim that internal nodes should be inaccessible from "outside". As another consequence of (5) we can state that a node with outdegree greater than one can never be an internal node of any gate, because all our patterns are trees.

Now we are ready to define what we mean by a realization of a subject graph over a library. The intuition is that all the nodes of the subject graph (except the primary outputs) are covered by gates. Figure 5 shows two different realizations of the same subject graph (an *exclusive or* constructed by means of our primitives). This example also shows how insertion of (inverting or non-inverting) buffers and automatic application of de Morgan's law can be enabled: one simply has to take care that in the subject graph there is at least one inverter between two nand primitives (and between a nand and a fanout point/primary input/primary output). This can easily achieved

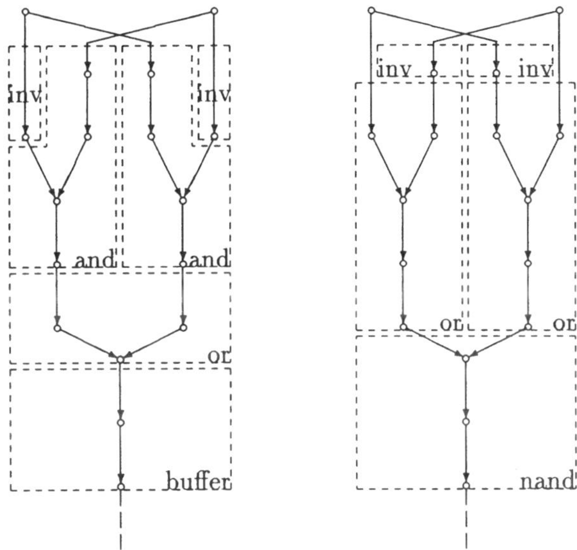

Figure 5: Different realizations of a subject graph

by inserting inverter pairs. Unnecessary inverter pairs inside[1] a fanout-free region will automatically removed during mapping if pattern trees with as well as pattern trees without inverter pairs at their root are considered.

In order to get a complete covering each node connected with a primary output v has to be realized by a gate $G \in \gamma(v)$; then of course each input $u \in inp(G)$ needs a realization by some gate $G' \in \gamma(u)$ and so on. Finally we require each internal[2] node to belong to exactly one gate, because we don't want parts of the circuit to be realized twice. To retain the analogy to definition 2 we will formally describe a realization of a subject graph as a (partial) mapping from nodes V_{pr} to the set of matching gates $\gamma(V_{pr})$. For all internal nodes this function will be undefined.

Definition 8 *Let $C = (V, E)$ be a subject graph and P be the set of pattern trees of a given cell library. A realization of C over P is a partial assignment*

$$\varrho : V_{pr} \longrightarrow \gamma(V_{pr})$$

iff (1) For all $v \in pred(PO)$ the value $\varrho(v)$ is defined.

---

[1] Inverter pairs at the leaves of a fanout-free region ask for a special treatment which can be done by similar techniques or by a subsequent fanout optimization pass.

[2] Notice that we identify an input node of a gate G with the output node of its driving gate; hence input and output nodes of G will belong to different gates (unless they represent primary inputs or primary outputs).

(2) If $\varrho(v)$ is defined and $G = \varrho(v)$
then (a) $\varrho(u)$ is defined for each input $u \in inp(G)\backslash PI$,
(b) $\varrho(u)$ is undefined for each internal node u of G.

The set of possible realizations of C will be denoted by R.

Before we extend our algorithms to optimize over the search space R instead of the set of sizings Σ, we want to give a rough sketch how to find the sets $\pi(v)$ of patterns matching at nodes v. The naive approach (to try out each pattern tree at each node) would result in a running time $O(n \cdot patsize)$, where n is the size of the subject graph and *patsize* denotes the size of all the pattern trees. Like [11, 2] we use a more sophisticated tree matching procedure derived from a well-known technique for code generation in compilers. Our matching procedure is derived from the *top down matching* algorithm of [2]. The idea is as follows:

1. Describe the pattern trees by a set of strings specifying the paths from the root to the leaves; we will call such a string s *selector of a leave* (if we want to emphasize its semantics) or *keyword* (if we are searching for occurences of s as an infix of another string s'). K be the set of all keywords. For example the pattern trees of figure 4 would be described by the keywords:

tree 1:	n1i1n1,	(1)	tree 2:	n1i1n1i1n1,	(5)
	n1i1n2,	(2)		n1i1n1i1n2,	(6)
	n2i1n1,	(3)		n1i1n2,	(7)
	n2i1n2,	(4)		n2	(8)

Herein the letters represent the type of the nodes (<u>i</u>nverters or <u>n</u>and primitives) and the digits represent edges (whether the first or the second edge ending in a node is to be selected).

2. Recursively traverse the subject tree in a depth first manner starting at the root going on contrary to the direction of the edges. At each node v consider the selector s of v in the subject tree (s being defined for subject trees in exactly the same way as for pattern trees). Find out whether a suffix of s is in the set of keywords. If this happens v is a candidate for an input of some gate G; (because each keyword describes a leaf of a pattern $p \in P$). In this case the potential output node u of G has to be informed. This can be achieved either by incrementing a counter $cnt[u, p]$ which is associated with pattern p at node u [11] or by coding this information into a bitstring $inf[v, p]$ and by transmitting this knowledge after returning from recursion to $inf[father(v), p]$ as it was done by [2] and also in our procedure. If a potential output node u has been informed that *all* the leafs of a pattern tree p actually match in the subject tree, then p matches at u.

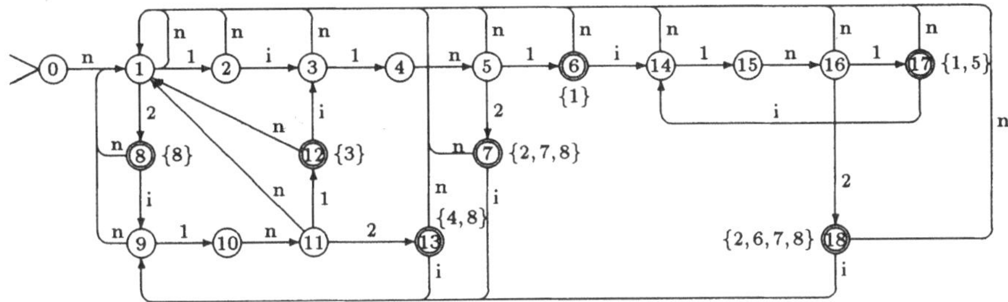

Figure 6: The automaton to the patterns

There remains the question of how to find out in (2.) which suffixes of the selector s are in K. As proposed by [11, 26] we use the Aho-Corasick algorithm for multiple keyword pattern matching [1] to fulfill this task; i.e. during a preprocessing phase we construct a complete deterministic finite automaton which is able to localize all occurences of each keyword as an infix of any string w. The output associated with each state of this automaton specifies the (index numbers) of all keywords which actually match at the end of the string consumed so far. Therefore during the depth first traversal of the subject graph the end of the current selector can be matched against the keyword set K by making the corresponding transitions in that automaton.

Figure 6 shows such an automaton for the keywords from the above example. State 0 represents the start state. The states 6,7,8,12,13,17 and 18 marked as accepting states are labeled with their output; i.e. (the indices of) the keywords they recognize. The numerous transitions back to the start state were omitted in this figure for the sake of a well-ordered arrangement. However since the automaton is complete by construction, every transition not appearing in the picture is to go back to state 0.

Time complexity for preprocessing is $O(\alpha \cdot m)$ where α denotes the size of the alphabet and and m is the number of states which is bounded by $1 + \sum_{k \in K} len(k)$. Matching time for p pattern trees in a subject graph of size n is $O(\alpha \cdot m + p \cdot n)$. The first term results from loading the automaton; the second term (which is in general more significant) is optimum, since the output may have size $O(p \cdot n)$. Notice that p denotes only the number of pattern trees and not their size as it would be in the case of a naive approach.

It is not difficult to see that all the methods for optimal sizing of tree-like circuits presented in section 3.1 can also be applied for finding optimal realizations $\varrho \in R$ of tree-like subject graphs. Instead of trying out each sizing x of a node v, now all the gates $G \in \gamma(v)$ have to be considered. Furthermore the inputs $inp(G)$ of each gate G take the place of the predecessors $u \in pred(v)$; finally the load of a node v is no longer given by the sizing of the successor node but by the gates $G' \in \lambda(v)$.

Contrary to the case of trees the heuristic for sizing of combinational circuits presented in section 3.2 is not directly transferable to the problem of optimally realizing an acyclic subject graph by tree patterns. The main reason is that we can no longer consider some fixed structure of gates as we could do in algorithm II. Instead each realization considered involves a different number of gates which embrace other nodes of the subject graph. As a first consequence the critical parts we consider should not be bordered by nodes which possibly could be internal nodes; i.e. the border of a critical part should consist of primary inputs, primary outputs and nodes v with $outdeg(v) > 1$. In other words: A fanout-free region of the subject graph should be considered either completely critical or completely uncritical.

Therefore it seems to be reasonable to consider each fanout-free region as a unity and to treat it as a big gate. The different implementations of such a *supergate* by means of the cell library can be used in the same way as we used the different sizings of a gate in our sizing algorithms. Therefore we can apply algorithm II to a circuit composed of supergates; the number of (super)gates in this "new" circuit is reduced to the number n' of fanout-free regions of the original subject graph. The different implementations of each supergate G can be computed dynamically during any step of iterative improvement by means of algorithm I.

We are now ready to estimate the computational complexity of this approach: When considering an investment of c area units $c+1$ implementations of any "critical" supergate are taken into consideration for each of l potential loads of that gate. An iteration step of step width w successively considers area investments $c = 0, 1, \ldots, w$. Therefore one iteration step costs time

$$(*) \qquad O(l \cdot n' \cdot w^2)$$

if we neglect at first the time for computing the different implementations of each supergate by means of algorithm I. Unfortunately this task turns out to corrupt the time complexity of the sizing heuristic: Let G be an arbitrary supergate, a_{min} the area of a minimum area realization of G and a its area at the beginning of the actual iteration step. It does not suffice to compute realizations of G with area $a, a + 1, \ldots, a + w$ out of some previously computed realizations because the arrival times at the inputs of the supergate may have changed. Instead we have to (re)compute optimum realizations of G with area $a_{min}, \ldots, a, a+1, \ldots, a+w$. Therefore computing the implementations of the supergates during a *single* iteration step may cost time

$$(**) \qquad O(p \cdot l \cdot n \cdot \Delta A)$$

where the number p of pattern trees results from estimating the number of cells that can be used to realize any node of the subject graph. By adding $(*)$ and $(**)$ and by summing up over all iteration steps we get the overall complexity

$$O(w_{max} \cdot l \cdot n' \cdot \Delta A + p \cdot l \cdot n \cdot \Delta A^2)$$

where w_{max} denotes the maximum step width chosen by our heuristic for step width adaptation. We started our first experiments on library mapping with a simplification of this approach which bounds step width to W and avoids the recomputation of realizations with area $a_{min}, \ldots, a - 1$ by confining to a restricted search space. Therefore this simplified heuristic can avoid the ΔA^2-term and manages with time $O(W \cdot p \cdot l \cdot n \cdot \Delta A)$; in our first trial we chosed step width being bounded by the area of the biggest cell.

5 Experimental results and concluding remarks

From the previous section we know that we can use the same approach for finding an optimal realization of a circuit over a library with cells of different types as we have proposed in [10] for sizing a circuit without changing the circuit structure. Obviously there are two alternatives: We can either determine the structure of the circuit and its sizing by means of two different passes or use a single optimization pass for both tasks. Both possibilities considerably differ in complexity: Remember that the running time of our basic algorithm is proportional to s^2, where s is the number of different implementations of any node; i.e. for the first pass of the first alternative s denotes the number of different pattern trees; for the second pass s is the number of different sizes for each cell type. However if we choose a single pass optimization two realizations of a node can differ in type or size; hence in this case s is the product of the number of pattern trees and the number of different sizes for each type. In order to examine whether the adaptation of our sizing heuristics to the library mapping problem is satisfactory and to try out whether single pass optimization is worth the money we have implemented our algorithms and made some experiments.

Our implementation was done in the programming language C++ on a SUN SPARC II workstation. For testing this implementation and for first experiments we chosed some circuits of modest size (up to about 100 gates, corresponding to subject graphs[3] of up to about 400 nodes) from the benchmark set of the Microelectronics Center of North Carolina [28]. In order to be able to consider different gate sizes for each cell type we have designed our own library. It consists of 38 cell implementations for 17 logical types (inverters, nands and nors of arities 2-4 and CMOS complex gates). Our delay model considers assymmetries between different cell inputs and differences between rise and fall transition times as we had proposed in [10]. The logical behaviour of the different cell types was described by 62 pattern trees.

Table 1 shows the results which we obtained by different optimization strategies. The leftmost column lists the number n of gates, area A and delay D of the unmapped circuits. The next column presents the improvement achieved by pure sizing. In the columns 3 and 4 we present the results of library mapping and sizing performed by two different passes: The experiments of column 3 started with a linear time pass to determine a minimum area mapping and performed sizind during a second pass, whereas pass 1 of column 4 performed a minimum delay mapping (over a library

---

[3]Insertion of inverter pairs as described in the previous section and decomposition of gates with more than two inputs increases the number of nodes in the subject graph on average by a factor of 3.3.

circuit	unmapped circuit			sizing of unmapped circ.			min area map. and sizing			min delay map. and sizing			single pass optimization		
	n	A	D	A	D	CPU	A	D	CPU	A	D	CPU	A	D	CPU
b1	15	42	5.56	45	5.03	0.7	41	5.21	2.8	61	4.48	3.6	42	5.02	8.5
C17	13	32	4.72	48	3.71	0.7	39	3.58	2.7	39	3.58	3.0	32	3.62	6.9
cm150a	70	201	16.83	297	15.33	3.8	183	13.48	5.6	253	12.38	43.5	187	13.77	73.3
cmb	54	156	10.60	214	8.96	2.4	218	6.62	4.5	234	6.63	11.1	169	7.06	42.5
majority	11	31	6.06	56	5.31	0.8	43	4.10	2.8	43	4.10	2.9	32	4.11	9.5
mux	90	266	16.87	384	13.43	5.4	239	11.81	6.7	281	11.96	22.0	239	11.81	66.7
parity	76	197	11.53	307	8.89	5.5	241	8.99	5.4	241	8.99	26.7	163	9.37	29.3
pcle	86	227	17.11	273	11.56	3.5	221	11.11	5.6	222	10.89	20.4	223	10.82	209.7
pcler8	103	277	18.65	354	12.90	6.8	296	12.82	7.5	285	11.85	12.2	271	12.02	61.2
tcon	72	176	5.26	176	5.26	1.7	128	5.31	3.5	132	5.27	10.2	132	5.27	23.7
unreg	113	338	10.50	343	9.47	4.8	276	8.13	8.4	292	8.02	25.9	286	8.05	101.8
z4ml	70	200	10.80	239	8.93	2.6	172	8.86	5.3	200	8.39	36.0	181	7.95	56.3

Table 1: Experimental results for the different optimization alternatives

containing only one implementation for each cell type). The last column shows the results of single pass optimization.

The series of experiments shown in the first column confirm that sizing is worth while: By pure sizing of the unmapped circuit, i.e. without any structural changes, we could achieve an average delay reduction of 16.8% by means of an area investment of 22.2%. It is true that this improvement is more moderate than that achieved by our previous experiments [10]; the reason for this is that our new cell library at present contains less realizations for most cell types. (So far we have four inverters, three 2-input-nands, three 2-input-nors but only two implementations for the other cell types.)

Let us look now at our first experiences with library mapping. The results achieved by starting with minimum area mapping and sizing during a second pass were amazingly good: We did not only need less area than by pure sizing but also delay was better for most benchmarks. Obviously minimum area mapping does not only reduce the size of a circuit but also its depth and thus delay. In fact for a quarter of the benchmarks minimum area mapping and minimum delay mapping resulted in the same realization. On average of course the experiments of column 3 yield realizations which are smaller, but a little slower than the results shown in column 4: Compared to the unmapped and unsized circuits of column 1 sizing after minimum area mapping resulted in a delay reduction of 23.2% by an area investment[4] of 3.8% whereas sizing after minimum delay mapping reduced delay by 25.8% and invested 14.3% area. Some particular cases, cmb and mux (pcler8), show however that starting with minimum delay mapping (minimum area mapping) has to result not necessarily in a fastest (smallest) realization after sizing. The reason for this surprising phenomenon is that our optimization algorithms consider only some discrete gate sizes. Thus a realization considered good by pass 1 can turn out less suitable for sizing in consequence

---

[4]I.e. the area gain achieved by minimum area mapping was just reinvested during the sizing pass.

of unavailable gate sizes (especially if there is a different number of implementations for each cell type). Computational complexity of minimum delay mapping is higher than of minimum area mapping: A minimum area mapping can be found by a linear time pass over the circuit, whereas delay is minimized by our (simplified) iterative heuristic.

The most expensive experiments performing library mapping and sizing during a single pass are presented in the last column. Delay reduction (23.9%) is comparable to the results of column 3 but the area values are better: In comparision with the unmapped and unsized circuit area even could be reduced by 6.4% on average. On the other hand the "optimum" of the unconstrained problem was not found in most cases. We believe that this is a consequence of bounding step width by the area of the biggest cell: To achieve a further improvement of delay it is often necessary to choose bigger realizations of more than one gate. Therefore step width bound of our actual mapping heuristic can become a limiting factor. In such cases better results would be possible by raising step width or using the more expensive approach sketched in section 4.

Let us conclude with some further interesting points which we would like to investigate in future.

- special cases of DAG-Matching

So far we have confined ourselves to cells which may be described by means of tree patterns for reasons of complexity [3]. This excludes cells as exors, multiplexers and multiple-output-cells from consideration. The question is whether to use heuristic approaches for DAG-matching if such cells are to be considered, or whether such cells should be installed first of all according to some peephole principle or whether even optimum matching is possible in some special cases.

- fanout optimization

Up to now we have adapted the gates at different load situations by sizing. This is the appropriate method if the capacitive load is moderate. However if large buses are to be driven or if fanout is very high chains or trees of buffers have to be inserted; (also look at our experimental results for the circuit tcon which is characterized by a gate with extremely high fanout). Several approaches have been proposed in literature to deal with this problem [12, 26, 25, 14]. We want to examine how these proposals harmonize with our algorithms. Moreover it is an open problem whether fanout optimization can be performed in connection with sizing or whether both tasks have to be solved by different passes: in both cases tree structures are considered, but direction of fanout trees is upside down. Moreover (pure) sizing considers a fixed tree whereas fanout optimization just is to determine an optimum structure.

- excluding false paths

Another weakness of our problem modelling is that we may overestimate delay by considering so-called "false paths", i.e. paths that can't be sensitized by any vector of boolean values at the primary inputs. It is known that considering static sensitization by a D-Algorithm approach may underestimate delay and that dynamic effects have to be taken into account for that reason [18, 17, 5]. Furthermore sensitizing conditions have to be "robust" against deviations of the manufactured circuit from idealized delay model [18, 17]. We want to examine methods to consider these aspects during optimization. Such methods may be possible in spite of problem complexity because we only need an upper bound for the delay of the longest sensitizable path and not the exact value for that delay.

These examples show that very different problems are connected with performance optimization in the field of VLSI-design. All those problems are very complex in themselves so that only in special cases optimum algorithms seem to be tractable. Therefore the first pleasant results should not be interpreted as ultimate solutions but as an encouragement for further investigations. Many parameters were not yet analysed sufficiently; it is possible that thoroughly new approaches could be found. Furthermore different optimization methods which were examined isolated so far should be attuned to each other and could even be integrated in a single tool. The main result of [24, 26] for example was that library mapping as proposed in Dagon and a cell based approach for sizing as presented in [9, 10] can be combined. In order to compare different strategies with regard to quality and complexity it is indispensable to make experiments. For that purpose we want to implement a software environment enabling the examination of different approaches concerning the domain of circuit analysis, synthesis and optimization.

Bibliography

[1] A.V. Aho and M.J. Corasick. Efficient string matching: An aid to bibliographic search. *Communications of the ACM, Vol. 18, No. 6*, pages 333–340, 1975.

[2] A.V. Aho, M. Ganapathi, and S.W.K. Tjiang. Code generation using tree matching and dynamic programming. *ACM Transactions on Programming Languages and Systems, Vol. 11, No. 4*, pages 491–516, 1989.

[3] A.V. Aho, S.C. Johnson, and J.D. Ullman. Code generation for expressions with common subexpressions. *Journal of the Association for Computing Machinery, Vol. 24, No. 1*, pages 146–160, 1977.

[4] M.R.C.M. Berkelaar and J.A.G. Jess. Gate sizing in mos digital circuits with linear programming. In *Proceedings of the European Design Automation Conference (EDAC90)*, pages 217–221, 1990.

[5] D.H.C. Du, S.H.C. Yen, and S. Ghanta. On the general false path problem in timing analysis. In *Proceedings of the 26th Design Automation Conference (DAC89)*, pages 555–560, 1989.

[6] L.A. Glasser and L.P.J. Hoyte. Delay and power optimization in vlsi circuits. In *Proceedings of the 21st Design Automation Conference (DAC84)*, pages 529–535, 1984.

[7] K.S. Hedlund. Aesop: A tool for automated transistor sizing. In *Proceedings of the 24th Design Automation Conference (DAC87)*, pages 114–120, 1987.

[8] U. Hinsberger. Zellenbasierte Dimensionierung kombinatorischer Schaltkreise. Master's thesis, Fachbereich Informatik, Universität des Saarlandes, Im Stadtwald, W–6600 Saarbrücken 11, FRG, 1990. 94 Seiten.

[9] U. Hinsberger and R. Kolla. A cell based approach to performance optimization of combinational circuits. Technical Report 14/1989, Sonderforschungsbereich 124 *VLSI Entwurfsmethoden und Parallelität*, Fachbereich Informatik, Universität des Saarlandes, Im Stadtwald, W–6600 Saarbrücken 11, FRG, 1989.

[10] U. Hinsberger and R. Kolla. Cell based performance optimization of combinational circuits. In *Proceedings of the 1st European Design Automation Conference (EDAC90)*, pages 594–599, 1990.

[11] C.M. Hoffmann and M.J. O'Donnell. Pattern matching in trees. *Journal of the Association for Computing Machinery, Vol. 29 , No. 1*, pages 68–95, 1982.

[12] H.J. Hoover, M.M. Klawe, and N.J. Pippenger. Bounding fan-out in logical networks. *Journal of the Association for Computing Machinery*, Vol. 31, No. 1:13–18, 1984.

[13] K. Keutzer. DAGON: Technology binding and local optimization by DAG matching. In *Proceedings of the 24th Design Automation Conference (DAC87)*, pages 341–347, June 1987.

[14] Shen Lin and M. Marek Sadowska. A fast and efficient agorithm for determining fanout trees in large networks. In *Proceedings of the 2nd European Design Automation Conference (EDAC91)*, pages 539–544, 1991.

[15] Shen Lin, M. Marek Sadowska, and E.S. Kuh. Delay and area optimization in standard-cell design. In *Proceedings of the 27th Design Automation Conference (DAC90)*, pages 349–352, 1991.

[16] D.P. Marple and A. El Gamal. Optimal selection of transistor sizes in digital VLSI circuits. In *Proceedings of Stanford Conference of Advanced Research in VLSI*, pages 151–172, 1987.

[17] P.C. McGeer and R.K. Brayton. Efficient algorithms for computing the longest viable path in a combinational network. In *Proceedings of the 26th Design Automation Conference (DAC89)*, pages 161–567, 1989.

[18] P.C. McGeer and R.K. Brayton. Provably corrext critical paths. In *Decennial Caltech Conference on VLSI*, 1989.

[19] G. De Micheli. Performance-oriented synthesis of large-scale domino MOS circuits. *IEEE Transactions on Computer Aided Design*, CAD-6(5):751–764, 1987.

[20] M.Matson. Optimization of digital MOS VLSI circuits. In *Proceedings of the Chapel Hill Conference on VLSI*, pages 109–126, May 1985.

[21] L. Nagel. A computer program to simulate semiconductor circuits. Technical Report ERL-520, University of California at Berkeley, 1975.

[22] F.W. Obermeier and R.H. Katz. An electrical optimizer that considers physical layout. In *Proceedings of the 25st Design Automation Conference (DAC88)*, pages 453–459, June 1988.

[23] J.K. Ousterhout. Crystal: A timing analyser for nMOS VLSI circuits. In R. Bryant, editor, *Proceedings of the Third Caltech Conference on VLSI*, pages 57–70. Computer Science Press, 1983.

[24] R. Rudell. *Logic Synthesis for VLSI Design*. PhD thesis, University of California, Berkeley, April 1989.

[25] K.J. Singh and A. Sangiovanni-Vincentelli. A heuristic algorithm for the fanout problem. In *Proceedings of the 27th Design Automation Conference (DAC90)*, pages 357–360, 1990.

[26] H. Touati, C. Moon, R. Brayton, and A. Wang. Performance-oriented technology mapping. In *Proceedings of the MIT VLSI Conference*, 1990.

[27] C.H.A. Wu, N. Vander Zanden, and D. Gajski. A new algorithm for transistor sizing in CMOS circuits. In *Proceedings of the European Design Automation Conference (EDAC90)*, pages 589–592, 1990.

[28] S. Yang. Logic synthesis and optimization benchmarks user guide (version 3.0). Technical report, Microelectronics Center of North Carolina, P.O. Box 12889, Research Triangle Park, NC 27709, January 1991.

An Algebraic Characterization of Context–Free Languages

Thomas Kretschmer

Auf dem Waas 16
D-6643 Borg
Germany

Abstract

We show that for any context–free language $L \subseteq T^+$ there is a length–preserving homomorphism μ and a homomorphism λ such that $L = (\overline{a}D(a,b))\lambda^{-1}\mu$, where $D(a,b)$ denotes the Dyck set over the alphabet $\{a, b\}$. λ and μ may be effectively constructed.

1   Introduction

Context–free languages may be characterized by various means. The classical way to specify a context–free language L is to give a push–down automata accepting L or a context–free grammar generating L. But there exist also algebraic characterizations of context–free languages , e.g., the theorem of Chomsky-Sch"utzenberger [2], the theorem of Greibach on a hardest context–free language [3] or the representation theorem of Shamir [5]. Here we will show that for any context–free language $L \subseteq T^+$ there is a length–preserving homomorphism μ and a homomorphism λ such that $L = (\overline{a}D(a,b))\lambda^{-1}\mu$, where $D(a,b)$ denotes the Dyck set over the alphabet $\{a,b\}$. λ and μ may be effectively constructed. The proof will make clear that this theorem is essentially just another way of expressing Shamir's theorem. Yokomori [6] proved independantly a weaker version of this result, but the connection to Shamir's result is not clear.

Notations Let T^* denote the free monoid over the finite alphabet T with identity 1. Define $T^+ := T^*\backslash\{1\}$. We write $P_f(T^*)$ for the set of finite subsets of T^*. We make $P_f(T^*)$ to a monoid with identity $\{1\}$ by introducing the following operation :

$$U \cdot V := \{uv \mid u \in U \text{ and } v \in V\}$$

for U, V finite subsets of T^*. If $\vartheta : X \to Y$ is a mapping, we write $x\vartheta$ for the image of x under ϑ. If $\vartheta' : Y \to Z$ is another mapping we denote the composition of ϑ and ϑ' by $\vartheta\vartheta'$. For a subset $U \subseteq Y$, $U\vartheta^{-1}$ is defined by $U\vartheta^{-1} := \{x \in X \mid x\vartheta \in U\}$.

A context–free grammar is denoted by $G = (V, T, P, S)$ where V is the set of variables, T is the set of terminals, $S \in V$ is the start symbol and

$$P \subset V \times (V \cup T)^*$$

is the finite set of productions. If $u \in (V \cup T)^*$ derives to $v \in (V \cup T)^*$, we write $u \xrightarrow{*} v$. G generates the language $L(G) := \{w \in T^* \mid S \xrightarrow{*} w\}$.

Now we want to define the Dyck set over a finite alphabet X. Let $\overline{X}$ be another alphabet such that $X \cap \overline{X} = \emptyset$ and such that there is a bijection $b : X \to \overline{X}$. We write $\overline{x}$ for xb $(x \in X)$. Then the Dyck set $D(X)$ over X is generated by $G = (\{S\}, X \cup \overline{X}, P, S)$ where

$$P = \{S \to SS, S \to 1\} \cup \{S \to xS\overline{x} \mid x \in X\}$$

(see e.g. [1]). For $u, v \in X^*$ we define $\overline{uv} = \overline{v} \cdot \overline{u}$.

A technical lemma We want to give an equivalent formulation for the fact that a language L is expressed as $L = L_0\lambda^{-1}\mu$.

Lemma 1 *Let $L \subseteq T^*$ and $L_0 \subseteq Y^*$ any languages. Then (a) and (b) are equivalent :*

(a) There is an alphabet X, a length–preserving homomorphism $\mu : X^* \to T^*$ and a homomorphism $\lambda : X^* \to Y^*$ such that $L = L_0\lambda^{-1}\mu$

(b) There is a homomorphism $\varphi : T^* \to P_f(Y^*)$ such that

$$w \in L \iff w\varphi \cap L_0 \neq \emptyset$$

Proof: Let us suppose that X, λ and μ are given such that $L = L_0\lambda^{-1}\mu$. We define φ by

$$t\varphi := t\mu^{-1}\lambda \quad (t \in T)$$

Then we get for any $w \in T^*$:

$$
\begin{aligned}
w \in L \iff & \; w \in L_0\lambda^{-1}\mu \\
\iff & \; \exists v \in L_0, u \in X^* : u\lambda = v \text{ and } u\mu = w \\
\iff & \; \exists v \in L_0 : v \in w\mu^{-1}\lambda = w\varphi \\
\iff & \; w\varphi \cap L_0 \neq \emptyset
\end{aligned}
$$

Now let us suppose that φ is given such that $w \in L \iff w\varphi \cap L_0 \neq \emptyset$. We have to define X, λ and μ. Let X be defined by

$$X := \{(t,m) \mid t \in T \text{ and } m \in t\varphi\} \subset T \times Y^*$$

and μ, λ by

$$
\begin{aligned}
(t,m)\mu & := t \\
(t,m)\lambda & := m
\end{aligned}
$$

for $(t,m) \in X$.

We will show that $L = L_0\lambda - 1\mu$. In the following the empty product is defined to be equal to 1.

"$\subseteq$" Let $w \in L$. We know that $w\varphi \cap L_0$ is not empty, i.e., there exists an $m \in w\varphi$ such that $m \in L_0$. Let $w = t_1 \cdot \ldots \cdot t_n$. Then for all $i \in \{1, \ldots, n\}$ there must be $m_i \in t_i\varphi$ such that $m_1 \cdot \ldots \cdot m_n = m$. Let u be equal to $(t_1, m_1) \cdot \ldots \cdot (t_n, m_n) \in X^*$. It is clear that $u\lambda = m$ and $u\mu = w$. Consequently $w \in L_0\lambda^{-1}\mu$.

"$\supseteq$" Let $w \in L_0\lambda^{-1}\mu$. Then there is a word $u \in X^*$ such that $u\mu = w$ and $u\lambda \in L_0$. Let $u = (t_1, m_1) \cdot \ldots \cdot (t_n, m_n)$. Clearly $m_i \in t_i\varphi$. λ, φ and μ being homomorphisms we obtain :

$$L_0 \ni u\lambda = m_1 \cdot \ldots \cdot m_n \in t_1\varphi \cdot \ldots \cdot t_n\varphi = (t_1 \cdot \ldots \cdot t_n)\varphi = u\mu\varphi = w\varphi$$

So $w\varphi \cap L_0$ is not empty and the supposition tells us that $w \in L$. ∎

2   A Shamir–like characterization of context–free languages

We will give a formulation of Shamir's theorem [5] that is suitable for proving our result. We fix the alphabet $A = \{a, b, \bar{a}, \bar{b}\}$.

                    Thomas Kretschmer

Theorem 1 *For any context–free language $L \subseteq T^+$ there is a monoid homomorphism*

$$\varphi : T^* \to P_f(A^*)$$

such that

$$w \in L \Longleftrightarrow w\varphi \cap \overline{a}D(a,b) \neq \emptyset$$

Proof: This result is well known, see e.g., [5] and [4]. To keep this article self–contained, we will sketch the proof.

Let $G = (V, T, P, S)$ be a context–free grammar for L that is in Greibach normal form, that is to say, $P \subset V \times TV^*$. First we show that there is a homomorphism

$$\tilde{\varphi} : T^* \to P_f((V \cup \overline{V})^*)$$

such that

$$(*) \quad w \in L \Longleftrightarrow w\tilde{\varphi} \cap \overline{S}D(V) \neq \emptyset$$

Define $\tilde{\varphi}$ by

$$t\tilde{\varphi} := \{\overline{v}u^\sigma \mid (v, tu) \in P\}$$

where $t \in T$ and u^σ denotes u mirrored.

Then the following statement holds for all $w \in T^+$ and $u \in V^*$:

There is a leftmost derivation $S \overset{*}{\to} wu$

$$\Longleftrightarrow$$

There is an $m \in w\tilde{\varphi}$ such that $m\overline{u}^\sigma \in \overline{S}D(V)$

This can be proved easily by induction on the length of w. Taking $u = 1$ we get $(*)$.

In case that V contains more than two elements we want to use $D(A)$ instead of $D(V)$. Let $V = \{v_0, \ldots, v_n\}$ where $v_0 = S$. We define a homomorphism

$$\pi : (V \cup \overline{V})^* \to A^*$$

by

$$\begin{aligned}
v_i\pi &:= ab^i \\
\overline{v_i}\pi &:= \overline{b}^i\overline{a}
\end{aligned}$$

and

$$\varphi : T^* \to P_f(A^*)$$

by $\varphi := \tilde{\varphi}\pi$. It is easy to check that

$$w \in L \Longleftrightarrow w\varphi \cap \overline{a}D(a,b) \neq \emptyset$$

$\blacksquare$

The main theorem If Lemma 1 and Theorem 1 are taken together, the following theorem is clear.

Theorem 2 *For any context–free language $L \subseteq T^+$ there is an alphabet X, a length–preserving homomorphism $\mu : X^* \to T^*$ and a homomorphism $\lambda : X^* \to A^*$ such that*

$$L = (\overline{a}D(a,b))\lambda^{-1}\mu$$

Remarks

- By applying the sketch of the proof of Theorem 1 and the proof of Lemma 1, we can effectively construct X, λ and μ.

- We will clarify the intuition behind the construction. Let $G = (V, T, P, S)$ be a context–free grammar in Greibach normal form such that $L = L(G)$. From the proof of Theorem 1 we get a homomorphism $\tilde{\varphi}$ such that

$$w \in L \iff w\tilde{\varphi} \cap \overline{S}D(V) \neq \emptyset$$

  If we look closer at the definition of X in the proof of Lemma 1, we see that we could take P instead of X and define λ and μ by $p\mu := t$ and $p\lambda := \overline{v}u^\sigma$ for $p = (v, tu) \in P$. A word u in $X^* = P^*$ corresponds to a sequence of productions. $u\mu$ is the corresponding terminal word and the condition $u\lambda \in \overline{S}D(V)$ is true if and only if this sequence of productions yields a valid leftmost derivation.

- Theorem 2 and Lemma 1 imply of course Theorem 1.

Bibliography

[1] J. Berstel. Transductions and Context–Free Languages. B.G. Teubner, Stuttgart 1979.

[2] N. Chomsky and M. Schützenberger. The algebraic theory of context–free languages , in : P. Braffort and D. Hirschberg (edts.). Computer programming and formal systems, North–Holland, Amsterdam 1963, 118–161.

[3] S. Greibach. The hardest context–free languages . SIAM Journal of Computing 2 (1973) 304–310.

[4] G. Hotz. A Representation Theorem of Infinite Dimensional Algebras and Applications to Language Theory. Journal of Computer and System Sciences 33 (1986) 423–455.

[5] E. Shamir. A Representation Theorem For Algebraic and Context–Free Power Series in Noncommuting Variables. Information and Control 11 (1967) 239–254.

[6] T. Yokomori. On Purely Morphic Characterizations of Context–Free Languages. Theoretical Computer Science 51 (1987) 301–308.

The Bisection Problem for Graphs of Degree 4 (Configuring Transputer Systems)

Juraj Hromkovič

Burkhard Monien

Universität Paderborn
4790 Paderborn
Germany

Abstract

It is well-known that for each $k \geq 3$ there exists a constant c_k and an infinite sequence $\{G_n\}_{n=8}^{\infty}$ of k-degree graphs (each G_n has exactly n vertices) that the bisection width of G_n is at least $c_k \cdot n$. It this paper some upper bounds on the c_k's are found. Let $\sigma_k(n)$ be the maximum of bisection widths of all k-degree graphs of n vertices. We prove that

$$\sigma_k(n) \leq \frac{(k-2)}{4} \cdot n + O(\sqrt{n})$$

for all even k. This result is improved for $k = 4$ by constructing two algorithms A and B, where for a given 4-degree graph G_n of n vertices

(i) A constructs a bisection of G_n involving at most $n/2 + 4$ edges for even $n \leq 60$ (i.e., $\sigma_4(n) \leq n/2 + 4$ for even $n \leq 60$)

(ii) B constructs a bisection of G_n involving at most $n/2 + 1$ edges for even $n \geq 350$ (i.e. $\sigma_4(n) \leq n/2 + 1$ for even $n \geq 350$).

The algorithms A and B run in $O(n^2)$ time on graphs of n vertices, and they are used to optimize hardware for building large transputer systems.

1   Introduction

The problem investigated here is related to the bisection problem, where for a given graph G a balanced partition with a minimal number of crossing edges has to be found. This problem is well-known and well-studied. It has many applications, especially in the field of VLSI layout. The problem is NP-complete [6, 3], and many heuristic algorithms have been proposed [8, 7, 3] to solve it. We do not deal directly with the classical problem of finding a minimal bisection of a given graph but with the question how large the minimal bisections of k-degree graphs (graphs whose degree is bounded by k) may be. Let $\sigma_k(n)$ be the maximum of bisection widths of all k-degree graphs of n vertices. Using expander graphs [1, 5, 4], we know that $\sigma_k(n) \in \Omega(n)$ for each $k \geq 3$. Namely, $\sigma_4(n) \geq n/7$ and $\sigma_3(n) \geq n/16$ follows from [1, 4]. Generally, the following lower bounds follow from [1, 2, 9]

$$\sigma_k(n) \geq (k - 2\sqrt{k-1})n/4$$

(We conjecture that these lower bounds are not very close to the real values $\sigma_k(n)$ because they are proved by probabilistic methods for almost all random k-regular graphs). The aim of this paper is to give some upper bounds on $\sigma_k(n)$ for $k \geq 3$.

In Section 2, we shall use a new combinatorial optimization technique to show that

$$\sigma_k(n) \leq (k - 2)n/4 + O(\sqrt{n}) \tag{1}$$

for each $k \in \{2 \cdot r \mid r \in N\}$, where N denotes the set of all nonnegative integers. Using this technique, we also get a more general result; Each k-regular graph of n vertices can be partitioned into d equal-sized components by removing at most

$$\frac{(k-2)}{2} \cdot n \cdot \frac{d-1}{d} + O\left(\sqrt{n}\right)$$

edges for k even. The proof of the above stated facts is constructive, because we give an algorithm C which finds for any given k-degree graph a bisection whose width is bounded by (1). Unfortunately, this algorithm C has an exponential time complexity and gives no good results for small n.

This unpleasant property of C will be overcome in Section 3 by constructing two algorithms A and B, where for a given 4-degree graph G_n of n vertices

  (i) A constructs a bisection of G_n involving at most $n/2 + 4$ edges for even $n \leq 60$, more precisely

$$\sigma_4(n) \;\leq\; n/2 + 4 \quad \text{for} \quad n \leq 60, n \equiv 0 \bmod 4$$
$$\sigma_4(n) \;\leq\; n/2 + 3 \quad \text{for} \quad n \leq 60, n \equiv 2 \bmod 4$$

These results are optimal for $n \leq 22$ and for $n = 26$.

  (ii) B constructs a bisection of G_n involving at most $n/2+1$ edges for even $n \geq 350$, i.e.

$$\sigma_4(n) \leq n/2 + 1 \quad \text{for even} \quad n \geq 350$$

Both algorithms A and B run in $O(n^2)$ time on graphs of n vertices.

In Section 4 we use the results of Section 3 to give an effective algorithm for configuring transputer systems. This new algorithm has led to an optimization of the use of hardware in the process of modular configuring of transputer systems. This optimization question was the original problem which has started the research presented in this paper.

2 Asymptotic Estimates

In this section we shall give an asymptotic estimate for the general task of partitioning k-degree graphs into $d \geq 2$ equal-sized components. This estimate implies that the bisection of 4-degree graphs can be done by removing $n/2 + o(n)$ edges, and that the 4-partition of 4-degree graphs can be done by removing $3n/4 + o(n)$ edges (note that these cases are the most important ones for the task of configuring transputer system). Now, let us start with partitioning regular graphs because these are the hardest ones among degree bounded graphs from the bisection width point of view.

Theorem 1 *Let d be some natural number, and let G be a 4-regular graph of n vertices, where $n = d \cdot h$ for some $h \in N$. Then one can partition G into d components, each of h vertices, by removing at most*

$$n \cdot \frac{d-1}{d} + O(\sqrt{n})$$

edges.

Proof: First, for any $m \in \mathcal{M} = \{m \in N \mid m \text{ divides } n/d\}$ we give an algorithm $d - PART(m, G)$ that finds a d-partitioning of G, and then we prove that this d-partitioning is achieved by removing at most $e(n, d, m) = (n + 2(md - 1)) \cdot ((d - 1))/d) \cdot (m/(m - 1/d))$ edges.

Algorithm 1 $d - PART(m, G)$

Input: *A 4-regular graph $G = (V, E)$ of n vertices, positive integers d, m with $n = dmb$ for some positive integer b*

Output: *d components of G : $D_1 = (V_1, E_1), \ldots, D_d = (V_d, E_d)$ and the set of removed edges $H_m = E - \bigcup_{i=1}^{d} E_d$ with the property $|H_m| \leq e(n, d, m)$.*

1. *Remove at most n edges from G in order to obtain a sub-graph $G' = (V, E')$ with the degree bounded by 2.*
   { This can be achieved by constructing an Eulerian Cycle in G and then removing each odd edge in the cycle.}

2. *Remove at most $2(md - 1)$ edges from G' in such a way that we obtain $m \cdot d$ components $C_i = (V_i', E_i'), i \in \{1, \ldots, md\}$, each exactly of n/md vertices. { So, we have found some md-partitioning by removing at most $n + 2(md - 1)$ edges }*

3. *Set $H := E - \bigcup_{i=1}^{md} E_i'$.*
   For each $i, j \in \{1, \ldots, md\}$ compute c_{ij} - the number of edges from H connecting the components C_i, and C_j.
   Divide the set $I = \{1, 2, \ldots, md\}$ into d disjoint sets $I_1, \ldots, I_d$, each of m elements, in such a way that the sum $\sum_{k=1}^{d} \sum_{i,j \in I_k} c_{ij}$ is maximal for all possible partitions of I.

4. *For each $k \in \{1, \ldots, d\}$ set $V_k = \bigcup_{i \in I_k} V_i'$ and set $D_k = (V_k, E_k)$ to be the induced graph of G by the set of vertices $V_k \subseteq V$.*
   { For each $k : |V_k| = m \cdot (n/md) = n/d \}$.
   Set $H_m := H - \bigcup_{k=1}^{d}(\bigcup_{i,j \in I_k}\{\{u, v\}, u \in V_i', v \in V_j'\}) = E - \bigcup_{k=1}^{d} E_k$

Now, let us prove that $|H_m| \leq e(n, d, m)$ for any $m \in \mathcal{M}$. Let, for any $i, j \in \{1, \ldots, md\}$, E_{ij} be the set of edges connecting the components $C_i = (V_i', E_i')$ and $C_j = (V_j', E_j')$, i.e., $E_{ij} = E \cap \{\{u, v\}| u \in V_i', v \in V_j'\}$. Clearly, there are exactly $\binom{md}{2}$ distinct sets E_{ij}. Let $a = \sum_{i<j} |E_{ij}| \leq |H| \leq n + 2(md - 1)$.

By constructing the components $D_1, \ldots D_d$ (each D_k as a union of m components $\bigcup_{i \in I_k} C_i$) we choose $d \cdot \binom{m}{2}$ sets E_{ij} from the set of edges H partitioning G into md components. Now we shall prove

$$|H_m| \leq a \cdot (1 - d\binom{m}{2}/\binom{md}{2}), \tag{2}$$

i.e., there exist $I_1, \ldots, I_d$ such that the number of edges removed from H is proportional to the number of sets E_{ij} removed from H.

Let there be z distinct partitions $\pi_1, \pi_2, \ldots, \pi_z$ of I into $I_1, I_2, \ldots, I_d$. Let for $i = 1, \ldots, z$, $H_m(\pi_i)$ be the subset of edges in H that divides G into d components. Since $|H_m| = min_{i=1,\ldots,z}\{|H_m(\pi_i)|\}$, to prove (2) it suffices to show that

$$z^{-1} \cdot \sum_{r=1}^{z} H_m(\pi_r) \leq a \cdot (1 - d\binom{md}{2}/\binom{md}{2}) \tag{3}.$$

Let l_{ij} be the number of occurences of E_{ij} in $H_m(\pi_1), \ldots, H_m(\pi_z)$. Since we consider all possible π_i's it is clear that $l_{ij} = l_{uv} = l$ for all $i, j, u, v \in \{1, \ldots, md\}$. Since each π_i contains exactly $\binom{md}{2} - d\binom{m}{2}$ E_{ij}'s, we have

$$l = z \cdot (1 - d\binom{m}{2}/\binom{md}{2}). \tag{4}$$

On the other hand, we have

$$\sum_{r=1}^{z} H_m(\pi_r) = \sum_{i<j} l_{ij}|E_{ij}| = l \cdot \sum_{i<j} |E_{ij}| = l \cdot a \tag{5}.$$

Now, following (4) and (5) we already obtain (3):

$$z^{-1} \cdot \sum_{r=1}^{z} H_m(\pi_r) \stackrel{(5)}{=} z^{-1} \cdot l \cdot a \stackrel{(4)}{=} a \cdot (1 - d\binom{m}{2}/\binom{md}{2}).$$

Thus, we have proved inequality (2) which implies

$$\begin{aligned}
|H_m| \;&\le\; a\Big(1 - \tfrac{d\cdot m(m-1)}{2}\big/\tfrac{md(md-1)}{2}\Big)\\
&\le\; (n + 2md - 2)\tfrac{md(md-1)-md(m-1)}{md(md-1)} =\\
&=\; (n + 2dm - 2)\tfrac{md-m}{md-1} = (n + 2dm - 2)\tfrac{d-1}{d}\cdot\tfrac{m}{m-1/d} =\\
&=\; e(n,d,m)
\end{aligned}$$

Clearly, we may choose m such that $e(n,d,m)$ is minimized. Thus,

$$min\{e(n,d,m)\,|\,m \in \mathcal{M}\} \le e\left(n,d,\lceil\sqrt{n}\,\rceil\right) \le n \cdot \tfrac{d-1}{d} + O(\sqrt{n}) \qquad\blacksquare$$

Concluding this section we shall still generalize our result for k-regular graphs where k is an even number.

Theorem 2 *Let k be an even integer. Let G be a k-regular graph of n vertices, where $n = d \cdot h$ for some $d, h \in N$. Then one can partition G into d components, each of h vertices, by removing at most*

$$\frac{(k-2)}{2}n \cdot \frac{d-1}{d} + O\left(\sqrt{n}\right)$$

edges.

Proof: Let k be an even, positive integer. To show our result, it is sufficent to change step 1 of the algorithm $d - PART(m, G)$ in the following way:

1': Remove at most $\frac{(k-2)}{2}n$ edges from G in order to obtain a graph $G = (V, E')$ with the degree bounded by 2.

This can be done because of Petersen 's lemma [12] for each even k and each k-regular graph of n vertices. Then the algorithm $d - PART(m, G)$ runs exactly in the way described in the previous proof. $\qquad\blacksquare$

Concluding Section 2, we note that Theorem 1 and Theorem 2 (and the same is true for the theorems proved in Section 3) hold also for arbitary graphs G of maximal degree 4 or of maximal degree k, respectively. Note that if we allow multiple edges, then by adding edges in an appropiate way we can easily construct a regular graph G' of degree 4 (or of degree k, respectively) such that G is a subgraph of G'. All the algorithms we are describing in this paper work also for regular graphs with multiple edges.

3 Improved Estimates for degree 4

In the previous section, we have presented some results regarding the partitioning of k-regular graphs into d equal-sized components for $d \ge 2$. In this section, we improve these results for $k = 4$ and $d = 2$. We note that we were not able to improve the asymptotic estimates established in Theorem 1 for $d \ge 3$, and we conjecture that the technique used there provides better estimates for larger d's than for smaller d's.

We shall present two algorithms for partitioning 4-regular graphs into two equal-sized components. The first one works very well for small numbers n, and the second one shows that $\sigma(n) \leq n/2 + 1$ for even $n \geq 350$. Before giving our algorithms we claim that small graphs can require the removal of more than $n/2$ edges in order to be 2-partitioned. To illustrate this fact we show the hardest graphs for $n = 12, 16$ and 20 in Figures 1, 2, and 3 respectively. Using a computer program we have shown that the bisection width of these graphs is equal to 10,12 and 14, respectively. Furthermore the $K_{4,4}$ (the complete bipartite graph of 8 vertices) has bisection width 8 and the bipartite graph G_{26}, defined by

$$
\begin{aligned}
G_{26} &= \left(\{u_i, v_i; 1 \leq i \leq 13\}, E_{26}\right), \\
E_{26} &= \left\{\{u_i, v_i\}, \{u_i, v_{i+1}\}, \{u_i, v_{i+2}\}, \{u_i, v_{i+3}\}; 1 \leq i \leq 13\right\} \\
&\qquad \text{(where the addition is performed modulo 13),}
\end{aligned}
$$

has bisection width 16.

These examples show that our Theorem 3 gives optimal results for $n \leq 22$ and for $n = 26$.

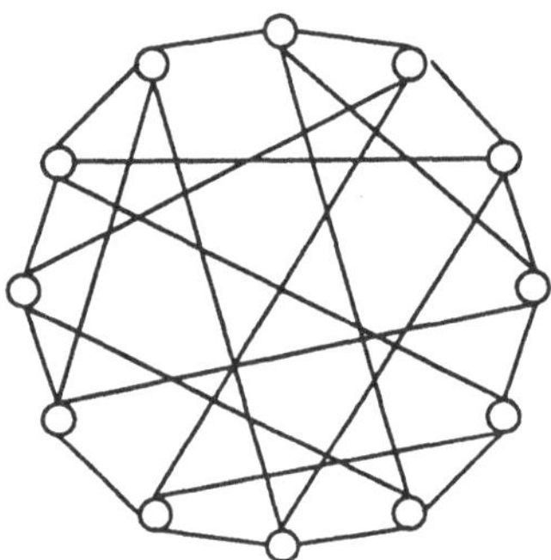

Figure 1, A graph with 12 nodes, $\sigma_4 = 10$

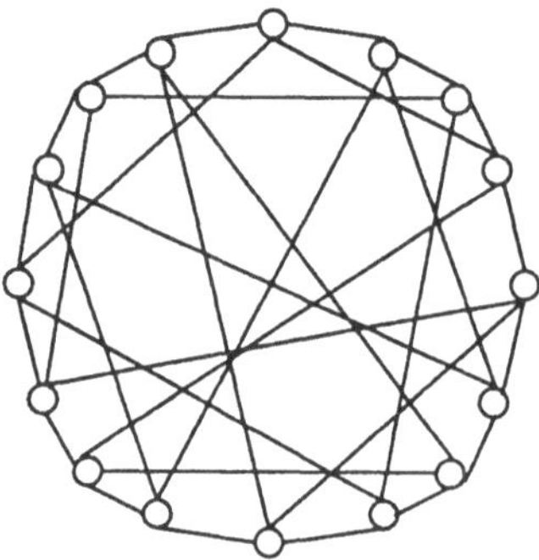

Figure 2, A graph with 16 nodes, $\sigma_4 = 12$

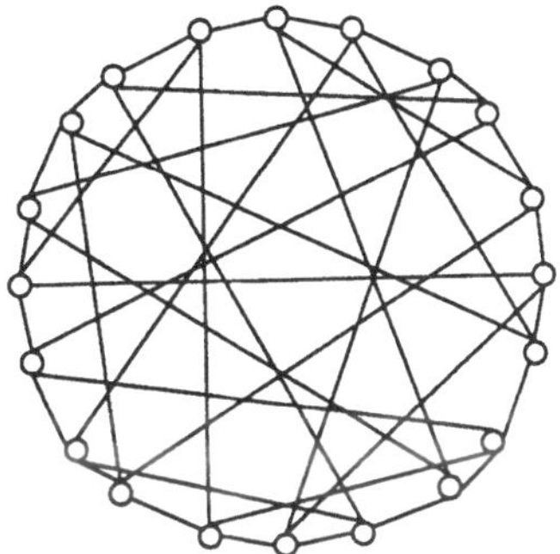

Figure 3, A graph with 20 nodes, $\sigma_4 = 14$

Both our techniques use the so-called "Balancing Lemma" established here, which shows that we are able to construct a balanced partition from an "almost balanced" (specified later) partition by increasing the number of edges between the two components at most by 2.

The first technique works very well for small $n \leq 60$, and it is used to prove that

$$\sigma_4(n) \leq n/2 + 4 \qquad \text{for all } n \equiv 0 \bmod 4,\ n \leq 60$$
$$\sigma_4(n) \leq n/2 + 3 \qquad \text{for all } n \equiv 2 \bmod 4,\ n \leq 60.$$

This technique is based on an iterative approach. Starting with one small component (the smallest cycle in the graph) on one side and one large component on the other side we iteratively increase the amount of vertices in the small component by transfering some small subgraphs from the large component to the small one. The small subgraphs removed are chosen in such a way that the number of its edges is as large as possible in the comparison with the number of its vertices. This iterative process stops after reaching an "almost balanced" partition. The Balancing Lemma is used to obtain the balanced partition of the given graph. Considering the examples stated above, we see that some small graphs also require $n/2 + 4$ edges to be removed in order to divide it into two equal-sized components.

The second technique is used to show that

$$\sigma_4(n) \leq n/2 + 1 \quad \text{for even} \quad n \geq 350.$$

This technique is based on local optimization, i. e. it uses local replacements between two given components. The algorithm based on this technique starts with a balanced partition of the given graph and decreases the number of edges connecting the two components by repeatedly replacing some subgraphs of the two components. The algorithm halts either in a balanced partition with at most $n/2 + 1$ edges or in an "almost balanced" partition with less than $n/2$ edges. Obviously, the Balancing Lemma can be used in the second case in order to obtain the required partition.

Both algorithms presented are simple to use and they work in $O(n^2)$ time. The main difficulty is of mathematical nature: to give the proofs showing that the resulting partitions have a small number of edges between the two components.

Now, we introduce our "Balancing Lemma". Before the formulation we need the following definitions.

Definition 1 *Let $G = (V, E)$ be an undirected regular graph of degree 4. Let $n = |V|$, n even. A <u>partition</u> $\pi = \pi(G)$ is a mapping $\pi : V \to \{1, 2\}$. Set $V_i = V_i(\pi) = \{u \in V; \pi(u) = i\}, i = 1, 2$, and $r(\pi) = |V_1(\pi)|$. For convenience and w.l.o.g., we will always assume that $r(\pi) \leq n/2$ holds. We call $bal(\pi) = n/2 - r(\pi)$ the balance of π, and we call the partition π <u>balanced</u> if $bal(\pi) = 0$ holds.*

Of course, we are not only interested in finding good balanced partitions, but, in constructing them, we have to consider also partitions which are not balanced. Let $Ext(\pi) = \{e \in E; e = \{u, v\} \text{ with } \pi(u) \neq \pi(v)\}, ext(\pi) = |Ext(\pi)|$ be the set of crossing edges and the number of crossing edges, respectively, and let $Int(\pi) = \{e \in E; e = \{u, v\} \text{ with } \pi(u) = \pi(v) = 1\}, int(\pi) = |Int(\pi)|$ be the set (respectively number) of internal edges of V_1. Note that $ext(\pi) = 4 \cdot r(\pi) - 2 \cdot int(\pi)$ holds.

Now, we are giving a definition of the "almost-balanced" partition. Note that the "almost-balanced" property is relative here because it is related also to the number $ext(\pi)$.

A partition π is called <u>almost balanced</u> if there exists some number z such that

(i) $ext(\pi) \geq z \cdot bal(\pi) = z(n/2 - r(\pi))$, and

(ii) $4 \cdot (n/2 + bal(\pi)) \leq 5 \cdot ext(\pi) + 2 \cdot z$.

We characterize the nodes according to the number of crossing edges incident to them. For $u \in V$ let $ext_\pi(u) = |\{v \in V; \{u, v\} \in E, \pi(u) \neq \pi(v)\}|$ be the number of crossing edges incident to u. We call a node u an A node, B node, C node or D node, respectively, if $ext(u) \geq 3$ or $ext(u) = 2, 1$ or 0.

We can now formulate and prove our main lemma which will be very useful for both bisection algorithms.

Lemma 1 *(Balancing Lemma) Let π be an almost balanced partition with $ext(\pi) \geq 4$. Then we can construct a balanced partition $\hat{\pi}$ with $ext(\hat{\pi}) \leq ext(\pi) + 2$.*

Proof: Consider the partition π. While $r(\pi) < \frac{n}{2}$ and there exists an A or B node u in V_2 shift this node u to V_1, i.e. redefine π by setting $\pi(u) = 1$ and leave $\pi(v)$ unchanged for $v \neq u$. Note that the number of crossing edges decreases by shifting an A node and remains unchanged by shifting a B node. If the number of crossing edges has decreased, then we can still shift C nodes or D nodes to V_1. In this way, we finally reach a balanced partition π' with $ext(\pi') \leq ext(\pi)$, or we reach a partition π' with $bal(\pi') \leq bal(\pi), r(\pi) \leq r(\pi') < n/2, ext(\pi') = ext(\pi)$ and $V_2(\pi')$ contains no A nodes and no B nodes. In the first case we have proved the lemma. In the second case the new partition π' (which of course may be the old partition π) still fulfills the conditions of the lemma, and additionally, we know that $V_2(\pi')$ contains no A nodes and no B nodes.

For the sake of convenience, we rename our partition and write π instead of π'. Consider $V_2 = V_2(\pi)$. Let F_C be the set of C nodes and F_D be the set of D nodes in V_2 and let $c = |F_C|$ and $d = |F_D|$. Then $c + d = |V_2| = n - r(\pi)$ and $c = ext(\pi)$.

Now let $U \subset F_C$ be a set of C nodes and let $Int(U) = \{e = \{u, v\}; e \in E, u, v \in U\}$ be its set of internal edges. Every node from U is connected via exactly one of its edges to a node from V_1. Therefore, there are $3 \cdot |U| - 2 \cdot |Int(U)|$ edges connecting U with nodes in $V_2 - U$. If we shift all nodes from U to V_1, then we have defined a new partition π' with $ext(\pi') = ext(\pi) + 2 \cdot (|U| - |Int(U)|)$.

Note that every connected graph of m nodes has at least $m - 1$ edges. We proceed as follows. We compute the connected components of the graph $H = (F_C, Int(F_C))$. If there exists a connected component of at least $bal(\pi) = n/2 - r(\pi)$ nodes, then there exists also a connected graph $(U, Int(U)), U \subset F_C$, with $|U| = bal(\pi)$. Since U is connected, we have $|Int(U)| \geq |U| - 1$, therefore, the balanced partition $\hat{\pi}$ obtained by shifting the nodes from U to V_1 fulfills $ext(\hat{\pi}) \leq ext(\pi) + 2$.

From now on, we assume that every connected component of H has less than $bal(\pi) = n/2 - r(\pi)$ nodes. We will show that in this case there exists a cycle consisting only of C nodes. Let x be the number of connected components of H. Then $ext(\pi) = c < x \cdot (\frac{n}{2} - r(\pi))$ holds, i.e. $x > z$, for any z satisfying conditions (i) of π. Now let us assume that H contains no cycles, i.e. H is a forest.

$\implies$ $|Int(F_C)| = c - x$

$\implies$ There are $3c - 2(c - x) = c + 2x$ edges connecting C nodes with D nodes within V_2 (see the fact stated above that $3 \cdot |U| - 2 \cdot |Int(U)|$ edges connect U with $V_2 - U$)

$\implies$ $4d \geq c + 2x$, since every D node is incident to at most 4 of these edges

$\implies$ $4 \cdot (n - r(\pi)) - 4 \cdot ext(\pi) = 4d \geq c + 2x = ext(\pi) + 2x$ since $d = n - r(\pi) - ext(\pi)$

This is a contradiction because $x > z$ for any z and each z satisfies (ii) $4(n - r(\pi)) \leq 5ext(\pi) + 2z$.

Therefore, we can assume now that there exists a cycle within F_C, i.e. there exists $U \subset F_C$ with $|Int(U)| \geq |U|$. Since U induces a connected graph, we also have that $|U| < bal(\pi)$. We can shift the nodes from U to V_1 and result in a new partition π' with $r(\pi) < r(\pi') < n/2$ (i.e., $bal(\pi') < bal(\pi)$) and $ext(\pi') = ext(\pi)$. With this partition fulfilling (i) and (ii), we can iteratively start the whole construction described in this proof. Finally, we will reach in this way the partition $\hat{\pi}$ we are aiming for. $\blacksquare$

Now, using two graph-theoretical approaches described in the introduction, we are able to prove the following results.

Theorem 3 *Let n be an even number, $n \leq 60$. Then*

$$\sigma_4(n) \leq n/2 + 4, \; \text{if } n \equiv 0 \; mod \; 4,$$
$$\sigma_4(n) \leq n/2 + 3, \; \text{if } n \equiv 2 \; mod \; 4.$$

Theorem 4 *Let n be an even number, $n \geq 350$, then $\sigma_4(n) \leq n/2 + 1$.*

Proof: of Theorem 3: Let $G = (V, E), |V| = n$, be some regular graph of degree 4. We will find the bisection by starting from some cycle U of small length, and then we will iteratively enlarge the set U by trying to get as many internal edges as possible.

Consider $U \subset V, |U| \leq \frac{n}{2}$. Let $int(U)$ be the number of internal edges and $ext(U)$ be the number of edges connecting nodes from U with nodes from $V - U$. Obviously, $ext(U) = 4 \cdot |U| - 2 \cdot int(U)$ holds.

Furthermore, let $C(U)$ be the nodes from $V - U$ which are connected by an edge to a node from U, i.e. $C(U) = \{v \in V; v \notin U, \exists u \in U \text{ with } \{u, v\} \in E\}$. Set $D(U) = V - U - C(U)$ and $\delta(U) = n - |U| - ext(U)$.

We will later use the following two observations:

(1) If $3 \cdot ext(U) > 4 \cdot \delta(U)$, then there exists $\hat{U}$ with $\hat{U} \supset U, |\hat{U}| = |U| + 2, int(\hat{U}) \geq int(U) + 3$.

(2) If $3 \cdot ext(U) > \delta(U)$, then there exists $\hat{U}$ with $\hat{U} \supset U, |\hat{U}| = |U| + 3, int(\hat{U}) \geq int(U) + 4$.

In order to show the validity of these observations, consider the nodes from $C(U)$. If there exists some node $v \in C(U)$ which is connected by at least two edges to nodes from U, then we have found $\hat{U}$ by adding v and one (or two, respectively) nodes from $C(U)$ to U. On the other hand, if there exist two nodes $v_1, v_2 \in C(U)$, which are connected by an edge $\{v_1, v_2\} \in E$, then we have found $\hat{U}$ by adding v_1 and v_2 (and one further node from $C(U)$, respectively) to U. Therefore we can assume now that every node from $C(U)$ is connected by exactly one edge to nodes from U and that there are no edges connecting nodes from $C(U)$. Then $|C(U)| = ext(U), |D(U)| = \delta(U)$ and there exist $3 \cdot |C(U)|$ edges connecting $C(U)$ with $D(U)$. This is not possible if $3 \cdot |C(U)| > 4 \cdot |D(U)|$ holds. Thus, we have shown (1). If $3 \cdot |C(U)| > |D(U)|$, then there exists some $w \in D(U)$ which is connected to two nodes $v_1, v_2 \in C(U)$. We have found $\hat{U}$ by adding v_1, v_2, w to U.

Now, we will prove our theorem by using the above observations and the Balancing Lemma. We will state the proof here only for $16 \leq n \leq 60$ and for $n \equiv 0 \; mod \; 4$. The other cases can be proved in the same way.

First, let $16 \leq n \leq 44$. Note that every regular graph of degree 4 has a cycle of length at most 4 if $n \leq 16$, of length at most 5 if $n \leq 24$ and of length at most 6 if $n \leq 52$. Thus, we know that G contains a cycle of at most $n/4$ nodes. Using this cycle, we can construct a set $U \subset V$ with $|U| = n/4$ and $int(U) \geq n/4$ (by adding some nodes connected to the nodes of the cycle to U).

We want to find $\hat{U}$ with $|\hat{U}| = \frac{n}{4} + 2$ and $int(\hat{U}) \geq \frac{n}{4} + 3$. If $int(U) = n/4$ then $ext(U) = 4 \cdot |U| - 2 \cdot int(U) = n/2$ and $\delta(U) = n - |U| - ext(U) = n/4$. Therefore, $3 \cdot ext(U) > 4 \cdot \delta(U)$ holds, and we get the set $\hat{U}$ from observation (1).

If $int(U) > n/4$ then we find $\hat{U}$ by adding arbitrary nodes from $C(U)$.

Note that $ext(\hat{U}) \leq \frac{n}{2} + 2$. While $ext(\hat{U}) \leq \frac{n}{2} + 2$, we can add arbitrary nodes from $C(\hat{U})$ to $\hat{U}$, till we have found a set $\hat{U}$ with $|\hat{U}| = \frac{n}{2}$ and $ext(\hat{U}) \leq \frac{n}{2} + 2$, or with $n/4 + 2 \leq |\hat{U}| \leq n/2$ and $ext(\hat{U}) = \frac{n}{2} + 2$. In the first case we have proved our theorem. Now, let $n/4 + 2 \leq |\hat{U}| \leq \frac{n}{2}$ and $ext(\hat{U}) = n/2 + 2$. Define π by $\pi(u) = 1$ iff $u \in U$. Then $bal(\pi) = n/2 - |U| \leq \frac{n}{4} - 2$ and $ext(\pi) = \frac{n}{2} + 2$. This partition is "almost balanced", since if we set $z = 2$, then

$$ext(\pi) = n/2 + 2 \geq 2(n/4 - 2) \geq 2 \cdot bal(\pi)$$
$$\text{and} \quad 4 \cdot (n/2 + bal(\pi)) \leq 3n - 8 \leq 5 \cdot n/2 + 14$$
$$= 5 \cdot ext(\pi) + 4 = 5 \cdot ext(\pi) + 2z \text{ if } n \leq 44.$$

Therefore, we find a balanced partititon π' with $ext(\pi') \leq ext(\pi) + 2 \leq \frac{n}{2} + 4$ by using Lemma 3.1 for $n \leq 44$.

Now, we have to consider the case $44 \leq n \leq 60$. G contains a cycle of at most $n/4 - 3$ nodes. Using this cycle, we can construct a set $U \subset V$ with $|U| = n/4 - 3$ and $int(U) \geq n/4 - 3$.

In the following we consider only the most difficult case that $int(U) = n/4 - 3$. Note that $ext(U) = n/2 - 6$ and $\delta(U) = n/4 + 9$. Therefore, $3 \cdot ext(U) > \delta(U)$ holds. Because of observation (2), we can find a set U' with $|U'| = n/4$ and $int(U') \geq n/4 + 1$.

Again, we consider in the following only the most difficult case $int(U') = n/4 + 1$. Note that $ext(U') = n/2 - 2$ and $\delta(U') = n/4 + 2$. Therefore, $3 \cdot ext(U) > 4 \cdot \delta(U)$ holds. It is easy to see that we can apply now observation (1) twice and this leads to a set U'' with $Ext(U'') \leq n/2 + 2$. And in the same way as above we can assume now that $\frac{n}{4} + 4 < |U''| < \frac{n}{2}$ and $ext(U'') = \frac{n}{2} + 2$.

In order to use Lemma 1, it remains to show that the partition π, defined by $\pi(u) = 1$ iff $u \in U''$, is an "almost balanced" partition. If we set $z = 2$, then

$$ext(\pi) = \frac{n}{2} + 2 \geq 2 \cdot (\frac{n}{4} - 4) \geq 2 \cdot bal(\pi) = z \, bal(\pi)$$
$$\text{and} \quad 4(\frac{n}{2} + bal(\pi)) \leq 3n - 16 \leq 5 \cdot \frac{n}{2} + 14$$
$$= 5 \cdot ext(\pi) + 4 = 5 \cdot ext(\pi) + 2z \text{ if } n \leq 60$$

∎

Our second technique uses local transformations. We start with an arbitrary partition. In each step we are looking for some set of nodes laying on one side of the partition whose shift to the other side reduces the number of crossing edges. We call a subset of nodes t-helpful, $t \in N$, if the number of crossing edges is decreased by $2t$ this way. A t-helpful graph is a graph $H = (S, E)$, where S is a t-helpful set and E contains all edges of the original graph G connecting nodes in S. Figure 4 shows two 1-helpful graphs.

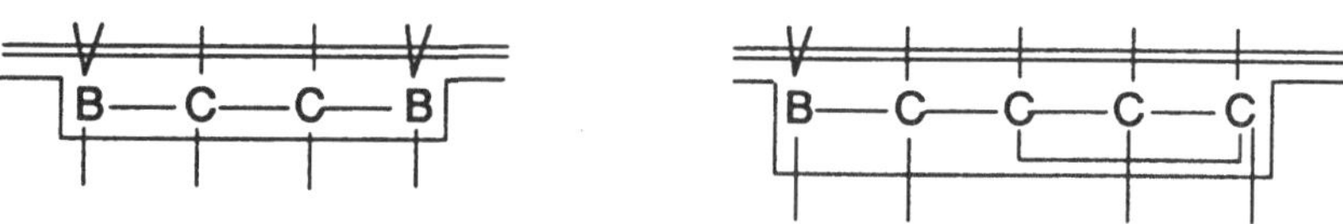

Figure 4: Two 1-helpful graphs

Our method now is the following. Let π be a partition. Determine a 2-helpful graph of small cardinality and define a new partition π' by shifting the nodes of this graph to the other side. Then π' is not balanced, but $ext(\pi') \leq ext(\pi) - 4$. Now compute a balanced partition $\hat{\pi}$ with $ext(\hat{\pi}) \leq ext(\pi') + 2 \leq ext(\pi) - 2$ by using the Balancing Lemma.

Lemma 2 *Let π be a balanced partition with $ext(\pi) > n/2 + 1$. Then we can find a balanced partition $\hat{\pi}$ with $ext(\hat{\pi}) < ext(\pi)$ by exchanging only two nodes or there exists a 2-helpful graph of cardinality at most $4 \cdot \lceil \log_2 n \rceil + 3$.*

Proof: Let us consider first the case that there exist some A nodes in V_1 and V_2 respectively.

If V_1 or V_2, respectively, contains an A node v with 4 external edges then this node v forms a 2-helpful graph. Likewise, if V_1 or V_2, respectively, contains two A nodes u and v (each with 3 external edges and 1 internal edge), then $\{u, v\}$ forms a 2-helpful graph.

On the other hand, if V_1 contains an A node u and V_2 contains one A node v (each with 3 external edges and 1 internal edge), then we define a new balanced partition $\hat{\pi}$ by shifting u to V_2 and v to V_1. Obviously, $ext(\hat{\pi}) = ext(\pi) - 4$ if $\{u, v\} \in E$ and $ext(\hat{\pi}) = ext(\pi) - 2$, otherwise.

Thus, we can assume that either V_1 or V_2 does not contain any A-node. W.l.o.g. we say that V_1 does not contain any A node. Using this assumption we show in what follows that V_1 contains a 2-helpful graph.

Since for any V_1 without A nodes $ext(\pi) = c + 2b$ holds, we obtain $ext(\pi) = c + 2b \geq n/2 + 2 = (b + c + d) + 2$, and thus $b \geq d + 2$ and $b > 1$ holds.

Let $B(V_1), C(V_1), D(V_1)$ be the sets of B nodes, C nodes and D nodes from V_1 and consider $v \in B(V_1)$. Denote by $R(v)$ the set of all nodes from V_1 which can be reached from v by a path of length at most $\lceil \log_2 n \rceil$ which does not contain any D node (Note that the final node of a path may be a D node). Let $G(v)$ be the graph induced by E and $R(v)$. Furthermore, let $s(v)$ be the number of edges in $G(v)$ which are incident to the D nodes and set $R = \cup_{v \in B(V_1)} R(v)$.

First we show, that if $s(v) \leq 1$ for any $v \in B(V_1)$, then $G(v)$ contains a 2-helpful graph of cardinality at most $4 \cdot \lceil \log_2 n \rceil - 1$. Note that $G(v)$ contains the B node v and therefore the above assumption is true if $G(v)$ contains two further B nodes, or one further B node and one cycle, or two cycles (note that such cycles can consist only of C nodes and B nodes). The corresponding 2-helpful graphs are shown in Figure 5.

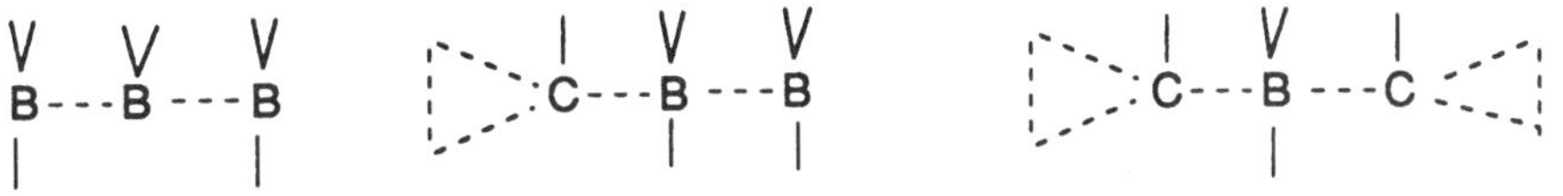

Figure 5: Three 2-helpful graphs

We will show that $G(v)$ contains at least one of these 2-helpful subgraphs, if $s(v) \leq 1$ holds. Assume on the contrary that $s(v) \leq 1$ and $G(v)$ contains only one further B node (in addition to node v) and no cycle, or one cycle and no further B node. We construct a new graph $G'(v)$ from $G(v)$ by erasing the edge to the D node (if it exists) and an edge of the cycle (if it exists) of largest distance from node v, and then by replacing any node of degree 2 (except the node v) by the corresponding edge. $G'(v)$ contains no cycles, it contains (except the node v) no further node of degree 2, and for every node $w \in R'(v)$, where $R'(v)$ is the node set of $G'(v)$, if there exists a path of length λ from v to w in $G(v)$ then there exists a path of length at least $\lambda - 2$ in $G'(v)$. Therefore, $G'(v)$ contains a complete binary tree of height $\lceil \log_2 n \rceil - 2$ and thus has at least $n/2 - 1$ nodes. This is a contradiction since we have erased (at least) two nodes and therefore, $R(v) > n/2$ would follow.

Thus, we can assume now that $s(v) \geq 2$ holds for all B nodes $v \in V_1$. We consider now the B nodes v with $s(v) = 2$. We show that in this case either $R(v) - D(v_1) = \{v\}$ (i.e. v is connected directly to two D nodes), or $G(v)$ contains a 1-helpful graph of cardinality at most $2 \cdot \lceil \log_2 n \rceil + 1$.

Note that $G(v)$ contains a 1-helpful graph of this size, if it contains a cycle or one further B node (in addition to node v). One can show in just the same way as above that under the assumption $s(v) = 2$ this is always the case.

Two 1-helpful graphs form a 2-helpful graph. Thus, we can assume now (otherwise we have proven our lemma) that for at most one node $v \in B(V_1)$ with $s(v) = 2$ the condition $R(v) - D(v_1) = \{v\}$ does not hold, and that at most one of the sets $R(v), v \in B(V_1)$, contains one further B node (in addition to v).

Let us call an edge incident to a D node active if it belongs to $R(v)$ for some $v \in B(V_1)$. Let $E(D)$ be the set of all active edges.

If there exists a D node all of whose 4 edges are active, then we have found a 2-helpful graph of cardinality at most $4 \cdot \lceil \log_2 n \rceil + 1$, since through all of these active edges some B node can be reached by a path of length $\leq \lceil \log_2 n \rceil$ consisting only of C nodes.

Therefore we can assume now, that for every D node at most three of its edges are active. We shall see that in this case there exist two D nodes w_1, w_2 with the following property:

w_1 and w_2 both have 3 active edges and at least two of these edges lead to B nodes v_1, v_2 (which may coincide) with $s(v_1) = s(v_2) = 2$.

In order to see this let d_i, $0 \leq i \leq 3$, be the number of D nodes with i active edges. Obviously, $d = d_0 + d_1 + d_2 + d_3$. Likewise let b_2 be the number of B nodes v with $s(v) = 2$ and let b_3 be the number of B nodes v with $s(v) \geq 3$. Then $b = b_2 + b_3$. Note, that we could assume that at most one of the sets $R(v), v \in V(B_1)$, contains one further B node. Therefore, there exist at most two B nodes w_1, w_2, whose sets $R(w_1)$ and $R(w_2)$ are not disjoint and, thus, may define the same active edges, i.e. $2b_2 + 3b_3 \leq |E(D)| + 3$ holds. Because of the definitions of the $d_i, 0 \leq i \leq 3$, $d_1 + 2d_2 + 3d_3 = |E(D)|$ holds and by subtracting these two relations we get

$$2(b_2 - d_2) + 3(b_3 - d_3) \leq d_1 + 3$$

We combine this with the relations mentioned above

$$b_2 + b_3 = b \geq d + 2 \geq d_1 + d_2 + d_3 + 2,$$

i.e. $(b_2 - d_2) + (b_3 - d_3) \geq d_1 + 2$.

Multiplying this inequality by 3 and subtracting the two inequalities we obtain

$$b_2 - d_2 \geq 2d_1 + 3.$$

i.e. $2b_2 \geq 2d_2 + d_1 + 6$

Thus, there are at least 6 active edges which are incident to B nodes with at most 2 active edges and which cannot be the active edges of any D node with at most 2 active edges. Then any two D nodes, w_1, w_2 each incident to one of these edges fulfill the above property.

We distinguish two cases:

1.) $R(v_1) - D(v_1) = \{v_1\}, R(v_2) - D(v_1) = \{v_2\}$

Then v_1 and v_2 are connected directly to the D nodes w_1 or w_2, respectively, and we have found a 2-helpful graph of cardinality $4\lceil \log_2 n \rceil + 3$.

The 2-helpful graph is shown in figure 6.

Figure 6: The 2-helpful graph for case 1

2.) $R(v_1) - D(v_1) = \{v_1\}, G(v_2)$ contains a helpful graph (and this helpful graph contains v_2 and has cardinality at most $2 \cdot \lceil \log_2 n \rceil + 1$).

Let v_1 be connected with w_1. Then w_1 together with the 3 paths connecting w_1 to the B nodes and together with the helpful graph from $G(v_2)$ form a 2-helpful graph of cardinality at most $4 \cdot \lceil \log_2 n \rceil + 2$. We have to distinguish the cases whether v_2 is associated with w_1 or with w_2, respectively. The 2-helpful graphs are shown in Figure 7.

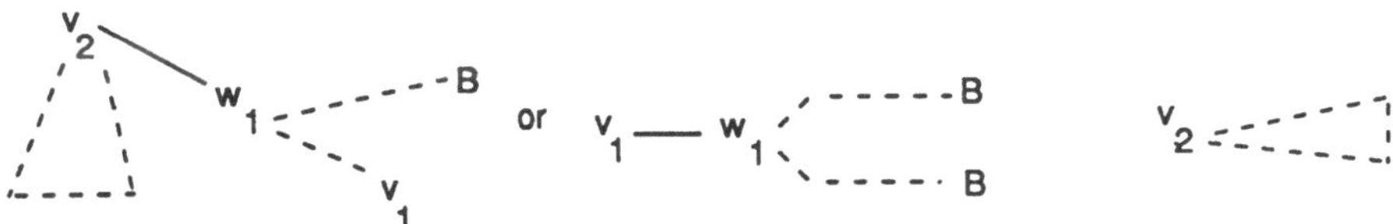

Figure 7: The 2-helpful graphs for case 2

Thus we have shown that we always can find a 2-helpful graph of cardinality at most $4 \cdot \lceil \log_2 n \rceil + 3$. ∎

Combining Lemma 1 and Lemma 2, we get

Proof: of Theorem 4: Let π be a balanced partition with $ext(\pi) > \frac{n}{2} + 1$. Because of Lemma 2 we can decrease the number of crossing edges by exchanging only two nodes, or we find a 2-helpful graph S of cardinality at most $4 \cdot \lceil \log_2 n \rceil + 3$.

Define now an (unbalanced) partition π' by shifting all nodes from S to the other side. Let us show that π' is almost balanced. Since $ext(\pi') = ext(\pi) - 4$ and $bal(\pi') = |S| \le 4\lceil \log_2 n \rceil + 3$ we have:

$$ext(\pi') \ge \frac{n}{2} - 2 \ge 2bal(\pi)$$

$$4\left(\frac{n}{2} + bal(\pi')\right) \le 4(\frac{n}{2} + 4\lceil \log_2 n \rceil + 3) \le 5 \cdot \left(\frac{n}{2} - 2\right) + 4 \le 5 \cdot ext(\pi') + 2 \cdot z$$

for $n \ge 350$. Therefore, by applying the Balancing Lemma we get a balanced partition $\hat{\pi}$ with $ext(\hat{\pi}) \le ext(\pi') + 2 \le ext(\pi) - 2$. While $ext(\pi) > \frac{n}{2} + 2$ we can proceed as above and we will finally get a balanced partition $\hat{\pi}$ with $ext(\hat{\pi}) \le \frac{n}{2} + 1$. ∎

We want to mention that we use essentially breadth first search in the proofs of Lemma 1 and Lemma 2. Thus, the final partitions can be computed within time $0(n^2)$.

4 Optimization of Hardware in Building Transputer Systems

The research presented in this paper is motivated by the problem of configuring transputer systems.

A transputer is a processor with 4 communication links. Therefore it is theoretically possible to use any graph of degree 4 as a communication structure. In a transputer system the physical interconnection network should allow to configure every communication graph of degree 4 in such a way that the edges of the graph can be mapped onto very short paths in the interconnection network. Of course, it would be optimal if we could guarantee mappings where these paths only have length 1. It turns out that the old edge partition theorem of J. Petersen [12] is very helpful in this context. This theorem states, as already mentioned before, that every regular graph of degree 4 can be partitioned into two graphs of degree 2. This result was rediscovered whithin the Esprit Project 1085 [11] and can be used to realize transputer systems as shown in Figure 8.

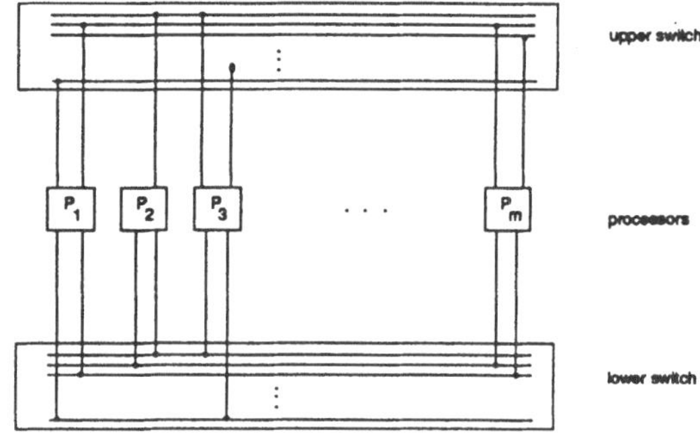

Figure 8: A transputer system with 2 switches

Note that with a switch of N entries transputer systems with at most $N/2$ processors can be built by using the above construction. In order to build systems with a larger number of processors it is quite natural to use a modular approach where two realisations of the above kind are connected by external edges as it is shown in Figure 9.

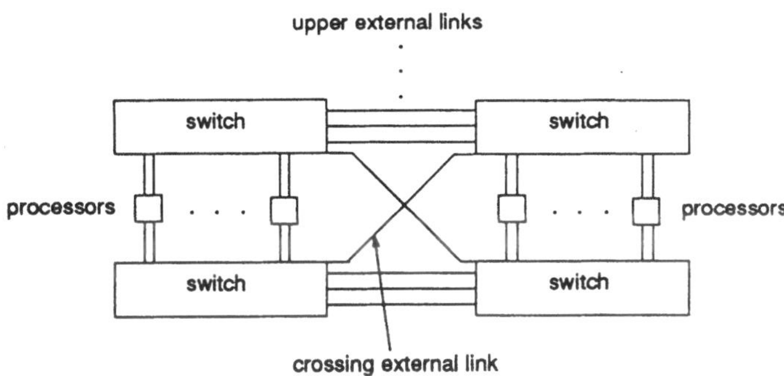

Figure 9: A modular parallel architecture

Note that in this construction all edges of the communication graph can be realized by connections passing at most two switches. The number of processors (we can connect in this way) is maximized by minimizing the number of external links. We will show in Theorem 4.1 that $\sigma_4(n) + 1$ external edges are sufficient where n is the number of processors. In order to guarantee this we need one pair of crossing external links connecting the upper and lower switches. Note that this "configuration problem" is somewhat more difficult than finding a good bisection since the external edges have to be associated with different classes of edges (upper external links, lower external links, crossing external links). We want to mention that transputer systems with 64 processors have been built by using our construction and switches with $N = 96$ entries.

Now, we are able to prove the following theorem which reduces this complex problem of configuring transputers to the problem of finding a reasonable upper bound on $\sigma_4(n)$.

Theorem 5 *For each partition of a 4-regular logical network of n processors into two equal sized parts there is an assignment of logical processors to the physical transputers with the properties:*

(1) the number of all external links is at most $\sigma_4(n) + 1$

(2) the number of upper external links is at most $\lfloor \sigma_4(n)/2 \rfloor$ and the number of lower external links is at most $\lfloor \sigma_4(n)/2 \rfloor - 1$.

(3) there are exactly two crossing links laid out as depicted in Fig. 9 .

Proof: Let G be a 4-regular graph of n vertices, and let π be a bisection of G with $ext(\pi) \leq \sigma_4(n)$. Let us assign the vertices of $V_1(\pi)[V_2(\pi)]$ to the processors of the left module [to the right module] in an arbitrary way. Now, we assign the edges of G

to the links of the transputer system without using any of the two crossing external links. If $ext(\pi) \geq 2$, then we define a new assignment of edges that uses both crossing external links and has the property (2).

Let P be a sequence of all edges $e_1, \ldots, e_m$ of G which is an Eulerian cycle in G. We define the first assignment in such a way that each odd edge in P is assigned to an upper link (internal or external) and each even edge in P is assigned to a lower link. In the correspondence to this assignment we can unambiguously assign a word $w_P = w = w_1 \ldots w_m \in \{u_i, u_e, l_i, l_e\}^m$ to P, where $w_r = u_i(u_e)$ if the r-th edge of P is assigned to an upper internal (external) link and $w_r = l_i(l_e)$ if the r-th edge of P is assigned to an lower internal (external) link. Let $c_1(c_2)$ be the external crossing link connecting the left lower switch (left upper switch) with the right upper (right lower) one.

Let $\#a(u)$ denote the number of occurrences of the symbol a in a word u. We shall call a word $x \in \{u_i, u_e, l_i, l_e\}^*$ a cyclic subword of w if $w = zxy$ for some $z, y \in \{u_i, u_e, l_i, l_e\}^*$, or $x = z_1 z_2$ and $w = z_2 v z_1$ for some $v \in \{u_i, u_e, l_i, l_e\}^+$. Let h be the homomorphism changing upper links to lower links and vice versa, i. e. $h(u_e) = l_e, h(u_i) = l_i, h(l_i) = u_i, h(l_e) = u_e, h(c_1) = c_2$, and $h(c_2) = c_1$.

Now, we shall distinguish two possibilities. If $|\#u_e(w) - \#l_e(w)| \leq 1$ then we change the assignment $A(w)$ correspoding to w in the following way. W. l. o. g. we may assume that $\#u_e(w) \geq \#l_e(w)$ (if not we consider the assignment $A(h(w))$ instead of $A(w)$) and that $\#l_e(w) \geq 1$ (if not, then $A(w)$ is already the assignment we look for). Obviously, there exists a subword $u_e x l_e$ (or $l_e x u_e$) of $w = y_1 u_e x l_e y_2$ where $x \in \{u_i, l_i\}^*$. Now, let us assign the crossing edges c_1 and c_2 to the edges corresponding to u_e and l_e in $u_e x l_e$ in the following way. If u_e corresponds to an edge leading from the right upper switch to the left upper switch then set $r = y_1 c_2 h(x) c_1 y_1$, otherwise set $r = y_1 c_1 h(x) c_2 y_2$. Obviously, the assignment $A(r)$ corresponding to r has the required properties (1),(2), and (3).

Let $|\#u_e(w) - \#l_e(w)| \geq 2$. W. l. o. g. we may assume that $\#u_e(w) = \#l_e(w) + k$ for some $k \geq 2$. Let $i = \lfloor k/2 \rfloor - 1$ if $\lfloor k/2 \rfloor$ is even, and $i = \lfloor k/2 \rfloor$ if $\lfloor k/2 \rfloor$ is odd. Then there exists a cyclic subword $u_e y u_e$ of w such that $\#u_e(y) = \#l_e(y) + i$. W. l. o. g. we may assume $w = z_1 u_e y u_e z_2$ for some $z_1, z_2 \in \{u_e, u_i, l_e, l_i\}^*$. We set either $r = z_1 c_1 h(y) c_2 z_2$ or $r = z_1 c_2 h(y) c_1 z_2$ depending on the direction of the edge corresponding to u_e on the $(|z_1| + 1)$-th position in w (as in the case stated above). Obviously, either $A(r)$ or $A(h(r))$ fulfils the properties (1), (2) and (3) of Theorem 5.

∎

Bibliography

[1] N. Alon Eigenvalues and expanders. Combinatorica 6 (1986), 85-95.

[2] N. Alon and V.D. Milman. λ_1, isometric inequalities for graphs, and superconcentrators. J. Combinatorial Theory B 38 (1985), 73-88.

[3] T.N. Bui, S. Chanduri, F.T. Leighton and M. Sipser. Graph bisection algorithms with good average case behavior. Combinatorica 7 (1987), 171-191.

[4] A. Broder and E. Shamir. On the second eigenvalue of random regular graphs. In: Proc. 28th Annual Symp. on FOCS, IEEE 1987, 286-294.

[5] Gabber and Z. Galil. Explicit constructions of linear-sized superconcentrators. J. Comput. Syst. Sci. 22 (1981), 407-420.

[6] M.R. Garey and D.S. Johnson. Some simplified NP-complete graph problems. Theor. Comp. Science 1 (1976), 237-267.

[7] D.S. Johnson, C.R. Aragon, L.A. Mc Geoch and C. Schevon. Optimization by simulated annealing: An experimental evaluation (Part I). Preprint, AT + T Bell Labs, Murray Hill, NY (1985).

[8] B.W. Kernighan and S. Lin. An efficient heuristic procedure for partitioning graphs. Bell Systems Techn. J. 49 (1970), 291-307.

[9] A. Lubotzky, R. Phillips and P. Sarnak. Ramanujan graphs. Combinatorica 8 (1988), No. 3, 261-277.

[10] H. Mühlenbein, O. Krämer, G. Peise and R. Rinn. The Megaframe Hypercluster - A reconfigurable architecture for massively parallel computers. IEEE Conference on Computer Architecture, Jerusalem 1989.

[11] D.A. Nicole. Esprit Project 1085, Reconfigurable Transputer Processor Architecture, Proc. CONPAR 88, 12-39.

[12] J. Petersen. Die Theorie der regulären Graphs. Acta Mathematica 15 (1891), 193-220.

Complexity of Closeness, Sparseness and Segment Equivalence for Context-Free and Regular Languages

Dung T. Huynh

Computer Science Program
University of Texas at Dallas
Richardson, Texas 75083
USA

Abstract

In this paper, we investigate the complexity of deciding closeness, segment equivalence and sparseness for context-free and regular languages. It will be shown that the closeness problem for context-free grammars (CFGs) is undecidable while it is **PSPACE**-complete for nondeterministic finite automata (NFAs) and **NL**-complete for deterministic finite automata (DFAs). The segment equivalence problems for CFGs and NFAs are co-**NP**-complete. It is **NL**-complete for DFAs. If encoded in binary, the segment equivalence problems for CFGs and NFAs are co-**NE**-complete and **PSPACE**-complete, respectively. The sparseness problems for NFAs and DFAs are **NL**-complete. We also show that the equivalence problems for CFGs and NFAs generating commutative languages are Π_2^{P}-complete and co-**NP**-complete, respectively. For trim DFAs generating commutative languages the equivalence problem is in **L**.

Keywords. Closeness, equivalence, sparseness, context-free language, regular language, complexity, grammar, automata, Turing machine.

1 Introduction

The equivalence problem is certainly an important decision problem that has been investigated extentively in the literature (cf. [4, 6, 7, 9, 18, 20, 21, 12, 13]), where numerous decidability and complexity results can be found. For the important class of context-free languages (CFLs), the equivalence problem is well known to be undecidable. In fact, the undecidability holds already for the class of linear CFLs. It is therefore interesting to develop necessary conditions for the equivalence problem for CFLs and subclasses. An remarkable result along this line is Parikh Theorem which states that the commutative images of CFLs are semilinear sets. This provides us a useful criterion to solve the equivalence problem for CFLs in certain cases: if the commutative images of two CFLs are not the same, they are not equivalent. However, as shown in [13], deciding whether the commutative images of two CFLs coincide is unfortunately Π_2^P-hard. Another nice invariant for CFLs can be found in an elegant theorem by Hotz [8]: the Hotz groups of two equivalent context-free grammars (CFGs) are isomorphic. Hotz's Theorem yields a simple criterion for CFG equivalence by considering the commutative images of Hotz groups. This gives us a polynomial-time algorithm to decide CFG equivalence in one direction ([10]).

In this paper, we study the complexity of two simple criteria for the equivalence problem for CFLs and regular sets. We also classify the complexity of the equivalence problem for the restricted class of commutative languages (in the sense of Ginsburg & Spanier [4]). For a language $L \subseteq \Sigma^*$ the *census function of L*, denoted by c_L, is a function $c_L : \{1\}^* \to \mathbf{N}$ defined by : $c_L(1^n) := \mathrm{Card}(L^{\leq n})$, where $L^{\leq n}$ denotes the set of strings of length less than or equal n in L. Following [1], L is said to be *sparse* if $c_L(1^n) \leq p(n)$ for all n, where $p(n)$ is some fixed polynomial. The two simple criteria for the equivalence problem are as follows. The first criterion involves the notion of closeness studied in structural complexity theory (cf. [22, 19, 14]). The similarity of two languages is measured in terms of the density of their symmetric difference: they are said to be *close* (or p-close) if their symmetric difference is sparse. Thus, in a sense two languages are essentially equivalent iff they differ only by a sparse set. The question we address here is whether deciding the closeness of two languages is substantially easier than deciding their equivalence. The second criterion for the equivalence problem is the segment equivalence problem. Given two languages L_1 and L_2, and an integer n, it is to determine whether $L_1^{\leq n} = L_2^{\leq n}$. We define these decision problems more precisely in the following. Let $\mathbf{C}$ denote a class of grammars or automata. For $M \in \mathbf{C}, L(M)$ denotes the language accepted (generated) by M.

The *closeness problem* for $\mathbf{C}$:

Input. $M_1, M_2 \in \mathbf{C}$

Question. Is $[L(M_1) - L(M_2)] \cup [L(M_2) - L(M_1)]$ a sparse set?

The *segment equivalence problem* for $\mathbf{C}$:

Input. $M_1, M_2 \in \mathbf{C}$ and 1^n.

Question. Is $L(M_1)^{\leq n} = L(M_2)^{\leq n}$?

If n is encoded in binary, it is called the binary-encoded segment equivalence problem. We will also study the complexity of testing sparseness.
The *sparseness problem* for **C**:

Input. $M \in$ **C**

Question. Is $L(M)$ sparse?

The results of this paper are as follows. The closeness problem for CFGs is undecidable while it is **PSPACE**-complete for NFAs. For DFAs the closeness problem is **NL**-complete. Thus, the closeness problem is as hard as the equivalence problem. The segment equivalence problem for CFGs is co-**NP**-complete, and this completeness holds for NFAs as well. For DFAs the segment equivalence problem is **NL**-complete (even with the restriction that the DFAs in the input contain only accessible states). We also show that the binary-encoded segment equivalence problem for CFGs is co-**NE**-complete while it is **PSPACE**-complete if NFAs are under consideration. (Note that the binary-encoded segment equivalence problem for DFAs is essentially the equivalence problem.) The sparseness problem for NFAs and DFAs will be shown to be **NL**-complete while the exact complexity of the sparseness problem for CFGs is presently unknown. As a matter of fact, the question of whether this problem can be solvable in polynomial time was posed as an open problem in [9] and [15]. Finally, we consider the class of commutative CFLs and show that the equivalence problem for this class of languages is Π_2^P-complete. The proof of this fact is based on earlier results by the author in [12], where we proved that the inequivalence problem for CFGs with a unary terminal alphabet is Σ_2^P-complete. We also show that the equivalence problem for NFAs generating commutative languages is co-**NP**-complete while it is **L** if trim DFAs are in the input instead. These results extend known results about the complexity of the equivalence problem for languages over a unary terminal alphabet.

This paper is organized as follows. Section 2 contains some basic definitions and common notations used later on. In Section 3 we study the closeness problem while Section 4 concerns the complexity of the segment equivalence problem. The sparseness problem will be investigated in Section 5. In Section 6 we derive complexity results for the equivalence problem for commutative languages. Section 7 contains some concluding remarks.

2 Preliminaries

In this section we briefly review some commonly known definitions and notations which will be used later on. **N** denotes the set of nonnegative integers. For a finite alphabet Σ, Σ^* denotes the set of all strings over Σ. For a string $w \in \Sigma^*$, $|w|$ denotes its length. A language is sparse if it has a polynomial-bounded census function. The complexity classes **L, NL, P, NP** are defined as usual. **NE** denotes the complexity class NTIME(2^{poly}). For **NL** completeness results $\mathbf{NC}^{(1)}$ reducibilities will be used, whereas logspace reducibilities are used for higher complexity classes (cf. [3] for further notions of parallel computation). $\Sigma_k^{\mathbf{P}}$ and $\mathbf{\Pi}_k^{\mathbf{P}}$ denote the k-th level of the polynomial time hierarchy. The reader is referred to [7] for further complexity theoretic notions.

In the following a nondeterministic finite automaton (NFA) is denoted by a 5-tuple $(Q, \Sigma, \delta, q_0, F)$, where $\delta : Q \times (\Sigma \cup \{\epsilon\}) \to 2^Q$ is the transition relation. (ϵ denotes the empty string.) If $\mathrm{Card}(\delta(q, a)) \leq 1$ for all $q \in Q$ and $a \in \Sigma$, then it is a deterministic finite automaton (DFA). In an FA a state q is said to be accessible if there is a path from the initial state to q. An FA is called accessible if it contains only accessible states. A state q is said to be coaccessible if there is a path from q to some final state in F. An FA is said to be trim if every state is accessible and coaccessible. (N, Σ, P, S) denotes a context-free grammar (CFG), where N is the set of nonterminals, Σ the set of terminals, $S \in N$ the initial symbol and $P \subseteq N \times (N \cup \Sigma)^*$ the finite set of productions. For a CFG G, $|G|$ denotes the size of G. Similarly, $|M|$ denotes the size of an FA M.

Let $L \subseteq \Sigma^*$ be a language, where $\Sigma = \{a_1, \ldots, a_k\}$ is some fixed alphabet. The *Parikh mapping* ψ on Σ^* is a morphism $\psi : \Sigma^* \to \mathbf{N}^k$ such that $\psi(uv) = \psi(u) + \psi(v)$ and for all $i = 1, \ldots, k$, $\psi(a_i)$ is the i-th unit vector in $\mathbf{N}^k$. A set $S \subseteq \mathbf{N}^k$ is called a *semilinear set* (cf. e.g.[6]) if there exist positive integers $n, m_1, \ldots, m_n$ and vectors $v_j^{(l)}$ in $\mathbf{N}^k$ for $l = 1, \ldots, n$, $j = 1, \ldots, m_l$ such that

$$S = \bigcup_{l=1}^n \{v_0^{(l)} + \sum_{j=1}^{m_l} \lambda_j v_j^{(l)} \mid \lambda_j \in \mathbf{N}\}.$$

For the special case $n = 1$, S is called a *linear set*.

A language $L \subseteq \Sigma^*$ is said to be *bounded* ([4]) if there exist strings $w_1, \ldots, w_n$ such that $L \subseteq w_1^* \ldots w_n^*$. L is said to be *commutative* ([4]) if for all $u, v \in L$, $uv = vu$. Note that every subset of $\{0\}^*$ is a commutative language and every commutative language is bounded.

3 Complexity of the Closeness Problem

In this section we classify the complexity of the closeness problem. We show that the closeness problem for CFGs is undecidable. We also show that for NFAs and DFAs it is **PSPACE**- and **NL**-complete, respectively. In particular, **NL** completeness holds even when the DFAs under consideration are accessible.

Theorem 1 *The closeness problem for CFGs is undecidable.*

Proof: We reduce the emptiness problem for Turing machines to the closeness problem for CFGs using the well-known proof technique of describing Turing machine computations by CFLs ([5]). To this end let $L \subseteq \{0, 1\}^*$ be a language accepted by a deterministic Turing machine $M = (Q, \{0, 1\}, \Gamma, \delta, q_0, B, F)$, where we may assume without loss of generality that M has only one semi-infinite tape left-end-marked by $\not\!c$ and $F = \{q_f\}$, and whenever M enters state q_f, it halts in q_f without changing the tape cell being scanned. We modify M slightly by adding a few transitions so that M, in state q_f, nondeterministically selects a symbol $X \in \Gamma$, prints X on the tape cell being scanned, and remains in state q_f without moving the head left or right. Let $M' = (Q, \{0, 1\}, \Gamma, \delta', q_0, B, F)$ be the resulting nondeterministic Turing machine.

Following [5], we construct a CFG G that describes the set of invalid computations of M' on all inputs $w \in \{0, 1\}^*$. Let $\Delta := Q \cup \Gamma \cup \{\#\}$, where $\#$ is a new symbol not in Γ. A valid computation of M on input w is encoded by a string $I_w \in \Delta^*$ defined by:

$$I_w := \alpha_0 \# \alpha_1^R \# \alpha_2 \# \alpha_3^R \# \ldots \# \alpha_f$$

such that

1. for all $i \geq 0$, α_i is a string not ending in B that encodes a configuration of M',

2. α_0 is the initial configuration with input w,

3. for all $j \geq 0$, α_j yields α_{j+1} by one move of M',

4. α_f is a final configuration of M'.

It can easily be shown ([5]) that the set $\text{VAL}_{M'} := \{I_w \mid w \in \{0, 1\}^*\}$ of all valid computations of M' on all inputs $w \in \{0, 1\}^*$ is the intersection of two deterministic CFLs. Moreover, the set $\text{INVAL}_{M'}$ of all invalid computations of M', the complement $\Delta^* - \text{VAL}_{M'}$, is even a linear CFL.

Now observe that $L(M') \neq \emptyset$ iff $\text{VAL}_{M'} \neq \emptyset$. We show further that if $\text{VAL}_{M'} \neq \emptyset$, then $\text{VAL}_{M'}$ is nonsparse. Indeed, suppose that $\text{VAL}_{M'} \neq \emptyset$. Let $I_w = \alpha_0 \# \alpha_1^R \# \ldots$ $\ldots \# \alpha_f \in \text{VAL}_{M'}$, where α_f is an accepting configuration of M'. Let $c := |I_w|$ and $d := |\alpha_f|$. From the construction of M', M' can make $|\Gamma|$ nondeterministic moves in states q_f and q'_f. Thus, for sufficiently large n, there are at least $|\Gamma|^{(n-c)/(d+1)}$ valid computations of M' on input w. Obviously, $\text{VAL}_{M'}$ is nonsparse. We have that $L(M') = \emptyset$ iff $\Delta^* - \text{INVAL}_{M'}$ is sparse iff Δ^* and $\text{INVAL}_{M'}$ are close. Since a CFG for $\text{INVAL}_{M'}$ can be effectively constructed from M', the emptiness problem for Turing machines is recursively reducible to the closeness problem for CFGs. This completes the proof of Theorem 1. ∎

We next show that the closeness problem for DFAs is **NL**-complete. The **NL** hardness holds even when the DFAs in the input are accessible. (Note that graph accessibility is already an **NL**-complete problem.)

Theorem 2 *The closeness problem for (accessible) DFAs is **NL**-complete under* **NC**$^{(1)}$ *reducibilities.*

Proof: We first show the **NL** upper bound. To this end, let $M_1 = (Q_1, \Sigma, \delta_1, q_1, F_1)$ and $M_2 = (Q_2, \Sigma, \delta_2, q_2, F_2)$ be the two DFAs in the input. To determine whether

$L(M_1)$ and $L(M_2)$ are close, we check whether $L(M_1) - L(M_2)$ and $L(M_2) - L(M_1)$ are sparse. Without loss of generality we only need to show how to verify that $L(M_1) - L(M_2)$ is sparse by a nondeterministic logspace-bounded Turing machine. We construct from M_1 and M_2 a DFA N accepting $L(M_1) - L(M_2)$ by an $\mathbf{NC}^{(1)}$ circuit. The DFA N is modified into an NFA N' with a unary input alphabet $\{0\}$ by relabeling all transitions in N by 0. Obviously, $L(N)$ is sparse iff the NFA N' is polynomially ambiguous. As shown in [2], the problem of determining whether a given NFA is polynomially ambiguous is $\mathbf{NL}$-complete under $\mathbf{NC}^{(1)}$ reducibilities. This fact together with the closure of $\mathbf{NL}$ under $\mathbf{NC}^{(1)}$ reducibilities ([16]) imply that the closeness problem for DFAs is in $\mathbf{NL}$. (Note that one can also use the $\mathbf{NL}$ upper bound for the sparseness problem for NFAs in Lemma 2 to obtain the same upper bound for testing whether $L(N)$ is sparse.)

We now show $\mathbf{NL}$ hardness. The idea is to reduce the accessibility problem for directed graphs, a problem well-known to be $\mathbf{NL}$-complete under $\mathbf{NC}^{(1)}$ reducibilities ([3]), to the closeness problem for DFAs. The graph accessibility problem (GAP) is the problem of deciding for a given (directed) graph G and two vertices s and g whether there is a path from s to g in G. A special case of GAP is the accessibility problem for graphs whose vertices have outdegrees ≤ 2, denoted by 2-GAP. An even more restricted version of 2-GAP is the monotone graph accessiblity problem for graphs whose vertices are numbered so that edges go from lower-numbered vertices to higher-numbered vertices. Let monotone 2-GAP denote this problem. It is known that monotone 2-GAP is also $\mathbf{NL}$-complete under $\mathbf{NC}^{(1)}$ reducibilities. We reduce monotone 2-GAP to the closeness problem for DFAs. To this end, let (G, s, g) be an instance of monotone 2-GAP, where $G = (V, E)$, s is the source vertex and g is the goal vertex. Without loss of generality we may assume that $indegree(s) = 0$ and $outdegree(g) = 0$, and the vertices in V are numbered so that $V = \{v_1, v_2, \ldots, v_n\}$ with $s = v_1$, $g = v_n$. We first modify G by adding vertices s', u, v and edges $(s', u), (u, v), (v, u)$ and (v, s). Let G' be the resulting graph with set of vertices $V' = V \cup \{s', u, v\}$. For the new instance (G', s', g) of 2-GAP (*not* an instance of monotone 2-GAP), it holds that there is a path from s' to g in G' iff there is a path from s to g in G. Obviously, G' can be constructed from G by an $\mathbf{NC}^{(1)}$ circuit. We now construct from G' an instance of the closeness problem for DFAs. We first construct a DFA M as follows. $M = (Q, \Sigma, \delta, s', \{g\})$, where $Q = V' \cup \{q_r\}$, $\Sigma = \{0, 1\}$, and δ is defined for each vertex v_i by:

1. if $outdegree(v_i) = 2$: let v_j, v_k be two vertices adjacent to v_i and define $\delta(v_i, 0) = v_j$ and $\delta(v_i, 1) = v_k$,

2. if $outdegree(v_i) = 1$: let v_j be the vertex adjacent to v_i and define $\delta(v_i, a) = v_j$ for all $a \in \Sigma$,

3. if $outdegree(v_i) = 0$: define $\delta(v_i, a) = q_r$ for all $a \in \Sigma$,

4. define $\delta(q_r, a) = q_r$ for all $a \in \Sigma$,

5. define $\delta(s', a) = u$, $\delta(u, a) = v$ for all $a \in \Sigma$, and $\delta(v, 0) = s$, $\delta(v, 1) = u$.

Clearly, the DFA M is well defined and it holds that there is a path from s' to g in G' iff $L(M) \neq \emptyset$. Observe that $L(M) \neq \emptyset$ iff $L(M)$ is nonsparse. Indeed suppose that $L(M) \neq \emptyset$. Let $w \in L(M)$. Obviously, w is of the form aw' for some $a \in \Sigma$ and $w' \in \Sigma^*$. Then the nonsparse set $a\{01, 11\}^*w'$ is contained in $L(M)$ which is therefore also nonsparse.

Since the DFA M may contain inaccessible states, we construct from M a DFA M_1 with only accessible states as follows. $M_1 = (Q_1, \Sigma, \delta_1, q_0, \{g\})$, where $Q_1 = Q \cup \{q_0, q_1, \ldots, q_n\}$ and δ_1 is defined by:

- $\delta_1(q, a) = \delta(q, a)$ for all $q \in Q$ and $a \in \Sigma$,

- $\delta_1(q_0, 0) = s'$

- $\delta_1(q_i, 0) = v_i$ for $1 \leq i \leq n$,

- $\delta_1(q_i, 1) = q_{i+1}$ for $0 \leq i \leq n - 1$,

- $\delta_1(q_n, 1) = g$.

It can easily be seen that M_1 contains only accessible states. Further, since the graph G is monotone, if there is no path from s to g in G, then $L(M_1)$ is finite. Thus, there is a path from s to g in G iff there is a path from s' to g in G' iff $L(M_1)$ is nonsparse. Now let M_2 be a DFA with only accessible states that accepts the set $\{\epsilon\}$. Then, $L(M_1)$ is sparse iff $L(M_1)$ and $L(M_2)$ are close. As the construction of M_1 and M_2 from (G, s, g) can be carried out by an $\mathbf{NC}^{(1)}$ circuit, we obtain an $\mathbf{NC}^{(1)}$ reduction of monotone 2-GAP to the closeness problem for DFAs. We conclude that the closeness problem for DFAs is $\mathbf{NL}$-complete under $\mathbf{NC}^{(1)}$ reducibilities. $\mathbf{NL}$ hardness holds even when the DFAs in the input are accessible. ∎

Finally we show the **PSPACE** completeness of the closeness problem for NFAs.

Theorem 3 *The closeness problem for NFAs is* **PSPACE**-*complete under logspace reducibilities.*

Proof: To show the **PSPACE** hardness, we reduce the computations of a polynomial space-bounded deterministic Turing machine to the closeness problem for NFAs. The reduction is essentially the generic reduction by Meyer & Stockmeyer in [18]. Let $N = (Q, \{0, 1\}, \Gamma, \delta, q_0, B, F)$ be a $p()$ space-bounded deterministic Turing machine, where $p()$ is some fixed polynomial. We assume without loss of generality that N has only one semi-infinite tape left-end-marked by $\notin$ and $F = \{q_f\}$, and whenever N enters state q_f, it halts in q_f without changing the content of the tape cell being scanned. As in the proof of Theorem 1, we modify N slightly by adding a few transitions so that N, in state q_f, nondeterministically selects a symbol $X \in \Gamma$, prints it on the tape cell being scanned, and remains in state q_f without moving the head left or right. Let N'

be the resulting $p()$ space-bounded nondeterministic Turing machine. Following [18] one constructs from an input string w of N' a regular expression α_w that describes the invalid computations of N' on w so that $w \in L(N')$ iff there exists a valid computation of N' on input w iff $\Delta^* - L(\alpha_w) \neq \emptyset$, where Δ is a suitable alphabet. Further, as argued in the proof of Theorem 1, the set $\Delta^* - L(\alpha_w) \neq \emptyset$ iff it is nonsparse. Now from α_w one can construct an equivalent NFA M_1. Let M_2 be an NFA accepting Δ^*. We have that $w \in L(N')$ iff $L(M_1)$ and $L(M_2)$ are close. Since the construction of the NFAs M_1 and M_2 from w can be carried out by a logspace-bounded Turing machine, we obtain a logspace reduction of $L(N)$ to the closeness problem for NFAs. This yields the **PSPACE** hardness.

To show the **PSPACE** upper bound we argue as in the proof of Theorem 2. Let M_1 and M_2 be the two NFAs in the input. From M_1 and M_2 one can construct two DFAs N_1 and N_2 which accept the sets $L(M_1) - L(M_2)$ and $L(M_2) - L(M_1)$, respectively. Note that the size of the DFAs N_1 and N_2 are exponential in terms of the size of the NFAs M_1 and M_2. However, N_1 and N_2 can be constructed from M_1, M_2 by a deterministic Turing machine using an amount of space polynomial in the size of M_1 and M_2. (Note that in measuring the space complexity of a Turing machine with output tape, the output tape is not counted.) We then relabel the transitions in N_1 and N_2 using the unary alphabet $\{0\}$ to obtain the NFAs N_1' and N_2'. Since the problem of determining whether a given NFA is polynomially ambiguous is **NL**-complete under $\mathbf{NC}^{(1)}$ reducibilities, we can test whether N_1' and N_2' are polynomially ambiguous using a amount of space polynomial in the size of M_1 and M_2. Using the familiar construction showing that logspace computations are closed under composition, we obtain a polynomial space-bounded deterministic Turing machine for the closeness problem for NFAs. This completes the proof of Theorem 3. ∎

4 Complexity of the Segment Equivalence Problem

In this section we study the complexity of the segment equivalence problem. We show that the segment equivalence problems for CFGs and NFAs are both co-**NP**-complete. For (accessible) DFAs it is **NL**-complete. The binary-encoded segment equivalence problems for CFGs and NFAs will be shown to be co-**NE**- and **PSPACE**-complete, respectively.

Theorem 4 *The segment equivalence problem for NFAs is co-**NP**-complete under logspace reducibilities.*

Proof: The co-**NP** upper bound can be shown easily as follows. Let $(M_1, M_2, 1^n)$ be an instance of the segment equivalence problem for NFAs. Clearly, $L(M_1)^{\leq n} = L(M_2)^{\leq n}$ iff for all $w \in \Sigma^{\leq n}$, $w \in L(M_1) \Leftrightarrow w \in L(M_2)$, where Σ is the input alphabet of M_1 and M_2. As the membership problem for NFAs is obviously solvable in polynomial time, the co-**NP** upper bound for the segment equivalence problem follows.

The co-**NP** hardness essentially follows from the co-**NP** hardness of the equivalence problem for regular expressions involving concatenation and union only. The co-**NP** hardness of the equivalence problem for such regular expressions is shown by constructing a generic reduction of a language $L \in$ co-**NP** to regular expression equivalence. Let L be the complement of a language accepted by a polynomial time-bounded nondeterministic Turing machine. The reduction computes for an input string w a regular expression α_w such that $w \in L$ iff $L(\alpha_w) = \Delta^{p(n)}$, where Δ is a suitable alphabet and $p()$ is some fixed polynomial. (Note that $L(\alpha_w)$ is contained in $\Delta^{p(n)}$.) This implies the co-**NP** hardness of the segment equivalence problem for NFAs.   ∎

Corollary 1 *The binary-encoded segment equivalence problem for NFAs is* **PSPACE**-*complete under logspace reducibilities.*

Proof: This follows easily from the **PSPACE** completeness of the equivalence problem for NFAs. Indeed, if the integer n in the input is encoded in binary, then the **PSPACE** hardness of the equivalence problem implies the **PSPACE** hardness of the segment equivalence. The **PSPACE** upper bound is obvious.   ∎

Theorem 5 *The segment equivalence problem for CFGs in co-**NP**-complete under logspace reducibilities.*

Proof: The co-**NP** hardness is a direct consequence of the co-**NP** hardness of the segment equivalence problem for NFAs. The co-**NP** upper bound follows from the fact that the membership problem for CFGs is in **P** by an application of the well-known algorithm by Earley ([6]). (Note that Earley algorithm does not require that the given CFG contain no ϵ-productions.)   ∎

Corollary 2 *The binary-encoded segment equivalence problem for CFGs is co-**NE**-complete under logspace reducibilities.*

Proof: The upper bound follows by an argument similar to the proof of the upper bound in Theorem 5. The lower bound follows from the co-**NE** hardness of the equivalence problem for CFGs generating finite languages ([9]).   ∎

Finally we show the **NL** completeness of the segment equivalence problem for DFAs.

Theorem 6 *The segment equivalence problem for (accessible) DFAs is* **NL**-*complete under* **NC**$^{(1)}$ *reducibilities.*

Proof: We first show the **NL** upper bound. To this end, let $(M_1, M_2, 1^n)$ be an instance of the segment equivalence problem for DFAs, where $M_1 = (Q_1, \Sigma, \delta_1, q_1, F_1)$ and $M_2 = (Q_2, \Sigma, \delta_2, q_2, F_2)$ are DFAs. From M_1 and M_2, we construct two DFAs N_1 and N_2 that accept $L(M_1) - L(M_2)$ and $L(M_2) - L(M_1)$, respectively. Clearly, $L(M_1)^{\leq n} = L(M_2)^{\leq n}$ iff $L(N_1)^{\leq n}$ and $L(N_2)^{\leq n}$ contains no strings of length $\leq n$. Thus, we only need to verify that in N_1 and N_2 there do not exist paths of length $\leq n$ from an intial state to some final state. As **NL** is closed under $\mathbf{NC}^{(1)}$ reducibilities, this task can be done in **NL**. This proves the upper bound.

We next show **NL** hardness. Note that in [2] we showed the **NL** completeness of the bounded non-universality for DFAs which would imply the **NL** hardness of the segment equivalence problem for DFAs since bounded universality is a special case of segment equivalence. However, the DFA constructed in [2] may contain inaccessible states. We therefore proceed as follows. Consider the DFA M_1 constructed in the proof of the **NL** hardness in Theorem 2 for a given instance (G, s, g) of monotone 2-GAP. Let M_2 be the DFA obtained from M_1 by removing the states s', u, v and adding a new transition $\delta_2(q_0, 0) = q_r$. The DFA M_2 contains only accessible states and accepts a subset of $L(M_1)$ which is contained in $\Sigma^{\leq n+1}$. It can easily be seen that $L(M_1)^{\leq n+3} = L(M_2)^{\leq n+3}$ iff there is no path from s to g in G. We conclude that the segment equivalence problem for DFAs containing only accessible states is **NL**-hard under $\mathbf{NC}^{(1)}$ reducibilities. This completes the proof of Theorem 6. ∎

5   Complexity of the Sparseness Problem

This section is devoted to the study of the sparseness problem. While the exact complexity of the sparseness problem for CFGs is presently unknown, we show that the sparseness problems for NFAs and DFAs are both **NL**-complete.

Lemma 1 *The sparseness problem for (accessible) DFAs in* **NL**-*hard under* $\mathbf{NC}^{(1)}$ *reducibilities.*

Proof: Recall the proof of the **NL** hardness of the closeness problem for DFAs in which we construct the DFA M_1 from an instance (G, s, g) of monotone 2-GAP. M_1 contains only accessible states. Further it holds that there is a path from s to g in G iff $L(M_1)$ is nonsparse. This completes the proof. ∎

Lemma 2 *The sparseness problem for NFAs is in* **NL**.

Proof: Let $M = (Q, \Sigma, \delta, q_0, F)$ be an NFA. Without loss of generality we assume that every state $q \in Q$ is accessible and coaccessible, i.e. q lies on some path from q_0 to some final state in F. We apply some results in [4] and [15].

Fact 1. ([4]) A language L is commutative iff there exists a string x such that $L \subseteq x^*$.

Fact 2. ([15]) A CFL L is bounded iff it is sparse.

For a state $q \in Q$, let M_q denote the NFA $(Q, \Sigma, \delta, q, \{q\})$. From Lemma 5.5.6 in [4] and Fact 1, we have

Fact 3. $L(M)$ is sparse iff for all $q \in Q$, $L(M_q)$ is commutative.

Let $q \in Q$ be some state. To test whether $L(M_q)$ is commutative we proceed as follows. From Fact 1 it follows that $L(M_q)$ is commutative iff $L(M_q) \subseteq x^*$, where x is a prefix of any string $w \in L(M_q)$. Such a string w of length $\leq |Q|$ can easily be computed by a nondeterministic logspace-bounded Turing machine. Let x be a prefix of w. We verify whether $L(M_q) \subseteq x^*$ by constructing a DFA $M_{\bar{x}}$ that accepts $\Sigma^* - x^*$, the complement of x^*, and then determining whether $L(M_q) \cap L(M_{\bar{x}}) = \emptyset$. This can be done on a nondeterministic logspace-bounded Turing machine as well. Noting that **NL** is closed under $\mathbf{NC}^{(1)}$ reducibilities, it is easy to see that testing whether $L(M_q)$ is commutative for all $q \in Q$ can be done in **NL**. This completes the proof of Lemma 2. ∎

We note that the proof of Lemma 2 can be slightly modified so that the **NL** upper bound holds for the sparseness problem for linear CFGs as well. From Lemmas 1 and 2 we obtain as corollary

Theorem 7 *The sparseness problems for NFAs and (accessible) DFAs are both* **NL**-*complete under* $\mathbf{NC}^{(1)}$ *reducibilities.*

Since a language $L \subseteq \Sigma^*$ is sparse iff Σ^* and $\bar{L}(= \Sigma^* - L)$ are close, we obtain the following

Corollary 3 *(1) Deciding for a given DFA M whether $\overline{L(M)}$ is sparse is* **NL**-*complete under* $\mathbf{NC}^{(1)}$ *reducibilities.*
(2) Deciding for a given NFA M whether $\overline{L(M)}$ is sparse is **PSPACE**-*complete under logspace reducibilities.*
(3) Deciding for a given CFG G whether $\overline{L(G)}$ is sparse is undecidable.

Proof: (1) follows immediately from Theorem 7 whereas (2) follows from the proof of Theorem 3. (3) follows from the proof of Theorem 1. ∎

6 Complexity of the Equivalence Problem for Commutative Languages

In this section we study the complexity of the equivalence problem for commutative languages. We show that the equivalence problems for CFGs and NFAs generating commutative languages are $\mathbf{\Pi}_2^\mathbf{P}$-complete and co-**NP**-complete, respectively. These results are generalizations of the same results for languages over a unary alphabet.

We also show that the equivalence problem for trim DFAs generating commutative languages is in **L**. We first consider commutative CFLs. In the following we will make use of a result in [11] about the complexity of the commutative (uniform) word problem for CFGs.

The *commutative word problem* for CFGs:

Input. A CFG $G = (N, \{a_1, \ldots, a_k\}, P, S)$ and a vector $v \in \mathbf{N}^k$.

Question. Is $v \in \psi(L(G))$?

Lemma 3 *The commutative word problem for CFGs is* **NP**-*complete under logspace reducibilities.*

Proof: See [11], where the proof was carried out even for commutative CFGs (in which commutative strings are encoded as binary-integer vetors). ∎

We now show the following

Theorem 8 *The equivalence problem for CFGs generating commutative languages is* $\Pi_2^{\mathbf{P}}$-*complete under logspace reducibilities.*

Proof: The $\Pi_2^{\mathbf{P}}$ hardness follows directly from the $\Pi_2^{\mathbf{P}}$ hardness of the equivalence problem for CFGs with a unary terminal alphabet since languages over a unary alphabet are obviously commutative. In the following we show the $\Pi_2^{\mathbf{P}}$ upper bound.

Let $G_1 = (N_1, \Sigma, P_1, S_1)$ and $G_2 = (N_2, \Sigma, P_2, S_2)$ be two CFGs, where $\Sigma = \{a_1, \ldots, a_k\}$. Assume that $L(G_1)$ and $L(G_2)$ are commutative, and, without loss of generality, that both contain some nonempty string. Since $L(G_1)$ and $L(G_2)$ are commutative, there exist strings w_1 and w_2 in Σ^* such that $L(G_1) \subseteq w_1^*$ and $L(G_2) \subseteq w_2^*$. Without loss of generality let $|w_1| \geq |w_2|$. It follows that $L(G_1) = L(G_2)$ implies that $L(G_1)$ and $L(G_2)$ are both contained in w_2^*. Let $w := w_2$. For $i = 1, 2$ define $S_i = \{m \mid w^m \in L(G_i)\}$. Then, $L(G_1) = L(G_2)$ iff $S_1 = S_2$.

Claim 1 *There exists a fixed polynomial* $p_1()$ *such that* $S_1 = S_2$ *iff for all* $m \leq 2^{p_1(|G_1|+|G_2|)}$, $m \in S_1 \Longleftrightarrow m \in S_2$.

Proof Sketch: (of Claim 1). First note that, by Parikh Theorem, $\psi(L(G_1))$ and $\psi(L(G_2))$ are semilinear sets in $\mathbf{N}^k$. Further, since $L(G_1)$ and $L(G_2)$ are subsets of w^*, $\psi(L(G_1))$ and $psi(L(G_2))$ are also subsets of the linear set

$$\mathrm{Lin}(0; \psi(w)) = \{\lambda\psi(w) \mid \lambda \in \mathbf{N}\}.$$

Now consider the morphism $\varphi : \mathbf{N} \to \mathbf{N}^k$ defined by $\varphi(1) := \psi(w)$. Obviously, φ is a bijection from $\mathbf{N}$ onto $\mathrm{Lin}(0; \psi(w))$. A careful analysis of the proof of Parikh Theorem shows that we can obtain semilinear set representations for M_1 and M_2 from the representations of $\psi(L(G_1))$ and $\psi(L(G_2))$ since the constants and periods in the representations of $\psi(L(G_1))$ and $\psi(L(G_2))$ are multiples of $\psi(w)$. In this manner, G_1

and G_2 may be regarded as CFGs generating languages over the unary alphabet $\{0\}$. We therefore can apply Proposition 4.6 in [12] and obtain Claim 1. ∎

For a vector $u = (u_1, \ldots, u_k) \in \mathbf{N}^k$, $|u|$ denotes $\sum_{i=1}^{k} u_i$. We prove

Claim 2 *There exists a fixed polynomial $p_2()$ such that $\psi(L(G_1)) = \psi(L(G_2))$ iff for all vector $v \in \mathbf{N}^k$ with $|v| \le 2^{p_2(|G_1|+|G_2|)}$, $v \in \psi(L(G_1)) \Longleftrightarrow v \in \psi(L(G_2))$.*

Proof: (of Claim 2). One simply notes that there exists a fixed polynomial $q()$ such that $|w| \le 2^{q(|G_1|+|G_2|)}$. ∎

We now show an effective characterization of the equivalence of two commutative CFLs.

Claim 3 *Let G_1 and G_2 be two CFGs such that $L(G_1)$ and $L(G_2)$ are commutative CFLs. Further for $i = 1, 2$ let $L(G_i) \cap \Sigma^+ \ne \emptyset$. Then $L(G_1) = L(G_2)$ iff the following conditions hold:*

(1) *For all vector $v \in \mathbf{N}^k$ with $|v| \le 2^{p_2(|G_1|+|G_2|)}$, $v \in \psi(L(G_1)) \Longleftrightarrow v \in \psi(L(G_2))$.*

(2) *For all positive integer $l \le 2^{p_2(|G_1|+|G_2|)}$, $\sigma \in \Sigma$, σ is the l-th symbol of some string in $L(G_1)$ iff σ is the l-th symbol of some string in $L(G_2)$.*

Proof: (of Claim 3). We only need to show that conditions (1) and (2) imply $L(G_1) = L(G_2)$. Recall that $L(G_1) \subseteq w_1^*$ and $L(G_2) \subseteq w_2^*$ for some strings w_1 and w_2 in Σ^*. Let $|w_1| \ge |w_2|$. Then for $i = 1, 2$, w_i is prefix of every string in $L(G_i)$. Therefore, condition (2) implies that w_1 is a prefix of w_2. Let w be a prefix of w_1, where $|w|$ is the greatest common divisor of $|w_1|$ and $|w_2|$. We claim that $L(G_1)$ and $L(G_2)$ are both contained in w^*. Indeed, let x be some string of length $\le 2^{p_2(|G_1|+|G_2|)}$ in $L(G_1)$. Then the commutativity of $L(G_1), L(G_2)$ and conditions (1) and (2) imply that x also belongs to $L(G_2)$. Thus, there exist integers $r, s \ge 1$ such that $w_1^r = x = w_2^s$. From Lemma 5.5.2 in [4] it follows that $w_1 w_2 = w_2 w_1$, which implies that $w_1 = w_2^j w_3$, where $j \ge 1$ and $|w_3| = |w_1| \bmod |w_2|$. If $|w_2|$ divides $|w_1|$, then $w_3 = \epsilon$ and $w = w_2$, and $L(G_1)$ and $L(G_2)$ are both contained in w^*. Otherwise, $0 < |w_3| < |w_2|$ and $w_2 w_3 = w_3 w_2$. Continuing this procedure we ultimately obtain a prefix w of w_1 and w_2 with $|w| = \gcd(|w_1|, |w_2|)$ such that $w_1 = w^{r'}$ and $w_2 = w^{s'}$ for some $r', s' \ge 1$. Obviously, $L(G_1)$ and $L(G_2)$ are both contained in w^*.

Observe that $|w|$ is bounded by $|w_1|$ and $|w_2|$, and hence by $2^{p_2(|G_1|+|G_2|)}$. Now consider the sets M_1 and M_2 defined above. From the proofs of Claims 1 and 2, it follows that condition (1) implies that $M_1 = M_2$, which in turn implies that $L(G_1) = L(G_2)$. This completes the proof of Claim 3. ∎

Applying Claim 3 we determine whether $L(G_1) = L(G_2)$ by verifying conditions (1) and (2). Obviously, condition (1) can be tested by a $\mathbf{\Pi}_2^{\mathbf{P}}$ machine since the commutative word problem for CFGs is, by Lemma 3, in **NP**. We next show that condition (2) can be tested by a $\mathbf{\Pi}_2^{\mathbf{P}}$ machine also. To this end, we only need to show

248   *Dung T. Huynh*

that the problem of determining for a positive integer $l \leq 2^{p_2(|G_1|+|G_2|)}$ and $\sigma \in \Sigma$ whether σ is the l-th symbol of some string in $L(G_1)$ (or $L(G_2)$) is in **NP**. Note that every string in $L(G_1)$ is of the form w^r for some $r \geq 1$. We construct from G_1 a CFG G_1' that generates the set

$$U_1 := \{0^n \sigma \mid \text{there exist} x, y \in \Sigma^* : |x| = n \text{ and } x\sigma y \in L(G_1)\},$$

where 0 is a new terminal symbol not in Σ. It is not hard to see that G_1' can be constructed from G_1 in polynomial time. Further, it holds that σ is the l-th symbol of some string in $L(G_1)$ iff $0^{l-1}\sigma \in L(G_1')$. Now consider the Parikh image $\psi(L(G_1')) \subseteq \mathbf{N}^{k+1}$. Since 0 is not in Σ, we have that $0^{l-1}\sigma \in L(G_1')$ iff $\psi(0^{l-1}\sigma) \in \psi(L(G_1'))$. Thus, the problem of determining for an integer $l \leq 2^{p_2(|G_1|+|G_2|)}$ and $\sigma \in \Sigma$ whether σ is the l-th symbol of some string in $L(G_1)$ is polynomial time reducible to the commutative word problem for CFGs, and hence belongs to **NP** also. Thus, we conclude that condition (2) of Claim 3 can be tested by a $\mathbf{\Pi_2^p}$ machine as well. Finally, to complete the proof of Theorem 8 one simply notes that conditions (1) and (2) can be combined into a single sentence in which all universal quantifiers precede all existential quantifiers. ∎

We next show that the equivalence problem for NFAs generating commutative languages is co-**NP**-complete.

Theorem 9 *The equivalence problem for NFAs generating commutative languages in co-**NP**-complete under logspace reducibilities.*

Proof: The co-**NP** hardness follows from the co-**NP** completeness of the equivalence problem for NFAs with a unary terminal alphabet ([20]). It remains to show the co-**NP** upper bound. To this end, let $M_i = (Q_i, \Sigma, \delta_i, q_i, F_i)$ be an NFA, where $i = 1, 2$. Assume that $L(M_i)$ is commutative for $i = 1, 2$. Consider M_1. As in the proof of Lemma 2, we find a string w such that $L(M_1) \subseteq w^*$ using a nondeterministic logspace-bounded Turing machine. We then check whether $L(M_2) \subseteq w^*$ in **NL**. If it is not the case, then $L(M_1) \neq L(M_2)$. Hence suppose that $L(M_2) \subseteq w^*$. Now let 0 be a new symbol not in Σ and define for $i = 1, 2$ the sets S_i as follows:

$$S_i := \{0^m \mid w^m \in L(M_i)\}.$$

Then $L(M_1) = L(M_2) \iff S_1 = S_2$. From M_i we construct in **NL** an NFA N_i accepting S_i by simply modifying the transition relation δ_i to β_i: $\beta_i(q, 0) := \delta_i(q, w)$. Thus, the equivalence of M_1 and M_2 is polynomial-time reducible to the equivalence of N_1 and N_2 which are NFAs with a unary terminal alphabet. Thus, the co-**NP** upper bound for the equivalence problem for NFAs with a unary terminal alphabet implies the same upper bound for the equivalence problem for NFAs generating commutative languages. This completes the proof of Theorem 9. ∎

Finally we consider the equivalence problem for DFAs generating commutative languages. As accessibility and coaccessibility in DFAs are **NL**-hard, we assume that the DFAs in the input are trim. We show that for trim DFAs generating commutative languages the equivalence problem is in **L**, and hence solvable in $O(\log n)$ steps on the EREW PRAM model.

Theorem 10 *The equivalence problem for trim DFAs is in* **L**.

Proof: Let $M = (Q, \Sigma, \delta, q_0, F)$ be a trim DFA accepting a commutative language. Then there exists a string $x \in \Sigma^*$ such that $L(M) \subseteq x^*$. Since M is deterministic and trim, the transition diagram of M is a path π of the form

$$q_0 \to^{a_0} p_1 \to^{a_1} p_2 \to^{a_2} \ldots \to^{a_{l-1}} p_l \to^{a_l} p_{l+1} \ldots \to^{a_{l+j-1}} p_{l+j} \to^{a_{l+j}} p_l$$

so that $a_0 \ldots a_l \ldots a_{l+j}$ is prefix of some string in x^* and $\{q_0, p_1, \ldots, p_l, \ldots, p_{l+j}\}$ the subpath of π from p_l back to p_l forms a simple loop. Let $\delta^\lambda(q_0)$ denote the state accessible from q_0 by a path of length $\lambda \geq 0$.

Now let $M_1 = (Q_1, \Sigma, \delta_1, q_1, F_1)$ and $M_2 = (Q_2, \Sigma, \delta_2, q_2, F_2)$ be two trim DFAs in the input. We verify whether $L(M_1) = L(M_2)$ as follows. For each $\lambda = 0, \ldots, |Q_1| \times |Q_2|$, check that the state $\delta_1^\lambda(q_1)$ is in F_1 iff the state $\delta_2^\lambda(q_2)$ is in F_2, and that there is a transition labeled a at $\delta_1^\lambda(q_1)$ in M_1 iff there is a transition labeled a at $\delta_2^\lambda(q_2)$ in M_2, $a \in \Sigma$. It is easily seen that this procedure can be carried out on a deterministic logspace-bounded Turing machine. ∎

7 Concluding Remarks

In this paper we have investigated the complexity of two simple criteria for the equivalence problem for CFLs and regular sets. We have also classified the complexity of the sparseness problem for CFLs and regular sets, and the equivalence problem for the restricted class of commutative languages. Our results regarding the closeness problem and the segment equivalence problem show that they remain computationally infeasible for the class of CFLs and the class of regular sets represented by NFAs. If DFAs are in the input instead, the closeness and segment equivalence problems are **NL**-complete and hence solvable in polynomial time (or even in $\mathbf{NC^{(2)}}$). Thus, for the class of CFLs it appears, to the author's knowledge, that the only invariant that provides a polynomial-time solvable criterion for the equivalence problem is the commutative Hotz group. It remains an interesting open problem to find further simple invariants for CFLs so that the equivalence problem can be efficiently solved. The exact complexity of the sparseness problem for CFGs remains open. It is not clear whether this problem is computationally hard. Finally we find it interesting to extend our results in Section 6 for bounded languages.

Acknowledgment. The author wishes to thank Lu Tian for a careful reading of the first version of this paper and several helpful remarks.

Bibliography

[1] Berman, L. and Hartmanis, J. On the Isomorphism and Density of NP and other Complete Sets. *SIAM J. of Computing*, pp. 305-322, 1977.

[2] Cho, S. and Huynh, D.T. The Parallel Complexity of Finite State Automata Problems. to appear in *Information & Computation*.

[3] Cook, S. A Taxonomy of Problems Which Have Fast Parallel Algorithms. *Information & Computation 64*, pp. 2-22, 1985.

[4] Ginsburg, S. *The Mathematical Theory of Context-Free Languages*. McGraw-Hill, 1966.

[5] Hartmanis, J. Context-Free Languages and Turing Machine Computations. *Proc. Symp. on Applied Mathematics 19*, pp. 42-51, 1967.

[6] Harrison, M. A. *Introduction to Formal Language Theory*. Addison-Wesley, 1978.

[7] Hopcroft, J. & Ullman, J. *Introduction to Automata Theory, Languages and Computations*. Addison-Wesley, 1979.

[8] Hotz, G. Eine Neue Invariante für Kontext-Freie Sprachen. *Theoretical Computer Science 11*, pp. 107-116, 1980.

[9] Hunt, H.B., Rosenkrantz, D.J. and Szymanski, T.G. On the Equivalence, Containment, and Covering Problems for the Regular and Context-Free Languages. *J. of Computer and System Science 12*, pp. 222-268, 1976.

[10] Huynh, D.T. Remarks on the Complexity of an Invariant of Context-Free Grammars. *Acta Informatica 17*, pp. 89-99, 1982.

[11] Huynh, D.T. Commutative Grammars: The Complexity of Uniform Word Problems. *Information & Computation 57*, pp. 21-39, 1983.

[12] Huynh, D.T. Deciding the Inequivalence of Context-Free Grammars with 1-Letter Terminal Alphabet Is Σ_2^p-Complete. *Theoretical Computer Science 33*, pp. 305-326, 1984.

[13] Huynh, D.T. The Complexity of Equivalence Problems for Commutative Grammars. *Information & Computation 66*, pp. 103-121, 1985.

[14] Huynh, D.T. Some Observations about the Randomness of Hard Problems. *SIAM J. on Computing 15*, pp. 1101-1105, 1986.

[15] Ibarra, O. and Ravikumar, B. On the Sparseness, Ambiguity and Other Decision Problems for Acceptors and Transducers. *Proc. 3rd Ann. Symposium on Theoretical Aspects of Computer Science*, LNCS 210, pp. 171-179, 1986.

[16] Immerman, N. Nondeterministic Space Is Closed under Complementation. *SIAM J. on Computing.* 17, pp. 935-938, 1988.

[17] Jones, N.D., Lien, Y.E. & Laaser, W.T. New Problems Complete for Nondeterministic Log Space. *Mathematical Systems Theory* 10, pp. 1-17, 1976.

[18] Meyer, A.R. and Stockmeyer, L.J. The Equivalence Problem for Regular Expressions with Squaring Requires Exponential Space. *Proc. 13th IEEE Symp. on Switching and Automata Theory*, pp. 125-129, 1972.

[19] Schöning, U. Complete Sets and Closeness to Complexity Classes. *Mathematical Systems Theory* 19, pp. 29-41, 1986.

[20] Stockmeyer, L. & Meyer, A.R. Word Problem Requiring Exponential Time: Preliminary Report. *Proc. 5th Ann. ACM Symp. on Theory of Computing*, pp. 1-9, 1973.

[21] Stockmeyer, L.J. The Complexity of Decision Problems in Automata Theory and Logic, Report TR-133, MIT Project MAC, Cambridge, Mass., 1974. by Circuits", *SIAM J. on Comput.* 13, pp. 409-422, 1984.

[22] Yesha, Y. On Certain Polynomial-Time Truth-Table Reducibilities of Complete Sets to Sparse Sets. *SIAM J. on Computing* 12, pp. 411-425, 1983.

Communication Complexity and lower bounds for sequential computation

Bala Kalyanasundaram
Dept. of Computer Science
University of Pittsburgh

Georg Schnitger
Dept. of Computer Science
Pennsylvania State University

Abstract

Information-theoretic approaches for lower bound problems are discussed and two applications of Communication Complexity are presented.

The first application concerns one-tape Turing machines with an additional one-way input tape. It is shown that lower bounds on the Communication Complexity of a given language immediately imply lower bounds on the running time for this Turing machine model. Consequently, lower bounds for the Turing machine complexity of specific languages are derived. Emphasis is given to bounded-error probabilistic Turing machines, since no previous lower bounds have been obtained for this computation mode.

The second application concerns a real-time comparison between Schoenhage's Storage Modification machines and the machine model of Kolmogorov and Uspenskii. A non-standard model of Communication Complexity is defined. It is shown that non-trivial lower bounds for this communication model will imply that Storage Modification machines cannot be simulated in real time by Kolmogorov-Uspenskii machines.

1   Introduction

Information-theoretic measures have turned out to be of crucial importance when establishing lower bounds on computational resources. *Entropy, Kolmogorov Complexity* and *Communication Complexity* are examples of measures of relevance to lower bound problems. We will restrict our attention to Kolmogorov Complexity and Communication Complexity which seem to be the two most successful tools.

We confine ourselves to a very sketchy discussion and refer to the survey [19] on Kolmogorov Complexity and its history. Informally, the Kolmogorov Complexity $K(u|v)$ measures the amount of randomness of u *given* v. On a more formal note, for a fixed *universal* Turing machine M, $K(u|v)$ is defined to be the minimal length of a string x such that U, on input $x\#v$, outputs u. Paul [24] was the first to utilize Kolmogorov Complexity for lower bound problems. In particular, a simplified proof of Hennie's theorem (a quadratic lower bound on time for one-tape Turing machines recognizing palindromes) is given as well as lower bounds on the time complexity of restricted sorting procedures. This approach was then extended by Paul, Seiferas and Simon [27] to lower bound problems for on-line computation. Again, we refer to [19] for a discussion of the now quite considerable literature on lower bounds via Kolmogorov Complexity.

One of the basic problems of Kolmogorov Complexity is the effect of *Magic* [24]. Magic occurs whenever two strings x and y have individually no information on a third string z, but combined have all the required information; in other words $K(z|x) = K(z|y) = |z|$ but $K(z|x,y) = O(1)$. (This phenomenon can happen for instance if $z_i = x_i \oplus y_i$ for all i.) Communication Complexity is one approach that tries to deal with the problem of Magic.

The two agent model of Communication Complexity was introduced by Abelson [1]. Yao [36] contains the first results for the case of binary input. In the two agent model an input x is partitioned into two substrings x_1 and x_2 of same length. The first agent P receives x_1 and the second agent Q receives x_2. P and Q will deterministically exchange messages in order to determine whether x belongs to a given language $L \subseteq \{0,1\}^{2n}$. The maximum over all inputs x of the number of bits exchanged by the best protocol is then defined to be the (deterministic) communication complexity of L, relative to the given input assignment. Various types of computational modes can be considered including among others nondeterministic and probabilistic computation. We refer to the surveys of Orlitsky and El Gamal [23] and Lovasz [16] for a discussion of further results.

The first application of Communication Complexity is due to Thompson [34] who considers lower bounds on the area and time for VLSI chips. It was also observed quite early that lower bounds on time can be obtained for one-tape Turing machines (see [26] for lower bounds on unbounded probabilistic one-tape Turing machines). In the next section we determine lower bounds on time for bounded error probabilistic one-tape Turing machines with an additional one-way input tape. We show that lower bounds on Communication Complexity immediately imply lower bounds for this Turing machine model (see Theorem 2). Moreover, a similar relationship can be established for deterministic and nondeterministic Turing machines [10]. These results improve on [13] and also present for the first time lower bounds for "natural" languages (see Theorem 1).

The most dramatic application of Communication Complexity has been obtained by Karchmer and Wigderson [15] and Raz and Wigderson [29]. Here the Communication Complexity of search problems is considered and a relationship to the depth of monotone and unrestricted circuits is described. Raz and Widgerson establish a linear lower bound on the depth of monotone circuits for *matching*; this is achieved by reducing the problem of determining the Communication Complexity of the associated search problem to the problem of determining the probabilistic Communication Complexity of the language *Set Disjointness*. The complexity of the latter problem was determined by Kalyanasundaram and Schnitger in [12]. The argument of [12] was later simplified by Razborov [28].

Multi-party Communication is a second model of importance for Communication Complexity [4]. In this model we assume that the input is partitioned into s blocks. s agents participate with each agent seeing all but one block. The communication proceeds with agents successively broadcasting messages.

The best lower bounds to date are due to Babai, Nisan and Szegedy [3]. Their results imply lower bounds for a wide variety of issues including pseudorandom generators, time-space tradeoffs and branching programs [3]. Finally, Hastad and Goldmann [7] use multi-party Communication Complexity to derive lower bounds on the size of depth 3 threshold circuits (with small bottom fan-in).

In the last section we define a third model of Communication Complexity (see Remark 1). This model is tailor-made to tackle the long outstanding problem of an on-line comparison between Schoenhage's Storage Modification machines [30] and the machine model of Kolmogorov and Uspenskii [13] (see Theorem 3).

2 Lower Bounds on Time for Turing Machines

Hartmanis and Stearns [9] show that a multi-tape Turing machine that runs in time $t(n)$ can be simulated by a one-tape Turing machine in time $O(t^2(n))$. This simulation result holds for both deterministic and nondeterministic Turing machines. On the other hand, it is well known that a two-tape Turing machine can simulate a multi-tape Turing machine with a loss of a factor of $\log(t(n))$. This shows that two-tape Turing machines are quite powerful. Thus, it is not surprising that no lower bounds on time are known for specific languages, although Paul *et al.* [25] were successful in showing that nondeterministic linear time properly contains deterministic linear time for multi-tape Turing machines. The strongest Turing machine model for which non-linear lower bounds have been established is the one-tape Turing machine with an additional read-only two-way input tape ([21, 22]).

We investigate one-tape Turing machines with a special one-way input tape (called

ITM's from now on). *ITM*'s are more powerful than the geometrically weak one-tape Turing machines since *ITM*'s can recognize palindromes in linear time. *ITM*'s were considered in [20, 18, 6]. Maass obtained an $\Omega(n^2)$ lower bound for deterministic *ITM*'s simulating deterministic two-tape Turing machines. For every constant k, Galil, Kannan and Szemeredi [6] showed an $\Omega(n^2/\log^{(k)} n)$ lower bound for nondeterministic *ITM*'s simulating deterministic two-tape Turing machines, extending Maass's result.

We obtain a lower bound of $\Omega(n^2/\log n)$ for *probabilistic ITM*'s recognizing the language *Element Distinctness* for inputs of size $n \log n$. Since the complement of *Element Distinctness* can be recognized by a nondeterministic *ITM* in time $O(nlogn)$, we therefore obtain an almost quadratic separation between probabilism and nondeterminism.

Paturi and Simon [26] consider bounded error probabilistic on-line computation for various machine models. Their arguments use Kolmogorov Complexity but heavily utilize that functions with *multiple* output are to be computed.

We still use the geometrical arguments of Maass [20], but we replace the conventional Kolmogorov-Complexity arguments by arguments from Communication Complexity. This enables us to give the first lower bound for the time complexity of probabilistic *ITM*'s. Also we introduce a large class of languages (defined by using concepts from Communication Complexity) and prove almost quadratic lower bounds for *probabilistic ITM*'s recognizing any language in this class.

2.1 Difficult Languages

Our first goal is to introduce a class of languages none of which are recognizable efficiently by probabilistic *ITM*'s. The definition of this class will be based on concepts of Communication Complexity. Informally, we demand that the languages are difficult for the two-agent model under "almost" any input assignment. Let x be a binary string of length $2nB$. We can partition x into $2n$ blocks (of consecutive bit positions) of length B each. We say that an input assignment P is a B-partition if and only P assigns only entire blocks to the two agents and both agents receive the same number of blocks.

Let $L = \{L_n : n \in N\}$ where $L_n \subseteq \{0,1\}^{2nB(n)}$ and let $t(n)$ and $B(n)$ be functions of n. $\#L_n$ will denote the cardinality of the set L_n.

For a given $B(n)$-partition π we can describe L by the following $0 - 1$ matrix $M(L_n)_\pi$. Its rows correspond to inputs received by agent P (relative to π) and its columns correspond to inputs received by agent Q (relative to π). An entry in the matrix is 1 if and only if the corresponding input belongs to the language.

We now introduce a notion of hardness for probabilistic communication modeled after Yao's distributional complexity [37].

Definition 1
(a) We say that L is $(B(n), t(n)) -$ hard for probabilistic protocols if and only if there is a positive constant δ such that for any $B(n)$-partition π and any submatrix of

$M(L_n)_\pi$ containing S $(S > \#L_n/2^{t(n)})$ entries of L_n, the submatrix must also contain $\delta S \frac{2^{2nB(n)} - \#L_n}{\#L_n}$ entries that do not belong to L_n.

(b) We say that L has probabilistic complexity $t(n)$ for block size $B(n)$ if and only if either L or its complement is $(B(n), t(n)) -$ hard for probabilistic protocols.

First we are interested in languages that allow us to concentrate on a single input assignment.

Definition 2 *We say that L is $B(n)$-symmetric if and only if for any string $x = x_1 \dots x_{2n}$ (where $x_i \in \{0,1\}^{B(n)}$) and any permutation π, $x \in L$ if and only if $x_{\pi(1)}, \dots, x_{\pi(2n)} \in L$.*

Obviously, for $B(n)$-symmetric languages, we only have to define hardness relative to a fixed input assignment. We will now give an example of a difficult symmetric language, namely *Element Distinctness*. Since the complement of element distinctness can be recognized by a nondeterministic *ITM* in linear time, our lower bound shows that nondeterminism can be more powerful than probabilism.

Element Distinctness : A string $x_1 x_2 \dots x_{2n}$ $(x_i \in \{0,1\}^{2\log n})$ of $2n$ words is accepted if and only if any two words are distinct.

Theorem 1 *The probabilistic complexity of Element Distinctness is $\Omega(n)$ for block size $B(n) = 2\log n$.*

Proof: *Element Distinctness* is $2\log n$-symmetric. Thus, it suffices to consider the $2\log n-$partition which assigns the first half of the string to agent P and the rest to agent Q. Since two n-element subsets of $\{0, \dots, n^2 - 1\}$ will be disjoint with probability approaching e^{-1}, the matrix of *Element Distinctness* has a constant proportion of 0's and 1's.

Consider a submatrix of size at least $(n!\binom{n^2}{n})^2 2^{-t(n)}$. Our goal is to show that a constant proportion of the entries of the submatrix are 0's.

The submatrix must have at least $n!\binom{n^2}{n} 2^{-t(n)}$ rows as well as columns. We represent each row by the n-subset of $\{0, \dots, n^2 - 1\}$ specified by the corresponding input of agent P. We will partition the rows into groups where each group has at least $\binom{n^2}{n} 2^{-t(n)-1}$ elements. Moreover, for any group, we demand that any two rows of the group define distinct subsets. A simple counting argument shows that at least half of the total number of rows under consideration can be identified with some group. The same construction will also be carried out for the columns.

Consider a column group and a row group. The two groups give rise to a *Set Disjointness* problem: each such entry can be encoded as a pair (x, y) of two strings of length n^2 (the incidence vectors of the row-set respectively the column-set) where each string has exactly n 1's. By a result of Babai, Frankl and Simon [3], a constant proportion of the entries don't belong to the language *Element Distinctness* unless

$t(n)$ is a sufficiently small constant fraction of n. This analysis can be applied to every pair of row/column groups.

As an immediate consequence, at least a constant proportion of entries covered by the submatrix are 0's. ♠

The following theorem states that ITM's spend almost quadratic time recognizing hard languages.

Theorem 2 *Assume that $B(n) = \Omega(\log n)$ and $t(n) = O(nB(n))$. Let $t(n)$ be the probabilistic complexity of a language L for block size $B(n)$. Then any (bounded error) probabilistic ITM, recognizing L, requires at least $\Omega(t^2(n)/B(n))$ steps.*

The following two corollaries are immediate consequences of the two theorems and the fact that the complement of *Element Distinctness* can be recognized by a nondeterministic ITM in linear time.

Corollary 1 *Any probabilistic ITM that recognizes Element Distinctness requires $\Omega(n^2/\log n)$ steps.*

Corollary 2 *There exists a language that can be recognized by a nondeterministic ITM in time $O(n \log n)$ but any probabilistic (bounded error) ITM requires $\Omega(n^2/\log n)$ steps.*

2.2 The Lower Bound

Proof of Theorem 2 We can view a probabilistic ITM as a deterministic ITM with one more special read only one-way random tape. The random tape is filled with bits and the deterministic ITM consults this random tape and moves appropriately. We fix the contents of the random tape such that the resulting deterministic ITM (say M) recognizes L with error probability at most ϵ, where ε will be determined later. We also demand that M computes for an expected number (over all inputs belonging to L) of at most $c\frac{t^2(n)}{4B(n)}$ steps, for a sufficiently small constant c which will be determined later as well. Observe that at least a fraction of 3/4'th of all inputs (belonging to L) requires less than $c\frac{t^2(n)}{3B(n)}$ steps each. From now on we only concentrate on the set I_1 of these inputs.

Let $x \in I_1 \subseteq L$ be such an input. Without loss of generality, let us assume that the work-tape head only visits cells numbered 1 through $c\frac{t^2(n)}{3B(n)}$. We can associate with input x the rightmost cell such that at most half of all bits of x are read for the first time when the work-tape head is to the left of the cell.

We say that cell r is a *boundary* if and only if at least $(1/2)^{\frac{\#I_1 \cdot 3B(n)}{c \cdot t^2(n)}}$ inputs are associated with the work-tape cell $r + 1$. Observe that $\Omega(\#I_1)$ inputs are associated with boundaries.

Consider a boundary r. Let $I_2(r)$ be the set of those inputs that are associated with r. For $x \in I_2(r)$ let q be the rightmost cell to the left of r and let s be the leftmost cell to the right of r such that the length of the induced crossing sequence between cells $q-1$ and q as well as the length of the induced crossing sequence between s and $s+1$ is less than $\sqrt{c\frac{t(n)}{B(n)}}$ each. Obviously $s-q+1 \leq \sqrt{c} \cdot t(n)$.

We call such a pair (q,s) of work-tape cells a *desert* if it is induced by at least $(1/2)\frac{\#I_2(r)}{c \cdot t^2(n)}$ inputs (which we collect in the subset $I_3(q,r,s) \subseteq I_2(r)$). Observe that $\Omega(\#I_2(r))$ inputs of $I_2(r)$ are "covered" by deserts.

Consider the crossing sequences induced by cells $q-1$ and q as well as by s and $s+1$. These two crossing sequences induce a partition of the cells of the input-tape. Let $Left_1$, $Right_1$ and $Middle$ be the set of those input-bit positions read for the first time while the work-tape head is to the left of q, in the interval $[q,s]$ and to the right of s respectively. Observe that both $Left_1$ and $Right_1$ will contain at most half of all bit positions. Let p denote the number of partitions of the input-tape into $Left_1$, $Right_1$ and $Middle$. Since the length of the induced crossing sequence is less than $\sqrt{c\frac{t(n)}{B(n)}}$, we obtain:

$$
\begin{aligned}
p &\leq \sum_{j=0}^{2\sqrt{c\frac{t(n)}{B(n)}}-1} \binom{2nB(n)}{j} 3^j \\
&\leq 2\sqrt{c\frac{t(n)}{B(n)}}\binom{2nB(n)}{2\sqrt{c\frac{t(n)}{B(n)}}-1} 3^{2\sqrt{c\cdot t(n)}/B(n)} \\
&\leq (2nB(n))^{2\sqrt{c\cdot t(n)}/B(n)}\ 3^{2\sqrt{c\cdot t(n)}/B(n)} \\
&= 2^{O(\sqrt{c\cdot t(n)})}
\end{aligned}
$$

where we assumed $B(n) \geq \log n$.

We call a triple $(Left_1, Right_1, Middle)$ *good* if and only if the number of inputs inducing the triple is at least $(1/2p)\#I_3(q,r,s)$. A simple counting argument shows that a constant fraction of inputs in $I_3(q,r,s)$ induce good triples.

It suffices to show that for each good triple, a constant proportion of the inputs inducing the triple is incorrectly recognized by the ITM. Now fix a good triple $T = (Left_1, Right_1, Middle)$ and let $I_4(q,r,s,T)$ be the set of inputs of $I_3(q,r,s)$ which induce the triple. Observe that, as defined by the triple, the input-tape is cut into $2\sqrt{c\frac{t(n)}{B(n)}}$ pieces where each piece is a set of consecutive bit positions. Consider the set C of those pieces that belong to $Middle$. Now we perform the following distribution procedure to assign the pieces in C to $Left_1$ and $Right_1$ such that the cumulative size of $Left_1$ as well as $Right_1$ is exactly $nB(n)$.

Keep adding the pieces in C to $Left_1$, proceeding on the input tape from left to right, until the cumulative size of $Left_1$ exceeds $nB(n)$. Say this happens when the piece B is added to $Left_1$. Now split B into two pieces B_{Left} and B_{Right} (each piece is a set of consecutive input-tape cells) such that when B_{Left} is added to $Left_1$, the cumulative size of $Left_1$ is exactly $nB(n)$. Assume that B is split between the input tape cells cut and $cut+1$. Let $Left_2$ be the superset of $Left_1$ obtained after the addition of the blocks including B_{Left}. Add the rest of the pieces, including B_{Right}, to $Right_1$ and obtain the superset $Right_2$.

We now fix some components of the computation such that the remaining computation is determined when formalized as a two-agent communication problem.

1. The crossing sequence between cells $q - 1$ and q and between cells s and $s + 1$ ($O(\sqrt{c} \cdot t(n))$ bits).

2. The triple $(Left_1, Right_1, Middle)$ ($O(\sqrt{c} \cdot t(n))$ bits).

3. The content of the input cells being scanned by the input head when the work-tape head crossed the boundaries between cells $q - 1$ and q and between cells s and $s + 1$ ($O(2\sqrt{c}\frac{t(n)}{B(n)})$ bits).

4. The position of the work-tape head when the input head crosses the boundary between cut and $cut + 1$ ($O(\log n)$ bits).

5. The content of the work-tape cells in the interval $[q, s]$, when the input head crosses the boundary between cut and $cut + 1$ ($O(\sqrt{c} \cdot t(n))$ bits).

6. The input consists of $2nB(n)$ bits which are partitioned to form $2n$ words where each word is exactly $B(n)$ bits long. Since the induced good triple $(Left_1, Right_1, Middle)$ can split at most $2\sqrt{c} \cdot t(n)/B(n)$ words, we fix the contents of those $2\sqrt{c} \cdot t(n)/B(n)$ words.   ($O(\sqrt{c} \cdot t(n))$ bits).

There are at most $a = 2^{O(\sqrt{c} \cdot t(n))}$ ways of assigning values to the above six components of the computation. We call such an assignment *good* if and only if at least $\frac{1}{2a}\#I_4(q, r, s, T)$ inputs in $I_4(q, r, s, T)$ induce this assignment. Observe that a constant fraction of inputs in $I_4(q, r, s, T)$ induce a good assignment.

Let A be a *good* assignment. According the procedure described earlier, let $Left_2$ and $Right_2$ be the two induced supersets for the good triple $(Left_1, Right_1, Middle)$. Observe that $(Left_2, Right_2)$ is almost a $B(n)-$
partition, since the length of the crossing sequences allows the split of at most $2t(n)/B(n)$ input words. Now, distribute those divided words among the two agents in such a way that both get an identical proportion of words. Assume that $Left_3$ and $Right_3$ corresponds to the new block preserving partition.

We will now prove that the inputs inducing the good assignment A form a submatrix under the $B(n)-$partition $(Left_3, Right_3)$. Assume that g and h are two inputs inducing the same good assignment. Let g_1 (resp. h_1) be the substring of g (resp. h) induced by the set $Left_3$. Analogously, the binary strings g_2 and h_2 are the substrings of g and h (respectively) induced by $Right_3$. We can combine g_1 and h_2 to form an input string, say g'. Analogously, we can combine h_1 and g_2 to form h'. It is fairly simple to verify that these two strings g' and h' also induce the same good assignment. Therefore, they form a submatrix under the partition $(Left_3, Right_3)$.

For assignment A consider the communication model, where the two agents partition the input according to $Left_3$ (belonging to agent P) and $Right_3$ (belonging to agent Q). Observe that inputs that induce this good assignment form a submatrix of size at least $s = \frac{\#L}{2^{O(\sqrt{c}\cdot t(n))}}$. Now choose c such that $s \geq \frac{\#L}{2^{t(n)}}$.

We know that good assignments cover a constant proportion of all inputs belonging to L. All we have to observe now is that the hardness of the language implies a fixed constant error probability ε'. Therefore, if we choose the error probability ϵ of the ITM M to be smaller than ε', we arrive at a contradiction.   ♠

3   A non-standard Model of Communication

Storage Modification Machines (SMM's) were introduced by Schoenhage [30] in order to provide a definition of low order time complexity. We give a brief description of SMM's, for more detailed information we refer the reader to [5, 30].

The storage of a SMM is a (modifiable) directed graph of (bounded)fan-out d but *unbounded* fan-in. For each vertex v of the graph, the edges with v as the tail are labeled injectively by numbers from $0,1, \ldots ,d$. A head (called *center*) is moving around in the graph, modifying it or retrieving information. At each step of the computation the center will "sit" on some vertex v. We first list the operations that the center can perform in one step.

(a) The center can create a new vertex w and insert the edge (v, w), provided that the fan-out does not exceed d.

(b) The center can replace an edge (v, u) by an edge (v, w). Here w must be specified within the program by a path originating in v.

(c) The center can move from v to any vertex u provided that the edge (v, u) belongs to the graph.

(d) Information can be retrieved by asking whether two paths (both originating in v and specified within the program) end at the same vertex.

Schoenhage [30] shows that SMM's can simulate multi-dimensional Turing Machines in real-time. He also designed a linear time integer multiplication algorithm for SMM's. Schnorr [31] proves that SMM's are real-time equivalent to "Successor RAM's". These are random access machines whose only non trivial operation is the computation of the successor function. Finally SMM's coincide with the pointer machines introduced by Knuth (called "linking automata" in [10]). Further results on Storage Modification machines include time-space tradeoff's [8], time and space hierarchies [16] as well as results on space complexity [16, 35].

The appealing simplicity of the SMM-operations as well as the different characterizations just mentioned can be interpreted as supporting the view that SMM's provide a satisfying definition of low order time complexity.

Another reason for the interest in Storage Modification machines is that many important data structures can be readily implemented on SMM's [33] (assuming each node is also equipped with a register storing an integer). For instance, SMM's can solve the Union-Find problem in almost linear time and nonlinear time is required [32].

Kolmogorov and Uspenskii [14] define a related machine model (which we will call KUM's). The major difference is that the KUM storage graph is an undirected graph of **bounded** degree. A real-time simulation of KUM's by SMM's is straight forward. On the other hand, it is quite likely that SMM's cannot be simulated by KUM's in real-time, but this problem remains unresolved.

Our goal is a formulation of the KUM *versus* SMM problem as a non-standard communication problem. We believe that the communication problem is of independent interest and hope that it contributes to an eventual solution of the KUM *versus* SMM problem.

The non-standard communication model consists of two agents (which we call P and Q). The (binary) input is disjointly distributed among the agents who then have to be prepared to compute each one of a given collection of functions. The computation for each of the functions has to proceed according to the rules of the conventional communication model.

Our communication model becomes non-standard in that agent Q is allowed to see the entire input of its partner before the computation starts. Q then has to condense this input into relatively few bits and can use only the condensed string during the computation.

3.1 KUM's, SMM's and Communication Complexity

Let us first describe a language L which is easy for SMM's but seemingly tough for KUM's. We describe L by describing a SMM S that recognizes L in real-time. S proceeds in phases.

Phase 1: *Preprocessing.*

Let w be the input string. S scans w from left to right until it finds a position n with $w_n = 0$. Now S starts to build a data structure consisting of two graphs C_n and W_n.

C_n consists of a complete binary tree of depth n whose leaves are replaced by chains of length n. The $m = n \cdot 2^n$ nodes belonging to one of the chains are called *citizens*. W_n is a complete binary tree of depth $\frac{n}{2}$. The leaves of W_n are called *countries*.

Phase 2. *Assigning countries to citizens.*

S reads the next $m\frac{n}{2}$ bits of w and interprets the corresponding bit string as the encoding of a vector $(c_1,..., c_m)$ of countries. More specifically, S performs a preorder traversal of C_n and connects the i'th citizen encountered during the traversal to country c_i.

Phase 3. *Black-white coloring of countries.*

S continues by reading the next $2^{n/2}$ bits while performing a preorder traversal of W_n. In particular, S colors the ith country black (white) if the ith bit is 0 (1). The black-white coloring can be implemented by either connecting the i'th country to itself or to a distinguished vertex.

Phase 4. *The query phase.*
S reads the remaining n bits which it interprets as describing a chain in C_n. S traverses this chain and prints for each traversed citizen the color of its country.

Obviously, S is able to recognize the induced language L in real-time. Let us also observe that phases 1 and 2 require time $O(n \cdot m)$. Phase 3 requires $O(2^{n/2})$ steps and the length of phase 4 is proportional to n. Finally, let us point out that S utilizes the unbounded fan-in for its country nodes. A simulating KUM has to encode this one-step access of a country node in its undirected storage graph of *bounded* degree.

We assume from now on that a KUM K recognizes L in real-time. Consider K during the query phase. We want to investigate the extent to which K is capable of quickly combining the information presented in phases 2 (assigning citizens to countries) and 3 (black-white coloring of countries) and apply it to a given query. We will therefore try to simulate K by a communication model.

Two agents called P and Q are participating. P knows the assignment of countries to citizens, whereas Q knows the coloring of the countries. So far it is not probable that a simulation of K will succeed. The reason is that K, while coloring W_n, will visit previously constructed portions of the storage graph. In information theoretic terms this means that Q can "steal" information belonging to P. We therefore extend the above conventional communication model by providing Q (without charge) with some information belonging to P:

Let w_A be the bit string specifying the assignment of countries and let w_C be the bit string defining the coloring of countries. Let $G(w_A)$ be the graph constructed by K after processing the country assignment and let $G(w_A, w_C)$ be the (connected) subgraph of $G(w_A)$ revisited during the coloring phase. We color the edges of $G(w_A)$ with $O(1)$ colors such that the color sequence induced by a simple path of length $\Omega(n)$ will identify the path (see Lemma 1). We provide agent P with the edge-colored graph $G(w_A)$ and w_A.

Let $G_h(w_A, w_C)$ be the subgraph of $G(w_A)$ consisting of all nodes of $G(w_A)$ which have distance at most h from a node of $G(w_A, w_C)$. We set $h := \varepsilon n$, where ε is chosen such that $G_h(w_A, w_C)$ contains at most $\frac{m^{3/4}}{n}$ nodes. Next we assign the edge-colored subgraph $G_h(w_A, w_C)$ and w_C to agent Q. Both agents are also provided with the original edge labels (used by K) for their respective edges.

Thus, agent Q can steal $O(m^{3/4})$ bits of information. The edge-coloring of $G(w_A)$ will be chosen according to the following lemma. We assume that the undirected $G(w_A)$ has M ($M = O(n \cdot m)$) nodes and is of degree D.

Lemma 1 *Let L be a positive integer. Set $k = 16(\frac{log_D(M)}{L} + 2)$ and $C = (D+1)D^k$.*
Then there exists an edge-coloring with at most C colors such that any two different simple paths of length at least L have different color sequences.

Proof: With Vizing's theorem we can color the edges of G with $D + 1$ colors so that no two edges incident with the same vertex have the same color. Let $c_1 : E \to \{1, \ldots, D + 1\}$ be the corresonding coloring. Next we will label the edges again but now with a random coloring $c_2 : E \to \{1, \ldots, C/(D+1)\}$. Finally, let $c = (c_1, c_2)$ be the combined coloring with C colors.

Now assume that there are two different simple paths p and q with the same color sequence, where colors are assigned according to c. Expressing paths as the sequence of their edges, we get $p = (e_1, \ldots, e_L)$ and $q = (e'_1, \ldots, e'_L)$. Then, for every $i < L$, $e_i \neq e'_i$ or $e_{i+1} \neq e'_{i+1}$. (Otherwise p and q would be identical, since their c_1-colorings coincides. Observe that two edges e_i and e'_i can be identical without p and q being forced to be identical, namely if p and q traverse e_i in opposite directions.)

Consider the following undirected graph H. The edges of p and q form the nodes of H. If edge e'_i does not occur as an edge of p, then we connect e_i and e'_i. Otherwise, if $e'_i = e_j$ and $i \neq j$, then we connect e_j and e_i. Observe that an edge in H expresses that its endpoints are forced to have the same color.

It is now quite obvious that H will have at least $L/4$ edges and is of degree at most 2. In other words, the colors of at least $L/8$ edges are determined once the remaining edges are colored. Thus, the probability that p and q possess the same color sequence is at most $(C/(D+1))^{-L/8}$.

But there are at most $M^2 D^{2L}$ different pairs of paths of length L and c_2 will "fail" with probability at most

$$M^2 D^{2L}(C/(D+1))^{-L/8} \ = \ M^2 D^{2L} D^{-kL/8} \ = \ M^2 D^{-L(k/8-2)}.$$

But this probability is less than one according to the choice of k. This proves the claim for paths of length exactly L and thus the claim is established for paths of length at least L. ♠

3.2 An Almost Real-Time Simulation by the Communication Model

We can now proceed with our simulation.

Theorem 3 *Assume that K computes for t steps when reading the query string. Then the non-standard communication model can simulate each step of K by exchanging at most $O(\log t)$ bits.*

Moreover, agent Q steals at most $O(m^{3/4})$ bits of information from its partner who possesses $\vartheta(n \cdot m)$ bits of information.

Proof Sketch: We set $L = \frac{\varepsilon}{2}n$. Then, according to Lemma 1, we can label the edges of $G(w_A)$ with $O(1)$ colors. Let G be the subgraph $G_L(w_A, w_C)$. Observe that Q knows $G_{2L}(w_A, w_C)$. Initially, we declare agent P "responsible" for all nodes outside of G and declare agent Q responsible for all nodes inside of G. ($G_{2L}(w_A, w_C) - G$ serves as a buffer zone for Q in the same way that $G - G(w_A, w_C)$ serves as a buffer zone for P. Both buffer zones will be used to decipher paths from their color sequences.

Throughout the simulation of K we will maintain the following invariant:

P and Q are aware of any changes (performed by K when reading the query) concerning nodes that they are responsible for.

Obviously, the invariant is satisfied initially. Also, the invariant implies that the proper output can be given.

In each step of the simulation, the responsible agent will first communicate a code of the operation to be simulated. Furthermore, the colors of all edges involved in an operation will be communicated and it will be announced *whether* and *when* nodes involved in the current operation were modified (resp. encountered)) before. (This feature enforces the log t simulation delay.) The communication is started by Q.

Let us now informally discuss how K can avoid a successful simulation by our communication model. When processing the query string K will traverse G. The first problem arises when K is about to leave G: agent Q will announce that the center is leaving her territory but agent P still has to determine the point of entry. K might try to modify $G - G(w_A, w_C)$ so that P will be unable to decode the "entry-path" from its color sequence. But this plot will be unsuccessful, since Q will announce if the currently traversed node was previously involved in a modification and if so, will also announce the time step in which this modification occured.

The second problem arises when P is about to leave his territory. Q will not get any advance notice from P, since P is unaware of Q's territory. Nevertheless, Q's decoding capability is sufficient for her to determine the intruding path (by taking into account P's announcements concerning modifications of $G(w_A)$). ♠

Remark 1 *We describe the obtained communication problem.*

(a) *For $m_1 = 2^n$ and $m_2 = 2^{n/2}$ we define the function family $(f_1, \ldots, f_{m_1})$ with $f_i : \{0,1\}^{\frac{n}{2} \cdot n \cdot m_1} \times \{0,1\}^{m_2} \to \{0,1\}^n$. In particular,*

$$f_i(c_1^1, \ldots, c_n^1, \ldots, c_1^{m_1}, \ldots c_n^{m_1}; b_0, \ldots, b_{m_2-1}) = (b_{c_1^i}, \ldots, b_{c_n^i})$$

where $c_j^i \in \{0,1\}^{\frac{n}{2}}$ is interpreted as an integer.

(b) *Agent P receives $w_A = (c_1^1, \ldots, c_n^{m_1})$ as input and agent Q receives $w_C = (b_0, \ldots, b_{m_2-1})$. Moreover, Q can steal $O(m_1^{3/4})$ bits of information from P; in other words, Q has access to $steal(w_A, w_C) \in \{0,1\}^{m_1^{3/4}}$.*

(c) *When presented with input i both agents compute $f_i(w_A, w_C)$ cooperatively by exchanging as few bits as possible. For a given protocol P let $B_P(n)$ be the maximum over all inputs (w_A, w_C) and over all i $(1 \le i \le m_1)$ of the number of bits exchanged by P. Finally, let $B(n) = \min\{B_P(n) : P\}$.*

(d) *If $B(n) = \omega(n \log n)$, then KUM's can not simulate SMM's in real-time. This is a consequence of Theorem 3, since a real-time simulation of S by K requires K to process the query phase in time $O(n)$.*

(e) *A trivial upper bound is $B(n) = O(n^2)$, which can be shown to be tight if no information is stolen.*

Bibliography

[1] H. Abelson, "Lower Bounds on Information Transfer in Distributed Computations", *Proc. 19th IEEE Symp. on Foundations of Computer Science*, 1978, pp. 151-158.

[2] L. Babai, P. Frankl and J. Simon, "BPP and the Polynomial Time Hierarchy in Communication Complexity Theory", *Proc. 27th Annual IEEE Symp. on Foundations of Computer Science*, 1986, pp. 337-347.

[3] L. Babai, N. Nisan and M. Szegedy, "Multiparty Protocols and Logspace-hard Pseudorandom Sequences", *Proc. 21st Annual ACM Symp. on Theory of Computing*, 1989, pp. 1-11.

[4] A.K. Chandra, M.L. Furst and R.J. Lipton, "Multi-party protocols", *Proc. 15th Annual ACM Symp. on Theory of Computing*, 1983, pp. 94-99.

[5] Y. Gurevich, "On Kolmogorov machines and related issues", *Bull. of EATCS*, 1988.

[6] Z. Galil, R. Kannan and E. Szemeredi, "On Nontrivial Separators for k-Page Graphs and Simulations by Nondeterministic One-Tape Turing Machines", *Proc. 18th Annual ACM Symp. on Theory of Computing*, 1986, pp. 39-49.

[7] J. Hastad and M. Goldmann, "On the Power of Small Depth Threshold Circuits", *Proc. 31st Annual IEEE Symp. on Foundations of Computer Science*, 1990, pp. 610-618.

[8] J.Y. Halpern, M.C. Loui, A.R. Meyer and D. Weise, "On Time Versus Space III", *Math. Systems Theory* 19, 1986, pp. 13-28.

[9] J. Hartmanis and R.E. Stearns, "On the Computational Complexity of Algorithms", *Trans. Amer. Math. Soc.* , Vol. 117, 1965, pp. 285-306.

[10] B. Kalyanasundaram, "Lower Bounds on Time, Space and Communication", Ph. D. *Dissertation*, The Pennsylvania State University, Dec. 1988.

[11] D. E. Knuth, "The Art of Computer Programming", vol. 1, *Addison-Wesley*, Reading Ma., 1968, pp. 462-463.

[12] B. Kalyanasundaram and G. Schnitger, "The Probabilistic Communication Complexity of Set Intersection", *Proc. 2nd Annual Conference on Structure in Complexity Theory*, 1987, pp. 41-47. To appear in *SIAM J. Disc. Math.*

[13] B. Kalyanasundaram and G. Schnitger, "On the Power of One-Tape Probabilistic Turing Machines", *Proc. 24th Annual Allerton Conference on Communication, Control and Computation*, 1986, pp. 749-757.

[14] A. N. Kolmogorov and V. A. Uspenskii, "On the Definition of an Algorithm", *AMS Transl.* 2nd series, vol. 29, 1963, pp. 217-245.

[15] M. Karchmer and A. Wigderson, "Monotone Circuits for Connectivity Require Superlogarithmic Depth", *SIAM J. Disc. Math* 3, 1990, pp. 255-265.

[16] L. Lovasz, "Communication Complexity: A Survey", *Tech. Report CS-TR-204-89*, Princeton University, 1989.

[17] D.R. Luginbuhl and M.C. Loui, "Hierarchies and Space measures for pointer machines", *University of Illinois at Urbana Champaign*, Tech. Rep. UILU-ENG-88-22445.

[18] M. Li, L. Longpre and P.M.B. Vitanyi, "On the Power of the Queue", Structure in Complexity Theory, *Lecture Notes in Computer Science,* Vol. 223, 1986, pp. 219-223.

[19] M. Li and P.M.B Vitanyi, "Kolmogorov Complexity and its Applications", in J. van Leeuwen ed., *Handbook of Theoretical Computer Science*, vol. A,

[20] W. Maass, "Quadratic Lower Bounds for Deterministic and Nondeterministic One-Tape Turing Machines", *Proc. 16th Annual ACM Symp. on Theory of Computing*, 1984, pp. 401-408.

[21] W. Maass, G. Schnitger and E. Szemeredi, "Two Tapes are Better than One for Off-line Turing Machines", *Proc. 19th Annual ACM Symp. on Theory of Computing*, 1987, 94-100.

[22] W. Maass, G. Schnitger, E. Szemeredi and G. Turan, "Two Tapes are Better than One for Off-line Turing Machines", *Tech. Report CS-90-30*, Department of Computer Science, The Pennsylvania State University, 1990.

[23] A. Orlitsky and A. El Gamal, "Communication Complexity", in Y. Abu-Mostafa ed. *Complexity in Information Theory*, Springer Verlag, 1988.

[24] W.J. Paul, "Kolmogorov's Complexity and Lower Bounds", in L. Budach ed., *Proc. 2nd Internat. Conf. on Fundamentals of Computation Theory*, Akademie Verlag, Berlin, 1979, pp. 325-334.

[25] W.J. Paul, N. Pippenger, E. Szemeredi and W.T. Trotter, "On Determinism versus Nondeterminism and related Problems", *Proc. 24th Annual IEEE Symp. on Foundations of Computer Science* 1983, pp. 429-438.

[26] R. Paturi and J. Simon, "Lower Bounds on the Time of Probabilistic On-line Simulations", *Proc. 24th Annual IEEE Symp. on Foundations of Computer Science*, 1983, pp. 343-350.

[27] W.J. Paul, J.I. Seiferas and J. Simon, "An Information Theoretic Approach to to Time Bounds for On-line Computation", *J. Comput. System Sci.* 23, 1981, pp. 108-126.

[28] A.A. Razborov, "On the Distributional Complexity of Disjointness", *Proc. 17th ICALP*, 1990, pp. 249-253.

[29] R. Raz and A. Wigderson, "Monotone Circuits for Matching Require Linear Depth", *Proc. 22nd Annual ACM Symp. on Theory of Computing*, 1990, pp. 287-292.

[30] A. Schoenhage, "Storage Modification Machines", *SIAM J. Comput.*, 9, 1980, pp. 490-508.

[31] C. P. Schnorr, "Rekursive Funktionen und ihre Komplexitaet", *Teubner*, Stuttgart, 1974.

[32] R. E. Tarjan, "A Class of Algorithms Which Require Nonlinear Time to Maintain Disjoint Sets", *J. Comput. System Sci.* 18, 1979, pp. 110-127.

[33] R. E. Tarjan, "Data Structures and Network Algorithms", *SIAM,* 1983, page 2.

[34] C. D. Thompson, "Area-Time Complexity for VLSI", *Proc. 11th Annual ACM Symp. on Theory of Computing* 1979, pp. 81-88.

[35] P. Van Emde Boas, "Space Measures for Storage Modification Machines", *University of Amsterdam*, Tech. Rep. FVI-UVA-87-16. To appear in *Information Processing Letters*.

[36] A.C. Yao, "Some Complexity Questions Related to Distributed Computing", *Proc. 11th Annual ACM Symp. on Theory of Computing*, 1979, pp. 209-213.

[37] A.C. Yao, "Lower Bounds by Probabilistic Arguments", *Proc. 24th Annual IEEE Symp. on Foundations of Computer Science*, 1983, pp. 420-428.

On the Stack Size of a Class of Backtrack Trees

Rainer Kemp

Johann Wolfgang Goethe-Universität
Fachbereich Informatik
6000 Frankfurt am Main
Germany

Abstract

We derive a lower and an upper bound for the average stack size of a tree contained in a family $\mathcal{F}_p(h)$ of non-regularly distributed binary trees introduced by P.W.Purdom for the purpose of modelling backtrack trees. The considered trees have a height less than or equal to h and their shapes are controlled by an external parameter $p \in [0, 1]$.

We show that the average stack size of a tree appearing in $\mathcal{F}_p(h)$ is bounded by a constant for $0 \leq p < \frac{1}{2}$, and that it grows at most logarithmically in h if $p = \frac{1}{2}$; for $\frac{1}{2} < p \leq 1$, the average stack size grows linearly in h.

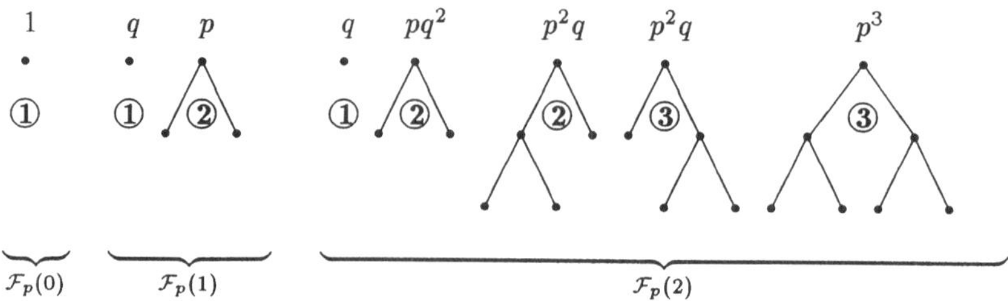

Figure 1. All trees $T \in \mathcal{F}_p(h)$ for $h \leq 2$. The root of a tree is marked by its probability. q stands for $(1 - p)$. The encircled numbers represent the stack size of the corresponding tree.

1   Introduction and Basic Definitions

Let T be an *extended binary tree* [5;p.399] with the set of internal nodes $I(T)$, the set of leaves $L(T)$ and the root $r(T) \in I(T) \cup L(T)$. We shall use the convention that the one-node tree has no internal nodes and exactly one leaf. For any two nodes $u, v \in I(T) \cup L(T)$, $d(u,v)$ denotes the *distance* from u to v defined as the length of the shortest path (= number of nodes on the path minus one) from u to v. A node $x \in I(T) \cup L(T)$ with $d(r(T), x) = \ell$ has the *level* ℓ. The set of all internal nodes and leaves appearing in T at level ℓ is denoted by $I_\ell(T)$ and $L_\ell(T)$, respectively. The tree T has the *height* h if the maximum level of a node is equal to h.

In [7] a family of trees has been introduced in order to estimate the number of nodes of a backtrack tree by doing partial backtrack search. This family of trees $\mathcal{F}_p(h)$, $p \in [0, 1]$, $h \in \mathbb{N}_0$, consists of all extended binary trees with height less than or equal to h, where each tree $T \in \mathcal{F}_p(h)$ is associated with a nonnegative real number $\varphi_{p,h}(T)$ recursively defined by:

(a) If T is the one-node tree then $\varphi_{p,h}(T) := p\,\delta_{h,0} + 1 - p, h \geq 0$;

(b) If T has the left subtree $T_1 \in \mathcal{F}_p(h - 1)$ and the right subtree $T_2 \in \mathcal{F}_p(h - 1)$ then $\varphi_{p,h}(T) := p\,\varphi_{p,h-1}(T_1)\,\varphi_{p,h-1}(T_2), h \geq 1$.

The trees $T \in \mathcal{F}_p(h)$ with $h \leq 2$ are drawn in Figure 1. It is not hard to show that for each $(p, h) \in [0, 1] \times \mathbb{N}_0$, the numbers $\varphi_{p,h}(T)$ define a probability distribution on the set $\mathcal{F}_p(h)$ [see [4], [7]; Lemma 1].

In [4] the average behaviour of *additive weights* defined on $\mathcal{F}_p(h)$ has been investigated. In the main, an additive weight $w_p(T)$ of a tree $T \in \mathcal{F}_p(h)$ with the left subtree T_1 and the right subtree T_2 is composed of the weighted sum $c_1 w_p(T_1) + c_2 w_p(T_2)$, $c_1, c_2 \in \mathbb{R}_+ \cup \{0\}$ of the additive weights of its subtrees and of some quantities which are described by the so-called weight functions; these functions are defined on the number of nodes appearing in the whole tree and in its subtrees. Choosing special weight functions, the corresponding additive weight yields a recursive definition of a characteristic parameter of the tree as say the internal (external) path length, the internal (external; internal-external) free path length, the left (right) branch length, the number of (root-free) paths between internal (external; internal-external) nodes etc.[1] In [4] a general approach to the computation of the average weight $\underline{w}_p(h)$ of a tree $T \in \mathcal{F}_p(h)$ has been presented. This approach yields *exact* expressions for many types of average weights and *exact asymptotic* equivalents to $\underline{w}_p(h)$ if the weight functions are arbitrary polynomials in the number of leaves. Generally, the average weight $\underline{w}_p(h)$ satisfies an inhomogeneous linear recurrence with constant coefficients. The results concerning the maximum growth of the average weight $\underline{w}_p(h)$ as a function of h can be summarized as follows:

	$p < 0.5$	$p = 0.5$	$p > 0.5$
$p(c_1 + c_2) < 1$	*constant*	*polynomially*	*exponentially*
$p(c_1 + c_2) = 1$	*linearly*	*polynomially*	*exponentially*
$p(c_1 + c_2) > 1$	*exponentially*	*exponentially*	*exponentially*

For example, if $c_1 = c_2 = 1$ then the left and right subtree of $T \in \mathcal{F}_p(h)$ contribute the same quota to the additive weight $w_p(T)$ of the whole tree. In this case, the average additive weight $\underline{w}_p(h)$ is a constant if $p < \frac{1}{2}$ and it grows at most polynomially if $p = \frac{1}{2}$; if $p > \frac{1}{2}$, then $\underline{w}_p(h)$ has an exponential growth in h.

In [8] these results have been generalized to simply generated trees ([6]) with a given finite set of allowed node degrees.

In this paper, we deal with the stack size $S_p(T)$ of a binary backtrack tree $T \in \mathcal{F}_p(h)$. Formally, the *stack size* $S_p(T)$ is recursively defined by ([1],[2])

$$S_p(T) := \mathbf{if} \ |I(T) \cup L(T)| = 1 \ \mathbf{then} \ 1$$
$$\mathbf{else} \ \mathbf{if} \ S_p(T_1) > S_p(T_2) \ \mathbf{then} \ S_p(T_1)$$
$$\mathbf{else} \ S_p(T_2) + 1;$$

where $T_1 \in \mathcal{F}_p(h-1)$ and $T_2 \in \mathcal{F}_p(h-1)$ are the left and right subtree of $T \in \mathcal{F}_p(h)$, respectively. $S_p(T)$ is the maximum number of nodes stored in the stack during postorder-traversing of $T \in \mathcal{F}_p(h)$ (traverse the left and right subtree from left to right, then visit the root). In Figure 1, the encircled numbers represent the value of the stack size of the corresponding tree. This important parameter ($\overset{\wedge}{=}$ the space requirements during backtracking by means of a stack) *cannot* be represented by an additive weight. In fact, the computation of the average stack size $\underline{S}_p(h)$ of a tree $T \in \mathcal{F}_p(h)$ leads to a nonlinear, double-recursive recurrence (Lemma 2). We are not able to derive its exact solution but some tricky estimations yield upper and lower bounds for the average stack size $\underline{S}_p(h)$. In the main, we are able to prove that (Theorem 2, 3, 4)

---

[1]An extended list of parameters which are describable by an additive weight can be found in [3].

- $\underline{S}_p(h)$ is bounded by a constant, if $0 \le p < \frac{1}{2}$;
  (more precisely: $\underline{S}_p(h) < \frac{1-2p}{1-p} C(\frac{p}{1-p})$, where $C(x) := \sum_{s \ge 0} d(s+1)x^s$; $d(n)$ denotes the number of positive divisors of the natural number $n \in \mathbb{N}$);

- $\underline{S}_p(h)$ is growing at most logarithmically in h, if $p = \frac{1}{2}$;
  (more precisely: $\underline{S}_p(h) \le H_h$; H_n denotes the n-th Harmonic number);

- $\underline{S}_p(h)$ is growing linearly in h, if $\frac{1}{2} < p \le 1$;
  (more precisely: $\frac{2p-1}{p}\left[1 - \frac{\ln(p)}{\ln(1-p)}\right] h + \mathcal{O}(1) < \underline{S}_p(h) < \frac{2p-1}{p} h + \mathcal{O}(1)$).

2   The Average Stack Size

In this section, we shall present a general approach to the computation of the average stack size $\underline{S}_p(h)$ of a tree $T \in \mathcal{F}_p(h)$. We start with the following lemma (cf.[4]) which can easily be proved by induction on h using the recursive definition of $\varphi_{p,h}(T)$ presented in the preceding section.

Lemma 1 *Let $h \in \mathbb{N}_0$ and $T \in \mathcal{F}_p(h)$.*

(a) *We have*

$$\varphi_{p,h}(T) = p^{|I(T)|} \, (1-p)^{|L(T)|-|L_h(T)|}$$

(b) *The sequence $\varphi_{p,h}(T)$, $T \in \mathcal{F}_p(h)$, is a probability distribution on $\mathcal{F}_p(h)$.*

An inspection of the preceding lemma shows that we obtain big bushy trees with high probability if p is near 1 and little short trees with high probability if p is near 0; for p near $\frac{1}{2}$, we obtain long skinny trees with high probability.

Now, let $\mathcal{F}_p^{(h)}(s) := \{T \in \mathcal{F}_p(h) | S(T) \le s\}$ be the set of all trees $T \in \mathcal{F}_p(h)$ with a stack size less than or equal s. The following lemma gives us information about the probability

$$f_{s,h}(p) := \sum_{T \in \mathcal{F}_p^{(h)}(s)} \varphi_{p,h}(T) \tag{1}$$

that a tree $T \in \mathcal{F}_p(h)$ has a stack size less than or equal to s.

Lemma 2 *Let $(s,h) \in \mathbb{N} \times \mathbb{N}_0$. We have*

$$f_{s,0}(p) = 1 \quad ; \quad f_{1,h}(p) = p\,\delta_{h,0} + 1 - p$$
$$f_{s,h}(p) = 1 - p + p\,f_{s,h-1}(p)\,f_{s-1,h-1}(p), \quad s \ge 2, \quad h \ge 1.$$

Proof: The one-node tree in $\mathcal{F}_p(h)$ appears with probability $p\,\delta_{h,0} + 1 - p$. It is the single tree with a stack size equal to one. Thus, the initial conditions for $f_{s,h}(p)$ are valid.

Now, let $h \geq 1$ and $T \in \mathcal{F}_p(h)$ with the left subtree $T_1 \in \mathcal{F}_p(h-1)$ and the right subtree $T_2 \in \mathcal{F}_p(h-1)$. The tree T has the stack size s if either T_1 has the stack size s and T_2 has a stack size less than s or T_1 has a stack size less than s and T_2 has a stack size equal to $s - 1$. Thus, by the recursive definition of $\varphi_{p,h}(T)$

$$f_{s,h}(p) - f_{s-1,h}(p) = \sum_{T \in \mathcal{F}_p^{(h)}(s) \backslash \mathcal{F}_p^{(h)}(s-1)} \varphi_{p,h}(T)$$

$$= \sum_{T_1 \in \mathcal{F}_p^{(h-1)}(s) \backslash \mathcal{F}_p^{(h-1)}(s-1)} \sum_{T_2 \in \mathcal{F}_p^{(h-1)}(s-1)} p\, \varphi_{p,h-1}(T_1)\, \varphi_{p,h-1}(T_2) +$$

$$+ \sum_{T_1 \in \mathcal{F}_p^{(h-1)}(s-1)} \sum_{T_2 \in \mathcal{F}_p^{(h-1)}(s-1) \backslash \mathcal{F}_p^{(h-1)}(s-2)} p\, \varphi_{p,h-1}(T_1)\, \varphi_{p,h-1}(T_2)$$

$$= p\left[f_{s,h-1}(p) - f_{s-1,h-1}(p)\right] f_{s-1,h-1}(p) +$$
$$+ p f_{s-1,h-1}(p) \left[f_{s-1,h-1}(p) - f_{s-2,h-1}(p)\right]$$

$$= p\left[f_{s,h-1}(p) f_{s-1,h-1}(p) - f_{s-1,h-1}(p) f_{s-2,h-1}(p)\right].$$

Hence, the sequence $u_s := f_{s,h}(p) - p\, f_{s,h-1}(p) f_{s-1,h-1}(p)$ is a constant sequence with the value $u_1 = 1 - p$ because $h \geq 1$ and $f_{0,h}(p) = 0$. This completes the proof of our lemma. $\blacksquare$

Thus, $f_{s,h}(p)$ satisfies a nonlinear, double-recursive recurrence. Since it seems to be hopeless to derive the exact solution of that recurrence, our main aim is to compute good upper and lower bounds for $f_{s,h}(p)$. Before we present such bounds we have to make some observations summarized in the following technical lemma.

Lemma 3 *The probability $f_{s,h}(p)$ that a tree $T \in \mathcal{F}_p(h)$ has a stack size less than or equal to s fulfills the following properties:*

(a) $f_{s,h}(p) = 1$ *for* $s \geq h + 1$*;*
(b) $f_{h,h}(p) = 1 - p^h$ *for* $h \in I\!N$*;*
(c) $f_{s,h}(p) \geq f_{s,h+1}(p)$ *for* $(s, h) \in I\!N \times I\!N_0$*;*
(d) $f_{s,h}(p) \leq f_{s+1,h+1}(p)$ *for* $(s, h) \in I\!N \times I\!N_0$*.*

Proof: (a) Since a binary tree of height h has always a stack size less than or equal to $h + 1$ we have $\mathcal{F}_p^{(h)}(s) = \mathcal{F}_p(h)$ for $s \geq h + 1$. Thus, (a) follows by (1) and by Lemma 1(b).
(b) Choosing $s = h$ in Lemma 2, we find by means of part (a)

$$f_{1,1}(p) = 1 - p$$
$$f_{h,h}(p) = 1 - p + p\, f_{h-1,h-1}(p), \quad h \geq 2.$$

Now, the result follows by iteration of this recurrence.

(c) Let $\Delta_{s,h}(p) = f_{s,h}(p) - f_{s,h+1}(p)$. By Lemma 2, we find

$$\begin{aligned}
\Delta_{s,h}(p) &= p\,[f_{s,h-1}(p)f_{s-1,h-1}(p) - f_{s,h}(p)f_{s-1,h}(p)] \\
&= p\,[\Delta_{s,h-1}(p)f_{s-1,h-1}(p) + \Delta_{s-1,h-1}(p)f_{s,h}(p)]
\end{aligned}$$

for $s \geq 2$, $h \geq 1$, and furthermore $\Delta_{1,h}(p) = 0$, $h \in \mathbb{N}$, and $\Delta_{s,0}(p) = p\,\delta_{s,1}$, $s \in \mathbb{N}$. Since the initial conditions for $\Delta_{s,h}(p)$ are nonnegative and $f_{s,h}(p) \in [0,1]$, the derived recurrence yields only nonnegative values for $\Delta_{s,h}(p)$. Hence, $\Delta_{s,h}(p) \geq 0$ for $(s,h) \in \mathbb{N} \times \mathbb{N}_0$.

(d) Let $\nabla_{s,h}(p) := f_{s+1,h+1}(p) - f_{s,h}(p)$. Again, we obtain by Lemma 2

$$\begin{aligned}
\nabla_{s,h}(p) &= p\,[f_{s+1,h}(p)f_{s,h}(p) - f_{s,h-1}(p)f_{s-1,h-1}(p)] \\
&= p\,[\nabla_{s,h-1}(p)f_{s,h}(p) + \nabla_{s-1,h-1}(p)f_{s,h-1}(p)]
\end{aligned}$$

for $s \geq 2$, $h \geq 1$. Thus, if we are able to show that the initial conditions $\nabla_{1,h}(p)$, $h \in \mathbb{N}_0$, and $\nabla_{s,0}(p)$, $s \in \mathbb{N}$, are nonnegative, the same argument as in the proof of part (c) yields to $\nabla_{s,h}(p) \geq 0$ for $(s,h) \in \mathbb{N} \times \mathbb{N}_0$. An application of Lemma 2 immediately leads to $\nabla_{s,0}(p) = 0$, $s \in \mathbb{N}$. It remains to prove that $\nabla_{1,h}(p) \geq 0$, $h \in \mathbb{N}_0$. Again, choosing $s = 2$ in Lemma 2, we find the recurrence

$$\begin{aligned}
f_{2,0}(p) &= 1 \\
f_{2,h}(p) &= 1 - p + p(p\,\delta_{h-1,0} + 1 - p)f_{2,h-1}(p), \quad h \geq 1.
\end{aligned}$$

The iteration of this recurrence yields the solution

$$\begin{aligned}
f_{2,h}(p) &= (1-p)\left[p^{h-1}(1-p)^{h-2} + \sum_{0 \leq i \leq h-2}\left(p(1-p)\right)^i\right] \\
&= \frac{1}{1-p(1-p)}\left[1 - p + p^{h+1}(1-p)^{h-1}\right]
\end{aligned}$$

for $h \geq 1$. Thus. $\nabla_{1,0}(p) = f_{2,1}(p) - f_{1,0}(p) = 1 - 1 = 0$ and for $h \geq 1$

$$\begin{aligned}
\nabla_{1,h}(p) &= f_{2,h+1}(p) - f_{1,h}(p) \\
&= \frac{p(1-p)^2}{1-p(1-p)}\left[1 + p^{h+1}(1-p)^{h-2}\right] \geq 0.
\end{aligned}$$

This completes the proof of our lemma. ∎

Now, we are ready to derive lower and upper bounds for the probability $f_{s,h}(p)$.

Theorem 1 *The probability $f_{s,h}(p)$ that a tree $T \in \mathcal{F}_p(h)$ has a stack size less than or equal to s satisfies the following inequalities for $1 \leq s \leq h$:*

(a) *If $p \in [0,1] \setminus \{\frac{1}{2}\}$ then*

$$f_{s,h}(p) \geq (1-p)\frac{p^s - (1-p)^s}{p^{s+1} - (1-p)^{s+1}}$$

$$f_{s,h}(p) \leq (1-p)\frac{p^h - (1 - 2p + p^{s+1})(1-p)^{h-s-1}}{p^{h+1} - (1 - 2p + p^{s+1})(1-p)^{h-s}}$$

(b) *If $p = \frac{1}{2}$ then*

$$\frac{s}{s+1} \leq f_{s,h}(p) \leq \frac{h - s - 1 + 2^s}{h - s + 2^s}$$

Proof: First, we shall prove the lower bound for $f_{s,h}(p)$. Using the inequality presented in part (c) of Lemma 3, we find by Lemma 2

$$f_{s,h}(p) \geq 1 - p + p\, f_{s,h}(p)\, f_{s-1,h-1}(p), \quad s \geq 2.$$

Iterating this recurrence, we obtain

$$f_{s,h}(p) \geq (1-p)\sum_{i\geq0} p^i f^i_{s-1.h-1}(p) = \frac{1-p}{1 - p\, f_{s-1,h-1}(p)}, \quad s \geq 2.$$

Furthermore, $f_{1,h}(p) = 1 - p$ by Lemma 2. Hence, $f_{s,h}(p) \geq F_s(p)$, where $F_s(p)$ has the continued fraction representation

$$F_1(p) = 1 - p$$

$$F_s(p) = \frac{1-p}{1 - p\, F_{s-1}(p)}, \quad s \geq 2.$$

Solving this recurrence by standard methods, we immediately find the lower bound for $f_{s,h}(p)$ stated in part (a) of our theorem. The lower bound presented in part (b) follows by an application of L'Hospital's rule to the lower bound given in part (a).

Next, let us turn to the upper bound for $f_{s,h}(p)$. Applying the inequality stated in part (d) of Lemma 3 to the recurrence given in Lemma 2 yields

$$f_{s,h}(p) \leq 1 - p + p\, f_{s,h-1}(p)\, f_{s,h}(p), \quad s \geq 2.$$

The iteration of this recurrence leads to

$$f_{s,h}(p) \leq (1-p)\sum_{i\geq0} p^i f^i_{s,h-1}(p) = \frac{1-p}{1 - p\, f_{s,h-1}(p)}, \quad s \geq 2.$$

Moreover, $f_{s,s}(p) = 1 - p^s$ by part (b) of Lemma 3. Thus, $f_{s,h}(p) \leq G_h(s,p)$, where $G_h(s,p)$ has the continued fraction representation

$$G_s(s,p) = 1 - p^s$$

$$G_h(s,p) = \frac{1-p}{1 - p\, G_{h-1}(s,p)}, \quad h > s \geq 1.$$

Again, using standard methods, we find the upper bound for $f_{s,h}(p)$ presented in part (a) of our theorem. The upper bound stated in part (b) follows by applying d'Hospital's rule to the upper bound given in part (a). This completes the proof of our theorem. ∎

The probability that a tree $T \in \mathcal{F}_p(h)$ has the stack size s is equal to $f_{s,h}(p) - f_{s-1,h}(p)$. Thus, the average stack size of such a tree is given by

$$\begin{aligned}
\underline{S}_p(h) &= \sum_{1 \le s \le h+1} s[f_{s,h}(p) - f_{s-1,h}(p)] \\
&= 1 + h - \sum_{1 \le s \le h} f_{s,h}(p). \tag{2}
\end{aligned}$$

The following theorem gives an upper bound for $\underline{S}_p(h)$.

Theorem 2 *Let $d(n)$ be the number of the positive divisors of $n \in I\!N$ and let $C(x)$ be the convergent series $C(x) := \sum_{s \ge 0} d(s+1)x^s$, $|x| < 1$. The average stack size $\underline{S}_p(h)$ of a tree $T \in \mathcal{F}_p(h)$ satisfies the inequalities:*

$$\underline{S}_p(h) \begin{cases} < \frac{1-2p}{1-p} C\left(\frac{p}{1-p}\right) & \text{if } 0 \le p < \frac{1}{2} \\[2mm] \le H_h & \text{if } p = \frac{1}{2} \\[2mm] < \frac{2p-1}{p} h + \frac{2p-1}{p}\left[1 + \frac{1-p}{p} C\left(\frac{1-p}{p}\right)\right] & \text{if } \frac{1}{2} < p \le 1 \end{cases}$$

Here, $H_n = \sum_{1 \le i \le n} \frac{1}{i}$ denotes the n-th Harmonic number.

Proof: Inserting the lower bound of $f_{s,h}(p)$ stated in Theorem 1(a) into (2), we find for $p \in [0,1] \setminus \{\frac{1}{2}\}$

$$\underline{S}_p(h) \le 1 + h - (1-p) \sum_{1 \le s \le h} \frac{p^s - (1-p)^s}{p^{s+1} - (1-p)^{s+1}}. \tag{3}$$

First, let us consider the case $0 \le p < \frac{1}{2}$, that is $x := \frac{p}{1-p} < 1$. We immediately obtain

$$\underline{S}_p(h) \le 1 + h - \sum_{1 \le s \le h} \frac{1 - x^s}{1 - x^{s+1}} = 1 + (1-x) \sum_{1 \le s \le h} \frac{x^s}{1 - x^{s+1}}.$$

The sum appearing in the last expression converges for $h \to \infty$. Since for $|z| < 1$

$$\sum_{s \ge 1} \frac{z^s}{1 - z^{s+1}} = \sum_{s \ge 1} \sum_{\lambda \ge 0} z^{(s+1)\lambda + s} = \sum_{s \ge 1} [d(s+1) - 1]z^s,$$

we finally find

$$\underline{S}_p(h) < 1 + (1-x) \sum_{s \ge 1} [d(s+1) - 1]x^s = (1-x) \sum_{s \ge 0} d(s+1)x^s.$$

This expression is identical to that given in the theorem for $0 \leq p < \frac{1}{2}$.

Next, let $\frac{1}{2} < p \leq 1$, that is $x := \frac{1-p}{p} < 1$. Here, we obtain in a similar way

$$\underline{S}_p(h) \leq 1 + (1-x)h + x(1-x) \sum_{1 \leq s \leq h} \frac{x^s}{1 - x^{s+1}}$$

and therefore

$$\underline{S}_p(h) < (1-x)h + (1-x)\left[1 + \sum_{s \geq 0} d(s+1)x^{s+1}\right].$$

For $p = \frac{1}{2}$, the lower bound for $f_{s,h}(p)$ presented in Theorem 1(b) leads directly to

$$\underline{S}_p(h) \leq 1 + h - \sum_{1 \leq s < h} \left[1 - \frac{1}{s+1}\right] = H_h.$$

This completes the proof of our theorem.   ∎

Since $H_n = \ln(n) + \gamma + \mathcal{O}(\frac{1}{n})$, where $\gamma = 0.577\,215\ldots$ is Euler's constant ([5]), the preceding theorem tells us that the average stack size $\underline{S}_p(h)$ of a tree $T \in \mathcal{F}_p(h)$ is $\mathcal{O}(1)$ for $0 \leq p < \frac{1}{2}$, $\mathcal{O}(\ln(h))$ for $p = \frac{1}{2}$ and $\mathcal{O}(h)$ for $\frac{1}{2} < p \leq 1$. Next let us derive a lower bound for the average stack size $\underline{S}_p(h)$.

Theorem 3 *The average stack size $\underline{S}_p(h)$ of a tree $T \in \mathcal{F}_p(h)$ satisfies the following inequalities:*

$$\underline{S}_p(h) \begin{cases} \geq 1 + o(1) & \text{if } 0 \leq p \leq \frac{1}{2} \\ > \frac{2p-1}{p}\left[1 - \frac{\ln(p)}{\ln(1-p)}\right] h + \mathcal{O}(1) & \text{if } \frac{1}{2} < p \leq 1 \end{cases}$$

Proof: Inserting the upper bound for $f_{s,h}(p)$ stated in Theorem 1(a) into (2), we obtain for $p \in [0,1] \setminus \{\frac{1}{2}\}$

$$\underline{S}_p(h) \geq 1 + h - (1-p) \sum_{1 \leq s \leq h} \frac{p^h - (1 - 2p + p^{s+1})(1-p)^{h-s-1}}{p^{h+1} - (1 - 2p + p^{s+1})(1-p)^{h-s}}$$

$$= 1 + \sum_{1 \leq s \leq h} \frac{(2p-1)p^h}{p^{h+1} - (1 - 2p + p^{s+1})(1-p)^{h-s}}. \tag{4}$$

First, let us consider the case $0 \leq p < \frac{1}{2}$, that is $x := \frac{p}{1-p} < 1$. In this case, we can write

$$\underline{S}_p(h) \geq 1 + \sum_{1 \leq s \leq h} \frac{(1-2p)p^h}{(1-p)^{h-s}(1 - 2p + p^{s+1}) - p^{h+1}},$$

where all terms appearing in the sum are nonnegative because $(1-p)^{h-s}(1 - 2p + p^{s+1}) - p^{h+1} > p^{s+1}(1-p)^{h-s} - p^{h+1} = p^{h+1}[x^{-(h-s)} - 1] \geq 0$. Since $p^s \leq p$, we immediately find $1 - 2p + p^{s+1} \geq (1 - 2p + p^s)(1 - p)$. Thus, all terms appearing in the above sum can be estimated by the term for $s = 1$ and we obtain

$$\sum_{1 \leq s \leq h} \frac{(1 - 2p)p^h}{(1 - p)^{h-s}(1 - 2p + p^{s+1}) - p^{h+1}} \leq \frac{1 - 2p}{p} h \frac{x^{h+1}}{1 - x^{h+1}} \xrightarrow{h \to \infty} 0.$$

Hence, the sum is convergent and we have proved our result for $0 \leq p < \frac{1}{2}$.

Next, let $\frac{1}{2} < p < 1$, that is $x := \frac{1-p}{p} < 1$. Here, we have $p^{h+1} - (1 - 2p + p^{s+1})(1 - p)^{h-s} > p^{h+1} - p^{s+1}(1 - p)^{h-s} = p^{h+1}[1 - x^{h-s}] \geq 0$. Thus, all terms appearing in the sum (4) are nonnegative. Introducing the function $\xi_{p,h}(s) := \frac{(2p-1)p^h}{p^{h+1} - (1-2p)(1-p)^s}$ and applying Euler's summation formula to (4), we obtain

$$\underline{S}_p(h) > 1 + \sum_{1 \leq s \leq h} \frac{(2p - 1)p^h}{p^{h+1} - (1 - 2p)(1 - p)^{h-s}}$$

$$= 1 + \sum_{0 \leq s < h} \frac{(2p - 1)p^h}{p^{h+1} - (1 - 2p)(1 - p)^s}$$

$$= 1 + \int_0^{h-1} \xi_{p,h}(s)\, ds + \frac{1}{2}\Big[\xi_{p,h}(0) + \xi_{p,h}(h - 1)\Big]$$

$$+ \int_0^{h-1} (s - \lfloor s \rfloor - \frac{1}{2})\xi'_{p,h}(s)\, ds.$$

Now, we successively find

$$\int_0^{h-1} \xi_{p,h}(s)\, ds = \frac{2p - 1}{p}\left[s - \frac{\ln\left(p^{h+1} + (2p - 1)(1 - p)^s\right)}{\ln(1 - p)}\right]\Bigg|_0^{h-1}$$

$$= \frac{2p - 1}{p}\left[1 - \frac{\ln(p)}{\ln(1 - p)}\right] h -$$

$$- \frac{2p - 1}{p\ln(1 - p)}\Big[\ln(1 - p) + \ln(p) - \ln(2p - 1) +$$

$$+ \ln\left(1 + \frac{2p - 1}{p(1 - p)}(\frac{1 - p}{p})^h\right) - \ln\left(1 + \frac{p^{h+1}}{2p - 1}\right)\Big]$$

$$= \frac{2p - 1}{p}\left[1 - \frac{\ln(p)}{\ln(1 - p)}\right] h + \mathcal{O}(1)$$

and

$$\frac{1}{2}\Big[\xi_{p,h}(0) + \xi_{p,h}(h - 1)\Big] = \frac{1}{2}\left[\frac{(2p - 1)p^h}{p^{h+1} + 2p - 1} + \frac{(2p - 1)p^h}{p^{h+1} - (1 - 2p)(1 - p)^{h-1}}\right]$$

$$= \frac{2p - 1}{2p} + \mathcal{O}(\max\{x^h, p^h\}) = \mathcal{O}(1)$$

and

$$\left| \int_0^{h-1} (s - \lfloor s \rfloor - \frac{1}{2}) \xi'_{p,h}(s) \, ds \right| \leq \frac{1}{2} \int_0^{h-1} \xi'_{p,h}(s) \, ds$$

$$= \frac{1}{2} \left[\xi_{p,h}(h-1) - \xi_{p,h}(0) \right]$$

$$= \frac{2p-1}{2p} + \mathcal{O}(\max\{x^h, p^h\}) = \mathcal{O}(1)$$

because $\xi'_{p,h}(s) = -\frac{(1-2p)^2(1-p)^s p^h \ln(1-p)}{\left[p^{h+1} - (1-2p)(1-p)^s \right]^2} \geq 0$. Thus, we have proved

$$\underline{S}_p(h) > \frac{2p-1}{p} \left[1 - \frac{\ln(p)}{\ln(1-p)} \right] h + \mathcal{O}(1).$$

An inspection of (4) shows that this lower bound is also valid for $p = 1$ because $\lim_{p \to 1} \frac{\ln(p)}{\ln(1-p)} = 0$. For $p = \frac{1}{2}$, the upper bound for $f_{s,h}(p)$ stated in Theorem 1(b) leads to

$$\underline{S}_p(h) \geq 1 + h - \sum_{1 \leq s \leq h} \left(1 - \frac{1}{h - s + 2^s} \right) = 1 + o(1).$$

This completes the proof of our theorem. ∎

The preceding theorem shows that the average stack size $\underline{S}_p(h)$ of a tree $T \in \mathcal{F}_p(h)$ is $\Omega(1)$ for $0 \leq p \leq \frac{1}{2}$ and that it grows at least linearly in h for $\frac{1}{2} < p \leq 1$. Combining Theorem 2 and Theorem 3, we have proved the following main result.

Theorem 4 *The average stack size $\underline{S}_p(h)$ of a tree $T \in \mathcal{F}_p(h)$ satisfies the following inequalities:*

(a) *If $0 \leq p < \frac{1}{2}$:*

$$1 + o(1) \leq \underline{S}_p(h) < \frac{1 - 2p}{1 - p} C \left(\frac{p}{1 - p} \right) = \mathcal{O}(1)$$

(b) *If $p = \frac{1}{2}$:*

$$1 + o(1) \leq \underline{S}_p(h) \leq H_h = \mathcal{O}(\ln(h))$$

(c) *If $\frac{1}{2} < p \leq 1$:*

$$\frac{2p-1}{p} \left[1 - \frac{\ln(p)}{\ln(1-p)} \right] h + \mathcal{O}(1) < \underline{S}_p(h) < \frac{2p-1}{p} h + \mathcal{O}(1).$$

Here, $C(x)$ denotes the function introduced in Theorem 2. □

Figure 2 shows the graph of the average stack size $\underline{S}_p(10)$ together with the lower and the upper bounds stated in the formulas (3) and (4), respectively.

For $0 \leq p < \frac{1}{2}$, the upper bound seems to be an excellent estimate of the exact values of $\underline{S}_p(h)$; if $\frac{1}{2} < p \leq 1$, the lower bound seems to be more appropriate to describe the exact values of $\underline{S}_p(h)$. For this choice of p, part (c) of Theorem 4 shows that the gap between the upper and the lower bound is $f(p)\,h$ for large h, where $f(p) := \frac{(2p-1)\ln(p)}{p\ln(1-p)}$. The graph of this function is drawn in Figure 3. The function $f(p)$ has a maximum at $p_0 \approx .629\,229\,279\ldots$. Since $f(p_0) \approx .191\,781\ldots$, the gap between the upper and the lower bound is less than $.192\,h$.

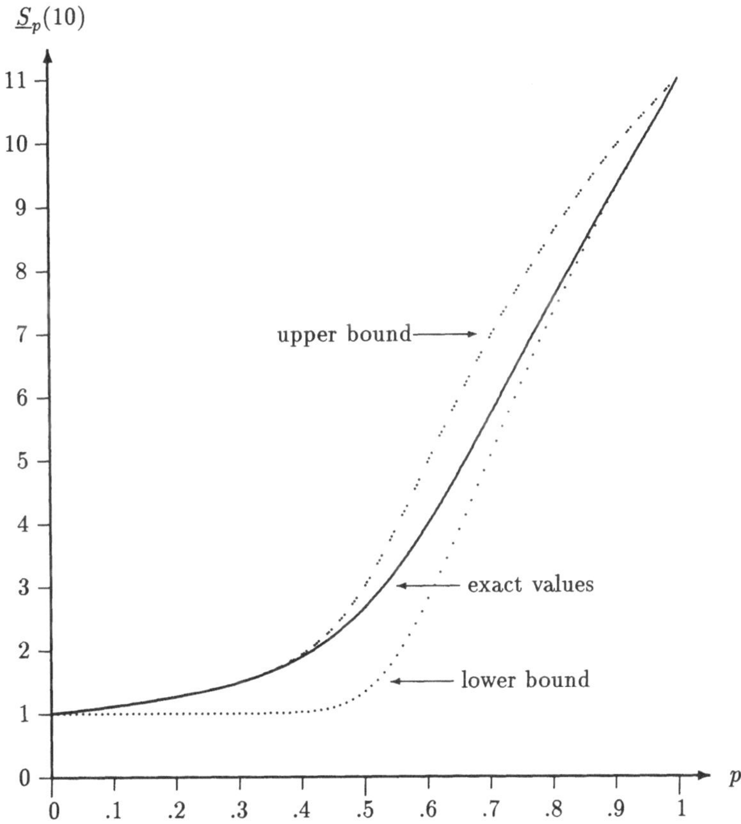

Figure 2. The graph of the average stack size $\underline{S}_p(10)$ together with the lower and the upper bound as a function of p.

3 Concluding Remarks

In this paper, we have derived a lower and an upper bound for the average stack size $\underline{S}_p(h)$ of a special class of binary backtrack trees. The derivation of these bounds requires only elementary methods. For $0 \le p < \frac{1}{2}$ and $\frac{1}{2} < p \le 1$ we have shown that $\underline{S}_p(h) = \Theta(1)$ and $\underline{S}_p(h) = \Theta(h)$, respectively. In the case $p = \frac{1}{2}$, we have only proved that $\underline{S}_p(h) = \mathcal{O}(\ln(h))$. It remains the question, whether this result can also be improved to $\underline{S}_p(h) = \Theta(\ln(h))$. Experimental results seem to confirm this conjecture. Of course, the central question goes unanswered: What is the *exact* (asymptotic) behaviour of the average stack size $\underline{S}_p(h)$?

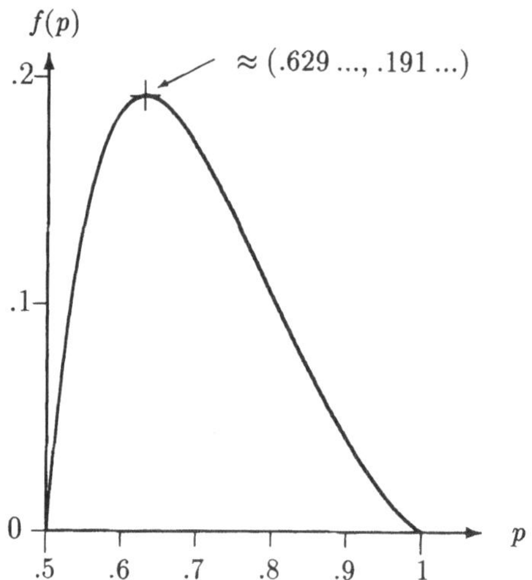

Figure 3. The graph of the function $f(p) = \frac{2p-1}{p}\,\frac{\ln(p)}{\ln(1-p)}$ for $\frac{1}{2} \le p \le 1$.

Bibliography

[1] Kemp, R.: A note on the stack size of regularly distributed binary trees, *BIT* **20** (1980), 157-163.

[2] Kemp, R.: On the average oscillation of a stack, *COMBINATORICA* **2**(2) (1982), 157-176.

[3] Kemp, R.: The expected additive weight of trees, *Acta Informatica* **26** (1989), 711-740.

[4] Kemp, R.: The analysis of a special class of backtrack trees, *preprint* Johann Wolfgang Goethe-Universität Frankfurt a.M., 1991.

[5] Knuth, D.E.: *The Art of Computer Programming,Vol. 1, 2nd ed.* Addison-Wesley, Reading, Mass. 1973.

[6] Meir, A., Moon, J.W.: On the altitude of nodes in random trees, *Can. J, Math.* **30** (1978), 997-1015.

[7] Purdom, P.W.: Tree size by partial backtracking, *SIAM J. Comput.* **4**(4) (1978), 481-491.

[8] Trier, U.: Additive weights of a special class of nonuniformly distributed backtrack trees, *preprint* Johann Wolfgang Goethe-Universität Frankfurt a.M., 1991.

Randomized Incremental Construction of Abstract Voronoi Diagrams[1]

Rolf Klein,

Fachbereich Mathematik
Universität–GHS–Essen
4300 Essen 1
Germany

Kurt Mehlhorn, Stefan Meiser

Max Planck Institut
für Informatik
6600 Saarbrücken
Germany

Abstract

Abstract Voronoi diagrams were introduced by R. Klein [14, 11, 12] as an axiomatic basis of Voronoi diagrams. We show how to construct abstract Voronoi diagrams in time $O(n \log n)$ by a randomized algorithm; the algorithm is based on Clarkson and Shor's randomized incremental construction technique [6]. The new algorithm has the following advantages over previous algorithms:

- It can handle a much wider class of abstract Voronoi diagrams than the algorithms presented in [14, 17].

- It can be adapted to a concrete kind of Voronoi diagram by providing a single basic operation, namely the construction of a Voronoi diagram of five sites. Moreover, all geometric decisions are confined to the basic operation, and using this operation, abstract Voronoi diagrams can be constructed in a purely combinatorial manner.

[1]This work was supported partially by the DFG, grants SPP Me 620/6 and Kl 655/2-1, and partially by the ESPRIT II Basic Research Actions Program of the EC under contract No. 3075 (project ALCOM). A preliminary version of this paper was presented at the SIGAL Symposium on Algorithms, Tokyo 1990.

1   Introduction

The Voronoi diagram of a set of sites in the plane partitions the plane into regions, called Voronoi regions, one to a site. The Voronoi region of a site s is the set of points in the plane for which s is the closest site among all the sites.

The Voronoi diagram has many applications in diverse fields, cf. Leven and Sharir [16] or Aurenhammer [1] for a list of applications and a history of Voronoi diagrams. Different types of diagrams result from considering different notions of distance, e. g. Euclidean or L_p-norm or convex distance functions, and different sorts of sites, e. g. points, line segments, or circles. For many types of diagrams efficient construction algorithms have been found; these are either based on the divide-and-conquer technique due to Shamos and Hoey [18], the sweepline technique due to Fortune [8], geometric transforms due to Brown [5] and Edelsbrunner and Seidel [7], or the randomized incremental construction technique due to Clarkson and Shor [6].

A unifying approach to Voronoi diagrams was proposed by Klein [11, 12, 13, 14], cf. [7] for a related approach. He does not use the concept of distance as the basic notion but rather the concept of bisecting curve, i. e. he assumes for each pair $\{p, q\}$ of sites the existence of a bisector $J(p, q)$, which is homeomorphic to a line and divides the plane into a p-region and a q-region. The intersection of all p-regions for different q's is then the Voronoi-region of site p. He also postulates that Voronoi-regions are simply-connected and partition the plane. He shows that abstract Voronoi diagrams have already many of the properties of concrete Voronoi diagrams, cf. Section 2.

At present there are two algorithms for the construction of abstract Voronoi diagrams. Both algorithms assume that certain elementary operations on bisecting curves, e. g. computation of the intersections, take $O(1)$ time, and both algorithms can handle only subclasses of abstract Voronoi diagrams.

Klein [14] presented an off-spring of the Shamos and Hoey divide-and-conquer algorithm. He has to assume that any set S of sites can be split in time $O(|S|)$ into about equal sized subsets L and R such that the bisector between L and R (= the common boundary of regions in L with regions in R) is acyclic and, under this assumption, constructs the Voronoi diagrams of n sites in time $O(n \log n)$. There are cases, e. g. points with additive weights in the Euclidean plane, where it is not known if such partitions exist.

Mehlhorn, Meiser and Ó' Dúnlaing [17] presented an off-spring of the Clarkson and Shor randomized incremental algorithm. They have to assume that the set of bisectors is regular, i. e. no four of them share a point and any point of intersection of two bisectors is a proper crossing of the bisectors. Under these assumptions, their algorithm runs in expected time $O(n \log n)$, the average being taken over all permutations of the input. There are cases, e. g. point sites in the Manhattan metric, where this assumption does not hold.

In this paper, we extend the randomized incremental algorithm and show that it can handle abstract Voronoi diagrams in (almost) their full generality; cf. the remark following Definition 1 in Section 2 for the minor restriction which we have to make. The algorithm runs in expected time $O(n \log n)$ and is as simple as the algorithm in [17], although correctness proof and running time analysis are more involved. The algorithm is uniform in the sense that only a single operation, namely the construction of a Voronoi diagram for 5 sites, depends on the specific type of Voronoi diagram and has to be programmed in order to adapt the algorithm to the type of the diagram. Moreover, all numerical operations take place within this particular operation.

In particular, comparisons only take place between objects which are related in the topology of the diagram. The incremental algorithm of Guibas and Stolfi [10] for Euclidean diagrams also has this property but neither the Plane-Sweep- nor the Divide-and-Conquer-algorithm do. Both algorithms need to sort the sites by x-coordinates. Moreover, the Plane-Sweep-algorithm sorts the computed events by x-coordinates; the Divide-and-Conquer-algorithm sorts the nodes of the diagram by y-coordinates in its merge step. In both cases, objects are compared to each other that are not at all related in their topology. Therefore, it may be difficult to make geometric decisions in a consistent manner. From a programmers point of view, concentrating the numerical computations inside a single operation may facilitate the handling of instable arithmetic.

As mentioned above, our algorithm is based on Clarkson and Shor's randomized incremental construction technique [6]. We make use of the refinement proposed in [9, 2, 4, 3], in particular, we use the notion of history graph instead of the original conflict graph.

The paper is organized as follows: In Section 2 we introduce abstract Voronoi diagrams; we give the definitions and state some properties. Section 3 deals with the notion of conflict that is basic to the randomized incremental construction scheme and introduces our basic operation. The algorithm is then given in Section 4.

Throughout the paper, we use the following notation: For a subset $X \subseteq \mathbb{R}^2$ the closure, boundary and interior of X are denoted by $\overline{X}$, *bd* X and *int* X, respectively. Instead of *int* X we also write X°.

2   Abstract Voronoi Diagrams

Let $n \in \mathbb{N}$, and for every pair of integers p, q such that $1 \leq p \neq q < n$ let $D(p, q)$ be either empty or an open unbounded subset of $\mathbb{R}^2$ and let $J(p, q)$ be the boundary of $D(p, q)$. We postulate:

1) $J(p, q) = J(q, p)$ and for each p, q such that $p \neq q$ the regions $D(p, q)$, $J(p, q)$ and $D(q, p)$ form a partition of $\mathbb{R}^2$ into three disjoint sets.

2) If $\emptyset \neq D(p, q) \neq \mathbb{R}^2$ then $J(p, q)$ is homeomorphic to the open interval $(0, 1)$.

We call $J(p,q)$ the bisecting curve for sites p and q and $D(p,q)$ the region of dominance of p over q. Following Klein [14], the abstract Voronoi diagram is now defined as follows:

Definition 1 *Let $S = \{1, \ldots, n-1\}$ and let $<$ be a linear order on S. Let*

$$R_<(p,q) := \begin{cases} D(p,q) \cup J(p,q) & \text{if } p < q \\ D(p,q) & \text{if } p > q \end{cases}$$

$$VR_<(p,S) := \bigcap_{\substack{q \in S \\ q \neq p}} R_<(p,q)$$

$$V_<(S) := \bigcup_{p \in S} bd\ VR_<(p,S)$$

$VR^o_<(p,S)$ *is called the* Voronoi region *of p or p-region w.r.t. to S and $<$, $VR_<(p,S)$ is called the* extended Voronoi region *of p w.r.t. to S and $<$, and $V_<(S)$ is called the* Voronoi diagram *of S with respect to $<$.* □

We require that the Voronoi regions and the bisecting curves satisfy the following two conditions:

1) Any two bisecting curves intersect in only a finite number of connected components.

2) For all non-empty subsets S' of S and all orderings $<$ of S

   A) if $VR_<(p,S')$ is non-empty then $VR_<(p,S')$ is path-connected and has non-empty interior for each $p \in S'$,

   B) $\mathbb{R}^2 = \bigcup_{p \in S'} VR_<(p,S')$

We will show that Voronoi diagrams can be constructed in time $O(n \log n)$ by a randomized algorithm provided that properties 1 and 2 hold and that the bisecting curves are computationally simple (cf. Section 3.2 for a precise definition).

Remarks:

a) Klein [14] has shown that Condition 2B is equivalent to the following transitivity property: $R_<(p,q) \cap R_<(q,r) \subseteq R_<(p,r)$ for any three pairwise distinct sites $p, q, r \in S$.

b) The union in 2B is disjoint by the definition of Voronoi regions.

c) Our Conditions 2A and 2B are slightly stronger than Klein's original definition in [14]. Klein considers a fixed ordering on S and requires that 2A and 2B hold for this fixed ordering. Figure 1 shows a system of bisectors for three sites p, q, r which satisfies 2A and 2B for the orderings $p < q < r$ and $p < r < q$, but not for any other ordering of these three sites. Note that p-region is disconnected if p is not the minimal site w.r.t. $<$. Our slightly stronger assumption excludes this degenerate case.

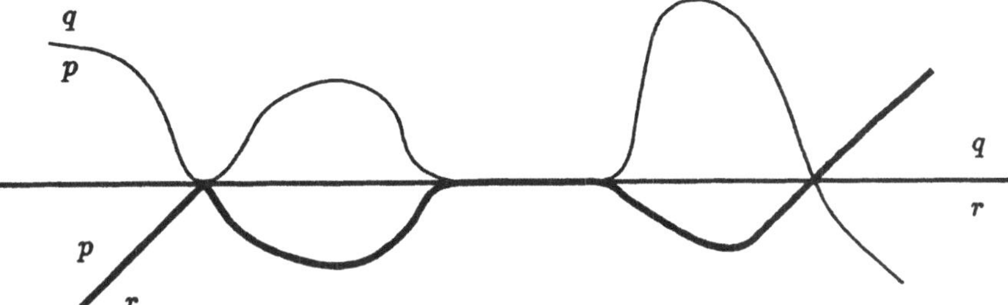

Figure 1: A degenerate case

d) Let p be a real number with $1 < p < \infty$ and let d denote the L_p-distance function. Let S be a set of points in the plane. For points $x, y \in S$ define $J(x, y) = \{z \mid d(x, z) = d(y, z)\}$. Then properties 1 and 2 hold.

e) Our algorithm works incrementally and considers the elements of S in random order. Let $\prec$ be the order in which the algorithm considers the elements of S. Then our algorithm constructs the diagram $V_\prec(S)$ provided that properties 1, 2A and 2B (with $<$ equal to $\prec$) and Facts 1 and 2 below hold. This observation can be used to construct Voronoi diagrams in some situations where properties 1 and 2 do not hold.

Let d be a *nice* metric on $\mathbb{R}^2$, cf. [12, 14] for a definition. All L_p-metrics, $1 \le p \le \infty$, are nice. Let S be a set of points in the plane. For points $x, y \in S$ with $x \prec y$ define $J(x, y) = J(y, x) = bd\,\{z \mid d(x, z) \le d(y, z)\} = bd\,\{z \mid d(y, z) < d(x, z)\}$, $D(x, y) = \{z \mid d(x, z) \le d(y, z)\} - J(x, y)$ and $D(y, x) = \{z \mid d(y, z) < d(x, z)\}$. In [14] it was shown that properties 1, 2A and 2B hold for the linear order $<$ equal to $\prec$. For the L_1- and the L_∞-metric Facts 1 and 2 also hold. Thus, our algorithm can be used to construct the diagram $V_\prec(S)$ in the L_1- and the L_∞-metric. The drawback of this approach is that the diagram depends on the insertion order. We do not know whether the bisecting curves in the L_1- and the L_∞-metric can be defined in such a way that properties 2A and 2B hold for all linear orders $<$; note that for points x and y lying on a line of slope ± 1 the set $\{z \mid d(x, z) = d(y, z)\}$ is not just a curve and hence there is a choice in the definition of the bisecting curve.

□

Our first Lemma shows that the Voronoi diagram does not depend on the particular linear order used in its definition.

Lemma 1 *Let $<_1$ and $<_2$ be two linear orderings of S and let $S' \subseteq S$*

a) int $VR_{<_1}(p, S') =$ int $VR_{<_2}(p, S')$ for all $p \in S'$.

b) $V_{<_1}(S') = V_{<_2}(S')$

Proof: a) Let $x \in int\ VR_{<_1}(p, S')$. Then $x \in D(p, q)$ for all $q \in S' - \{p\}$. Thus $x \in int\ VR_{<_2}(p, S')$, i.e. $int\ VR_{<_1}(p, S') \subseteq int\ VR_{<_2}(p, S')$. Equality follows by symmetry.

b) Let $x \in V_{<_1}(S')$. Then $x \in bd\ int\ VR_{<_1}(p, S')$ for some $p \in S'$ by Theorem 2.3.5 of [14]. Thus $x \in V_{<_2}(S')$ by part a), i.e. $V_{<_1}(S') \subseteq V_{<_2}(S')$. Equality follows by symmetry. ∎

In light of Lemma 1, we will write $V(S)$ instead of $V_<(S)$ and $int\ VR(p, S)$ instead of $int\ VR_<(p, S)$ from now on. We will also write $VR(p, S)$ instead of $VR_<(p, S)$ when the linear order $<$ is clear from the context.

Definition 2 *An edge e of $V(S)$ is a maximal connected subset of $V(S)$ such that every point $x \in e$ lies on $bd\ int\ VR(p, S)$ for exactly two sites p of S. The edge is said to separate the regions of these two sites. A vertex v of $V(S)$ is a point $x \in V(S)$ which lies on $bd\ int\ VR(p, S)$ for at least three sites p of S.*

The following Facts characterize the topology of abstract Voronoi diagrams. Let $<$ be a linear order on S and define

$$\min(p, q) = \begin{cases} p & \text{if } p < q \\ q & \text{if } q < p \end{cases}$$

Fact 1 *a) All but finitely many points of $V(S)$ belong to an edge of $V(S)$.*
b) For each point $x \in V(S)$ there are arbitrarily small neighborhoods U of x having the following properties: $V(S) \cap bd\ U$ is finite. Let $w_1, \ldots, w_h$ be the points in $V(S) \cap bd\ U$ as encountered in a clockwise traversal of $bd\ U$. Then $h \geq 2$ and $V(S) \cap U$ is the union of curve segments $\beta_1, \ldots, \beta_h$ where β_i connects x to w_i and the β_i's are disjoint except at their common endpoint x. For each i, $1 \leq i \leq h$, there is a site $p_i \in S$ such that the open "piece of pie" bordered by β_i, β_{i+1} (read indices mod h) is contained in $VR^{\circ}(p_i, S)$. Then $p_i \neq p_j$ for $i \neq j$. Let $q_i \in S$ be the site such that $\beta_i - x \subseteq VR(q_i, S)$. We have $q_i \leq \min\{p_{i-1}, p_i\}$. The point x belongs to $VR(p, S)$, where $p = \min\{p_1, \ldots, p_h, q_1, \ldots, q_h\}$. Also, only the site p can occur more than once among $p_1, \ldots, p_h, q_1, \ldots, q_h$. □

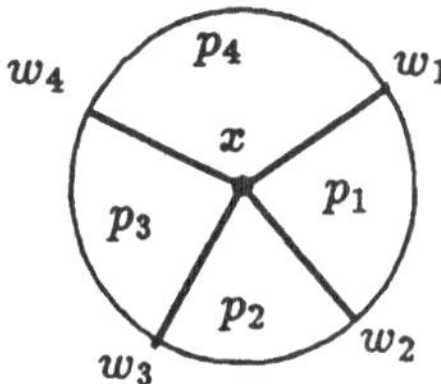

a vertex x

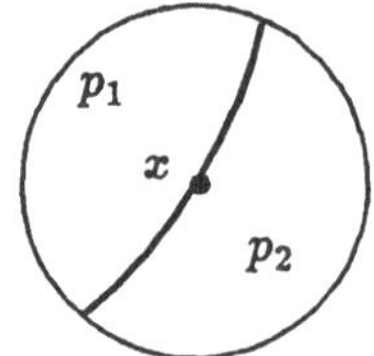

a point x on an edge
$x \in VR(\min\{p_1, p_2\}, S)$

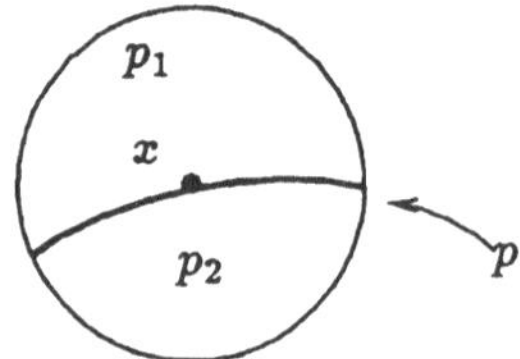

$x \in VR(p, S)$ and
$p < \min\{p_1, p_2\}$

Figure 2: Illustration of Fact 1.

Figure 2 illustrates Fact 1. Fact 1 is an easy consequence of Theorem 2.3.5 of [14]. In Theorem 2.3.5 of [14] it is allowed that p (but only p!) owns several open pieces of pie. Since the open pieces of pie are invariant with respect to the ordering $<$ by Lemma 1, p can own at most one open piece of pie under our strengthened conditions.

For the sequel, it is helpful to restrict attention to the "finite part" of $V(S)$. Let Γ be a simple closed curve such that in the outer domain of Γ any two bisectors are either disjoint or identical. We add a site ∞ to S, define $J(p,\infty) = J(\infty,p) = \Gamma$ for all p, $1 \leq p < n$, and $D(\infty,p)$ to be the outer domain of Γ for each p, $1 \leq p < n$.

Fact 2 *$V(S)$ is connected. The extended Voronoi region of a site $p \in S - \{\infty\}$ is simply-connected, each non-empty Voronoi region $VR^\circ(p,S)$, $p \in S - \{\infty\}$, is homeomorphic to an open disc and its boundary is a simple closed curve. The Voronoi region of site ∞ is not simply connected but it has only one hole being the inner domain of Γ. A Voronoi diagram can be represented as a planar graph in a natural way. The vertices and edges of the graph are the vertices and edges of $V(S)$, respectively; the faces of the graph correspond to the non-empty Voronoi regions. We use $V(S)$ to also denote this graph.* $\square$

For a proof of Fact 2 see Lemma 2.2.4 and Theorems 2.3.5 and 2.5.5 [14].

The extended Voronoi region $VR(p,S)$ for a site $p \in S - \{\infty\}$, consists of its Voronoi region, some vertices and edges on the boundary of $int\ VR(p,S)$, and some other vertices and edges of $V(S)$. The other edges and vertices form trees rooted at $bd\ int\ VR(p,S)$, cf. Figure 3.

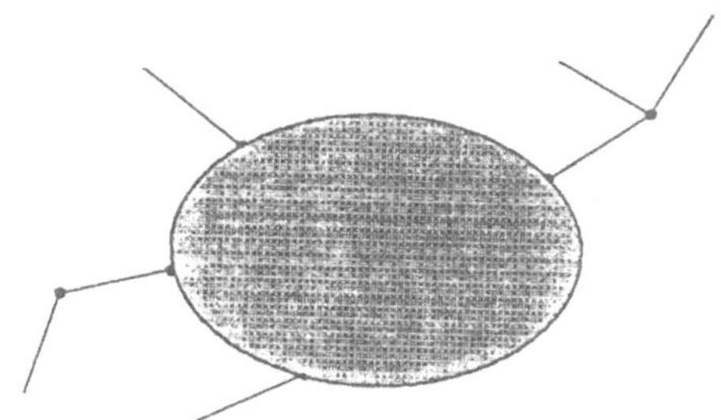

Figure 3: An extended Voronoi region $VR_<(p,S)$.

We next illustrate the concepts introduced so far on an example which we will use as our running example throughout the paper. Figure 4 shows $V(\{a,b,\infty\})$ and $V(\{a,b,c,\infty\})$. Assume $a < b < c < \infty$. Then edges e_2 and e_3 belong to $VR_<(a,\{a,b,\infty\})$ and edges e_4, e_5, e_7 and e_8 belong to $VR_<(a,\{a,b,c,\infty\})$.

For the remainder of the section, let $R \subseteq S$, $\infty \in R$, $s \in S - R$, $\mathcal{S} = VR^\circ(s, R \cup \{s\})$, and $\mathcal{E} = V(R) \cap \overline{\mathcal{S}}$. Then $bd\ \mathcal{S} = bd\ \overline{\mathcal{S}}$ is a simple closed curve.

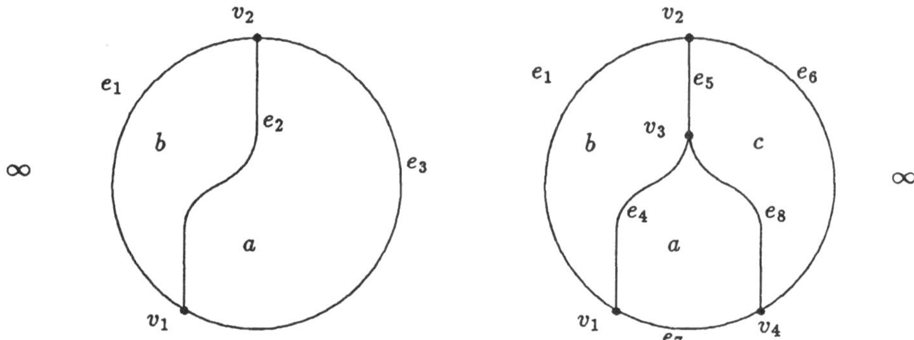

Figure 4: The enclosing circle represents Γ. Edges e_1 and e_3 are part of $J(b,\infty) = \Gamma$ and $J(a,\infty) = \Gamma$, respectively. The bisector between sites b and c is not shown.

Lemma 2 *If $S \neq \emptyset$ then $\mathcal{E}$ is a non-empty connected set which intersects bd S. Moreover, $\mathcal{E}$ is not just a single point.*

Proof: If $\mathcal{E}$ were empty, then $\overline{S} \subseteq VR^o(p, R)$ for some $p \in R - \{\infty\}$. Consequently, $VR(p, R \cup \{s\})$ would not be simply connected. Now let $\mathcal{E}_1, \mathcal{E}_2, \ldots, \mathcal{E}_k$ be the connected components of $\mathcal{E}$ for some k. Observe that no $\mathcal{E}_j$ can be entirely contained inside S for else $V(R)$ would not be connected, a contradiction.

Assume $k \geq 2$. Then a path $\mathcal{P} \subseteq \overline{S} - \mathcal{E}$ exists, connecting two points x and y on the boundary of S and separating $\mathcal{E}_1$ from $\mathcal{E}_2$, cf. Figure 5. From $\mathcal{P} \cap \mathcal{E} = \emptyset$ we have

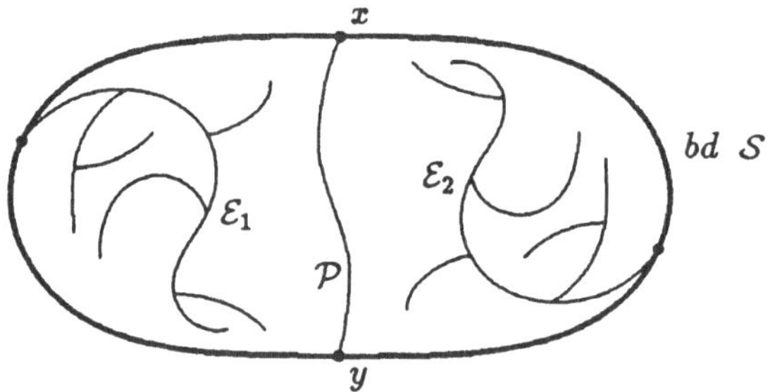

Figure 5:

$\mathcal{P} \cap V(R) = \emptyset$ and thus $\mathcal{P} \subseteq int\ VR(r, R)$ for a site $r \in R - \{\infty\}$. $x, y \in \mathcal{P}$ implies that all sufficiently small neighborhoods $U(x)$ and $U(y)$ are entirely contained in $VR(r, R)$. The points in the intersection of these neighborhoods with the complement of $\overline{S}$ thus lie in $VR(r, R \cup \{s\})$ and can be connected by a path $\mathcal{Q} \subseteq int\ VR(r, R \cup \{s\}) \subseteq int\ VR(r, R)$. The cycle $\mathcal{P} \circ \mathcal{Q}$ is therefore entirely contained in $int\ VR(r, R)$ and contains $\mathcal{E}_1$ or $\mathcal{E}_2$ in its interior. This is a contradiction.

At this point we have shown that $\mathcal{E}$ is a non-empty connected set which intersects bd S. Assume now that $\mathcal{E}$ is a single point. This point, say v, is either a vertex of

$V(R)$ or lies on an edge of $V(R)$. In either case, one of the regions of $V(R)$ incident to v in $V(R)$ is split by $\mathcal{S}$ in a neigborhood of v and hence represented twice at v in $VR(R \cup \{s\})$, cf. Figure 6, a contradiction to Fact 1.   ∎

Figure 6:

Note that Lemma 2 implies in particular, that if $\mathcal{S} \neq \emptyset$ then $\overline{\mathcal{S}}$ intersects an edge of $V(R)$. Lemma 3 discusses the various forms which an intersection between $\overline{\mathcal{S}}$ and an edge e of $V(R)$ can have.

Lemma 3 *Let e be an edge of $V(R)$. If $e \cap \overline{\mathcal{S}} \neq \emptyset$, then either $e \cap \overline{\mathcal{S}} = V(R) \cap \overline{\mathcal{S}}$ and $e \cap \overline{\mathcal{S}}$ is a single component or $e - \overline{\mathcal{S}}$ is a single component (possibly empty).*

Proof: Assume first that $e \cap \overline{\mathcal{S}} = V(R) \cap \overline{\mathcal{S}}$. Because $V(R) \cap \overline{\mathcal{S}}$ is connected by Lemma 2, $e \cap \overline{\mathcal{S}}$ is connected. Assume next that $e \cap \overline{\mathcal{S}} \neq V(R) \cap \overline{\mathcal{S}}$. Then for every point $x \in e \cap \overline{\mathcal{S}}$ one of the subpaths of e connecting x to an endpoint of e must be contained in $\overline{\mathcal{S}}$, since $V(R) \cap \mathcal{S}$ is a connected set. Hence $e - \overline{\mathcal{S}}$ is a single component. ∎

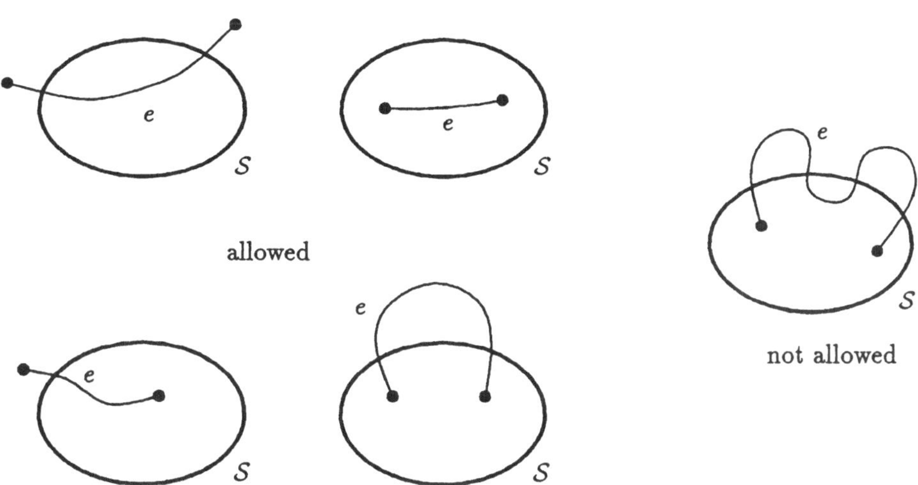

Figure 7: Four situations as allowed by Lemma 3 and one impossible.

3   Conflicts and the Basic Operation

The notion of conflict is central for the randomized incremental construction scheme of Clarkson and Shor. For the construction of Voronoi diagrams the appropriate notion is the conflict between edges and sites. We define this notion in Section 3.1. For the analysis of the algorithm it is important that edges of a Voronoi diagram are treated as purely combinatorial objects. Such a characterization is also derived in Section 3.1. In Section 3.2 we show how to decide the existence of conflicts by looking at diagrams of at most five sites.

3.1   Conflicts and a Combinatorial Characterization of Edges

For a set R of sites we use $E(R)$ to denote the set of bounded edges of the Voronoi diagram of R.

Definition 3 *Let $e \in E(R)$ separate p-region and q-region, and let $t \in S - R$ be a site.*

    a) t conflicts with e with respect to R iff for all neigborhoods U of e:

$$\left(\overline{VR^{\circ}(p, R)} \cup \overline{VR^{\circ}(q, R)}\right) \cap U \cap \overline{VR^{\circ}(t, R \cup \{t\})} - \{v, w\} \neq \emptyset$$

    where v and w are the endpoints of e.

    b) t intersects e with respect to R iff $e \cap \overline{VR^{\circ}(t, R \cup \{t\})} \neq \emptyset$.

    c) Let v be an endpoint of e. t clips e at v with respect to R iff for all neigborhoods U of v:

$$e \cap U \cap \overline{VR^{\circ}(t, R \cup \{t\})} \neq \emptyset.$$

Remarks:

    a) Figure 8 illustrates Definition 3.

    b) When R is clear from the context, the phrase "with respect to" is omitted.

    c) In Lemma 6 we will show that t conflicts with e iff either t intersects e or clips one of the four edges "neighboring" e.

$\square$

Definition 4 *Let $R \subseteq S$.*

    a) A description D over R is a set $\{(r_p, p, q, r_q), (s_q, q, p, s_p)\}$, where $\{p, q, r_p, r_q, s_p, s_q\} \subseteq R$, $p \neq q$, and $\{p, q\} \cap \{r_p, r_q, s_p, s_q\} = \emptyset$. $\mathcal{F}(R)$ denotes the set of descriptions over R and set(D) denotes the set $\{p, q, r_p, r_q, s_p, s_q\}$.

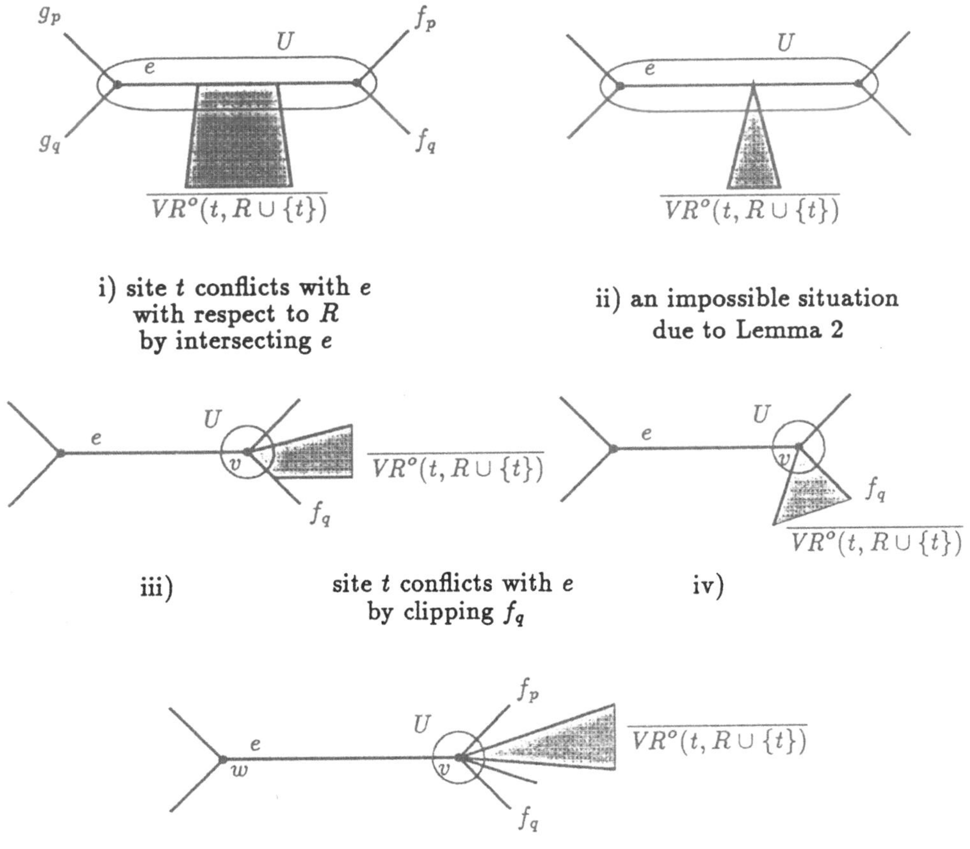

Figure 8: An illustration of Definition 3.

b) Let $e \in E(R)$ separate p- and q-region. Let f_p and g_p be the edges preceding and following e in a clockwise traversal of the boundary of $VR^\circ(p, R)$, and let g_q and f_q be the edges preceding and following e in a clockwise traversal of the boundary of $VR^\circ(q, R)$, cf. Figure 9. We call f_p, g_p, f_q, g_q the edges neighboring e in $V(R)$. Assume further that f_p separates p- and s_p-region, g_p separates p- and r_p-region, and g_q separates q- and r_q-region, and f_q separates q- and s_q-region. Then $D_R(e) = \{(r_p, p, q, r_q), (s_q, q, p, s_p)\} \in \mathcal{F}(R)$ is the description of e with respect to R.

$\square$

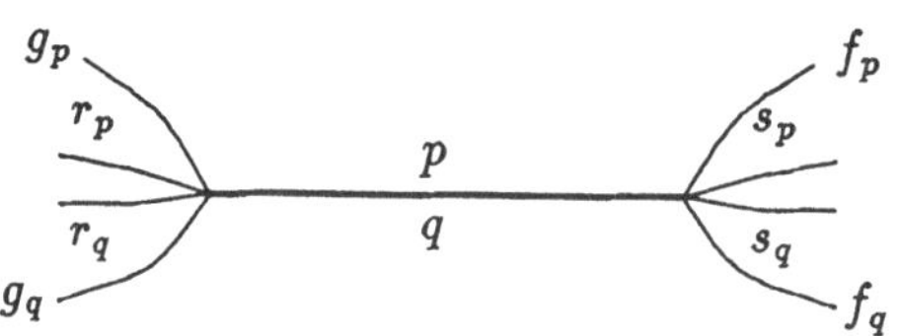

Figure 9:

Lemma 4 *Let e be a bounded edge of $V(R)$ and let $set(D_R(e)) \subseteq R' \subseteq R$. Then e is an edge of $V(R')$ and $D_R(e) = D_{R'}(e)$.*

Proof: Clearly $VR(t, R') \supseteq VR(t, R)$ for every $t \in R'$. Thus e is an edge of $V(R')$ and $D_R(e) = D_{R'}(e)$. $\blacksquare$

Lemma 5 *Let e and f be bounded edges of $V(R)$ with $e \neq f$. Then $D_R(e) \neq D_R(f)$.*

Proof: The existence of two distinct edges with the same description clearly contradicts the fact that all regions are path-connected. $\blacksquare$

Definition 5 *Let D be a description over S and let $t \in S - set(D)$ be a site. t conflicts with D iff there is no bounded edge in $V(set(D) \cup \{t\})$ with decription D.* $\square$

The two definitions of conflict are reconciled by the following lemma.

Lemma 6 *Let $R \subseteq S$, $e \in E(R)$, and $t \in S - R$. The following three statements are equivalent:*

a) t conflicts with e with respect to R.

b) t intersects e with respect to R or clips one of the four edges neighboring e at an endpoint common with e.

c) t conflicts with $D_R(e)$.

Proof: Let e separate p- and q-region, let v and w be the endpoints of e, let f_p, g_p, g_q, f_q, s_p, s_q, r_p, and r_q be defined as in Definition 4, and let $D = D_R(e)$.

a) $\Rightarrow$ b): If t conflicts with e, i.e. $\left(\overline{VR^\circ(p, R)} \cup \overline{VR^\circ(q, R)}\right) \cap U \cap \overline{VR^\circ(t, R \cup \{t\})} - \{v, w\} \neq \emptyset$ for all neighborhoods U of e, then $\overline{e} \cap \overline{VR^\circ(t, R \cup \{t\})} \neq \emptyset$, i.e. either $e \cap \overline{VR^\circ(t, R \cup \{t\})} \neq \emptyset$ or $x \in \overline{VR^\circ(t, R \cup \{t\})}$ where x is an endpoint of e. In the former case, t intersects e and we are done. In the latter case, t must clip one of the edges adjacent to x by Lemma 2. Since $\left(\overline{VR^\circ(p, R)} \cup \overline{VR^\circ(q, R)}\right) \cap U \cap \overline{VR^\circ(t, R \cup \{t\})} - \{v, w\} \neq \emptyset$ for all neighborhoods U of e, t must clip either e or one of the two edges neighboring e at x.

b) $\Rightarrow$ c): By Fact 1 and since $VR^\circ(x, R) \subseteq VR^\circ(x, set(D))$ for $x \in set(D)$, we have $VR^\circ(y, R) \cap U = VR^\circ(y, set(D)) \cap U$ for $y \in \{p, q\}$ and all sufficiently small neighborhoods U of e. In particular, e is an edge of $V(set(D))$ and there are edges f_p', g_p', f_q', g_q' of $V(set(D))$ with $f_p \subseteq f_p'$, $g_p \subseteq g_p'$, $f_q \subseteq f_q'$, and $g_q \subseteq g_q'$.

Since t intersects e or clips one of the edges f_p, g_p, f_q or g_q at their common endpoint with e, we have $U \cap \overline{VR^\circ(t, R \cup \{t\})} \cap (e \cup g_p \cup g_q \cup f_p \cup f_q) \neq \emptyset$ for all neighborhoods U of e. Since $VR^\circ(t, set(D) \cup \{t\}) \supseteq VR^\circ(t, R \cup \{t\})$ this implies $U \cap \overline{VR^\circ(t, set(D) \cup \{t\})} \cap (e \cup g_p' \cup g_q' \cup f_p' \cup f_q') \neq \emptyset$ for all neighborhoods U of e, i.e. either e is not an edge of $V(set(D) \cup \{t\})$ or e is an edge but has a description different from D with respect to $V(set(D) \cup \{t\})$. Suppose now, that there were an edge f of $V(set(D) \cup \{t\})$ with $D_{set(D) \cup \{t\}}(f) = D$. Then certainly, $e \neq f$. By Lemma 4, f would also exist in $V(set(D))$ and would have description D there. This contradicts Lemma 5. Thus t conflicts with D.

c) $\Rightarrow$ a): We show that non-a) implies non-c). So assume that there is a neighborhood U of e such that $\left(\overline{VR^\circ(p, R)} \cup \overline{VR^\circ(q, R)}\right) \cap U \cap \overline{VR^\circ(t, R \cup \{t\})} \subseteq \{v, w\}$. Thus $\overline{VR^\circ(x, R \cup \{t\})} \cap U \supseteq \left(\overline{VR^\circ(x, R)} \cap U\right) - \{v, w\}$ for $x \in \{p, q\}$ and hence $\left(\overline{VR^\circ(p, R \cup \{t\})} \cup \overline{VR^\circ(q, R \cup \{t\})}\right) \cap U \cap \overline{VR^\circ(t, R \cup \{t\})} \subseteq \{v, w\}$. Thus e also exists with description D in $V(R \cup \{t\})$ and also in $V(set(D) \cup \{t\})$ by Lemma 4. Hence t does not conflict with $D_R(e)$. $\blacksquare$

Remark: Lemma 6 establishes the equivalence of three different notions of conflict. The first notion (statement a)) is the most intuitive, the second notion (statement b)) is the key for the algorithmic treatment of conflicts, and the third notion (statement c)) is crucial for the analysis of the incremental algorithm. The following Lemma 7 completes our combinatorial characterization of edges. $\square$

Definition 6 *Let $R \subseteq S$. Define $\mathcal{F}_0(R) = \{D \in \mathcal{F}(R) \mid V(set(D))$ contains an edge with description D and D does not conflict with any site $t \in R - set(D)\}$.*

Lemma 7 *Let $\infty \in R \subseteq S$. Then $e \mapsto D_R(e)$ is a bijection between the edges of $V(R)$ and $\mathcal{F}_0(R)$.*

Proof: Note first that all edges of $V(R)$ are bounded since $\infty \in R$. Let $e \in E(R)$ be an edge separating p- and q-region, let $D = D_R(e)$, and let $t \in R - set(D)$. By

Lemma 4, e with description D is also edge of $V(set(D))$ and $V(set(D) \cup \{t\})$. In particular t does not conflict with D, i.e. $D \in \mathcal{F}_0(R)$. We have now shown that the mapping $e \mapsto D_R(e)$ maps $E(R)$ into $\mathcal{F}_0(R)$. By Lemma 5, the mapping is injective. It remains to show surjectivity.

Let $D \in \mathcal{F}_0(R)$ be arbitrary. Then there is an edge with description D in $V(set(D))$, and there is an edge with description D in $V(set(D) \cup \{t\})$ for all $t \in R - set(D)$. If $R = set(D)$, this shows the existence of $e \in V(R)$ with $D_R(e) = D$. So assume that $R \neq set(D)$. If $V(R)$ contains no edge with $D_R(e) = D$, then there must be a set R' and a site $t \notin R'$ such that $set(D) \subseteq R' \subseteq R' \cup \{t\} \subseteq R$, $V(R')$ contains an edge e with description D, and $V(R' \cup \{t\})$ contains no edge e with description D. Thus t conflicts with e with respect to R' and hence t conflicts with D by Lemma 6, a contradiction. Thus $e \in E(R)$ with $D_R(e) = D$ exists in either case. ∎

Example (continued): We continue the example from the end of Section 2. Using v_1, v_2, e_1, e_2, and e_3 as defined by Figure 4, the site c intersects edges e_2 and e_3 and clips edges e_2 and e_3 at their endpoint v_2 with respect to $V(R)$, $R = \{a, b, \infty\}$. It conflicts with all three edges of the diagram; note that e_1 separates b-region and ∞-region and that $(\overline{VR^\circ(b, R)} \cup \overline{VR^\circ(\infty, R)}) \cap U \cap \overline{VR^\circ(c, R \cup \{c\})} \nsubseteq \{v_1, v_2\}$ for all neighborhoods U of e_1.

We have $D_R(e_1) = \{(a, \infty, b, a), (a, b, \infty, a)\}$, $D_R(e_2) = \{(\infty, b, a, \infty), (\infty, a, b, \infty)\}$, and $D_R(e_3) = \{(b, a, \infty, b), (b, \infty, a, b)\}$. Finally, $\mathcal{F}_0(R) = \{D_R(e_i) \mid 1 \le i \le 3\}$.

3.2   Conflicts and the Voronoi Diagram of Five Sites

We show how conflicts between edges and sites can be decided by the computation of diagrams of five sites.

Definition 7 *Let $R \subseteq S$ and let p, q, and r be three distinct sites in R. A vertex v of $V(R)$ is called a (p, q, r)-vertex, if v is incident to p-, q-, and r-region, and there is a clockwise traversal of the regions incident to v which encounters p-region before q-region before r-region.*

The following Lemma strengthens Lemma 5.

Lemma 8 *Let $R \subseteq S$ and let p, q, and r be three distinct sites in R. Then $V(R)$ contains at most one (p, q, r)-vertex and at most one edge separating p- and q-region incident to that vertex.*

Proof: Assume otherwise, say v and w are two distinct (p, q, r)-vertices. Since $VR^\circ(p, R)$ and $VR^\circ(q, R)$ are path-connected there are paths $\mathcal{P}$ and $\mathcal{Q}$ connecting v and w and running completely (except at their endpoints) inside p- and q-region respectively. The cycle $\mathcal{P} \circ \mathcal{Q}$ then contains r-region in its interior and its exterior, a contradiction to the fact, that $int\, VR(r, R)$ is homeomorphic to a disc. Thus there is at most one (p, q, r)-vertex, say v. By Fact 1 there is exactly one edge separating p- and q-region incident to v. ∎

Remark: A Voronoi diagram can contain more than one edge separating p- and q-region as Figure 10 shows. □

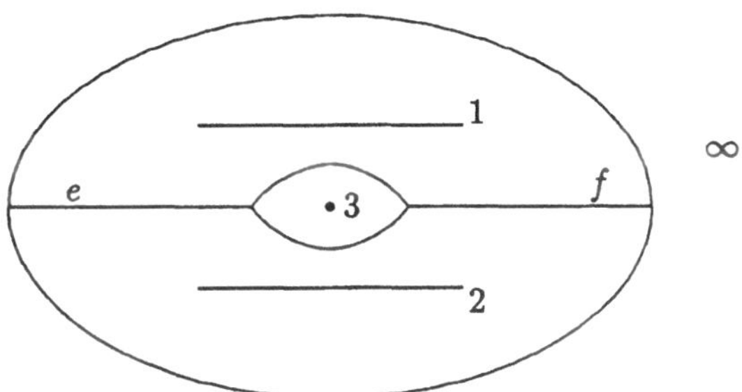

Figure 10: A Voronoi diagram of two line segment sites 1 and 2, a point site 3, and the site ∞.

We can now define the basic operation of our algorithm.

Basic Operation

Input: 5 sites p, q, r, s, t such that
      1) $p \neq q$, $\{p, q\} \cap \{r, s\} = \emptyset$, $t \notin \{p, q, r, s\}$, and
      2) $V(\{p, q, r, s\})$ contains an edge e separating p- and q-region
         and having a (p, q, r)- and a (q, p, s)-endpoint.

Remark: e is unique by Lemma 8.

Output: The combinatorial structure of $e \cap \overline{VR^o(t, \{p, q, r, s, t\})}$,
      i. e. one of the following:
      1) intersection is empty
      2) intersection is non-empty and consists of a single component:
         a) e itself
         b) a segment of e adjacent to the (p, q, r)-endpoint
         c) a segment of e adjacent to the (q, p, s)-endpoint
         d) a segment not adjacent to any endpoint of e
      3) intersection is non-empty and consists of exactly two components

The basic operation will be used for deciding if a given edge of $V(R)$ conflicts with a site $t \notin R$ and if so, for determining the combinatorial type of the conflict. Each call of *basic_op* will be charged one time unit. Note that the input to the basic operation is a combinatorial object, namely the 5-tuple (p, q, r, s, t), and that the output is a combinatorial object, namely a symbol in $\{1, 2a, 2b, 2c, 2d, 3\}$. Also note, that the six cases specified exhaust all possible cases by Lemma 3 and that in case 3 the two components are incident to one endpoint of e each. We use $basic\_op(p, q, r, s, t)$ to denote the output of the basic operation on input (p, q, r, s, t).

Lemma 9 *Let* $R \subseteq S$, $e \in E(R)$, $D_R(e) = \{(r_p, p, q, r_q), (s_q, q, p, s_p)\}$, $r \in \{r_p, r_q\}$, $s \in \{s_p, s_q\}$, $R' = \{p, q, r, s\}$, *and* $t \in S - R$.

   a) *e is an edge of* $V(R')$ *and* $e \cap \overline{VR^\circ(t, R \cup \{t\})} = e \cap \overline{VR^\circ(t, R' \cup \{t\})}$.

   b) *t intersects e with respect to R iff* $basic\_op(p, q, r, s, t) \in \{2a, 2b, 2c, 2d, 3\}$.

   c) *t clips e at the (p, q, r)-endpoint with respect to R iff* $basic\_op(p, q, r, s, t) \in \{2b, 3\}$.

Proof: a) Since $VR^\circ(x, R') \supseteq VR^\circ(x, R)$ for $x \in R'$, the point set e is an edge of $V(R')$. Also $e \cap \overline{VR^\circ(t, R \cup \{t\})} \subseteq e \cap \overline{VR^\circ(t, R' \cup \{t\})}$ since $VR^\circ(t, R \cup \{t\}) \subseteq VR^\circ(t, R' \cup \{t\})$. For the converse, let $x \in e \cap \overline{VR^\circ(t, R' \cup \{t\})}$ be arbitrary. Since $x \in e$, we have $U \cap VR^\circ(o, R) = \emptyset$ for all sufficiently small neighborhoods U of x and all $o \in R - \{p, q\}$. Since $x \in \overline{VR^\circ(t, R' \cup \{t\})}$, we have $x \notin VR^\circ(p, R' \cup \{t\}) \cup VR^\circ(q, R' \cup \{t\})$ and hence $x \notin VR^\circ(p, R \cup \{t\}) \cup VR^\circ(q, R \cup \{t\})$. Thus $x \in e \cap \overline{VR^\circ(t, R \cup \{t\})}$. b) and c) are immediate consequences of part a). ∎

Example (continued):   In $V(R)$, $R = \{a, b, \infty\}$, of Figure 4 the vertex v_2 is a (a, b, ∞)-vertex and the vertex v_1 is a (b, a, ∞)-vertex. We have $basic\_op(b, a, \infty, \infty, c) = 2c$ and $basic\_op(\infty, a, b, b, c) = 2b$. □

4   The Incremental Algorithm

In this section, we describe the incremental construction algorithm. The algorithm starts with set $R_3 = \{\infty, p, q\}$, where p and q are chosen uniformly at random from $S - \{\infty\}$, and then adds the remaining sites in random order, i. e. $R_{k+1} = R_k \cup \{s\}$, where s is chosen uniformly at random from $S - R_k$. The following data structures are maintained for the current set $R = R_k$ of sites:

1) The Voronoi diagram $V(R)$: It is stored as a planar map; with every face of $V(R)$ the corresponding site in R is stored.

2) The history graph $\mathcal{H}(R)$: It is a directed acyclic graph with a single source. Its vertex set is given by

$$\{source\} \cup \bigcup_{3 \leq i \leq k} \{D_{R_i}(e) \mid e \text{ is an edge of } V(R_i)\}.$$

The following history-graph invariants are maintained:

a) Every vertex of $\mathcal{H}(R)$ has outdegree at most 5 and the vertices in $\{D_R(e) \mid e$ edge of $V(R)\}$ have outdegree 0.

b) Every edge e of $V(R)$ is linked to its description $D_R(e)$ in $\mathcal{H}(R)$ and vice versa.

c) For every site $s \in S - R$ and every edge e of $V(R)$, such that s conflicts with $D_R(e)$, there is a path from *source* to $D_R(e)$ that visits only descriptions in conflict with s.

We now discuss how to construct $V(R \cup \{s\})$ and $\mathcal{H}(R \cup \{s\})$ from $V(R)$ and $\mathcal{H}(R)$. Let $E_s = \{e \mid e$ is an edge of $V(R)$ and $D_R(e)$ conflicts with $s\}$. We first show how to construct E_s (Step 1), then how to construct $V(R \cup \{s\})$ (Step 2), and finally how to construct $\mathcal{H}(R \cup \{s\})$ (Step 3). All of this will take time proportional to the number c of vertices of $\mathcal{H}(R)$ which are in conflict with s (note that vertices of $\mathcal{H}(R)$ are descriptions).

Step 1: Construction of E_s

Starting at the source of $\mathcal{H}(R)$ we explore all vertices of $\mathcal{H}(R)$ which are in conflict with s. Since the outdegree of $\mathcal{H}(R)$ is bounded by 5, this takes time proportional to c. Also, the search identifies the edge set E_s (by the third history graph invariant).

Lemma 10 *The set E_s can be computed in time $O(c)$.*

Step 2: Construction of $V(R \cup \{s\})$

As above, let $\mathcal{S} = VR^\circ(s, R \cup \{s\})$. We know from Lemma 2 that $\mathcal{S} \neq \emptyset$ iff $E_s \neq \emptyset$. So, $V(R \cup \{s\}) = V(R)$ and $\mathcal{H}(R \cup \{s\}) = \mathcal{H}(R)$ if $E_s = \emptyset$. We therefore assume $E_s \neq \emptyset$ from now on. For an edge $e \in E_s$, $e - \overline{\mathcal{S}}$ consists of at most two subsegments of e. Also, if $D_R(e) = \{(r_p, p, q, r_q), (t_q, q, p, t_p)\}$, $r \in \{r_p, r_q\}$, $t \in \{t_p, t_q\}$ then $basic\_op(p, q, r, t, s)$ tells us the structure of $e - \overline{\mathcal{S}}$. We call a point v an endpoint of $e - \overline{\mathcal{S}}$ if it is an endpoint of one of the subsegments of e. In this way, $e - \overline{\mathcal{S}}$ may have 0, 2, or 4 endpoints. These endpoints are pairwise distinct by Lemma 3. We first characterize the vertices of $V(R \cup \{s\})$. Let V be the set of vertices of $V(R)$ and let

$V_{del} = \{v \mid v \in V$ and all edges incident to v are clipped at v by $s\}$
$V_{unch} = \{v \mid v \in V$ and no edge incident to v is clipped at v by $s\}$
$V_{chang} = \{v \mid v \in V$ and some but not all edges incident to v are clipped at v by $s\}$
$V_{new} = \{v \mid v \notin V$ and v is endpoint of $e - \overline{\mathcal{S}}$ for some $e \in E_s\}$

Example (continued):   Let $R = \{a, b, \infty\}$ and $s = c$. Then $V_{del} = \emptyset$, $V_{unch} = \{v_1\}$, $V_{chang} = \{v_2\}$, and $V_{new} = \{v_3, v_4\}$. Note that our basic operation tells us that $e_2 - \overline{\mathcal{S}}$ connects v_1 and v_3 and $e_3 - \overline{\mathcal{S}}$ connects v_1 and v_4, cf. Figure 11.

Lemma 11 *Every vertex v of $V(R \cup \{s\})$ is contained in $V_{unch} \cup V_{chang} \cup V_{new}$.*

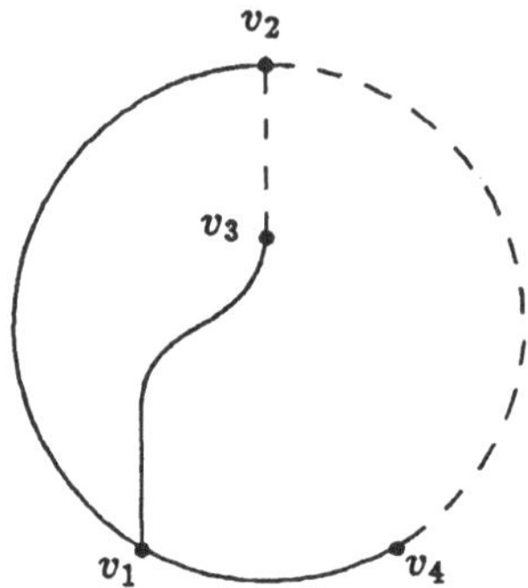

Figure 11: $V(R) \cap \overline{S}$ is shown dashed.

Proof: For every vertex v of $V(R \cup \{s\})$ there are two sites p and q different from s such that an edge e' of $V(R \cup \{s\})$ separating p- and q-region is incident to v. Also, there is an edge e of $V(R)$ with $e' \subseteq e$. Thus v is either a vertex of $V(R)$ or v lies on edge e of $V(R)$. In the latter case, v is an endpoint of $e - \overline{S}$ and hence $v \in V_{new}$. In the former case, e is not clipped at v by s and hence $v \in V_{unch} \cup V_{chang}$. ∎

Lemma 12 *Let $v \in V_{unch}$. Then $U \cap V(R) = U \cap V(R \cup \{s\})$ for all sufficiently small neighborhoods of v; in particular, v is a vertex of $V(R \cup \{s\})$.*

Proof: If $v \in V_{unch}$ then no edge of $V(R)$ incident to v is clipped at v by s. Thus $U \cap V(R) \cap \overline{S} \subseteq v$ for all sufficiently small neighborhoods of v. Thus $U \cap V(R) \cap \overline{S} = \emptyset$ by Lemma 2 and hence $U \cap V(R) = U \cap V(R \cup \{s\})$. ∎

Lemma 13 *Let $v \in V_{chang}$.*

> *a) In the clockwise ordering of edges of $V(R)$ around v there are edges f'' and f' ($f' = f''$ is possible) such that all edges between f'' and f' (inclusive) are not clipped at v by s and all edges between f' and f'' (exclusive) are clipped at v by s.*

> *b) Let e' be the edge following f' and let e'' be the edge preceding f'' in the clockwise ordering of edges of $V(R)$ around v, cf. Figure 12. Let f' and e' border p-region and e'' and f'' border q-region. Then v is a vertex of $V(R \cup \{s\})$ incident to the following edges: all edges between f'' and f' (inclusive), an edge separating p- and s-region, and an edge separating s- and q-region.*

Proof: a) Since some but not all edges incident to v are clipped at v by s, v must lie on *bd* S. Since *bd* S is a simple closed curve passing through v, the edges clipped at v by s and the edges not clipped at v by s must form contiguous subsequences in the clockwise ordering of edges around v. This proves a).

b) This is an immediate consequence of part a) and Fact 1. ∎

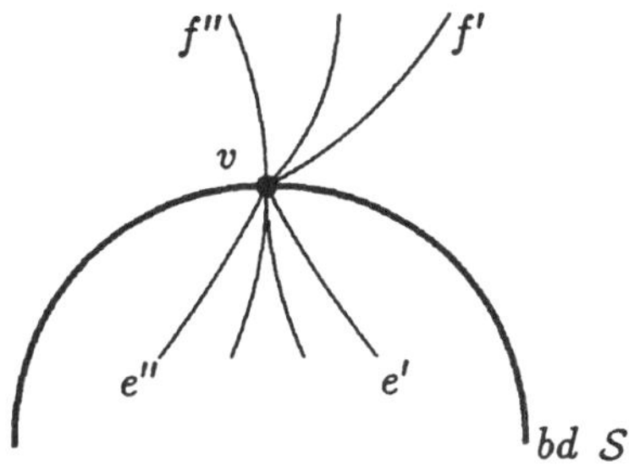

Figure 12:

Lemma 14 *Let $v \in V_{new}$ and let v lie on an edge e of $V(R)$ separating p- and q-region. Then v is a vertex of degree three in $V(R \cup \{s\})$. The three edges incident to v separate p- and q-, q- and s-, and s- and p-region respectively.*

Proof: obvious ∎

Example (continued): By Lemmas 11 to 14, the vertices v_2, v_3 and v_4 become incident to two new edges each, cf. Figure 13.

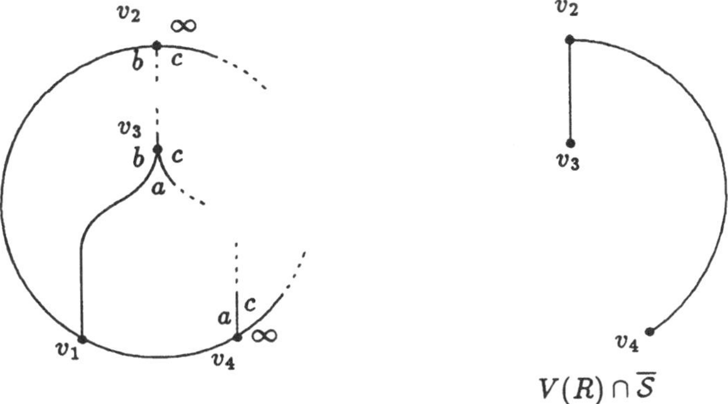

Figure 13: On the left, for each vertex $v \in V_{chang} \cup V_{new}$ the two new edges incident to v are shown. The deleted part of $V(R)$ is shown on the right.

At this point, we have characterized the vertex set of $V(R \cup \{s\})$ and also the set of edges incident to each vertex of $V(R \cup \{s\})$ in their clockwise ordering around v. It remains to link the two occurences of each edge. As above, let $\mathcal{E} = V(R) \cap \overline{S}$.

We know an embedding of $\mathcal{E}$ into the plane. The boundary of the outer face of $\mathcal{E}$ is a closed curve since $\mathcal{E}$ is connected. Also, the vertices on $bd\ S$ lie on $\mathcal{E}$ and $bd\ S$ is a simple closed curve. Hence a clockwise traversal of the boundary of the outer face of $\mathcal{E}$ yields the cyclic ordering of the "half-edges" of $bd\ S$, cf. Figure 14. This allows us to link the two occurrences of each edge. We conclude:

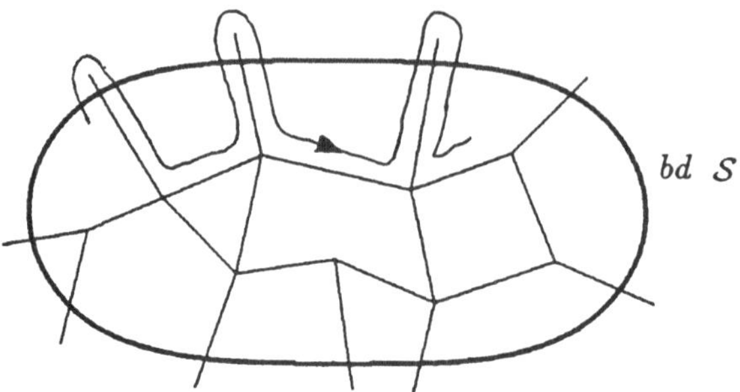

Figure 14:

Lemma 15 *Given E_s, $V(R \cup \{s\})$ can be constructed from $V(R)$ in time $O(|E_s|)$.* $\square$

Proof: Given E_s, one can determine the sets V_{del}, V_{chang}, and V_{new} in time $O(|E_s|)$. In the same time bound one can update the cyclic adjacency lists of these vertices. Finally the traversal of $\mathcal{E}$ takes time $O(|E_s|)$.   ∎

Example (continued): The clockwise march around $\mathcal{E}$ and joining the two occurences of each edge yields $V(R \cup \{c\})$ as shown in Figure 4.

Step 3: Computation of $\mathcal{H}(R \cup \{s\})$

We first characterize the set of vertices $\mathcal{H}(R \cup \{s\})$ which are not already vertices of $\mathcal{H}(R)$. Call an edge e of $V(R \cup \{s\})$ *new* if it is not a subset of an edge of $V(R)$, *shortened* if it is a proper subset of an edge of $V(R)$, *affected* if e is an edge of $V(R)$ and there is a vertex $v \in V_{chang}$ such that e is one of the edges f' or f'' defined in Lemma 13, cf. Figure 15.

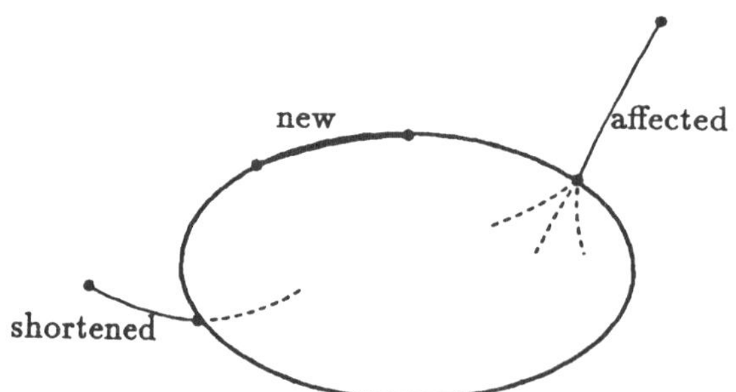

Figure 15: A characterization of edges

Lemma 16 *Let $V_{\mathcal{H}}(R)$ and $V_{\mathcal{H}}(R \cup \{s\})$ be the vertex set of $\mathcal{H}(R)$ and $\mathcal{H}(R \cup \{s\})$ respectively. Then $V_{\mathcal{H}}(R \cup \{s\}) - V_{\mathcal{H}}(R) = \{D_{R \cup \{s\}}(e) \mid e \text{ is a new, shortened, or affected edge of } V(R \cup \{s\})\}$.*

Proof: Let e be an edge of $V(R \cup \{s\})$ which is neither new, shortened, nor affected. Then e was already an edge of $V(R)$ and hence its endpoints must lie in $V_{unch} \cup V_{chang}$. Also if an endpoint v of e belongs to V_{chang} then e lies strictly between the edges f'' and f' defined in Lemma 13. Thus e's descriptions with respect to $V(R)$ and $V(R \cup \{s\})$ are identical.

Conversely, if e is new or shortened then e did not exist in $V(R)$, and if e is affected then e's description with respect to R and $R \cup \{s\}$ are different. ∎

We next discuss which edges have to be added to $\mathcal{H}(R \cup \{s\})$ in order to maintain the history-graph invariants.

Lemma 17 *Let e be a shortened or affected edge of $V(R \cup \{s\})$, let e' be the edge of $V(R)$ with $e \subseteq e'$, and let $t \in S - R - \{s\}$ be in conflict with $D_{R \cup \{s\}}(e)$. Then t is in conflict with $D_R(e')$.*

Proof: Let e separate p- and q-region, let v and w be the endpoints of e, let v' and w' be the endpoints of e', and let U' be any neighborhood of e'. Since $e \subseteq e'$ there is a neighborhood U of e with $U \subseteq U'$ and $U \cap \{v', w'\} \subseteq \{v, w\}$. Since t conflicts with $D_{R \cup \{s\}}(e)$, we have $(\overline{VR^\circ(p, R \cup \{s\})} \cup \overline{VR^\circ(q, R \cup \{s\})}) \cap U \cap \overline{VR^\circ(t, R \cup \{s, t\})} - \{v, w\} \neq \emptyset$ by Lemma 6. Thus $(\overline{VR^\circ(p, R)} \cup \overline{VR^\circ(q, R)}) \cap U \cap \overline{VR^\circ(t, R \cup \{t\})} - \{v, w\} \neq \emptyset$ and hence $(\overline{VR^\circ(p, R)} \cup \overline{VR^\circ(q, R)}) \cap U' \cap \overline{VR^\circ(t, R \cup \{t\})} - \{v', w'\} \neq \emptyset$ Thus t conflicts with $D_R(e')$ by Lemma 6. ∎

Lemma 18 *Let e be a new edge of $V(R \cup \{s\})$ with endpoints x_1 and x_2. Then $x_1, x_2 \in V_{new} \cup V_{chang}$. Let $p \in R$ be such that e separates p- and s-region in $V(R \cup \{s\})$ and let $\mathcal{P}$ be the part of bd $VR^\circ(p, R)$ between x_1 and x_2 such that in all sufficiently small neighborhoods of x_1 and x_2 the path $\mathcal{P}$ is contained in $\overline{S}$, as illustrated in Figure 16. Let $E_{\mathcal{P}}$ be the set of edges contained in $\mathcal{P}$ and let E_1 and E_2 be defined as follows. If $x_1 \in V_{new}$ then E_1 consists of the edge of $V(R)$ containing x_1. If $x_1 \in V_{chang}$ then let f', f'', e', and e'' be defined as in Lemma 13 where we also assume that f'' and e'' bound p-region in $V(R)$, cf. Figure 16. Then $e'' \in E_{\mathcal{P}}$ and E_1 consists of edges f', f'' and e'. E_2 is defined analogously. Finally, let $t \in S - R - \{s\}$ be in conflict with $D_{R \cup \{s\}}(e)$. Then t is in conflict with $D_R(h)$ for some $h \in E_{\mathcal{P}} \cup E_1 \cup E_2$.*

Proof: We assume for the sake of a contradiction that t does not conflict with $D_R(h)$ for any $h \in E_1 \cup E_2 \cup E_{\mathcal{P}}$

Claim 1 *t does not clip e or one of the neighboring edges of e in $V(R \cup \{s\})$ at x_1 or x_2.*

Proof: Assume otherwise, say t clips e or a neighboring edge of e at x_1. Assume first that $x_1 \in V_{new}$ and let e' be the edge of $V(R)$ containing x_1. Then $x_1 \in e' \cap \overline{VR^\circ(t, R \cup \{s, t\})} \subseteq e' \cap \overline{VR^\circ(t, R \cup \{t\})}$ and hence t conflicts with e', a contradiction. Assume next that $x_1 \in V_{chang}$. In $V(R \cup \{s\})$, t clips either f'' or one of the two new

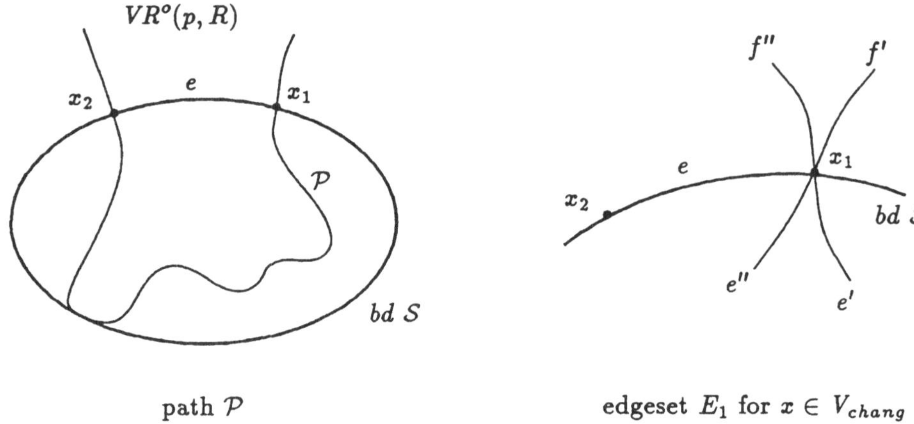

Figure 16: The definitions of Lemma 18

edges incident to x_1. In the former case, t must clip f'' at x_1 in $V(R)$, in the latter case, t must clip one of the edges f', f'', e' or e'' in $V(R)$. In either case, we have a contradiction.   ■

Claim 2 $\mathcal{P} \subseteq \overline{S}$ *and there is a point* $x \in e \cap \overline{VR^\circ(t, R \cup \{s, t\})}$ *which does not lie on* $\mathcal{P}$.

Proof: $\mathcal{P} \subseteq \overline{S}$ holds, because if $\mathcal{P}$ crossed $bd\ S - e$, then $V(R) \cap \overline{S}$ would not be connected, contradicting Lemma 2. Since t conflicts with e but does not clip e or a neighbor of e at x_1 or x_2, we have that $e \cap \overline{VR^\circ(t, R \cup \{s, t\})}$ is a non-empty subsegment of e not extending to either endpoint of e. This subsegment must contain a point x not in $\mathcal{P}$ since t does not conflict with any edge h, $h \in E_\mathcal{P} \cup E_1 \cup E_2$.   ■

Since t does not clip e or one of the neighboring edges of e at x_1 or x_2 in $V(R \cup \{s\})$, all points in the wedge at x_1 formed by e and the part of $bd\ VR^\circ(p, R)$ outside S belong to $VR^\circ(p, R \cup \{s, t\})$. The same holds true for the wedge at x_2. Since $VR^\circ(p, R \cup \{s, t\})$ is connected, there is a path $\mathcal{Q}$ from x_1 to x_2 running completely inside $VR^\circ(p, R \cup \{s, t\}) \subseteq VR^\circ(p, R \cup \{t\})$ except at the endpoints, cf. Figure 17. We may assume that $\mathcal{Q}$ does not touch $bd\ S$ (and therefore x does not lie on $\mathcal{Q}$). Thus x lies in the interior of the cycle $x_1 \circ \mathcal{P} \circ x_2 \circ \mathcal{Q}$; otherwise $VR^\circ(p, R)$ would not be simply connected. The point x belongs to $\overline{VR^\circ(t, R \cup \{t\})}$. Since $VR^\circ(p, R \cup \{t\})$ is simply connected, the region $\overline{VR^\circ(t, R \cup \{t\})}$ cannot be contained in the cycle $x_1 \circ \mathcal{P} \circ x_2 \circ \mathcal{Q}$. Since $\mathcal{Q} \cap \overline{VR^\circ(t, R \cup \{t\})} = \emptyset$, we conclude $\mathcal{P} \cap \overline{VR^\circ(t, R \cup \{t\})} \neq \emptyset$. The intersection cannot consist of a single point and hence t must conflict with an edge $h \in E_\mathcal{P} \cup E_1 \cup E_2$, a contradiction. This completes the proof of Lemma 18.   ■

In view of Lemmas 17 and 18 the following edges are added to the history graph.

For each shortened or affected edge e we add $(D(e'), D(e))$ where e' is a edge of $V(R)$ with $e \subseteq e'$.

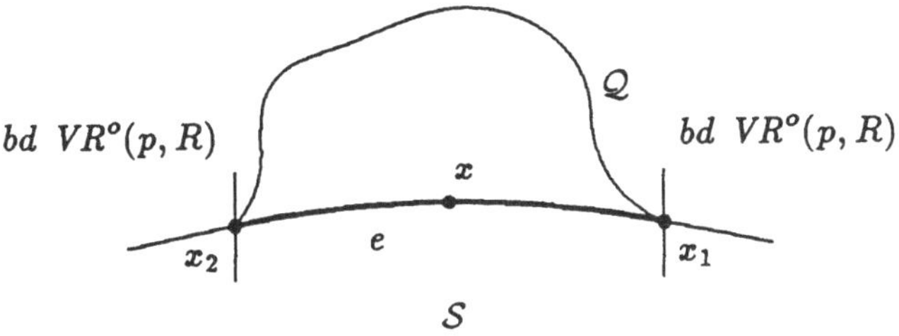

Figure 17:

For each new edge e, we add edges $(D(e'), D(e))$ for all $e' \in E_{\mathcal{P}} \cup E_1 \cup E_2$ where $E_{\mathcal{P}}$, E_1 and E_2 are defined as in Lemma 18.

In this way, part c) of the history-graph invariant is maintained by Lemmas 17 and 18. Part b) is trivial. For part a), we distinguish cases. Let e' be an edge of $V(R)$. If e' is also an edge of $V(R \cup \{s\})$ and $D_R(e') = D_{R \cup \{s\}}(e')$ then no edges out of $D_R(e')$ were added to the history graph. Otherwise, either $e' \subseteq \overline{S}$ or there is a shortened or affected edge e of $V(R \cup \{s\})$ with $e \subseteq e'$. In the former case, e' belongs to at most four sets E_1 or E_2, or two sets $E_{\mathcal{P}}$, i.e. the outdegree of $D_R(e')$ is at most four. In the latter case, e' belongs to at most four sets E_1 or E_2 and also has an edge to e, i.e. the outdegree of e' is at most 5. Figure 18 shows a situation where the outdegree is actually 5.

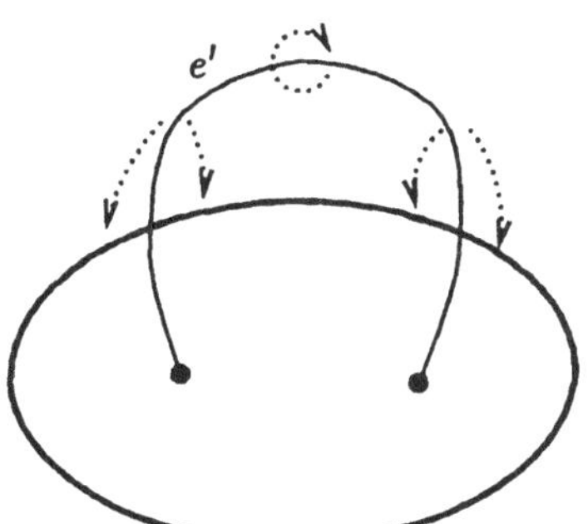

Figure 18: $D_R(e')$ has 5 children

Lemma 19 *Given E_s, $\mathcal{H}(R \cup \{s\})$ can be constructed from $V(R)$ and $\mathcal{H}(R)$ in time* $O(|E_s|)$.

Proof: By the discussion above. ∎

We summarize in:

Theorem 1 *a) Let $\infty \in R$ and $s \in S - R$. Then $V(R \cup \{s\})$ and $\mathcal{H}(R \cup \{s\})$ can be constructed from $V(R)$ and $\mathcal{H}(R)$ in time $O(c)$ where c is the number of vertices of*

$\mathcal{H}(R)$ which conflict with s.

b) For $R \subseteq S$, $|R| = 3$ and $\infty \in R$, the data structures $V(R)$ and $\mathcal{H}(R)$ can be set up in time $O(1)$.

Proof: a) That is the contents of Lemmas 10, 15 and 19.

b) The Voronoi diagram $V(R)$ for three sites ∞, p and q has the structure shown in Figure 19. The history graph $\mathcal{H}(\{\infty, p, q\})$ for these three sites simply consists of a node for the source and each of the three edges of $V(\{\infty, p, q\})$. The descriptions of the 3 edges of $V(\{\infty, p, q\})$ are made children of the source. Both structures can certainly be set up in time $O(1)$. ∎

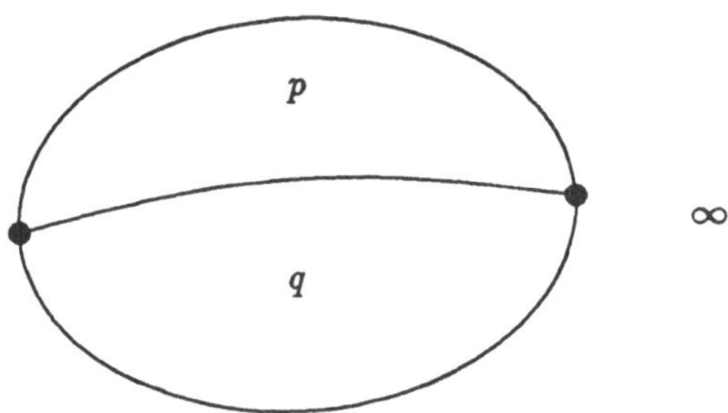

Figure 19:

Theorem 2 *An abstract Voronoi diagram of n sites can be computed by a randomized algorithm in expected time $O(n \log n)$ and expected space $O(n)$. Moreover, the expected time for inserting the n-th object is $O(\log n)$. Randomization here only concerns the order in which the sites are inserted.*

Proof: Let $\tau_0(r, S)$ be the expected cardinality of $\mathcal{F}_0(R)$ where R is a random subset of S of size r. Theorem 1 and the results of [3] imply that the expected space complexity of our algorithm is

$$O\left(\sum_{j=1}^{n} \frac{\tau_0(\lfloor \frac{n}{j} \rfloor, S)}{j} \right),$$

that the expected running time is

$$O\left(\sum_{j=1}^{n} \tau_0(\lfloor \frac{n}{j} \rfloor, S) \right)$$

and the expected time for inserting the last site is

$$O\left(\frac{1}{n} \sum_{j=1}^{n} \tau_0(\lfloor \frac{n}{j} \rfloor, S) \right)$$

In our case $\tau_0(r, S) \leq 3r$ since the Voronoi diagram of r sites has at most $3r$ edges by Fact 2. Thus we get expected space $O(n)$, expected construction time $O(n \log n)$ and expected insertion time $O(\log n)$. ∎

5 Conclusion

We showed that the construction of abstract Voronoi diagrams can be reduced efficiently and purely combinatorially to the construction of abstract Voronoi diagrams for five sites. Many previously considered types of Voronoi diagrams can thus be handled by the same simple algorithm. Further research will be done on the question of how the construction of the 5-site-diagram can be carried out efficiently dependent on the type of bisecting curves. Our goal is a "universal" algorithm for Voronoi diagrams.

Bibliography

[1] F. Aurenhammer. Voronoi Diagrams — A Survey. Tech. report 263, Institutes for Information Processing, Graz Technical University, Austria (1988), to appear in ACM Computing Surveys.

[2] J. D. Boissonnat, M. Devillers-Teillaud. On the Randomized Construction of the Delaunay Tree. Tech. report 1140, INRIA Sophia-Antipolis (1989).

[3] J. D. Boissonnat, O. Devillers, R. Schott, M. Teillaud, M. Yvinec. Applications of Random Sampling to On-line Algorithms in Computational Geometry. Tech. report 1285, INRIA Sophia-Antipolis (1990).

[4] J. D. Boissonnat, O. Devillers, M. Teillaud. A Dynamic Construction of Higher Order Voronoi Diagrams and its Randomized Analysis. Tech. report 1207, INRIA Sophia-Antipolis (1990).

[5] K. Q. Brown. Voronoi Diagrams from Convex Hulls. *IPL 9 (1979)*, pp. 223–228.

[6] K. L. Clarkson, P. W. Shor. Applications of Random Sampling in Computational Geometry, II. *Discrete & Computational Geometry 4 (1989)*, pp. 387–421.

[7] H. Edelsbrunner, R. Seidel. Voronoi Diagrams and Arrangements. *Discrete & Computational Geometry 1 (1986)*, pp. 25–44.

[8] S. Fortune. A Sweepline Algorithm for Voronoi Diagrams. *Algorithmica 2 (1987)*, pp. 153–174.

[9] L. J. Guibas, D. E. Knuth, M. Sharir. Randomized Incremental Construction of Delaunay and Voronoi Diagrams. *Proc. 17th Int. Colloq. Automata, Languages and Programming (ICALP), Warwick (1990)*, LNCS 443, pp. 414–431.

[10] L. Guibas, J. Stolfi. Primitives for the Manipulation of General Subdivisions and the Computation of Voronoi Diagrams. *ACM Transactions on Graphics 4 (1985)*, pp. 74–123.

[11] R. Klein. Abstract Voronoi Diagrams and their Applications (extended abstract). In: *H. Noltemeier (Ed.), Computational Geometry and its Applications (CG '88), Würzburg (1988)*, LNCS 333, pp. 148–157.

[12] R. Klein. Voronoi Diagrams in the Moscow Metric (extended abstract). In: *J. van Leeuwen (Ed.), Graphtheoretic Concepts in Computer Science (WG '88), Amsterdam (1988)*, LNCS 344, pp. 434–441.

[13] R. Klein. Combinatorial Properties of Abstract Voronoi Diagrams. In: *M. Nagl (Ed.), Graphtheoretic Concepts in Computer Science (WG '89), Rolduc (1989)*, LNCS 411, pp. 356–369.

[14] R. Klein. *Concrete and Abstract Voronoi Diagrams.* LNCS 400 (1989).

[15] R. Klein, K. Mehlhorn, S. Meiser. On the Construction of Abstract Voronoi Diagrams, part II. *SIGAL Symp. on Algorithms, Tokyo (1990)*, LNCS 450, pp. 138-154.

[16] D. Leven, M. Sharir. Intersection and Proximity Problems and Voronoi Diagrams. In: *J. Schwartz and C. K. Yap (Eds.), Advances in Robotics, Vol. 1, Lawrence Erlbaum (1986)*, pp. 187–228.

[17] K. Mehlhorn, S. Meiser, C. Ó' Dúnlaing. On the Construction of Abstract Voronoi Diagrams. *Discrete & Computational Geometry 6 (1991)*, pp. 211–224.

[18] M. I. Shamos, D. Hoey. Closest Point Problems. *Proc. 16th IEEE Symp. on Foundations of Computer Science (1975)*, pp. 151–162.

Über die relativistische Struktur logischer Zeit in verteilten Systemen

Friedemann Mattern

FB Informatik
Universität des Saarlandes
6600 Saarbrücken
Germany

Zusammenfassung

Ein verteiltes System, welches aus n autonomen Prozessen besteht, die ausschließlich über Nachrichten unbestimmter Laufzeit miteinander kommunizieren, ist dadurch charakterisiert, daß a priori kein Prozeß eine konsistente Sicht des globalen Zustandes besitzt und eine gemeinsame Zeitbasis innerhalb des Systems nicht existiert. Interpretiert man nun den Zeitbegriff neu – als eine halbgeordnete Menge n−dimensionaler Vektoren mit Verbandstruktur – so erhält man einen logischen Zeitbegriff, der in einem solchen Systems ohne synchronisierte Uhren realisierbar ist und dennoch die durch die Nachrichten vermittelte Kausalität zwischen Ereignissen in isomorpher Weise repräsentiert. Damit bekommen auch die Relationen "später" und "gleichzeitig" eine neue, verallgemeinerte, jedoch adäquate und anschauliche Interpretation. Interessanterweise zeigt sich, daß der dadurch festgelegte Zeitbegriff eine relativistische Struktur aufweist und in enger Analogie zum Minkowski'schen Raumzeit-Modell steht – so entsprechen beispielsweise die konsistenten Schnitte verteilter Berechnungen den Lichtkegeln im Raumzeit-Modell. Im vorliegenden Beitrag wird das Konzept der "Vektorzeit" motiviert und hergeleitet sowie die Analogie zur relativistischen Raumzeit diskutiert.

1   Zeit

Was also ist die Zeit?
Wenn niemand mich danach fragt,
weiß ich's,
will ich's aber einem Fragenden erklären,
weiß ich's nicht.

    Augustinus, Bekenntnisse

Zeit ist ein flüchtiges Phänomen, sie vergeht lautlos und unaufhaltsam.[1] Dennoch spielt Zeit eine wichtige Rolle im täglichen Leben – so ist "Zeit" auch das am häufigsten verwendete Substantiv der deutschen Sprache [15]. Wie sehr die Zeit unser Handeln und Denken prägt, indem sie fast alle Lebensbereiche erfaßt, wird schon an einer willkürlichen Auswahl von "Zeitbegriffen" aus einem Wörterbuch deutlich, die in dieser Aneinanderreihung fast etwas skuril wirkt: *Zeitalter, Zeitarbeit, Zeitbombe, Zeitdruck, Zeitfrage, Zeitgeber, Zeitgeist, Zeitgenosse, Zeitgeschmack, Zeitkarte, Zeitlauf, Zeitlupe, Zeitnehmer, Zeitnot, Zeitpolitik, Zeitpunkt, Zeitraffer, Zeitraum, Zeitschloß, Zeitwende, Zeitwert, Zeitzeichen.* Tritt das bestimmende Element "Zeit" als Suffix auf, dann läßt sich eine nicht minder skurile Wortliste angeben: *Atomzeit, Endzeit, Freizeit, Friedenszeit, Frühzeit, Gleitzeit, Gründerzeit, Halbwertszeit, Jahreszeit, Lebenszeit, Neuzeit, Ortszeit, Raumzeit, Spätzeit, Steinzeit, Tageszeit, Teilzeit, Uhrzeit, Vorwarnzeit, Zwischenzeit.* Auch Redewendungen und Metaphern zeugen vom facettenreichen Erscheinungsbild der Zeit, so kann man Zeit nicht nur sparen, verlieren, absitzen, verbrauchen oder verschwenden, sondern auch herausschlagen, schinden, rauben, vertreiben und sogar totschlagen! Schließlich kann Zeit schön, gut, knapp, närrisch oder nachtschlafend sein.

    Längst schon hat man sich auf eine gemeinsame, vom zu ungenau gewordenen Verlauf der Gestirne unabhängige, "Weltzeit" geeinigt – der zeitkompakte Globus ist nicht nur für weltweit operierende Börsenmakler eine Realität geworden: Vielfältige globale Steuerungsmechanismen beruhen auf der Existenz einer gemeinsamen Zeit, "materialisiert" und implementiert in Form von genau synchronisierten und oft kunstvoll konstruierten Uhren.

    Was aber wäre, wenn es keine gemeinsame Zeitbasis gäbe? Wenn sich diese als bloße Fiktion herausstellen sollte? Oder wenn sie zwar theoretisch existiert, praktisch jedoch nicht erfahrbar wäre? Könnte man in einer solchen Welt überhaupt noch konsistente Aussagen über den globalen Zustand erhalten? Diese Fragestellungen sind nicht aus der Luft gegriffen, sie haben in mindestens zweierlei Hinsicht einen realen Hintergrund: Zum einen relativiert die moderne Physik den Begriff der Gleichzeitigkeit, Zeit ist also eigentlich kein absolutes und überall gleich verfügbares "Ding", zum zweiten spiegelt die Abwesenheit einer gemeinsamen Zeitbasis die Situation in verteilten Systemen wider, bei denen autonome Prozesse ausschließlich über Nachrichten mit nicht vernachlässigbaren (und oft nicht genau bekannten) Übertragungszeiten mitein-

---

[1] *Time goes, you say? Ah no! Alas, time stays, we go.* (Austin Dobson, The Paradox of Time.)

ander kommunizieren. Ausreichend genau synchronisierte Uhren stehen in solchen Systemen oft nicht zur Verfügung, in jedem Fall ist aber aus theoretischer Sicht die Fragestellung interessant, welche den globalen Zustand betreffenden Probleme unter diesen Bedingungen noch gelöst werden können: Wie führt man etwa eine korrekte Volkszählung ohne Uhr, Kalender und Stichdatum durch?

Das Phänomen Zeit wurde von Philosophen, Physikern, Logikern, Psychologen, Biologen und anderen Wissenschaftlern zu ergründen versucht.[2] Obwohl Zeitphänomene auch in der Informatik eine wichtige Rolle spielen – die dort konstruierten und analysierten Systeme laufen in der Zeit ab, Berechnungen sind Zustandsfolgen über einer globalen Zeit und Algorithmen sollten stets so wenig Zeit wie möglich verbrauchen – war die Frage nach dem Wesen der Zeit für diese Disziplin nie von Bedeutung. Um diesem Defizit ein wenig abzuhelfen, soll in diesem Beitrag das Konzept der Zeit, welches verteilten Systemen zugrundeliegt, untersucht werden. Dabei wird sich herausstellen, daß in solchen Systemen Zeit einen der bekannten Minkowski'schen Raumzeit vergleichbaren Charakter hat. Die Darstellung dieser "relativistischen Struktur" der Zeit ist unser Hauptthema.

2 Verteilte Berechnungen und Zeitdiagramme

Diese Art der Veranschaulichung des Zeitablaufs
durch ein zeichnerisches Bild im Raum
kann sehr nützlich sein;
für sie war jedoch die Erfindung
der Relativitätstheorie nicht einmal notwendig,
denn das hat jeder graphische Eisenbahnfahrplan
schon ebenso gemacht.

Hans Reichenbach, Philosophie der Raum-Zeit-Lehre

Ein *verteiltes System* besteht aus mehreren sequentiellen Prozessen, die ausschließlich über *Nachrichten* miteinander kommunizieren. Die Nachrichtenübertragungszeit kann dabei i.a. nicht vernachlässigt werden, sie soll hier als unbestimmt (jedoch endlich) angenommen werden. Von einem einheitlichen Zeitbegriff wird zunächst abstrahiert – zumindest in dem Sinne, daß die verschiedenen Prozesse eine gemeinsame Zeit, etwa durch Zugriff auf eine globale Uhr, nicht erfahren können. Unter diesen Voraussetzungen hat *a priori* kein Prozeß eine konsistente Sicht des globalen Zustandes. Diese für verteilte Systeme charakteristische Eigenschaft verleiht solchen Systemen gegenüber klassischen zentralistischen Systemen eine neue Qualität mit vielen interessanten und nicht einfach zu lösenden Problemen [18].

Das Verhalten jedes Prozesses im System wird durch einen Algorithmus bestimmt, der die Folge lokaler Aktionen sowie die Reaktion des Prozesses auf eintreffende Nach-

---

[2]Der am interdisziplinären oder eher "multikulturellen" Charakter dieser Untersuchungen interessierte Leser sei z.B. auf die Tagungsbände der "International Society for the Study of Time" verwiesen [11], deren erste Konferenz 1969 am Mathematischen Forschungsinstitut Oberwolfach stattfand.

                                  Friedemann Mattern

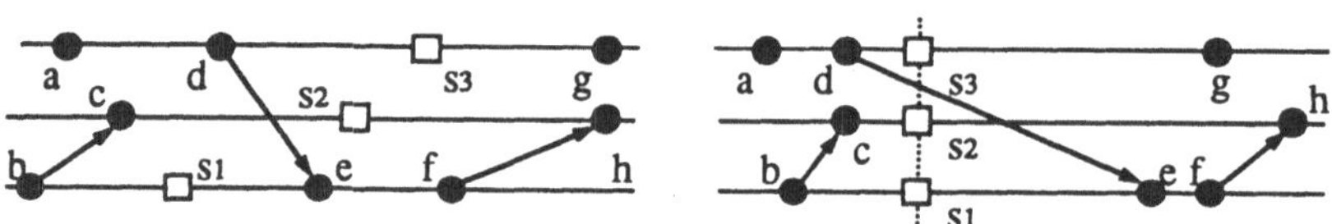

Abbildung 1: Zwei äquivalente Zeitdiagramme.

richten festlegt. Die mittels Nachrichtenaustausch koordinierte gleichzeitige Ausführung der lokalen Algorithmen wird als *verteilte Berechnung* bezeichnet. Formal kann das Stattfinden von Aktionen zu atomaren *Ereignissen* abstrahiert werden; man unterscheidet dabei zwischen *Sendeereignissen, Empfangsereignissen* und *internen Ereignissen*. Eine solchermaßen auf das von außen beobachtbare Verhalten reduzierte verteilte Berechnung läßt sich an einem *Zeitdiagramm* veranschaulichen (vgl. Abb. 1), bei dem die Ereignisse eines Prozesses in der Reihenfolge ihres Auftretens auf einer Prozeßachse angeordnet werden und man sich vorstellt, daß eine (fiktive) globale Zeit von links nach rechts fließt.[3] Nachrichten werden als von links nach rechts laufende Pfeile dargestellt; Aktionen, bzw. die durch sie induzierten Ereignisse, werden aufgrund ihrer Atomarität durch Punkte (oder kleine Quadrate) symbolisiert.

In Zeitdiagrammen wird der Ablauf einer verteilten Berechnung in einer Weise repräsentiert, wie sie ein idealer externer Beobachter sehen würde, der über jedes Stattfinden eines Ereignisses sofort informiert wird. Offenbar ist es für den Verlauf der Berechnung jedoch unerheblich, zu welchem exakten globalen Zeitpunkt ein Ereignis stattfindet, vorausgesetzt, die lokale Reihenfolge der Ereignisse wird nicht verändert und Nachrichtenpfeile verlaufen immer von links nach rechts. Daher stellt ein Zeitdiagramm stets eine ganze Klasse in einem später zu präzisierenden Sinne äquivalenter Abläufe dar. In dieser Hinsicht ist das linke Zeitdiagramm aus Abb. 1 äquivalent zum rechten Zeitdiagramm. Salopp ausgedrückt konnte das linke Diagramm durch Stauchen und Dehnen der lokalen Prozeßachsen in das rechte Diagramm überführt werden, wobei hier allerdings die Ereignisse s_1, s_2, s_3 senkrecht übereinander angeordnet wurden und durch eine vertikale Linie verbunden wurden. Derartige "Schnittlinien" werden im folgenden eine wichtige Rolle spielen.

Zur präzisen Formulierung der Begriffe treffen wir einige Definitionen, wobei wir zunächst von einem gegebenen Zeitdiagramm und der dadurch repräsentierten verteilten Berechnung mit einer Menge E von Ereignissen, die bei den Prozessen $P_1, ..., P_n$ stattfinden, ausgehen.

---

[3] *"It is true that there are certain implicit dangers in using such graphical representations, because in every geometrical diagram time appears to be misleadingly spatialized. On the other hand, such diagrams, provided we do not forget their symbolic nature, have a definite advantage..."* (Aus Milič Čapek's philosophischer Kritik am Minkowski'schen Raumzeit-Konzept [25] und dessen vereinfachten Modellvorstellungen – wie später gezeigt wird, sind unsere Zeitdiagramme den bei der Darstellung des Minkowski'schen Raumzeit-Konzepts verwendeten Diagrammen in der Tat sehr ähnlich.)

Definition 1 (Globale Ereignisordnung '$\prec$').
Es bezeichne '$\prec$' die kleinste transitive Relation auf der Ereignismenge E, so daß für $e, e' \in E$ die Beziehung $e \prec e'$ gilt, wenn

1. *e und e' im gleichen Prozeß stattfinden und e der unmittelbare Vorgänger von e' ist oder*

2. *e' durch den Empfang einer Nachricht ausgelöst wird, die von Ereignis e ausgesendet wurde (d.h. e' das zum Sendeereignis e gehörige Empfangsereignis ist).*

Die reflexive Hülle von $\prec$ wird mit $\preceq$ bezeichnet. Da in Zeitdiagrammen Nachrichten voraussetzungsgemäß nur von links nach rechts laufen, enthält die '$\prec$'-Relation keine Zyklen. Bei ihr handelt es sich um die von Lamport [16] definierte "happens before"-Relation; $e \prec e$ bedeutet, daß e vor e' stattfindet. Hierbei ist "vor" sowohl räumlich zu interpretieren (e liegt links von e') als auch zeitlich im Sinne der (fiktiven) Globalzeit. Man beachte, daß die Umkehrung dieses Sachverhalts nicht gilt, d.h. es ist nicht in jedem Fall $e \prec e'$, wenn in einem Zeitdiagramm e links von e' liegt (vgl. dazu beispielsweise die Ereignisse a und c in Abbildung 1).[4] Offenbar dient die Halbordnung '$\prec$' zur Abstraktion von der genauen zeitlichen Lage der Ereignisse, die beiden Diagramme aus Abb. 1 beschreiben somit die gleiche Halbordnung. Die '$\prec$'-Relation kann als (potentielle) *Kausalität* interpretiert werden und läßt sich an Zeitdiagrammen einfach als von links nach rechts laufende Pfade veranschaulichen; wir halten dies in einem Theorem fest:

Theorem 1 (Kausalkette).
In einem Zeitdiagramm gilt für zwei Ereignisse e, e' die Relation $e \prec e'$ genau dann, wenn es (unter Einbeziehung der von links nach rechts laufenden Prozeßachsen) einen Pfad von e nach e' gibt.

Die Korrektheit der Aussage ist insbesondere dann evident, wenn $e \prec e'$ als "e kann e' beeinflussen" interpretiert wird. Formal läßt sich das Theorem durch Induktion über die Pfadlänge unter Ausnutzung der Transitivität der '$\prec$'-Relation beweisen. In Abb. 1 ist beispielsweise $b \prec c$ (b ist ein direkter Vorgänger von c) und $a \prec h$ (Pfad von a über d, e, f zu h). Ein Pfad im Zeitdiagramm ist also nichts anderes als eine graphische Veranschaulichung einer (potentiellen) *Kausalkette*.

Für spätere Überlegungen ist die symmetrische Relation "kausal unabhängig" von Interesse. Wir treffen dazu folgende Definition:

Definition 2 ('$||$'-Relation).
Für zwei Ereignisse $e, e' \in E$ gilt $e||e'$ genau dann, wenn $\neg(e \prec e') \wedge \neg(e' \prec e)$.

Offensichtlich gilt somit für je zwei Ereignisse e, e' entweder $e \prec e'$ oder $e' \prec e$ oder $e||e'$. Bisher sind wir von einer gegebenen verteilten Berechnung ausgegangen, haben den Begriff der verteilten Berechnung selbst jedoch nur informell eingeführt. Die formale Definition soll nun konform zum bisher Gesagten nachgeholt werden.

---

[4]*Denn nicht alles, was aufeinander folgt, folgt auseinander.* (Ferdinand Seibt in "Die Zeit als Kategorie der Geschichte und als Kondition des historischen Sinns" [21].)

Definition 3 (Verteilte Berechnung).

Eine (n-fach) verteilte Berechnung über einer Ereignismenge E ist ein n-Tupel $(E_1, \dots$
$\dots, E_n)$ mit einer Menge $\Gamma \subseteq E \times E$ von korrespondierenden Sende- und Empfangser-
eignissen, wobei jedes $E_i \subseteq E$ vermöge einer Relation $\prec_i$ linear geordnet ist und
folgende drei Bedingungen erfüllt sein müssen:

1. *Die Ereignismengen $E_1, \dots, E_n$ sind paarweise disjunkt.*
2. *Γ ist links- und rechtseindeutig.*
3. *Die kleinste transitive Relation $\prec$, die die zwei Axiome*
   A1. $a \prec_i b \;\Rightarrow\; a \prec b$
   A2. $(a, b) \in \Gamma \;\Rightarrow\; a \prec b$
   erfüllt, ist eine (irreflexive) Halbordnung.

Offenbar stellen die E_i die lokalen Berechnungen der Prozesse dar, Γ die Nachrichten, und '$\prec$' ist die weiter oben bereits eingeführte "happens before"-Relation, welche die Kausalität vermittelt. Bedingung 3 der Definition besagt, daß diese Relation zyklenfrei sein muß – eine intuitiv einleuchtende Forderung. Die Zyklenfreiheit garantiert, daß sich zu jeder verteilten Berechnung ein Zeitdiagramm angeben läßt, bei dem die Nachrichtenpfeile stets von links nach rechts laufen. Ordnet man die Ereignisse auf den Prozeßachsen entsprechend der '$\prec_i$'-Relation von links nach rechts an, so ergibt sich damit bei einer ebenfalls von links nach rechts laufenden globalen Zeit[5] die beruhigende Erkenntnis, daß die Zukunft die Vergangenheit nicht beeinflussen kann.

3   Schnitte und Verbände

... ist das ganze Dasein
ein ewiges Trennen und Verbinden.

J.W. Goethe, Maximen und Reflexionen

Unser Anliegen ist die Definition eines adäquaten Zeitbegriffs für verteilte Systeme. Zwar haben wir vorausgesetzt, daß eine gemeinsame globale Zeit innerhalb eines verteilten Systems nicht erfahrbar ist, doch dürfen wir als Ausgangspunkt unserer Überlegungen Zeitdiagramme mit ihren fiktiven und dennoch suggestiven globalen Zeitachsen[6] betrachten.

In einem derartigen Diagramm sollten sich globale Zeitpunkte als senkrechte Linien darstellen, die alle Prozeßachsen "zur gleichen Zeit" schneiden. Da es zu einer verteilten Berechnung kein kanonisches Zeitdiagramm gibt, scheint dieser Ansatz zunächst problematisch zu sein. Nun lassen sich jedoch alle zueinander äquivalenten

---

[5] *"Everybody knows that time flows from left to right unless you are left-handed"* (Anonymer Graffito, der auf dem elektronischen "Bulletin Board" des internationalen Usenet-Netzes zu lesen war.)

[6] *"Eine geometrische Gerade ist ein sehr schlechtes Bild für die Zeit, auch wenn sie sich als Zeitachse bei der graphischen Darstellung größerer Verkaufszahlen und abnehmender Profite bewährt hat."* (J.T. Fraser in [10].)

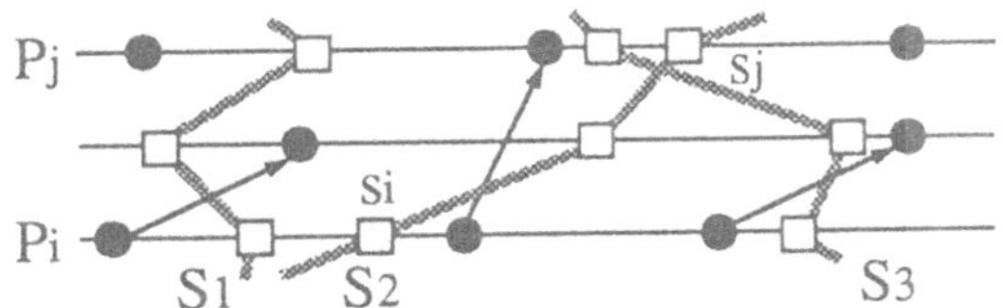

Abbildung 2: Konsistente und inkonsistente Schnitte.

Zeitdiagramme (d.h. alle Zeitdiagramme, die die gleiche Berechnung repräsentieren), durch Stauchen und Dehnen der lokalen Prozeßachsen ineinander überführen. Diese "Gummiband-Äquivalenztransformationen" sind damit genau diejenigen Transformationen auf Zeitdiagrammen, die die Kausalstruktur invariant lassen. Fixiert man daher eine (dehnbare) senkrechte Schnittlinie an den Prozeßachsen, indem man dort "Scharniere" in Form von Schnittereignissen $s_1, ..., s_n$ anbringt (vgl. Abb. 1), so wird durch die Gummibandtransformation die senkrechte Schnittlinie in eine schiefe oder gar zickzackförmige Linie transformiert. Unabhängig von ihrer Verformung teilt die Schnittlinie die Menge der Ereignisse jedoch in die gleichen zwei disjunkte Mengen, die "Vergangenheit" (die Ereignisse links der Schnittlinie) und die "Zukunft" (die restlichen Ereignisse). Damit qualifizieren sich Schnittlinien von Zeitdiagrammen als "Ersatz" für globale Zeitpunkte. Dieser Sachverhalt motiviert die Definition eines *Schnittes* einer verteilten Berechnung über einer Ereignismenge E als eine nach unten abgeschlossene Teilmenge von E:

Definition 4 (Lokale Ereignisordnung '$\prec_l$').
Die lokale Ereignisordnung *'$\prec_l$' ist definiert durch: $e \prec_l e'$ genau dann, wenn $e \prec e'$ und wenn e und e' Ereignisse des gleichen Prozesses sind.*

Definition 5 (Schnitt).
Eine endliche Teilmenge $S \subseteq E$ heißt Schnitt *von E, falls $(e \in S \wedge e' \prec_l e) \Rightarrow e' \in S$.*

Man beachte, daß eine Schnittlinie lediglich ein geometrisches Gebilde ist, das ein Zeitdiagramm durchtrennt, während ein Schnitt formal eine Menge von Ereignissen ist. Jeder Schnittlinie ist allerdings durch die Menge der "linken" Ereignisse ein Schnitt assoziiert, und oft ist es bequem, sich unter einem Schnitt anschaulich eine damit assoziierte Schnittlinie in einem Zeitdiagramm vorzustellen. Von besonderer Bedeutung ist eine spezielle Klasse von Schnitten:

Definition 6 (Konsistenter Schnitt).
Eine endliche Teilmenge $S \subseteq E$ heißt konsistenter Schnitt *von E, falls $(e \in S \wedge e' \prec e) \Rightarrow e' \in S$.*

Wie man leicht sieht, ist wegen $\prec_l \subseteq \prec$ jeder konsistente Schnitt ein Schnitt entsprechend Definition 5. Schnitte, die keine konsistenten Schnitte sind, heißen *inkonsistent*.

Abbildung 2 zeigt Schnittlinien zu konsistenten (S_1, S_3) und zu inkonsistenten (S_2) Schnitten. Da für konsistente Schnitte mit einem Empfangsereignis definitionsgemäß auch das zugehörige Sendeereignis zum Schnitt gehört, läßt sich eine inkonsistente Schnittlinie daran erkennen, daß eine Nachricht "aus der Zukunft" existiert, also eine Nachricht, deren Sendeereignis rechts der Schnittlinie liegt, obwohl sich das zugehörige Empfangsereignis links davon befindet. Solche Nachrichten, die die Schnittlinie "in falscher Richtung" überqueren, bewirken, daß mindestens zwei Schnittereignisse s_i, s_j kausal abhängig voneinander sind – offenbar existiert eine Kausalkette vom Schnittereignis s_i des Sendeprozesses P_i zum Schnittereignis s_j des Empfangsprozesses P_j (vgl. Abb. 2). Für konsistente Schnitte ist dies nicht der Fall, hier sind alle Schnittereignisse paarweise kausal unabhängig voneinander.

Intuitiv sollte in verteilten Systemen die Rolle von Zeitpunkten einer globalen Zeit von senkrechten Schnittlinien übernommen werden – daher sind Schnitte zu solchen Schnittlinien, die sich "geradebiegen" lassen, besonders interessant. Das folgende Theorem macht hierzu eine Aussage:

Theorem 2 (Gummiband-Konsistenzkriterium).
Eine Schnittlinie eines Zeitdiagramms repräsentiert genau dann einen konsistenten Schnitt, wenn sie sich mit der Gummibandtransformation in eine senkrechte Schnittlinie transformieren läßt.

Die nachfolgend skizzierte Überlegung zeigt, wie sich für das Theorem ein Beweis nach dem Do-it-yourself-Prinzip – im wörtlichen Sinne – "basteln" läßt: Ein gegebenes Zeitdiagramm mit einer konsistenten Schnittlinie schneide man entlang dieser Linie in zwei Hälften. Die rechte Hälfte schiebe man dann soweit nach rechts, bis sich die beiden Hälften nicht mehr überlappen und durch einen (senkrechten) Spalt getrennt sind. Durchtrennte Nachrichtenpfeile müssen wieder verbunden werden, sie verlaufen weiterhin von links nach rechts, da lediglich die Empfangsereignisse weiter in die Zukunft gerückt wurden. Die Schnittlinie läßt sich nun senkrecht in den Spalt zwischen den beiden Hälften legen. Umgekehrt ist es einleuchtend, daß eine für die Inkonsistenz des Schnittes verantwortliche Nachricht "aus der Zukunft" nicht in einem Zeitdiagramm mit senkrechter Schnittlinie dargestellt werden kann.

Konsistente Schnitte lassen sich als Mengen bereits geschehener Ereignisse auffassen – mit dieser temporalen Charakterisierung nähern wir uns wieder unserem Ziel der Definition eines Zeitbegriffs für verteilte Systeme. Da sich rechts einer Schnittlinie bei dieser Interpretation typischerweise mehrere Ereignisse befinden, die bei einem gegebenen Schnitt (d.h. in einem momentanen globalen Zustand) als "nächste" ausgeführt werden können, ist der jeweils "nächste Schnitt" i.a. nicht eindeutig bestimmt. Es läßt sich aber auf der Menge aller Schnitte eine *partielle* Ordnung "später" definieren, wobei anschaulich ein Schnitt S_2 *später* als ein Schnitt S_1 ist, wenn die Schnittlinie von S_2 rechts von derjenigen zu S_1 liegt (vgl. Abb. 2).

Definition 7 (Späterer Schnitt).
Ein Schnitt S_2 heißt später *als ein Schnitt S_1, wenn $S_1 \subseteq S_2$.*

Durch die mengentheoretische Festlegung von Schnitten nach Definition 5 ergibt sich folgendes:

Theorem 3 (Schnittverband).
Die Menge der Schnitte eines Zeitdiagramms bildet bezüglich "später" einen Verband.

Beweis: Folgt unmittelbar aus der Definition eines Verbandes als halbgeordnete Menge, bei der für je zwei Elemente eine größte untere Schranke inf und eine kleinste obere Schranke sup existiert. Offenbar gilt bezüglich "später" für je zwei Schnitte S_1, S_2: inf $= S_1 \cap S_2$ und sup $= S_1 \cup S_2$, wobei inf und sup bezüglich $\prec_l$ nach unten abgeschlossen sind. ∎

Auch die konsistenten Schnitte, die für die Festlegung eines logischen Zeitbegriffs alleine von Interesse sind, bilden einen Verband:

Theorem 4 (Unterverband konsistenter Schnitte).
Die Menge der konsistenten Schnitte ist ein Unterverband des Verbandes aller Schnitte.

Beweis. Es ist zu zeigen, daß die konsistenten Schnitte unter $\cup$ und $\cap$ abgeschlossen sind.

(1) Seien S_1, S_2 zwei konsistente Schnitte. Es seien ferner $x \in S_1 \cap S_2$ und $y \prec x$. Dann ist $y \in S_1$, da S_1 konsistent ist, und $y \in S_2$, da S_2 konsistent ist. Daher ist $y \in S_1 \cap S_2$, d.h. $S_1 \cap S_2$ ist konsistent.

(2) Sei $x \in S_1 \cup S_2$ und $y \prec x$. Man betrachtet zwei Fälle:

    (a) $x \in S_1$. Dann ist $y \in S_1$, weil S_1 konsistent ist.

    (b) $x \in S_2$. Dann ist $y \in S_2$, weil S_2 konsistent ist.

    Also ist $y \in S_1 \cup S_2$, d.h. $S_1 \cup S_2$ ist konsistent. □

Die Verbandstruktur konsistenter Schnitte garantiert, daß es für je zwei konsistente Schnitte S_1 und S_2 stets sowohl einen konsistenten Schnitt gibt, der später als beide ist, als auch einen konsistenten Schnitt, der früher als beide ist. Dies läßt sich auf eine beliebige endliche Menge konsistenter Schnitte erweitern: $\sup(S_1, ..., S_k) = S_1 \cup ... \cup S_k$ ist später als $S_1, ... S_k$; entsprechendes gilt für inf. Interpretiert man Schnitte als Zeit"punkte" in einem verteilten System, so heißt dies, daß eine Menge solcher Zeitpunkt stets eine gemeinsame ferne Zukunft und eine gemeinsame ferne Vergangenheit haben. Dies ist gewissermaßen ein Ersatz für den Verlust der Linearität, den die Zeit in den folgenden Abschnitten erleiden wird.

4   Logische Zeit

Ein jegliches hat seine Zeit.

Prediger Salomo 3,1

Unser Ziel besteht darin, für verteilte Systeme einen Zeitbegriff zu definieren, der einerseits innerhalb eines solchen Systems realisierbar ist (insbesondere also nicht auf globalen Uhren beruht), andererseits aber gewisse nützliche Eigenschaften aufweist, so daß auch die Bezeichnung "Zeit" gerechtfertigt ist. Sinnvoll wäre es beispielsweise, wenn der Zeitbegriff es gestatten würde, Ereignissen Zeitpunkte zuzuordnen und es damit ermöglichen würde, die potentielle Kausalität zwischen Ereignissen in dem Sinne zu bestimmen oder auszuschließen, daß ein "späteres" Ereignis ein "früheres" garantiert nicht beeinflussen kann. Ferner wäre es wünschenswert, daß, wenn auf allen Prozessen (im Sinne des Zeitbegriffs) "gleichzeitig" ein Schnappschuß des lokalen Zustandes gezogen wird, der dadurch gegebene globale Zustand in einer Weise konsistent ist, als wären alle lokalen Schnappschüsse zum gleichen Realzeitpunkt getätigt worden. Tatsächlich wird sich zeigen, daß der im nachfolgenden Abschnitt entwickelte vektorielle Zeitbegriff diesen Forderungen genügt.

Die Tatsache, daß sich in unserer "realen verteilten Welt" ein System von autonomen Realzeituhren konstruieren läßt, welche (ausreichende Synchronität vorausgesetzt) einen Zeitbegriff implementieren, der diese Eigenschaften ebenfalls aufweist und die Kausalbeziehung zwischen räumlich entfernten Ereignissen respektiert, ist – so betrachtet – das eigentlich Erstaunliche; Lamport bezeichnet dies sogar als ein "Mysterium des Universums" [16]. Faßt man *Realzeit* als eine Menge von Zeitpunkten auf, so scheint sie sich in ihrer formalen Struktur als eine irreflexive lineare Ordnung darzustellen, welche nach oben und unten unbeschränkt ist und wo die Zeitpunkte dicht liegen (vgl. dazu z.B. [24]). Typische Modelle hierfür sind die rationalen oder die reellen Zahlen.

Etwas anders stellt sich dies für die *logische Zeit* eines verteilten Systems dar. Bei einer verteilten Berechnung, wie wir sie in abstrakter Form definiert haben, geschieht zwischen zwei aufeinanderfolgenden Ereignissen nichts. Daher braucht in einem solchen Modell das Verstreichen von Zeit nur an das Stattfinden von Ereignissen gekoppelt zu werden. Dies ist eine in der Informatik durchaus übliche Modellvorstellung, so wird beispielsweise bei der ereignisgesteuerten Simulationsmethode das Fortschalten der Simulationsuhr durch die jeweilige Eintrittszeit des nächsten auszuführenden Ereignisses bestimmt. Da Zeit in diesem Fall nur als Folge stattgefundener Ereignisse wahrgenommen wird, ist logische Zeit nicht dicht, sondern *diskret*. Beschränkt man sich auf endliche verteilte Berechnungen, so darf Zeit auch *beschränkt* sein. Es wird sich im folgenden sogar herausstellen, daß auf ein weiteres Axiom verzichtet werden muß, möchte man einen adäquaten Zeitbegriff definieren, welcher die Kausalstruktur einer verteilten Berechnung widerspiegelt. Tatsächlich ist unser im folgenden entwickelte Zeitbegriff *nicht-linear* und Zeit in ihrer formalen Struktur somit nur eine *partielle* Ordnung.

Die Festlegung eines Zeitbegriffs für verteilte Systeme bedeutet, daß jedem Ereignis "der" Zeitpunkt zugeordnet werden muß, zu dem es stattfand. Aus dem Vergleich der Zeitpunkte sollten sich dann Rückschlüsse über die Beziehung der Ereignisse untereinander ergeben. Formal ist damit eine Abbildung $C : E \longrightarrow T$ gesucht, die jedem Ereignis $e \in E$ einen *Zeitstempel* $C(e)$ aus einer geeigneten Menge T zuordnet.

T sollte zumindest die Formulierung der Begriffe "früher" oder "später" erlauben, so daß wir bei T von einer Halbordnung $(T, <)$ ausgehen. Die Abbildung C läßt sich formal als *logische Uhr* auffassen. Eine sinnvolle Anforderung an C ist die sogenannte *Uhrenbedingung*

$$\forall e, e' \in E : e \prec e' \implies C(e) < C(e').$$

Sie besagt, daß ein Ereignis e einen kleineren Zeitstempel als ein Ereignis e' haben soll, wenn e das Ereignis e' kausal beeinflussen kann.

Lamport gab 1978 in seinem oft zitierten Artikel "Time, Clocks, and the Ordering of Events in a Distributed System" [16] ein Verfahren an, das auf den natürlichen Zahlen als Zeitbereich T beruht, und bei dem die virtuelle Uhr C durch ein System von Zählern (je einer für jeden Prozeß) und ein einfaches Nachrichtenprotokoll implementiert wird. Allerdings fehlt diesem Konzept eine an sich wünschenswerte Eigenschaft: Bei der Abbildung der Ereignisse in die linear geordneten natürlichen Zahlen geht Struktur verloren; Ereignisse, die unabhängig voneinander sind, bekommen Zeitstempel zugeordnet, als ob sie in einer bestimmten Reihenfolge stattfinden würden. Insbesondere läßt sich durch bloßes Überprüfen der Zeitstempel nicht nachweisen, daß ein Ereignis ein anderes Ereignis *nicht* kausal beeinflussen kann. Zu diesem Zweck sollte die Zeitstruktur T die Ereignisstruktur E *isomorph* repräsentieren, so daß die Implikation in der Uhrenbedingung in beide Richtungen gültig ist. Dies leistet die im nächsten Abschnitt beschriebene "Vektorzeit".

5 Vektorzeit

Andere Zeiten, andere Sitten.

Sprichwort

Zur Motivation der im folgenden betrachteten Vektorzeit geht man zunächst davon aus, daß jeder Prozeß P_i eine einfache, durch einen Zähler realisierte, logische Uhr besitzt, die bei jedem seiner Ereignisse um 1 erhöht wird. Ein idealisierter externer Beobachter, der unmittelbaren Zugriff auf alle lokalen Uhren hat, wüßte dann zu jedem Augenblick die jeweiligen lokalen Zeiten. Eine geeignete Datenstruktur zur Speicherung dieses globalen Zeitwissens stellt ein Vektor dar, welcher für jeden Prozeß eine Komponente besitzt. Das Beispiel aus Abb. 3 illustriert die Idee.

Das unmittelbare Zeitwissen des idealisierten Beobachters läßt sich innerhalb des Systems aufgrund der Nachrichtenverzögerung nicht realisieren. Unser Ziel besteht nun darin, ein Schema zu konstruieren, das jedem Prozeß eine *optimale Approximation* dieser idealen globalen Zeit ermöglicht, indem ein Prozeß zum frühest möglichen Zeitpunkt, allerdings ohne zusätzlichen Nachrichtenaufwand, über alle bereits geschehenen Ereignisse informiert wird. Zu diesem Zweck wird jeder Prozeß P_i mit einer Uhr C_i ausgestattet, die einen *Vektor* der Länge n aufnehmen kann, wobei n die Gesamtzahl der Prozesse bezeichnet. Eine Uhr C_i ist mit dem Nullvektor initialisiert; sie

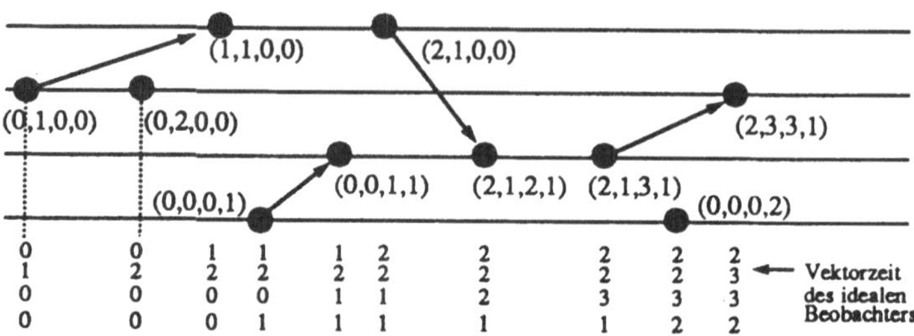

Abbildung 3: Fortpflanzung des Zeitwissens.

"tickt" jedesmal unmittelbar bevor auf Prozeß P_i ein Ereignis ausgeführt wird, indem der Wert der eigenen Komponente inkrementiert wird:

$$C_i[i] \;:=\; C_i[i] + 1.$$

Jede Nachricht enthält einen vektoriellen Zeitstempel t, wobei t der Zeitvektor der lokalen Uhr des Senders beim Absenden der Nachricht ist. Indem ein Prozeß mit einer Nachricht einen Zeitstempel t zugeschickt bekommt, erfährt er etwas über die Approximation der globalen Zeit anderer Prozesse. Der Empfänger kombiniert sein eigenes Zeitwissen C_i mit der Approximation t, die er zusammen mit der Nachricht erhält, mittels

$$C_i \;:=\; \sup(C_i, t),$$

wobei sup für das komponentenweise Maximum steht. Der gesuchte Zeitstempel $C(e)$ eines Ereignisses e von Prozeß P_i ist der Wert der Uhr C_i bei Ausführung von e (bei Empfangsereignissen nach der Aktualisierung durch den Nachrichten-Zeitstempel). Abbildung 3 illustriert die Fortpflanzung des Zeitwissens und die Aktualisierung der Vektoruhren an einem Beispiel.

Offensichtlich werden die Ereignisse eines Prozesses P_i durch die i-te Komponente der Uhr C_i durchnumeriert, d.h. vor einem Ereignis e geschahen bereits $|C(e)[i]| - 1$ andere Ereignisse auf dem gleichen Prozeß. Tatsächlich enthält der vektorielle Zeitstempel $C(e)$ eines Ereignisses e in kompakter Form sogar das gesamte Wissen darüber, von welchen anderen Ereignissen das Ereignis e kausal abhängig ist. So bedeutet beispielsweise $C(e)[k] = j$, daß Ereignis e vom ersten, zweiten, ..., j-ten Ereignis des Prozesses P_k abhängig ist, jedoch von keinem späteren Ereignis des Prozesses P_k. Alternativ zu der oben getroffenen operationalen Festlegung der Vektorzeit $C(e)$ eines Ereignisses e läßt diese sich daher wie folgt definieren:

Definition 8 (Vektorzeit eines Ereignisses).
Der vektorielle Zeitstempel $C(e)$ eines Ereignisses e ist derjenige Vektor aus $\mathbf{N}^n$, für dessen i-te Komponente $C(e)[i]$ gilt:

$$C(e)[i] \;=\; |\{e' \,|\, e' \text{ ist ein Ereignis von Prozeß } P_i \,\wedge\, e' \preceq e\}|.$$

Man macht sich leicht klar, daß diese Festlegung durch die oben genannten Regeln realisiert wird.

6 Die Struktur der Vektorzeit

Und überall hingen, lagen und standen Uhren.
Da gab es auch Weltzeituhren in Kugelform,
welche die Zeit für jeden Punkt der Erde anzeigten...
"Vielleicht", meinte Momo,
"braucht man dazu eben so eine Uhr."
Meister Hora schüttelte lächelnd den Kopf.
"Die Uhr allein würde niemand nützen.
Man muß sie auch lesen können."

Michael Ende, Momo

Um verschiedene Zeitvektoren miteinander vergleichen zu können, treffen wir zunächst folgende Definition:

Definition 9 (Zeitvektorordnung).
Für zwei Zeitvektoren u, v ist
$$u \leq v \ :\Leftrightarrow \forall i: \ u[i] \leq v[i],$$
$$u < v \ :\Leftrightarrow \ u \leq v \ \wedge \ u \neq v,$$
$$u \| v \ :\Leftrightarrow \ \neg(u < v) \ \wedge \ \neg(v < u).$$

Man beachte, daß '$\leq$' und '$<$' *partielle* Ordnungen sind. Die symmetrische und reflexive '$\|$'-Relation kann als eine Verallgemeinerung der *Gleichzeitigkeit* der Realzeit aufgefaßt werden.[7] Während allerdings der Begriff der Gegenwart in der Realzeit auf einen quasi existenzlosen Schnittpunkt zwischen Vergangenheit und Zukunft reduziert wird,[8] kann die durch '$\|$' induzierte "Gleichzeitigkeit" bei der Vektorzeit einen größeren Umfang annehmen – wobei jedoch zu beachten ist, daß die '$\|$'-Relation nicht transitiv ist!

Durch Definition 8 bzw. durch das im vorherigen Abschnitt beschriebene Verfahren wird einem Ereignis ein Zeitvektor zugeordnet. In natürlicher Weise kann jedoch auch einem *Schnitt S* ein Zeitvektor $\tau(S)$ zugeordnet werden, indem als Wert für die i-te Komponente von $\tau(S)$ die Anzahl der Ereignisse von S auf Prozeß P_i festgelegt wird:

Definition 10 (Zeitvektor eines Schnittes).
Sei S ein Schnitt. Der durch

$$\tau(S)[i] \ = \ |\{e \in S \,|\, e \text{ ist ein Ereignis von Prozeß } P_i\}|$$

definierte Vektor τ heißt Zeitvektor des Schnittes S.

---

[7]Bereits 1928 formulierte dies der Philosoph Hans Reichenbach so: *"Zwei gleichzeitige Ereignisse sind derart gelegen, daß weder von einem zum andern noch umgekehrt eine Wirkungskette zu eilen vermag. Die Ereignisse, die jetzt, in diesem Augenblick, in einem fernen Lande stattfinden, können von mir nicht mehr beeinflußt werden, auch nicht durch Telegramme; und umgekehrt können sie auf das, was gegenwärtig hier geschieht, keinen Einfluß mehr haben. Gleichzeitigkeit heißt Ausgeschaltetheit des Wirkungzusammenhangs."* [20]

[8] *"Denn wäre da eine Ausdehnung, müßte sie wiederum in Vergangenheit und Zukunft geteilt werden. Für die Gegenwart aber bliebe kein Raum."* (Augustinus, Bekenntnisse)

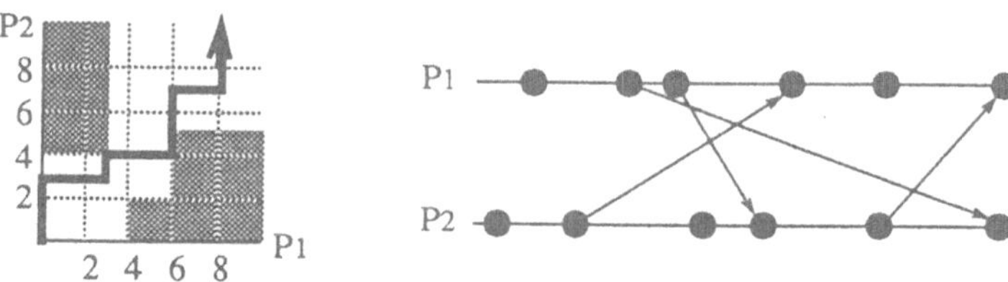

Abbildung 4: Verbandsgitter und zugehöriges Zeitdiagramm.

Aus Definition 10 und Definition 7 ergibt sich unmittelbar:

Korollar 1 (Späterer Zeitvektor eines Schnittes).
S_1 ist später als S_2 genau dann, wenn $\tau(S_2) \leq \tau(S_1)$.

Indem Schnitten Zeitvektoren zugeordnet werden, überträgt sich die Verbandstruktur der Schnitte bzw. der konsistenten Schnitte auf die entsprechenden Mengen von Zeitvektoren. Es gilt daher:

Theorem 5 (Zeitverband).
Die Zeitvektoren der Schnitte bilden bezüglich der Halbordnung '$\leq$' einen Verband, die Zeitvektoren der konsistenten Schnitte einen Unterverband.

Beweis: Zu je zwei Zeitvektoren $\tau(S_1), \tau(S_2)$ existieren Infimum $\tau(S_1 \cap S_2)$ und Supremum $\tau(S_1 \cup S_2)$. ∎

Da Zeitvektoren als Elemente aus $\mathbf{N}^n$ eine operationale Struktur darstellen, ist hier die Interpretation des Verbandes als algebraische Struktur mit den beiden Verknüpfungen *inf* und *sup* interessant. Es seien $\inf(x,y) = u$ mit $u[i] = \min(x[i], y[i])$ und $\sup(x,y) = v$ mit $v[i] = \max(x[i], y[i])$, d.h. *inf* und *sup* sind als die komponentenweise Minimum- bzw. Maximumoperation definiert. Man zeigt leicht einerseits den Zusammenhang zur ordnungstheoretischen Interpretation vermöge $x \leq y \Leftrightarrow x = \inf(x,y)$ und andererseits die nach der algebraischen Interpretation geforderten Eigenschaften Kommutatitivität und Assoziativität für beide Operationen sowie die Absorptionsgesetze $\sup(x, \inf(x,y)) = x$ und $\inf(x, \sup(x,y)) = x$.

Die Zeitvektorverbände sind somit ein besonders hübsches Beispiel für die duale Auffassung von Verbänden als Vermittler zwischen ordnungstheoretischer und algebraischer Struktur: Um Aussagen über frühere oder spätere Zeitpunkte in Form von Schnittmengen zu bekommen, bedient man sich der algebraischen Struktur, mit der es sich einfacher und effizienter rechnen läßt.

Zeitvektoren lassen sich als Koordinatenpunkte in einem n-dimensionalen Gitter darstellen, welches gleichzeitig die Verbandstruktur illustriert. Betrachtet man nur die zu konsistenten Schnitten gehörenden Vektoren – nur solche können, wie in Theorem 6 und 7 gezeigt wird, überhaupt als Zeitstempel von Ereignissen und Nachrichten auftreten – so ergeben sich insbesondere für $n = 3$ interessante und ästhetisch reizvolle geometrische Gebilde, da solche Verbandsgitter von den Seiten her "ausgefranst" sind. Abbildung 4 zeigt ein Beispiel für $n = 2$. In einem solchen Verbandsgitter bestimmt jeder vom Nullvektor wegführende Pfad, der sich in dem schlauchförmigen Gebilde in Richtung der "Raumdiagonalen" bewegt, eine Folge von Ereignissen, die eine Linearisierung der '$\prec$'-Kausalbeziehung darstellt. Ein solcher Pfad läßt sich daher als eine *konsistente Beobachtung* der verteilten Berechnung auffassen. Im allgemeinen existieren mehrere verschiedene konsistente Beobachtungen einer verteilten Berechnung; da nach einem bekannten mathematischen Theorem jedoch der Schnitt aller Linearisierungen einer Halbordnung die Halbordnung selbst ergibt, ist das, was allen Beobachtungen gemein ist, gerade die zugrundeliegende Kausalstruktur. Die ausgefranste Gestalt der Verbandsgitter hat ihre Ursache darin, daß sich die Zeit nicht immer frei in alle Richtungen ausbreiten kann, einige Dimensionen können vorübergehend gesperrt sein, da vor der Wahrnehmung eines Empfangsereignisses das zugehörige Sendeereignis wahrgenommen werden muß.

Durch Definition 8 wurden Ereignissen Zeitvektoren zugeordnet, während durch Definition 10 Schnitten Zeitvektoren zugeordnet wurden. Der Zusammenhang zwischen Zeitvektoren von Ereignissen und Schnitten ergibt sich daraus, daß die "kausale Vergangenheit" eines Ereignisses, d.h. die Menge all derjenigen Ereignisse, die ein Ereignis kausal beeinflussen können, einen konsistenten Schnitt bilden:

Definition 11 (Kausale Vergangenheit $\downarrow e$).
Die kausale Vergangenheit $\downarrow e$ eines Ereignisses e ist definiert als $\downarrow e = \{e'|e' \preceq e\}$.

Theorem 6 ($\downarrow e$ ist konsistent).
Für jedes Ereignis $e \in E$ ist $\downarrow e$ ein konsistenter Schnitt von E.

Der Beweis ergibt sich unmittelbar aus den Definitionen 6 und 11 sowie aus der Transitivität von '$\preceq$'. Nun wird auch klar, daß der Zeitstempel eines Ereignisses nichts anderes ist als der Zeitvektor seiner kausalen Vergangenheit:

Theorem 7 (Zeit ist kausale Vergangenheit).

$$C(e) = \tau(\downarrow e).$$

Beweis: $\tau(\downarrow e)[i] \stackrel{def}{=} |\{e' \in \downarrow e \mid e'$ ist ein Ereignis von Prozeß $P_i\}| = |\{e' \preceq e \mid e'$ ist ein Ereignis von Prozeß $P_i\}| \stackrel{def}{=} C(e)[i].$ ∎

Über den Zeitvektor wird vermöge Theorem 7 ein Ereignis mit seiner kausalen Vergangenheit "identifiziert". So gesehen stellt sich die zunächst etwas geheimnisvolle Vektorzeit als etwas relativ natürliches heraus: Das "Jetzt"ist eindeutig durch die Menge der in diesem Augenblick vergangenen Ereignisse bestimmt; Zeit *ist* die Menge der vergangenen Ereignisse! Die Zeitvektoren ermöglichen lediglich eine kompakte Repräsentation und eine algebraische Fassung – und damit rechnerisch einfache Handhabung – von Mengenoperationen auf den kausalen Vergangenheiten der Ereignisse, die den eigentlichen Zeitbegriff definieren.

Für den Beweis des nachfolgenden zentralen Theorems 8 benötigen wir noch ein Ergebnis, das konsistente Schnitte bezüglich der kausalen Vergangenheit als abgeschlossen charakterisiert:

Lemma 1 *Sei S ein konsistenter Schnitt. Dann gilt $e \in S \iff \downarrow e \subseteq S$.*

Beweis:
(1) "$\Rightarrow$": Aus $e \in S$ folgt wegen der Konsistenzeigenschaft von S mit $x \preceq e$ auch $x \in S$ für beliebige x. Insbesondere gilt daher $x \in \downarrow e \Rightarrow x \preceq e \Rightarrow x \in S$, also $\downarrow e \subseteq S$.
(2) "$\Leftarrow$": Aus $\downarrow e \subseteq S$ folgt $x \in \downarrow e \Rightarrow x \in S$. Wegen $e \in \downarrow e$ ist daher $e \in S$. ∎

Wir können nun zeigen, daß sich mit Vektoruhren eine verschärfte Form der Uhrenbedingung realisieren läßt; genauer besagt das folgende Theorem, daß die Kausalstruktur durch die Vektorzeit isomorph repräsentiert wird:

Theorem 8 (Isomorphie von Kausal- und Zeitstruktur).

$$\forall e, e' \in E : e \prec e' \iff C(e) < C(e').$$

Beweis: $e \preceq e' \overset{11}{\iff} e \in \downarrow e' \overset{6,1}{\iff} \downarrow e \subseteq \downarrow e' \overset{7,1}{\iff} \tau(\downarrow e) \leq \tau(\downarrow e') \overset{7}{\iff} C(e) \leq C(e')$. Da jedem Ereignis eindeutig ein Zeitstempel zugeordnet ist und verschiedene Ereignisse verschiedene Zeitstempel haben, folgt die Behauptung. ∎

Theorem 8 läßt sich an Zeitdiagrammen anschaulich interpretieren: Ein Ereignis e' hat einen größeren Zeitstempel als ein Ereignis e genau dann, wenn eine Kausalkette in Form eines Pfades von e nach e' existiert. Offensichtlich kann entlang eines solchen Pfades der Wert von Vektorkomponenten höchstens größer werden. Hat umgekehrt ein Ereignis e' einen größeren Zeitstempel als ein anderes Ereignis e, dann muß es einen Pfad von e nach e' geben, entlang welchem das "Zeitwissen" bezüglich $C(e)$ propagiert wurde. Aus Theorem 8 ergibt sich unmittelbar eine Möglichkeit, Ereignisse anhand ihres Zeitstempels auf kausale Unabhängigkeit zu überprüfen:

Korollar 2 $\forall e, e' \in E : e \| e' \iff C(e) \| C(e')$.

Etwas salopp ausgedrückt besagt das Korollar, daß es genau die "gleichzeitigen" Ereignisse sind, die sich gegenseitig nicht beeinflussen können.

7 Minkowski's relativistische Raumzeit

Von Stund an
sollen Raum für sich
und Zeit für sich
völlig zu Schatten herabsinken
und nur noch eine Art Union der beiden
soll Selbständigkeit bewahren.

Hermann Minkowski, Raum und Zeit

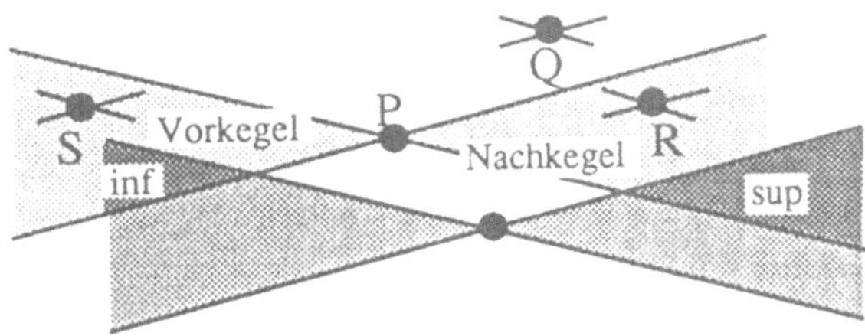

Abbildung 5: Lichtkegel in Minkowski's Raumzeit.

In diesem Abschnitt zeigen wir, daß eine prinzipielle Analogie zwischen dem Minkowski'schen Raumzeit-Modell und unseren Zeitdiagrammen bzw. den durch die Kausalbeziehung halbgeordneten Ereignisstrukturen existiert. Sie ergibt sich aus dem relativistischen Effekt, der immer dann auftritt, wenn die Laufzeiten von Signalen nicht vernachlässigbar sind.[9] Aufgrund der Begrenzung durch die Lichtgeschwindigkeit[10] stellt sich die Realzeit – analog zu unserer Vektorzeit – bei genauer Betrachtung nicht mehr als lineare Ordnung, sondern nur noch als eine *Halbordnung* dar. Dies läßt sich als *Relativierung der Gleichzeitigkeit* auffassen; es ist eine Konsequenz, die sich bereits aus den Grundannahmen der Relativitätstheorie ergibt. In seinen "Bemerkungen über die Beziehungen zwischen der Relativitätstheorie und der idealistischen Philosophie" formuliert dies Kurt Gödel so: "Die Behauptung, die Ereignisse A und B hätten gleichzeitig stattgefunden (und für eine große Gruppe von Ereignispaaren auch die Behauptung, A habe vor B stattgefunden), verliert ihren objektiven[11] Sinn insofern, als ein anderer Beobachter mit dem gleichen Anspruch auf Richtigkeit behaupten kann, A und B hätten nicht gleichzeitig stattgefunden (oder B habe vor A stattgefunden)" [12]. Selbst Max Born gibt allerdings zu, daß sich diese Relativität der Gleichzeitigkeit schwer erfassen läßt: "Daß dasselbe, was das Ich als zugleich empfindet, ein anderer als nacheinander bezeichnen soll, das läßt sich in der Tat durch das Zeiterlebnis nicht begreifen" [3].

Das geeignete Modell zur Darstellung der relativistischen Realität ist Minkowski's Raumzeit-Modell, bei dem ein $n-1$-dimensionaler Raum mit der eindimensionalen Zeit zu einem n-dimensionalen Bild der Welt verknüpft wird. Aufgrund der begrenzten Signallaufzeiten kann in diesem Modell ein bei einem bestimmten Raumzeitpunkt stattfindendes Ereignis nur dann ein anderes Ereignis beeinflussen, wenn es in dessen *Lichtkegel* liegt. Auf dem Kegelmantel liegen die Ereignisse (bzw. Raumzeitpunkte), die "gerade noch" beeinflußt werden können; der Öffnungswinkel des Kegels ist durch die Grenzgeschwindigkeit der Signalausbreitung bestimmt.

---

[9]Im Rahmen der Petrinetz-Theorie wurde diese Analogie von Budde in [4, 5] diskutiert.

[10]Der französische Physiker Paul Langevin bezeichnet in einer englischen Übersetzung seiner Werke die Lichtgeschwindigkeit daher konsequenterweise als "speed limit of causality".

[11]Für einige Ereignispaare ist allerdings die zeitliche Reihenfolge objektiv (d.h. für alle möglichen Beobachter gleich), nämlich für diejenigen Ereignispaare, die kausal voneinander abhängig sind. Ursache und Wirkung finden niemals gleichzeitig oder gar in umgekehrter Reihenfolge statt!

Abbildung 5 veranschaulicht diese Situation für $n = 2$. Ereignisse P und Q sind kausal unabhängig voneinander, während S Ereignis P beeinflussen kann (da es in dessen *Vorkegel* liegt) und P Ereignis R beeinflussen kann (R liegt im *Nachkegel* von P). Bezüglich der Kausalbeziehung gilt die Transitivität (wenn X im Nachkegel von Y liegt und Y im Nachkegel von Z, dann liegt auch X im Nachkegel von Z); man sieht jedoch, daß bezüglich der Unabhängigkeit die Transitivität nicht gilt: Offensichtlich sind sowohl P und Q als auch Q und R paarweise unabhängig, nicht jedoch P und R.

Analog zu den Schnitten unserer Zeitdiagramme bilden für $n = 2$ die Lichtkegel bezüglich der Inklusion sogar einen *Verband*. Abbildung 5 veranschaulicht die Konstruktion. Der Nachkegel des Supremums wird durch den Schnitt der beiden Nachkegel bestimmt, der Schnitt der beiden Vorkegel liefert den Vorkegel des Infimums. Durch die Identifikation der Kegel mit ihrer Spitze überträgt sich so die Verbandstruktur auf die Raumzeitpunkte. Eine analoge Identifikation haben wir bezüglich der Zeitdiagramme bei den Schnitten und den zugehörigen Schnittlinien vorgenommen.

Unsere Gummibandtransformationen aus Abschnitt 3, die die metrische, nicht jedoch die topologische Struktur verändern und damit die '$\prec$'-Relation invariant lassen, finden ihre Entsprechung bei der Minkowski'schen Raumzeit in den bekannten *Lorentz-Transformationen*. Diese stellen als orthogonale Drehungen des n-dimensionalen Raumes kausalerhaltende Koordinatentransformationen dar, die den Lichtkegel invariant lassen.

Bei der Vektorzeit erlaubte es die verschärfte Form der Uhrenbedingung (vgl. Theorem 8 und Korollar 2) zu überprüfen, ob zwei Ereignisse mit gegebenen Zeitstempeln in kausaler Beziehung zueinander stehen oder nicht. Ein rechnerisch ähnlich einfaches Kausalitätskriterium liefert die Kegelstruktur der Raumzeit: Bezeichnet c die Grenzgeschwindigkeit, also den Tangens des halben Öffnungswinkels des Kegels, dann sind zwei Ereignisse e_1 und e_2 mit den Koordinaten (x_1, t_1) bzw. (x_2, t_2) voneinander abhängig, wenn eines im Lichtkegel des anderen liegt, wenn also $c^2 (t_2 - t_1)^2 - (x_2 - x_1)^2 \geq 0$ gilt. Die Kegel-Inklusionsordnung läßt sich auf n-dimensionale reelle Räume ($n \geq 2$) verallgemeinern, indem für zwei beliebige Raumzeitpunkte $U = (u_1, ..., u_n)$ und $V = (v_1, ..., v_n)$ die Relation

$$U \leq V \quad :\Longleftrightarrow \quad \sum_{i=1}^{n-1} (u_i - v_i)^2 \leq c^2 (u_n - v_n)^2 \ \wedge \ u_n \leq v_n$$

definiert wird. Üblicherweise wird der Parameter c dabei zu 1 normiert; die n-te Koordinate übernimmt die Rolle einer gerichteten Zeit. (Die Normierung von c bedeutet lediglich eine bequeme Festlegung der Maßeinheiten – ist c die Lichtgeschwindigkeit, dann würde man die n-te Dimension beispielsweise in Jahren, die $n - 1$ ersten räumlichen Dimensionen in "Lichtjahren" messen). Man kann zeigen, daß die so definierte "Minkowski'sche Kausalordnung" eine Halbordnung ist und daß zu je zwei Punkten ein gemeinsamer größerer existiert.

Für $n = 2$ läßt sich aus der Minkowski'schen Kausalordnung sogar leicht unsere Zeitvektorordnung gewinnen: Betrachtet man die mit $c = 1$ normierten Nachkegel

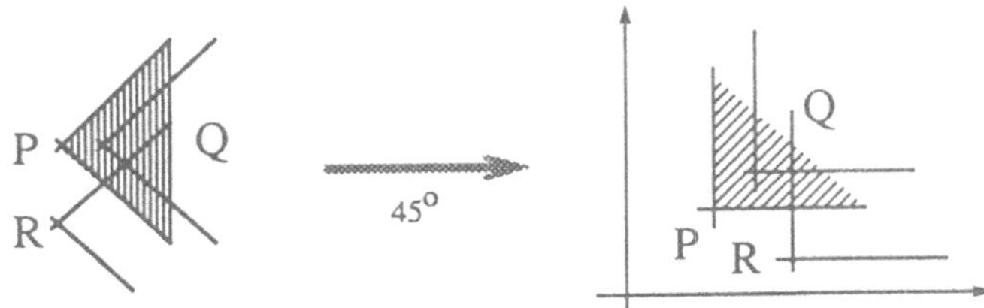

Abbildung 6: Isomorphie der Vektorzeit- und Lichtkegelordnung.

und dreht das Koordinatensystem um 45° nach links (vgl. Abb. 6), dann erkennt man, daß nun die "Zukunft" eines Punktes P aus denjenigen Punkten besteht, die rechts und oberhalb von P liegen. Für zwei Punkte $P = (p_1, p_2)$ und $Q = (q_1, q_2)$ gilt also $P \leq Q$ genau dann, wenn $p_1 \leq q_1$ und $p_2 \leq q_2$. Dies läßt sich entsprechend Definition 9 in vektorieller Form als $(p_1, p_2) \leq (q_1, q_2)$ schreiben. Salopp formuliert unterscheidet sich für n=2 die Minkowski'sche Raumzeit von der Vektorzeit nur durch eine 45°-Drehung des Raumes – die beiden Strukturen sind also praktisch identisch!

Das Minkowski'sche Konzept des Lichtkegels eines Ereignisses läßt sich prinzipiell auf unsere Zeitdiagramme übertragen: Der Vorkegel $\widehat{P}$ eines Raumzeitpunktes P enthält alle Punkte Q, die P beeinflussen können, d.h. $\widehat{P} = \{Q \mid Q \leq P\}$, wobei '$\leq$' die oben definierte Minkowski'sche Kausalordnung ist. Dies entspricht formal unserer Definition $\downarrow e = \{e' \mid e' \preceq e\}$ der kausalen Vergangenheit eines Ereignisses (vgl. Definition 11). Entsprechend ließe sich auch eine "kausale Zukunft" $\uparrow e = \{e' \mid e \preceq e'\}$ definieren, die dem Nachkegel entspricht. Da bei einem Zeitdiagramm kein Ereignis aus $\downarrow e$ rechts von e liegen kann, befindet sich der "Vorkegel" $\downarrow e$ eines Ereignisses e tatsächlich auch links des Ereignisses – allerdings hat ein solcher abstrakter Kegel i.a. keine geometrisch perfekte Kegelform, sondern "krumme" Begrenzungslinien. Da die abstrakten Kegel $\downarrow e$ und $\uparrow e$ jedoch konsistente Schnitte darstellen (vgl. Theorem 6), läßt sich in der gleichen konstruktiven Weise wie im skizzierten Beweis zum Gummibandkriterium (Theorem 2) der Kegelmantel "geradebiegen".

Für unsere reale Welt liefert die relativistische Raumzeit ein exakteres Bild der Wirklichkeit als die gemeinhin als linear angesehene "Standardzeit". Die Analogie zur geometrisch etwas anders gearteten Vektorzeit (halbgeordnete Mengen mit Verbandstruktur, kausalerhaltende Transformationenen, Ähnlichkeit der Zeitdiagramme) ist kein Zufall, sie beruht auf der beiden Modellen gleichermaßen zugrundeliegenden Voraussetzung, daß Ursache und Wirkung zeitlich getrennt sind und damit in der durch sie begrenzten "Zwischenzeit" kausal unabhängige Ereignisse stattfinden können. Die Analogie ist ein starkes Indiz dafür, daß Vektorzeit ein "richtiges" Zeitmodell für verteilte Systeme darstellt – daß die Vektorzeit zudem auch nützlich ist, wird im nächsten Abschnitt kurz skizziert.

8 Anwendungen

Kommt Zeit, kommt Rat.

Sprichwort

Da mit Zeitvektoren die Kausalstruktur verteilter Berechnungen – und damit die potentielle Abhängigkeit jedes Ereignisses von jedem anderen – einfach festgestellt werden kann, ergeben sich einige interessante Anwendungen dieses Konzeptes. Nachteilig für die praktische Verwendbarkeit ist lediglich der durch die Länge der Vektoren verursachte Aufwand. Wie Charron-Bost in [7] zeigt, existiert i.a. jedoch leider keine kompaktere Repräsentation der Kausalstruktur. Es helfen lediglich Optimierungstechniken, die im Einzelfall allerdings den Speicher- bzw. Kommunikationsaufwand deutlich reduzieren können: In [22] geben Singhal und Kshemkalyani ein effizientes Verfahren an, bei dem nur diejenigen Komponenten eines Vektors versendet werden, die sich seit der letzten Nachricht an den selben Empfänger geändert haben. Meldal et al. beschreiben in [19], wie auf das Speichern einiger Vektorkomponenten verzichtet werden kann, wenn die Verbindungstopologie der Prozesse bestimmte Eigenschaften aufweist.

Eine direkte Anwendung findet das Konzept der vektoriellen Zeit beim Testen verteilter Systeme. Wird jedes Ereignis mit einem vektoriellen Zeitstempel versehen, so ist es beispielsweise möglich, potentielle Race-Bedingungen – hervorgerufen durch zeitlich in inkorrekter Reihenfolge stattfindende Ereignisse, die (fälschlicherweise) in keiner kausalen Beziehung zueinander stehen – festzustellen. Vermöge Theorem 8 ist es gegebenenfalls auch möglich auszuschließen, daß ein Ereignis e die Ursache für ein fehlerhaftes Ereignis e' sein kann (nämlich dann, wenn $\neg(e \prec e'$ gilt). Die Verwendung von Vektoruhren für das Testen verteilter Systeme wird von Haban und Weigel [13] sowie von Fidge [8] beschrieben.

Zum Zweck der Leistungsanalyse verteilter Systeme ist es wichtig zu wissen, welche Ereignisse potentiell gleichzeitig ausgeführt werden können. Offenbar ist dies zumindest dann möglich, wenn die Ereignisse kausal unabhängig voneinander sind, wenn ihre Zeitstempel also in der '||'-Relation zueinander stehen. Eine Analyse der vektoriellen Zeitstempel von Ereignissen kann daher helfen, den Grad der Parallelität zu bestimmen [6].

Das Nachhalten von kausalen Abhängigkeiten zwischen Ereignissen spielt bei verteilten Datenbanken eine große Rolle, hier müssen z.B. die lokalen Sicherungskopien zueinander konsistent sein. Wird eine Teildatenbank (etwa nach einem Fehler) auf einen früheren Zustand zurückgesetzt, so hat dies evtl. Einfluß auf den Zustand anderer Teildatenbanken. Strom und Yemini [23], Johnson und Zwaenepoel [14] sowie Venkatesh et al. [26] zeigen, wie im Rahmen dieses Problembereichs Zeitvektoren verwendet werden können. Ebenfalls aus dem Bereich verteilter Datenbanken stammt das "distributed dictionary"-Problem: Die Sicht jedes beteiligten Prozesses auf die verteilte Datenbank soll konsistent in dem Sinne sein, daß der Prozeß die Einfüge- und Löschoperationen in einer Reihenfolge wahrnimmt, die mit der Kausalordnung verträglich ist. Die Verwendung von Zeitvektoren zur Lösung dieses Problems wird von Fischer und Michael [9][12] sowie Liskov und Ladin [17] diskutiert.

Weitere interessante Anwendungen des Zeitvektorkonzeptes stellen Algorithmen zur Berechnung konsistenter globaler Zustände [18], zur Realisierung von "causal memory" als eine schwach-synchrone Variante globalen Speichers [1] sowie zur Implementierung kausal geordneter Broadcasts [2] dar. Da an dieser Stelle auf diese Aspekte nicht weiter eingegangen werde kann, sei statt dessen auf die zitierte Literatur verwiesen.

9 Epilog

Oh! Mit der Zeit wären wir fertig.

  Schiller, Kabale und Liebe

Zeit hat viele Facetten – in diesem Beitrag haben wir uns notgedrungen auf einen einzigen Aspekt, die formale Struktur logischer Zeit in verteilten Systemen, beschränkt. So interessant solche Untersuchungen auch sein mögen – wir abstrahieren dabei doch von fast allem, was Zeit für uns bedeutet: für gute und schlechte Zeiten, glückliche und weniger glückliche Ereignisse ist kein Platz in dieser Theorie. Was bleibt, ist die Sehnsucht nach erfüllter Zeit – *echte Zeit* muß man leben!

Dank gebührt Bernadette Charron-Bost für die anregenden Diskussionen über das Konzept der Vektorzeit sowie Reinhard Schwarz für seine Anmerkungen und Verbesserungsvorschläge zum vorliegenden Beitrag.

Literaturverzeichnis

[1] M. AHAMAD, P. HUTTO, R. JOHN. Implementing and Programming Causal Distributed Shared Memory. Technical Report GIT-CC-90-49, College of Computing, Georgia Institute of Technology, 1990.

[2] K. BIRMAN, A. SCHIPER, P. STEPHENSON. Lightweight Causal and Atomic Group Multicast. Technical Report TR 91-1192, Computer Science Department, Cornell University, 1991.

[3] M. BORN. Die Relativitätstheorie Einsteins. *Springer-Verlag*, 1964.

---

[12]Dieser Konferenzbeitrag von Michael Fischer und Alan Michael stellt nach Kenntnis des Autors die früheste Veröffentlichung dar, bei der Zeitvektoren zum Nachhalten der kausalen Abhängigkeiten zwischen Ereignissen verwendet werden. Später wurde das Zeitvektorkonzept ("kausal unabhängig" von dieser Veröffentlichung) mehrfach wiederentdeckt.

[4] R. BUDDE. Einige Bemerkungen zum Verständnis nebenläufiger Prozesse und Systeme. *In: W. Brauer (Hg.): 11. GI-Jahrestagung, Springer-Verlag, Informatik-Fachberichte 50, pp. 448-459*, 1981.

[5] R. BUDDE, H. NIETERS. Einführung in die Netztheorie. *Regelungstechnik 32, pp. 76-80*, 1984.

[6] B. CHARRON-BOST. Combinatorics and Geometry of Consistent Cuts: Application to Concurrency Theory. *In: J-C Bermond, M. Raynal (eds) Proc. of the 3rd International Workshop on Distributed Algorithms, Springer-Verlag LNCS 392, pp. 45-56*, 1989.

[7] B. CHARRON-BOST. Concerning the Size of Logical Clocks in Distributed Systems. *Information Processing Letters*, 39:11–16, 1991.

[8] J. FIDGE. *Dynamic Analysis of Event Orderings in Message-Passing Systems.* PhD thesis, Department of Computer Science, The Australian National University, 1989.

[9] M. FISCHER, A. MICHAEL. Sacrificing Serializability to Attain High Availability of Data in an Unreliable Network. In *ACM SIGACT-SIGOPS Symp. on Principles of Database Systems*, pages 70–75, 1982.

[10] J. T. FRASER. *Die Zeit – vertraut und fremd.* Birkhäuser Verlag, 1988.

[11] J.T FRASER, F.C. HABER, G.H. MÜLLER. *The Study of Time.* Springer-Verlag, 1972.

[12] K. GÖDEL. Einige Bemerkungen über die Beziehungen zwischen der Relativitätstheorie und der idealistischen Philosophie. *In: Albert Einstein als Philosoph und Naturforscher (Hg.: P.A. Schlipp), Vieweg (Original: "A Remark," In: A. Einstein, Philosopher-Scientist, ed. by P.A. Schlipp, 1949)*, 1979.

[13] D. HABAN, W. WEIGEL. Global Events and Global Breakpoints in Distributed Systems. *Proc. 21st Hawaii International Conference on System Sciences, Vol. II, pp. 166-175*, 1988.

[14] D.B. JOHNSON, W. ZWAENEPOEL. Recovery in Distributed Systems Using Optimistic Message Logging and Checkpointing. Technical Report TR88-68, Department of Computer Science, Rice University, 1988.

[15] W. KÖNIG. DTV-Atlas zur Deutschen Sprache. *Deutscher Taschenbuch Verlag*, 1978.

[16] L. LAMPORT. Time, Clocks, and the Ordering of Events in a Distributed System. *Comm. of the ACM 21:7, pp. 558-565*, 1978.

[17] B. LISKOV, R. LADIN. Highly-Available Distributed Services and Fault-Tolerant Distributed Garbage Collection. *Proc. of the 5th ACM Symposium on Principles of Distributed Computing*, pages 29–39, 1986.

[18] F. MATTERN. Verteilte Basisalgorithmen. *Springer-Verlag, Informatik-Fachberichte Bd. 226*, 1989.

[19] S. MELDAL, S. SANKAR, J. VERA. Exploting Locality in Maintaining Potential Causality. Technical Report CSL-TR-91-466, Computer Systems Laboratory, Stanford University, 1991.

[20] H. REICHENBACH. Philosophie der Raum-Zeit-Lehre. *Walter de Gruyter*, 1928.

[21] F. SEIBT. *Die Zeit als Kategorie der Geschichte und als Kondition des historischen Sinns, In: Die Zeit – Dauer und Augenblick.* (Hg.: A. Peisl, A. Mohler), Oldenbourg Verlag (Schriften der Carl Friedrich von Siemens Stiftung), 1983.

[22] M. SINGHAL, A. KSHEMKALYANI. An Efficient Implementation of Vector Clocks. Technical report, Dept. of Computer and Information Science, The Ohio State University, 1991.

[23] R. STROM, S. YEMINI. Optimistic Recovery in Distributed Systems. *ACM Transactions on Computer Systems*, 3(3):204–226, 1985.

[24] J.F.A.K. VAN BENTHEM. The Logic of Time. *D. Reidel Publishing Company*, 1983.

[25] M. ČAPEK. Time–Space Rather than Space–Time. *Diogenes*, (123):30–49, 1983.

[26] K. VENKATESH, T. RADHAKRISHNAN, H.F. LI. Optimal Checkpointing and Local Recording for Domino-Free Rollback Recovery. *Information Processing Letters 25, pp. 295-303*, 1987.

A Hierarchy Preserving Hierarchical Bottom-Up 2-layer Wiring Algorithm with Respect to Via Minimization

Paul Molitor

Fachbereich 14 – Informatik
Universität des Saarlandes
6600 Saarbrücken
Germany

Abstract

Here, the following hierarchical constrained via minimization problem (HCVM) is examined: "Let Q be a circuit with a given layout hierarchy. Find a 2-layer wiring of Q which needs a number of vias minimal with respect to the preservation of hierarchy, i.e., on condition that the description of the result is (nearly) as short as the description of Q before the 2-layer wiring". The problem arises in connection with hierarchical physical synthesis, which is not highly developed yet although absolutely essential for the design of ULSI circuits. The computational complexity of HCVM is unknown till today. This paper presents a hierarchical bottom-up algorithm to (a variant of) this problem which is locally optimal, i.e., given the 2-layer wirings of the subcircuits of level i, it computes the optimal (partially induced by the wirings of the subcircuits) 2-layer wiring of the circuits of level $i + 1$. The algorithm's running time is $O(n^3)$ where n is the size of the hierarchical description of Q.

1 Introduction

Todays integrated circuits have up to several thousand transistors. Processing such designs in a naive manner requires very large internal representations with some millions of data so that even linear space (time) optimization algorithms would be too expensive. However, large designs have a regular structure. In such arrangements, there are lots of identical subcircuits so that they can be described by small hierarchical representations. During synthesis, it is desirable to handle all instances of a subcircuit identically in order to guarantee identical electrical behavior of all instances, to allow further hierarchical processing and to decrease the running time of the synthesis.

A leading feature of VLSI design systems is the placement and routing aspect. Typically, routing consists of two steps: *wire layout* and *wiring (layer assignment)*. The first step determines the placement of the routing segments. A 'planar' wire layout is obtained. In the wiring step, this layout is converted into an actual three-dimensional configuration of wires. Formally, wiring is formulated as follows: Given a wire layout L and 2 physical layers, assign each routing segment of L a unique physical layer in a way that the segments are electrically connected in the proper way. Adjacent edges of the same wire lying on different layers are connected with vertical connections, called *vias*. Segments of different wires crossing each other have to be assigned to different layers. A problem arising in this context is the *constrained via minimization problem* (CVM), i.e., computing a 2-layer wiring with a minimal number of vias. Minimizing the number of vias is important because too many vias lead to a decreased performance of the electrical circuit, a decreased yield of the manufacturing process and an increased amount of area required for interconnections. Thus, it is a step towards optimizing the performance of the circuit and possibly minimizing its manufacturing costs. A survey on the k-layer wiring problem, especially the constrained via minimization problem (for non-hierarchical circuits) is given in [13].

In this paper, we will first present a specification technique, namely (graphical) recursive net equations, producing compact hierarchical representations of large regular circuits. This technique is the kernel of the CADIC system. CADIC is a VLSI design system based on an algebraic approach well suited for the hierarchical design of integrated circuits. Main features of this system are the compact internal representation of large regular design objects and the hierarchical processing of each individual step of the automated synthesis. More details can be found in [1, 2, 5, 7, 10, 12]. Section 4 will introduce the problem of 2-layer wiring with respect to via minimization. Finally, the last two sections will concentrate on hierarchical layer assigment.

2 CADIC's design level

Boolean algebra is the classical basis for a calculus allowing the handling of logical circuits. This calculus was sufficient as long as the geometrical placement of the basic cells and the routing of the interconnection wires played a minor role for the design of a boolean circuit. With increasing chip complexity this is no longer the case. Now, a calculus based on simple operations and representing the logical function as well as some information about the geometrical arrangement of cells and wires is required. Nevertheless, neither the geometrical part must be too detailed because it would strongly depend on the fabrication process, nor the logical part should be on the lowest level, e.g., transistor design level, but on the level of basic cells computing digital values.

A first generalization of boolean algebra in this direction was given by Günter Hotz [4] in 1965 with the introduction of the $\times$-category. Unfortunately, the $\times$-categories only handle cells having their input ports on their northern boundaries and their output ports on their southern boundaries. For the purposes of VLSI design we need a generalization of this concept.

We consider circuits laid out into a rectangle R. The external connectors (input- and output ports) are placed on the boundary of R. To suppress geometrical and physical details of the manufacturing process and to become independent of technology, we omit the width and the layers of the wires. (Thus the designer has not to specify the layer in which a wire segment is embedded. The layer assignment will be generated later on during physical synthesis.) Furthermore, we ignore the internal structures and the sizes of the basic cells and only maintain the order of external connectors on their boundary. By considering crossings and junctions of wires as (passive) cells performing crossings and junctions of signals, this abstraction results in a planar arrangement of cells in the plane whose interconnections consist of crossing-free non-overlapping lines. We call such a structure *logic topographical net* or *lg-net*.

To suppress precise geometrical relations of this abstract layout (which is not a precise specification of the layout of an integrated circuit because of the abstractions made above) we consider two planar layouts of this kind to be equivalent if and only if they can be transformed into each other by a sequence of deformations such as deformation of wires, translation and stretching of cells, and order preserving translation of connectors along the side of a rectangle, i.e., deformations which maintain the planar topological structure of the layout. A maximal set of equivalent layouts is called *logic topological net* or *lt-net*.

Each wire segment is related to a type $t \in O$. For an lt-net α, we denote the sequence of the types of the connectors on the northern and southern side of α (from left to right) by $N(\alpha)$ and $S(\alpha)$, respectively, the sequences on the eastern and western side (from top to bottom) are denoted by $E(\alpha)$ and $W(\alpha)$, respectively. A type is an attribute of a wire, e.g., a wire of type $\underline{7}$ could represent 7 parallel wires (of type $\underline{1}$).

Let A and B be two lg-nets. If the eastern side of A geometrically matches the western side of B, i.e., both lg-nets have same height and the connectors on the eastern side of A match those on the western side of B, then we can construct a new lg-net $C = A \ominus B$ by superposing the eastern side of A and the western side of B, identifying the matching connectors and erasing the superposed borders. Analogously, the composition $A \oplus B$ is defined if and only if the southern border of A geometrically matches the northern one of B. It turns out that the operations $\ominus$ and $\oplus$ can be extended to lt-nets in a natural way. Consider two lt-nets α and β. If there exist two representatives $A \in \alpha$ and $B \in \beta$ for which the $\ominus$-composition ($\oplus$-composition) is defined, perform the operation on the lg-nets. Then, the class of the result is the composition of the lt-nets α and β, i.e. $\alpha \ominus \beta = [A \ominus B]$ and $\alpha \oplus \beta = [A \oplus B]$, respectively.

The corresponding algebraic structure is called *bicategory*. The exact definition can be found in [12].

3   Hierarchical design with CADIC

By allowing simultaneous cell– and edge-replacement operations, large regular lt-nets can be described in a way that can be easily surveyed. These refinements are homomorphisms on bicategories which are called *bifunctors*.

It is obvious that the parallel replacement of cells and edges is not allowed in any case. The necessary conditions which must be fulfilled can easily be drawn from the algebraic base and are fixed by the formal definition of bifunctors. Let $NET(\mathcal{A}, O)$ be the set the lt-nets composed of cells from cell library $\mathcal{A}$ and wire segments of types from O. Then, a refinement of an lt-net formally corresponds to a homomorphism $\varphi_2 : NET(\mathcal{A}, O) \rightarrow NET(\mathcal{A}, O)$ with $\varphi_2(\alpha \circ \beta) = \varphi_2(\alpha) \circ \varphi_2(\beta)$ ($\circ \in \{ \ominus , \oplus \}$) and a monoidhomomorphism $\varphi_1 : O^* \rightarrow O^*$. φ_1 defines the refinement of the signal types, φ_2 the refinement of the lt-nets themselves. The condition which has to be fulfilled is that the refinement of each cell $x \in A$ is compatible with the refinement of the wire types, i.e., $N(\varphi_2(x)) = \varphi_1(N(x))$, $S(\varphi_2(x)) = \varphi_1(S(x))$, $W(\varphi_2(x)) = \varphi_1(W(x))$ and $E(\varphi_2(x)) = \varphi_1(E(x))$. It follows that the refinement of the signal types already determines the refinement of the passive cells, namely the refinement of knees, junctions and cross-overs. Figure 1 shows the refinement of knees and junctions for $\varphi_1(t) = t_1 t_2$ $(t, t_1, t_2 \in O)$.

In the remainder of this section, we will demonstrate how to specify large regular lt-nets with the help of bifunctors by specifying a family of partial multipliers (see also [1, 7]). This type of multiplier is a tree-multiplier, which is a modified version of a Wallace tree multiplier [14] made suitable for VLSI design by Luk and Vuillemin [8].

The first step in the design is the following one: Let $a = (a_{n-1}, \ldots, a_0)$, $b = (b_{n-1}, \ldots, b_0)$ be two binary numbers. The product of a and b is equal to the sum of the n binary numbers of length $2n$ represented by the n lines of the following matrix P_n, where $a_i b_j$ denotes the *logical and* of the bits a_i and b_j, if $a_{2n-1} = a_{2n-2} = \ldots = a_n = a_{-1} = a_{-2} = \ldots = a_{-n+1} = 0$:

$$\begin{pmatrix} a_{2n-1}b_0 & a_{2n-2}b_0 & \cdots & a_0 b_0 \\ a_{2n-2}b_1 & a_{2n-3}b_1 & \cdots & a_{-1}b_1 \\ \vdots & \vdots & \vdots & \vdots \\ a_n b_{n-1} & a_{n-1}b_{n-1} & \cdots & a_{-n+1}b_{n-1} \end{pmatrix}$$

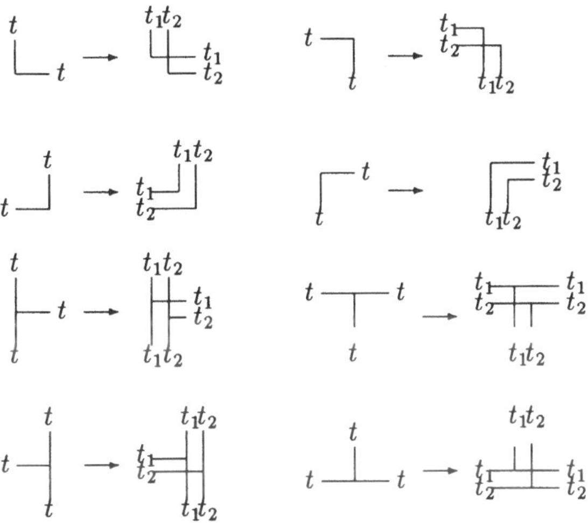

Figure 1: Refinement of the passive cells

Matrix P_n gives us the idea to compose an n-bit multiplier by $2n$ identical columns (or n identical rows). In this way, we obtain the first three recursive equations. Note that all the equations are described by figures and can be fed as figures in the computer.

(1) An n-bit multiplier is given by the circuit $c[n, 2n]$ (see figure 2), where $c[n, i]$ consists of i columns of an n-bit multiplier.

(2) To simplify the presentation, we assume that $n = 2^k$ for some $k \geq 1$. Then, $c[n, i]$ can be composed of two instances of $c[n, \frac{i}{2}]$ and is graphically described in figure 3. It follows that, parallel to the replacement of cell $c[n, i]$ in a circuit, each wire segment of type $k[i]$ is replaced by two wire segments of type $k[\frac{i}{2}]$.

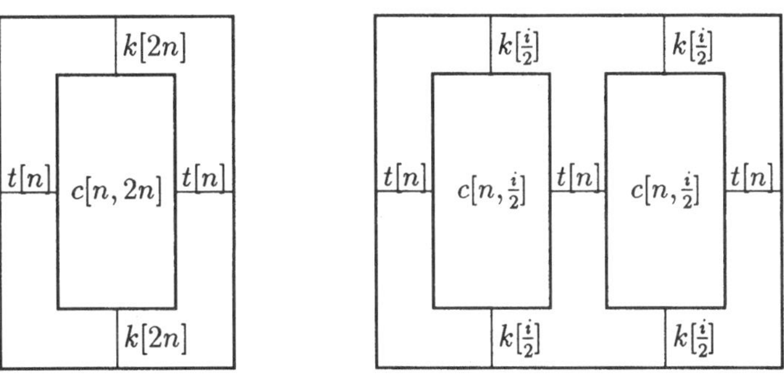

Figure 2   Figure 3

(3) The equation that $c[n, 2]$ consists of 2 columns of an n-bit multiplier which are denoted by $col[n]$ serves as base of the induction.

To complete the specification, circuit *col*[n] has to be specified recursively. This can be done by 3 further equations (see [1, 7]).

Following the graphical design of such a family of circuits, e.g., the family of the n-bit tree multipliers, the designer tells the system the circuit he wants to construct, e.g., the 128-bit multiplier. The system expands the special circuit following the recursive net equations. The result is a hierarchical logic topological net.

Notice that each multiplier *multiplier*[n] (e.g., $n = 128$) consists of many identical subcircuits due to the compactness of the specification. In order to take advantage of this fact, we introduce the notion of a *box*. Each variable on the left side of a net equation can represent a box. In the above example, *col*[2], *col*[3], ..., $c[2,2]$, $c[2,4]$, ... are such candidates. Now, if such a variable is declared to be a box, it occurs only once in the internal representation of the whole circuit, e.g. if $c[128,16]$ is a box, the internal representation of the 128-bit multiplier does not contain the $c[128,16]$-subcircuit 16 times but only once. Then, *hierarchical processing* means that attempts are made to handle all occurrences of a box in the same manner, e.g., that they are identically optimized. By this means, the internal representation only increases by a constant factor during synthesis which allows (further) hierarchical processing.

It is obvious that the quality of the optimizations during synthesis is impaired by hierarchical processing. There are trade-offs between 'compactness of the hierarchy' and 'quality of the result', i.e., the compacter the representation of the circuit is, the worse the quality of the hierarchical optimizations is. This trade-off has been investigated for the constrained via minimization problem during layer assignment [6].

In the following, we will present a hierarchical bottom-up 2-layer wiring algorithm which preserves the hierarchy and which takes advantage of not only the refinement of cells but also of the refinement of the signal types. To this, we need the following definition:

Definition 1 *A box is said to be of* level 0, *if it does not contain any other box; it is of* level $i + 1$, *if it contains at least one box of level i and no box of level greater than* i.

4   An approach to 2-layer wiring

An example shall illustrate the non-hierarchical 2-layer wiring problem. Formal definitions of the notions can be found in [7, 11].

Let L be the wire layout shown in figure 4. L divides the plane in elementary regions (*faces*), 28 inner faces and one outer face. The inner faces are enumerated in figure 4. Each face r has a boundary. We call a face r *odd*, if its boundary cannot be embedded into two layers without using a via. Otherwise, the face is called *even*. (See figure 5 where the odd faces of L are marked by $\star$.)

Obviously, it is impossible to find a 2-layer wiring without using a via if there is an odd face. The converse is true as well.

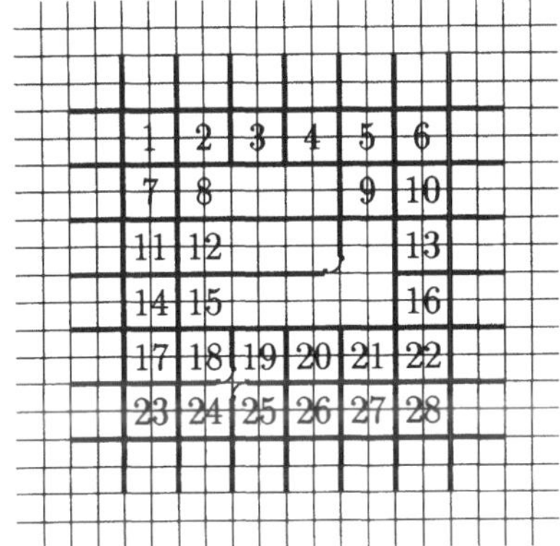

Figure 4: Wire layout L. The thin lines represent the square grid.

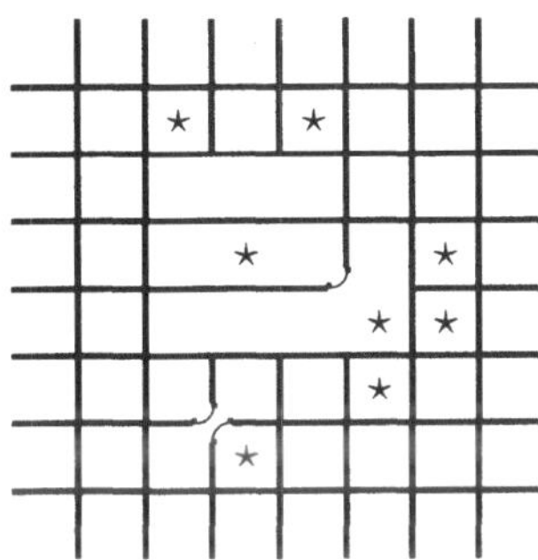

Figure 5: The odd faces of L are marked by $\star$.

Lemma 1 [11] *Let L be a wire layout. There is a 2-layer wiring of L needing no (further) via if and only if each face of L is even.*

Therefore, in order to obtain a 2-layer wiring, the odd faces have to be transformed into even ones by inserting vias on their boundaries. (We assume that a via is a vertex where the layer has to be changed.) Here, a via may only be located on a grid vertex which is touched by at most one signal net. In our example, if a via is placed, e.g., on the vertical wire segment (on the grid vertex) between face 2 and face 3, face 2 which was odd becomes even and face 3 which was even becomes odd. The insertion of a via between face 3 and 4 transforms these two odd faces into even ones. To sum it up, we have transformed the two odd faces 2 and 4 into even ones by joining them by a path of vias. We have *married* face 2 to face 4.

Lemma 2 [11] *In any layout L, there is an even number of odd faces.*

So, marrying each odd face to exactly one other odd face results in a layout where each face is even. By lemma 1, this new layout is 2-layer wirable without further vias. Conversely, any 2-layer wiring induces such a marriage of the odd faces. Figure 6 shows a 2-layer wiring of L. Here, faces 2 and 4, 13 and 16, 12 and 15, and 21 and 25 are joined. Note that the faces 21 and 25 are joined by a path of vias which runs across the outer face.

Now, let us concentrate on the dual graph of a wire layout.

Definition 2 *Let L be a wire layout. The* dual graph $G = (V, E)$ *of L is defined as follows. The set V of vertices consists of the faces of L and $\{f, f'\}$ is an edge of E, if the faces f and f' are adjacent to a common wire segment onto which a via may be placed.*

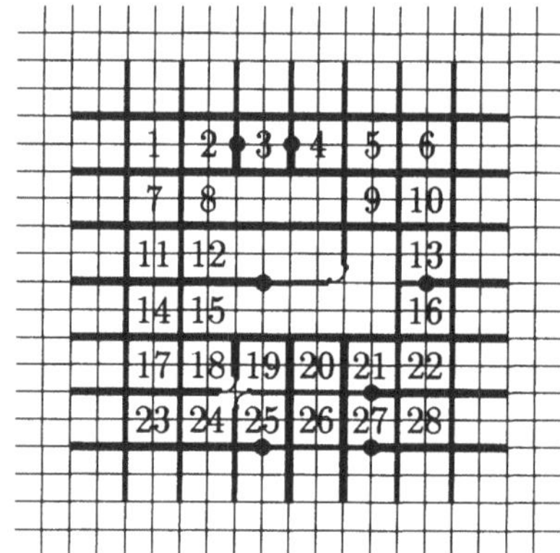

Figure 6: Two-layer wiring of L. The thick lines represent the one layer, the extra thick lines represent the other one.

The dual graph of the layout L of figure 4 is obviously connected (in the sense of graph theory [9]). Therefore, it is easy to find a perfect matching of the odd faces which determines the marriages. The situation is more complex in the case of dual graphs consisting of more than one connected component.

Definition 3 *A connected component of G is called* odd *if it contains an odd number of odd faces. Otherwise it is called* even.

Because of the statements above, it is easy to see that the following lemma holds:

Lemma 3 [11] *Let L be a wire layout and let G be its dual graph. There is a 2-layer wiring of L if and only if each connected component of G is even.*

In the following, we confine ourselves to 2-wirable wire layouts. To find a solution to the constrained via minimization problem, one has to transform all the odd faces into even ones by using a minimal number of vias. Thus, the following matching problem gives us the solution:

Lemma 4 *Let ODD be the set of the odd faces of the wire layout L. Then the via minimization problem is equivalent to the problem of finding a (disjoint) partition of ODD into pairs $\{f_{1_1}, f_{1_2}\}, \ldots, \{f_{k_1}, f_{k_2}\}$ such that*

$$\sum_{i=1}^{k} d_G(f_{i_1}, f_{i_2})$$

is minimal, where $d_G(g, g')$ is the length of the minimum length path connecting face g and face g' in the dual graph G.

This minimum weighted matching problem can be solved in running time $O(n^3)$ [3], where n is the amount of the odd faces. Note that all these statements hold as well for the case that the wire layout is not grid-based where there are two classes of edges, namely edges onto which vias are allowed and edges onto which vias must not be placed.

5 A step towards HCVM

In this section we confine ourselves to the case that the refinement of the wires (signal types) is the identity, i.e., $\varphi_1(t) = t$ ($\forall t \in O$). Before formulating the problem exactly, we need another definition.

Definition 4 *Let A be a box of level i composed of cells and boxes B, C, D, ... (of levels less than i) which are interconnected by some wire layout. Let δ_B, δ_C, δ_D, ... be 2-layer assignments of the boxes B, C, D, ..., respectively. A layer assignment δ_A of A is said to be induced by δ_B, δ_C, δ_D, ..., if δ_A is given by applying δ_X on all occurrences of box X ($X \in \{B, C, D, ..., \}$) and then realizing the 2-layer assignment of the interconnecting wire layout.*

Then, the problems of hierarchical layer assignment with respect to via minimization can be formulated in the following manner.
Given a box A of level i which is composed of cells and boxes B, C, D, ...
Problem 1
Given 2-layer assignments δ_B, δ_C, δ_D, ... of the boxes B, C, D, ..., respectively.
Find a (by δ_B, δ_C, δ_D, ...) induced 2-layer assignment δ_A of box A which has less (or equal) vias than any other (by δ_B, δ_C, δ_D, ...) induced 2-layer assignment δ'_A.
Problem 2
Find 2-layer assignments δ_B, δ_C, δ_D, ... of the boxes B, C, D, ... such that there is a (by δ_B, δ_C, δ_D, ...) induced 2-layer wiring δ_A of box A with a minimal number of vias compared to any other 2-layer wiring δ'_A of box A induced by any other 2-layer assignments δ'_B, δ'_C, δ'_D, ...
The computational complexity of both problems is unknown till today. In the following, we will present a solution to a variant of problem 1.

Obviously, problem 1 is equivalent to the pin preassignment problem, i.e., to the problem that pins are preassigned to some layers. Now, assume that two metal layers (ALU1 and ALU2) are available for wiring, and that the logical basic cells are already designed in such a way that there are two designs for every basic cell. Further, assume that the two designs $d_1(x)$ and $d_2(x)$ of each basic cell x are dual, i.e., that the ith pin of $d_1(x)$ is in layer ALU1 if and only if the ith pin of $d_2(x)$ is in ALU2. Then, two pins i and j are said to be in the same class, if i and j are preassigned to the same layer (in each of the two designs).

First, we consider a circuit of level 1, i.e., it only contains basic cells and wire segments. Figure 7 shows such a circuit composed of AND– and OR–cells. Assume

that the OR–cell is designed such that all pins are in the same class, i.e., that they
are preassigned to the same layer, either ALU1 or ALU2, and that the AND–cell is
designed in such a way that the left pin in the north and the pin in the south belong
to one class and the right pin in the north to the other class.

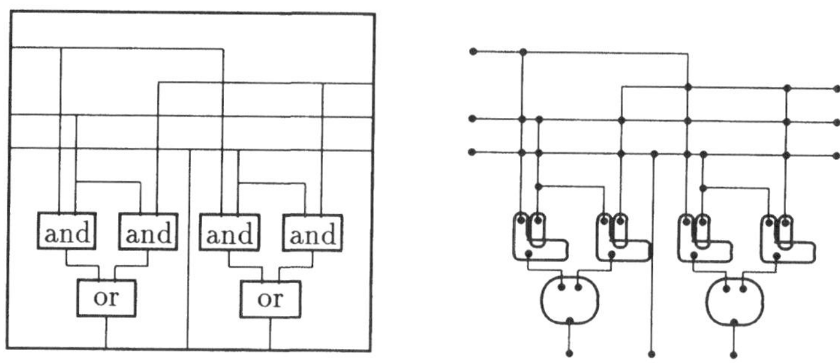

Figure 7: Transformation of a circuit of level 1

By making the following (mental) transformations of the circuit, problem 1 is
solved under the assumptions made above by applying the results of the previous
section.

We substitute each occurrence of the OR–cell by a new wire on which it is forbidden
to place a via and which crosses all the pins of this occurrence (see figure 7). Because
no via may be placed on this new wire, all the pins have to be assigned to the same
layer. (Note that the inner face encircled by the new wire is even.) Each occurrence
of the AND–cell is substituted by two new wires (general case) which cross each other
and on which it is forbidden to place a via. One of these wires crosses all pins of the
first class, the other one crosses all the pins of the second class. By this, the pins of
different classes are assigned to different layers. The right picture of figure 7 shows
the transformed circuit.

After this transformation step, the results of the previous section are applied, i.e.,
the odd faces and an optimal wedding of these faces are computed. Obviously, there
exist two different 2-layer assignments to the optimal wedding which are dual in the
above sense. Thus, the algorithm can be applied bottom up to boxes of level 2, 3, 4
and so on. Figure 8 shows the odd faces of the transformed circuit and one of the two
dual 2-layer wirings. Note that the outer face is odd, too.

6   Hierarchical layer assignment

In this section, we investigate the general case, i.e., the case where the refinement φ_1
of the signal types is not the identity. Without loss of generality we confine ourselves

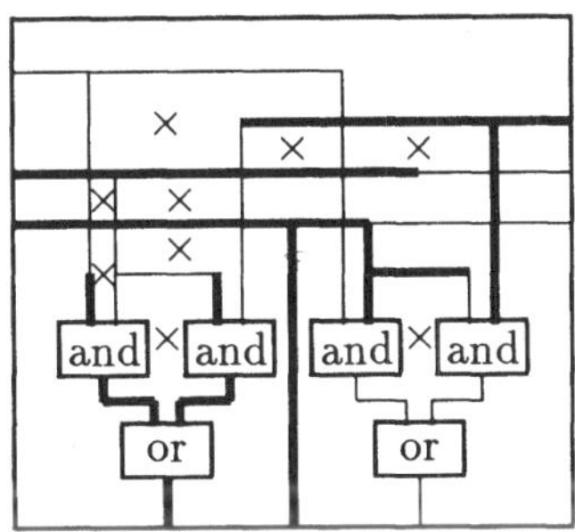

Figure 8: Odd faces and corresponding layer assigment of the above circuit

to the case that all the wires are busses. The types quote the bus width, i.e., the number of lines which are represented by the bus. We denote a bus width of n lines by signal type $\underline{n}$.

Surely, the bottom-up algorithm described above can be applied by expanding the busses in each step first. In the following, we present a method to take advantage of the busses during hierarchical via minimization. It results in a smaller number of odd faces which have to be handled by the minimum weighted matching problem actually computing the layer assignment (which has running time $O(n^3)$ where n is the number of odd faces to handle). Nevertheless, there will be a preprocessing and a postprocessing step which both have running time linear in the number of faces of the circuit with expanded busses.

In the following, we relate a non-negative weight to each edge of the dual graph. The distance between two faces is defined by the minimum weighted path of the dual graph connecting them. Wire segments of type $\underline{1}$ obtain the weight 1. (We identify the wire segments of the wire layout with the edges of the corresponding dual graph.)

Now, look at the situation illustrated in figure 9. The bus of type $\underline{n}$ is connected to occurrences of two boxes A and B. Assume that the layer assignments of A and B, respectively, have already been computed by our algorithm. Thus, there is a preassignment of the pins of A and B, and faces may be odd inside of the bus. For example, let the preassignment of the pins of the eastern side of A (from top to bottom) be $(c_0^A, c_1^A, c_1^A, c_1^A, c_0^A, c_0^A, c_1^A)$ (for $n = 7$) and of the pins of the western side of B (from top to bottom) be $(c_1^B, c_1^B, c_0^B, c_1^B, c_0^B, c_1^B, c_0^B)$. Here, c_0^A, c_1^A, and c_0^B, c_1^B denote the two pin classes of A and B, respectively. Obviously, there are four odd faces inside of the bus (see the right picture of figure 9 and the previous section).

To marry these four odd faces, there are two possibilities. The first one is to marry them to each other inside of the bus, i.e., the topmost odd face to the second one, and the third one to the bottom one (see the dashed lines in figure 9). For this, three

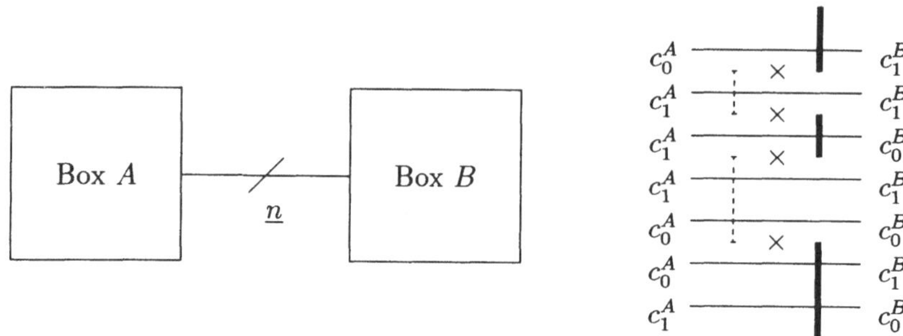

Figure 9: Illustration of the problem. At the right, the bus is expanded ($n = 7$) with a given preassignment of A and B. The marked faces are odd.

vias have to be inserted. The second possiblity is to marry the two middle ones to each other and the other two odd faces by pathes going outside of the bus (see the bold lines in figure 9). This possibility requires four vias to be inserted.

This consideration leads to the following construction. We substitute each bus of type $\underline{n}$ ($n \geq 4$) by three wire segments (see figure 10). The weight of these segments (note that we identify the wire segments of the wire layout with the edges of the corresponding dual graph) are defined in the following way.

If there is an even number (≥ 2) of odd faces inside of the bus, there are two possibilities to marry them, namely to marry them to each other inside of the bus and to marry two of them by pathes going outside of the bus (and the remainder to each other inside of the bus). This is modelled by defining the faces i_1 and i_2 to be odd and the weight of edge $\{i_1, i_2\}$ (which is adjacent to faces i_1 and i_2) to be equal to the number of vias required to marry all the odd faces inside of the bus, the weight of edge $\{o_1, i_1\}$ to be the number of vias required, if the two outermost odd faces are married through pathes going outside of the bus, and the weight of the third edge to be 0. Because an optimal wedding of the odd faces contains either edge $\{i_1, i_2\}$ or the two edges $\{o_1, i_1\}$ and $\{i_2, o_2\}$, the situation is (compactly) specified in a correct manner.

If there is an odd number of odd faces inside of the bus, all but one odd faces have to be married inside of the bus by an optimal wedding. The remaining odd face has to be married through a path leaving the inside of the bus either through the upper line or through the lower line. This case is illustrated by figure 11. This is modelled by defining either face i_1 or face i_2 to be odd and the weight of edge $\{i_1, i_2\}$ to be 0 (i.e., we identify face i_1 with face i_2), the weight of edge $\{o_1, i_1\}$ to be the number of vias required, if a wedding path leaves the inside of the bus by the upper line, and the weight of edge $\{i_2, o_2\}$ to be the number of vias required, if a wedding path leaves the inside of the bus by the lower line. Note that an optimal wedding of the odd faces contains either edge $\{o_1, i_1\}$ or edge $\{i_2, o_2\}$.

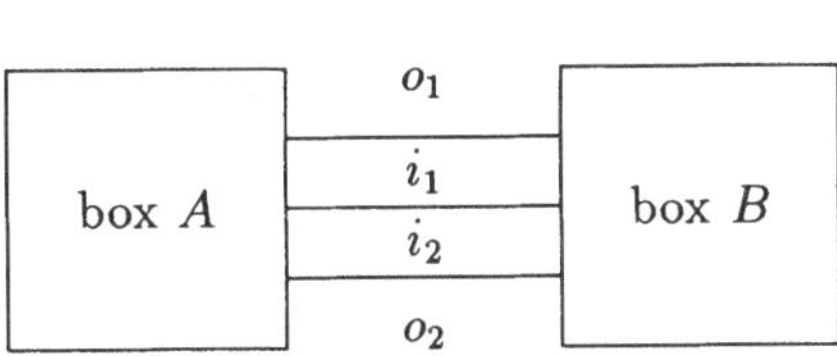

Figure 10: Transformed circuit

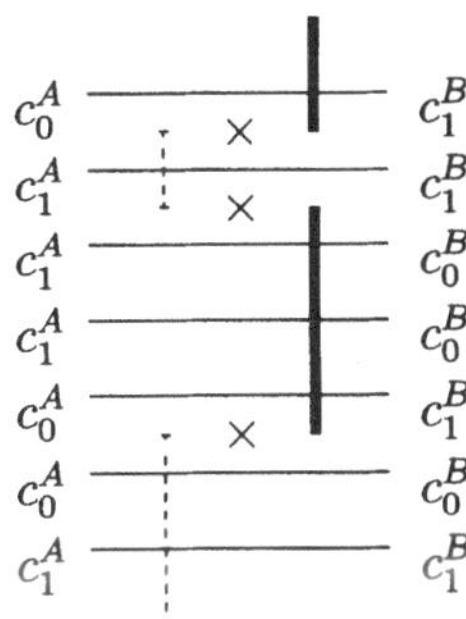

Figure 11: The dashed and bold lines denote the two possible weddings of the odd faces.

If each face inside of the bus is even, we define the faces i_1 and i_2 to be even, the weight of (any) two of the three edges to be 0 (i.e., we identify the three edges with each other), and the third one to be equal n, because n vias have to be inserted if a wedding path cross the bus.

Busses of type $\underline{2}$ or $\underline{3}$ will be expanded.

Apart from the situation just discussed where pin preassignment plays the major role, we have to determine how to perform layer assignment inside the passive cells of type $\underline{n}$. Considering figure 1, one notes that the refinements of the north-west-knee and south-east-knee are trivial with respect to layer assignment. The remaining refinements are a bit more complicate.

The refinement of a north-east-knee (south-west-knee) generates new odd faces inside of the bus, namely those marked in figure 12. In order to transform the odd faces on the inside into even ones, the algorithm inserts a via onto each of the $n - 2$ inner knees of type $\underline{1}$ as shown in figure 12, in any case. (Note that this is nearly optimal. At least $n - 5$ of these $n - 2$ odd faces cannot be married to another odd face through a path of length 1 because at most three faces adjacent to them can be odd.) Further vias, which probably are needed, may only be placed outside of the refinements.

Thus, a north-east-knee of type $\underline{n}$ can be represented by three wire segments crossing each other and one via arranged as shown in figure 13.

The weights of the three horizontal and the three vertical edges on the outside depend on the adjacent cells and boxes. This is done by the way discussed above. Figure 14 illustrates the situation of a north-east-knee connected to a box (where one face inside of the bus is odd).

The refinements of the junctions are handled in an analogous manner. Here, the algorithm places vias onto each of the $n - 2$ inner junctions of type $\underline{1}$, in any case. The representation of a junction is as shown in figure 15. The refinements of cross-overs do

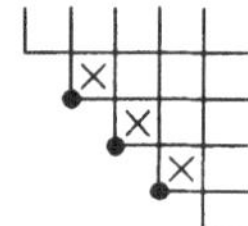

Figure 12: Refinement of a north-east-knee of type $\underline{5}$.

Figure 13: Representation of a north-east-knee of type $\underline{n}$

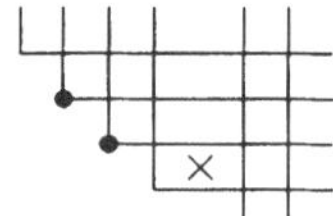

Figure 14: A knee of type $\underline{4}$ connected to a cross over.

Figure 15: Representation of a junction of type $\underline{n}$

not generate inner odd faces such that no via is placed on the inside. The remainder is handled analogously to the above discussion.

By this construction the faces inside of the busses are handled in the right way.

Finally, we only have to determine whether a face outside of a bus is even or odd. This can easily be done in the following way. Crossings and vias are points where the layer has to be changed even if they are part of the representation of an expanded passive cell. Pin preassignments at the northern or southern (western or eastern) side of a box are handled in the way described in the previous section by associating the pin class of the leftmost (topmost) line of a bus to the leftmost (topmost) of the three wires, and the pin class of the rightmost (bottommost) line of a bus to the rightmost (bottommost) of the three wires.

This completes the hierarchy preserving bottom up algorithm for 2-layer assignment with respect to via minimization. The algorithm's running time is $O(n^3 + m)$ where n is the size of the hierarchical description and m is less than n multiplied by the largest bus width. The last term is due to the preprocessing and postprocessing step at each hierarchy-level.

7   Concluding Remarks

The trade-off between 'compactness of the hierarchy' and 'quality of the result' (without the assumptions made) has been investigated by case studies in [6]. It has turned out that the quality of the result is not inversely proportional to the size of the hierarchical representation, that the trade-off is 'discontinuous', i.e., it can happen that the hierarchical optimization works considerably better after having manipulated the hierarchical representation only a little bit, e.g., by taking one variable out of the set of boxes, and that there often exists a very compact representation requiring only a few more vias than the 'flat circuit' does.

Acknowledgements

I express my deepest appreciations to Professor Dr. Günter Hotz, my advisor, who has had (and will have) a profound influence on the development of my work.

Bibliography

[1] B. Becker, Th. Burch, G. Hotz, D. Kiel, R. Kolla, P. Molitor, H. G. Osthof, G. Pitsch, and U. Sparmann. A graphical system for hierarchical specifications and checkups of VLSI circuits. In *Proceedings of EDAC90*, pages 174–179, 1990.

[2] B. Becker, G. Hotz, R. Kolla, P. Molitor, and H.G. Osthof. Hierarchical design based on a calculus of nets. In *Proceedings of DAC87*, pages 649–653, 1987.

[3] H. Gabow. *Implementation of algorithms for maximum matching on non-bipartite graphs*. PhD thesis, Stanford University, 1973.

[4] G. Hotz. Eine Algebraisierung des Syntheseproblems für Schaltkreise. *EIK*, 1:185–205,209–231, 1965.

[5] G. Hotz, R. Kolla, and P. Molitor. On network algebras and recursive functions. In *Proceedings of GRAGRA86*, LNCS 291, pages 250–261, 1986.

[6] R. Kolla and P. Molitor. A note on hierarchical layer assignment. *INTEGRA-TION, the VLSI Journal*, 7:213–230, 1989.

[7] R. Kolla, P. Molitor, and H.G. Osthof. *Einführung in den VLSI-Entwurf*. Leitfäden und Monographien der Informatik. B.G. Teubner Verlag, Stuttgart, 1989. 352 Seiten.

[8] W.K. Luk and J. Vuillemin. Recursive implementation of optimal time VLSI integer multipliers. In *Proceedings IFIP Congress 83*, pages 155–168, 1983.

[9] K. Mehlhorn. *Data Structures and Algorithms 2: Graph Algorithms and NP-Completeness*. EATCS Monographs on Theoretical Computer Science, 1984.

[10] P. Molitor. *Über die Bikategorie der logisch topologischen Netze und ihre Semantik*. PhD thesis, Universität des Saarlandes, 1986.

 Paul Molitor

[11] P. Molitor. On the contact minimization problem. In *Proceedings of STACS87*, pages 420–431, 1987.

[12] P. Molitor. Free net algebras in VLSI-theory. *FUNDAMENTA INFORMATI-CAE*, XI:117–142, 1988.

[13] P. Molitor. A survey on wiring. *EIK*, 27(1):3–19, 1991.

[14] C.S. Wallace. A suggestion for a fast multiplier. *IEEE-EC*, 13:14–17, 1964.

Eine $O(e \log e)$ - Heuristik für ein Flußproblem

Hans Georg Osthof

Saarbrücker Zeitung
D-6600 Saarbrücken

Zusammenfassung

In diesem Beitrag wird ein neuer Zugang zur Lösung eines speziellen Flußproblems aufgezeigt. Dieser ermöglicht die Berechnung einer oberen Schranke für den maximalen Flß in Zeit $O(e \log e)$, wohingegen die Algorithmen zur Bestimmung der exakten Lösung in Zeit $O\left(\frac{n^3}{\log n}\right)$ bzw. $O\left(e \cdot n \cdot \log \frac{n^2}{e}\right)$ arbeiten. Bemerkenswert ist, daß die Heuristik auf den mehr als 200.000 gerechneten Testbeispielen immer die exakte Lösung erzielte. Weiterhin existiert bisher auch auch kein (konstruiertes) Gegenbeispiel für die Heuristik.

1 Einführung

Gegeben sei ein ungerichteter, zusammenhängender Graph $G = (V, E)$ mit $V = V_{in} \dot\cup V_c$. Den Punkten aus V_c werden beliebige, feste Koordinaten in der euklidischen Ebene zugeordnet. Wir bezeichnen bezeichnen solche Graphen mit *Graphen mit festem Rand*. Gesucht wird eine Auslegung der Punktmenge V_{in}, so daß folgende gewichtete Kantensumme minimiert wird:

$$\sum_{(v_i,v_j)\in E} (w_{ij} \cdot |e_{ij}|)^p \text{ mit } 1 \le p \le \infty, \ w_{ij} > 0, \ |e_{ij}| = \text{ euklid. Kantenlänge}$$

Für ein festes p nennen wir eine solche Grapheinbettung l_p–optimales Layout. Naheliegende Anwendungen ergeben sich bei Verteilungsproblemen z.B. bei der Höchstintegration: Faßt man V_c als die Pins auf dem Chiprand und V_{in} als die auf der Chipfläche zu plazierenden Elemente auf, so werden für $p = \infty$ die längsten Kanten minimiert, für $p = 2$ ein Energieminimum angestrebt und für $p = 1$ die Summe der Kantenlängen minimiert.

Grundlegende Untersuchungen bzgl. Existenz, Eigenschaften und Berechnung solcher l_p-optimalen Einbettungen wurden in den Arbeiten von Becker, Becker-Groh, Hotz und Osthof [2, 1, 3, 10] durchgeführt.

Überraschender Weise ergeben sich aber auch Anwendungen in anderen Bereichen. Man kann die Ergebnisse und Algorithmen für Heuristiken zur Generierung von Minimalflächen, euklidischen Steinerbäumen und zur Lösung von Flußproblemen einsetzen. Auf letzteres soll hier eingegangen werden.
Wir betrachten folgendes spezielle Flußproblem:

Gegeben sei ein ungerichteter Graph $G = (V, E)$ mit zwei ausgezeichneten Punkten Q und S. Weiterhin seien alle Kanten mit positiven Gewichten versehen. Gesucht wird der maximale Fluß zwischen Quelle Q und Senke S. Diese Aufgabe läßt sich auch als Schnittproblem formulieren: Trenne Q und S durch Aufschneiden von Kanten so, daß die Summe der Gewichte dieser Kanten minimal ist. Diese Äquivalenz folgt aus dem berühmten *MaxFlow-MinCut Theorem* [11].

Für dieses klassische Problem sind kombinatorische Algorithmen bekannt, die im worst case in Zeit $O\left(\frac{n^3}{\log n}\right)$ bzw. $O\left(e \cdot n \cdot \log \frac{n^2}{e}\right)$ arbeiten ($|V| = n$, $|E| = e$)[4, 5, 6]. Wir zeigen im folgenden einen nichtkombinatorischen Weg auf. Dazu geben wir eine andere Fassung der Problemstellung.

2 Eine äquivalente Problemstellung

Wir fassen Q und S als Punktmenge V_c mit den festen Koordinaten $x_0 = 0$ bzw. $x_{n+1} = 1$ auf. Die Gewichte der Kanten bezeichnen wir wieder mit w_{ij}, wenn $(v_i, v_j) \in E$ ist. Im folgenden betrachten wir nur das eindimensionale Problem, d.h. $y_i = 0$, $i = 0, \ldots, n + 1$. Aufgabe ist es, jedem Punkt v_i der Punktmenge V_{in} eine Koordinate x_i so zu geben, daß folgende Kosten eines l_1-optimalen Layouts minimiert werden:

$$\sum_{(v_i, v_j) \in E} w_{ij} \cdot |x_i - x_j|$$

Folgender Zusammenhang besteht zwischen einem solchen optimalen Layout und dem *MaxFlow-MinCut* Problem:

Satz 1 *Die Kosten eines l_1-optimalen Layouts dieses speziellen Graphen* mit festem Rand *entsprechen dem maximalen Fluß bzw. minimalen Schnitt des Graphen.*

Für den Beweis benötigen wir noch einige Vorbetrachtungen. Im folgenden betrachten wir nur noch diese speziellen Graphen mit dem festen Rand Q und S. Da offensichtlich in einem optimalen Layout alle Punkte zwischen Q und S liegen, werden wir die Begriffe 'links' und 'rechts' bzgl. eines Punktes verwenden.

Als erstes schauen wir uns die l_1-optimale Auslegung eines Punktes an, wobei die Koordinaten der Nachbarknoten fest sind. Wir können folgende einfache Beobachtung machen:

Lemma 1 *Ein Punkt liegt lokal l_1-optimal zu seinen Nachbarn, wenn die Summe der Gewichte der nach links ausgehenden Kanten gleich der Summe der Gewichte der nach rechts ausgehenden Kanten ist. Liegen Einheitsgewichte vor, so betrachten wir die Anzahl der ausgehenden Kanten.*

Ist die optimale Lage eindeutig, so ist sie identisch zu einem Nachbarknoten, andernfalls bewegt sie sich in einem Intervall zwischen zwei Nachbarknoten.

Beweis: Für die l_1-optimale Lage eines Punktes sind nicht die Längen der anliegenden Kanten, sondern die Anzahl der Kanten wesentlich, da die 'Kraft', die jede auf den Punkt ausübt, konstant ist. Dies gilt auch für die Gewichtung, da diese nur Einfluß auf die Konstante hat. Ließen wir nur Gewichte $w \in \mathbf{N}$ zu, so entspräche die Gewichtung nur einer Vervielfachung von Kanten.

Die Nachbarn des zu optimierenden Punktes seien in aufsteigender Reihenfolge ihrer Koordinaten indiziert: $x_1 \leq x_2 \leq \ldots \leq x_n$ Betrachten wir die Einheitsgewichtung, so suchen wir das mittlere Element:

$$x_{opt} \begin{cases} = & x_{\frac{n+1}{2}} & \text{wenn } n \text{ ungerade} \\ \in & [x_{\frac{n}{2}}, x_{\frac{n}{2}+1}] & \text{wenn } n \text{ gerade} \end{cases}$$

Nehmen wir unterschiedliche Kantengewichte, ist es das Element, das das Gesamtgewicht halbiert:

$$x_{opt} \begin{cases} = & x_k & \text{wenn } \left(\sum_{i=1}^{k} w_i > \sum_{i=k+1}^{n} w_i \right) \wedge \left(\sum_{i=1}^{k-1} w_i < \sum_{i=k}^{n} w_i \right) \\ \in & [x_k, x_{k+1}] & \text{wenn } \sum_{i=1}^{k} w_i = \sum_{i=k+1}^{n} w_i \end{cases}$$

$\blacksquare$

Mit diesem Merkmal der l_1-optimalen Lage eines Punktes, können wir eine spezielle Eigenschaft für die l_1-optimale Einbettung eines Graphen beweisen:

Lemma 2 *Zu jedem Graphen existiert eine l_1-optimale Einbettung, wobei die inneren Punkte entweder auf Q oder auf S liegen.*

Beweis: Wir führen einen Induktionsbeweis.

Mit Lemma 1 gilt die Aussage für die optimale Lage eines inneren Punktes. Entweder liegt er auf Q (x_0) oder S (x_2) oder jede Lage in dem Intervall $[x_0, x_2]$ ist optimal.

Haben wir ein optimales Layout mit $n + 1$ inneren Punkten, so nehmen wir uns einen beliebigen Punkt vor. Dieser liegt wieder mit Lemma 1 auf einem Nachbarn oder er befindet sich in einem Intervall zwischen zwei Nachbarn. Dann können wir ihn auf einen der Nachbarn verschieben. Verschmelzen wir ihn nun mit einem Nachbarn, d.h. die beiden Knoten werden zu einem zusammengefaßt und die Nachbarschaftslisten vereinigt, so ändert sich nichts an den Kosten, jedoch gibt es nur noch n Knoten. Für diesen gilt nach Induktionsannahme die Voraussetzung. Folglich haben wir auch die angestrebte Einbettung für unseren Graphen mit den $n + 1$ inneren Punkten. ∎

Kommen wir nun zum Beweis des Satzes.

Beweis: Wir beweisen die Äquivalenz zum minimalen Schnitt im Graphen. Der minimale gewichtete Schnitt sei mit $mc \in \mathbf{R}^+$ und die Kosten eines minimalen gewichteten l_1-optimalen Layouts seien mit $k_1 \in \mathbf{R}^+$ bezeichnet.

$k_1 \leq mc$ Gegeben sei also ein minimaler Schnitt durch den Graphen. Wir konstruieren dann folgende Einbettung des Graphen. Alle links vom Schnitt liegenden Knoten werden auf Q und alle rechts liegenden Knoten auf S gelegt. Die Kosten dieser Einbettung sind gleich der Summe der Gewichte der Schnittkanten, da die den Schnittkanten entsprechenden Kanten im Layout die Länge 1 haben und die sonstigen Kanten Nullkanten sind.

Eine l_1-optimale Einbettung hat nun kleinere oder dieselben Kosten wie diese Einbettung.

$mc \leq k_1$ Gegeben sei ein l_1-optimales Layout eines Graphen. Nach Lemma 2 gibt es dann auch eine l_1-optimale Einbettung, deren Knoten entweder auf Q oder S liegen.

Ein minimaler Schnitt durch den Graphen muß nun kleiner gleich dem Schnitt sein, der durch Auftrennen der Kanten ensteht, die von Q nach S in diesem l_1-optimalen Layout führen. ∎

Wir erhalten also mit einer gewichteten l_1-optimalen Auslegung den maximalen Fluß bzw. minimalen Schnitt. Wie kann nun eine solche Einbettung schnell berechnet werden?

In den oben zitierten Arbeiten kommen Einzelschrittverfahren zum Einsatz, die sukzessive alle Punkte lokal optimal auslegen und dadurch die optimalen Einbettungen approximieren. Daß dies im vorliegenden Fall unter Umständen zu suboptimalen

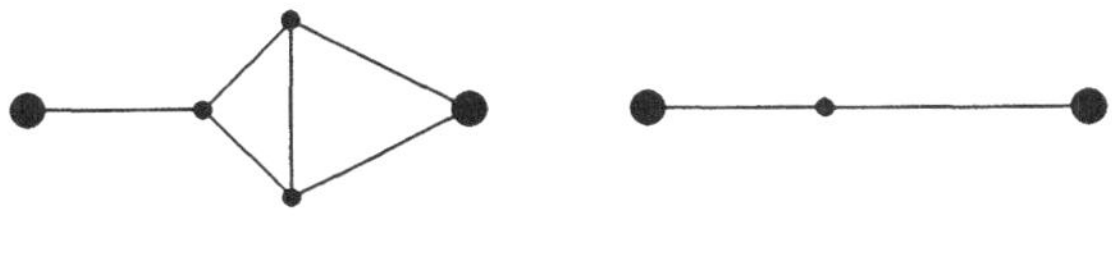

Abbildung 1:

Lösungen führen kann, soll an einem Beispiel aufgezeigt werden. Dazu benötigen wir noch folgende Definition:

Definition 1 *Wir bezeichnen eine Einbettung als lokal l_1-optimiert, wenn jeder Punkt der Einbettung lokal l_1-optimal bzgl. seiner Nachbarn liegt.*

In Abbildung 1 sehen wir einen Graphen mit drei inneren Knoten.

Liegen diese Punkte aufeinander wie in der rechten Figur, so liegt jeder Punkt lokal l_1-optimal, da jeder Knoten durch zwei Nachbarn in dieser Lage festgehalten und nur durch einen 'nach außen' gezogen wird. Diese Einbettung ist folglich lokal l_1-optimiert, jedoch nicht l_1-optimal. Denn im optimalen Layout liegen die inneren Knoten auf dem rechten Randpunkt. Da das Einzelschrittverfahren nur lokal l_1-optimierte Einbettungen approximieren kann, erhalten wir somit auch suboptimale Lösungen.

Wir können nun die optimale Einbettung über den Grenzwert der l_p-optimalen Einbettungen für $p \to 1+$ approximieren. Man kann aber auch versuchen, die lokalen Minima aufzulösen, in die sich das Einzelschrittverfahren verfängt. Im folgenden stellen wir eine Heuristik zur Approximation eines l_1-optimalen Layouts an, wodurch wir eine obere Schranke für die optimalen Kosten und damit eine obere Schranke für den maximalen Fluß erhalten.

3 Die Heuristik

Die Heuristik basiert darauf, daß die lokal optimierten Lagen dadurch entstehen, daß zueinander benachbarte Punkte bei der Optimierung aufeinander zu liegen kommen und danach lokal optimal eingebettet sind. Betrachtet man die Punkte jedoch als einen Knoten, so liegt dieser nicht mehr optimal, wie wir in Abbildung 1 gesehen haben. Vereinigt man in diesem Beispiel die Knoten und startet wieder das Einzelschrittverfahren, erhält man in diesem Fall das Optimum. Daß eine solche Strategie auch scheitern kann, zeigt Abbildung 2[1].

Liegen alle inneren Knoten wie in der unteren Figur aufeinander, so liegt jeder Knoten lokal optimiert, das Layout ist jedoch nicht optimal ausgelegt, da der minimale

---

[1]Der Einfachheit halber gehen wir bei den Beispielen immer von einer Einheitsgewichtung der Kanten aus.

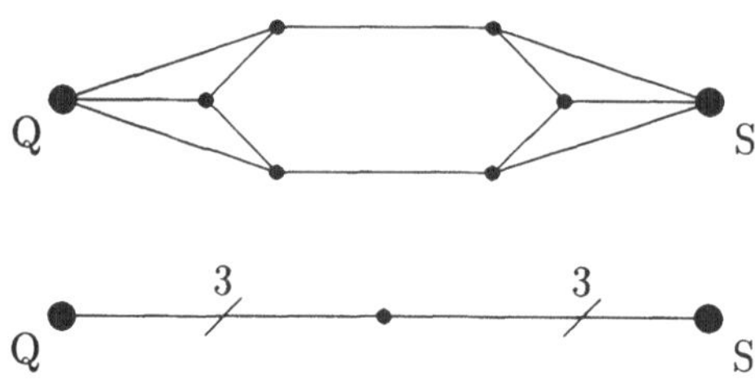

Abbildung 2:

Schnitt aus zwei Kanten besteht. Eine Vereinigung der Knoten führt folglich zu einem Fehler, da es dann drei kantendisjunkte Wege werden.

Wir können nun nicht ausschließen, daß eine solche Situation, in der strategiegemäß Punkte fälschlicherweise miteinander verschmolzen werden, im Laufe der Optimierung auftreten. Es hat sich jedoch in vielen Tests gezeigt, daß die Ausgangsplazierung für die Güte der Heuristiken eminent wichtig sind.

Wir wählen als Startkonfiguration für die Heuristiken die gewichtete l_2-optimale Einbettung der Punkte zwischen Q und S. In diesem ist die Summe der gewichteten Kantenquadrate minimiert.

Bevor wir auf die Heuristiken eingehen, wollen wir das Verfahren zur Bestimmung der l_1-optimalen Lage eines einzelnen Punktes angeben. In Lemma 1 ist die zu berechnende Lage angegeben. Da in den Heuristiken die Nachbarn nicht in sortierter Reihenfolge vorliegen werden, müssen diese sortiert werden und dann das (gewichtete) mittlere Element berechnet werden. Für eine sehr große Anzahl von Nachbarn lohnt es sich, den linearen Median-Algorithmus zu verwenden. Sind nämlich die Kanten mit Einheitsgewichten versehen, so haben wir gerade den Median [8] der Nachbarn zu bestimmen. Gewichtete Kanten können wir auch so interpretieren, daß die Nachbarn gewichtet sind. Dann genügt eine kleine Änderung im Algorithmus zur Bestimmung des *gewichteten Medians*.

Lemma 3 *Der gewichtete Median kann mit Hilfe des linearen Median Algorithmus berechnet werden.*

Beweis: Eine einfache Version verwendet wiederholt den einfachen Median zur Halbierung der Datenmenge.

Man bestimmt den Median $x_{\frac{n}{2}}$. Ist die Summe der Gewichte links des Medians größer, sucht man rekursiv in dieser um die Hälfte reduzierten Datenmenge weiter, ist sie kleiner, sucht man entsprechend in der rechten Datenmenge, andernfalls ist der gewichtete Median gefunden. Zu beachten ist, daß bei weiterer Suche in der rechten Datenmenge der Schwellenwert 'Gesamtgewichthalbe' um das Gewicht der linken Menge verringert wird.

Die Laufzeit ist offensichtlich linear.

Man kann den gewichteten Median aber auch direkt durch eine einfache Abänderung des Median Algorithmus erhalten.

Dieser reduziert seine Datenmenge dadurch, daß er zuerst in linearer Zeit den 'Median der Mediane' bestimmt und dann mit dessen Hilfe die Menge der Kandidaten für den Median bestimmt. Dazu werden die Elemente in zwei Mengen aufgeteilt: Die eine enthält die Elemente, die größer als der Median der Mediane sind, die andere die kleineren. Die Kandidatenmenge ist dann die bzgl. der *Anzahl der Elemente* größere Menge.

Rekursiv wird dann dieses Vorgehen auf diese Menge angewendet, um so schließlich den Median zu bestimmen.

Für den gewichteten Median richten wir uns einfach bei der Bestimmung der Kandidatenmenge nicht nach der *Anzahl der Elemente*, sondern nach der *Summe der Gewichte der Elemente*. ∎

Bemerkung: In einer Implementierung wird man den *randomisierten* Median Algorithmus vorziehen, da der 'normale' Algorithmus eine zu große Laufzeitkonstante aufweist. Dort wird der 'Median der Mediane' zufällig bestimmt und die Laufzeit ist dann im Mittel linear [8].

Wir diskutieren nun die Heuristik. Vorauszuschicken ist noch eine Vereinbarung bei dem lokalen Schritt des Einzelschrittverfahrens:

Ist die optimale Lage nicht eindeutig, setzen wir sie als linke Intervallgrenze fest.

Das bedeutet, daß sich die lokal optimale Lage eines Punktes immer auf einem Nachbarn befindet. Ein wichtiges Merkmal der Heuristik notieren wir in folgender Bemerkung:

Bemerkung: Die Heuristik endet damit, daß es keine inneren Knoten mehr gibt.

```
begin
    berechne die l₂-optimale Vorplazierung
    solange es noch einen inneren Knoten v gibt
        lege v l₁-optimal aus
        vereinige ihn mit dem Nachbarn
end
```

Satz 2 *Die Heuristik benötigt eine Laufzeit von $O(e \log e)$.*

Beweis: Offensichtlich terminiert dieses Verfahren.

Das Auffinden der lokal optimalen Lage eines Punktes mit der anschließenden Vereinigung mit dem entsprechenden Nachbarn stellt eine *Union-Find*-Operation dar. Es gibt nun mehrer effiziente Varianten zur Realisierung dieser Operation [8], die sich hauptsächlich in der Behandlung des *Trade-offs*, der bei der effizienten Realisierung der Unions und Finds existiert, unterscheiden.

Wir verwenden die Variante, die für m Finds und n Unions eine Zeit von $O(m + n \log n)$ benötigt. Diese ist am günstigsten, da wir in einem lokalen Schritt (Auffinden des Medians) alle Nachbarn eines Punktes durchmustern müssen, folglich Kanten innerhalb der Gesamtheuristik öfters durchmustern, was den *Finds* entspricht. Vereinigungen oder *Unions* von Knoten führen wir dagegen nur n-mal durch.

Als Datenstruktur verwenden wir eine Knotenliste und eine Kantenliste. Jeder Knoten einschließlich Q und S hat eine Liste von Zeigern auf seine zugehörigen Kanten und jede Kante zeigt auf ihre beiden Endpunkte. Außerdem speichert sie ihr Gewicht. Wichtig bei einem *Union* von zwei Knoten v_i und v_j ist, daß wir die Änderungen der Daten bei dem mit kleinerem Grad vornehmen; sei dies v_j. Wir hängen die Liste von Kantenzeigern von v_j an v_i. Wir durchmustern diese Kanten und lassen diese nun auf v_i anstatt auf v_j zeigen.

Wir können somit bei einer Medianberechnung immer direkt auf alle Nachbarn zugreifen, womit wir linear im Grad eines Knotens bei einer lokalen Operation bleiben.

Um die Laufzeit abzuschätzen, überlegen wir uns, daß wir jede Kante höchstens $O(\log e)$-mal betrachten. Wir nehmen bei den Unions immer die Änderungen an dem Knoten mit kleinerem Grad vor und zwar in linearer Zeit bzgl. des Grades. Nach der Union hat der Gesamtknoten dann mindestens doppelt so großen Grad. Da nach den n Unions der Grad nicht gößer als die Kantenmenge E sein kann, nimmt jede Kante höchstens $O(\log e)$-mal an einem Union teil.

Ist e_j der Grad des Knotens v_j, so finden wir die Nachbarknoten und den gewichteten Median v_i in Zeit $O(e_j)$ und bilden $Union(v_i, v_j)$ in Zeit $O(\min\{e_i, e_j\})$ Wir können wegen obiger Argumentation lokal abrechnen. Wir summieren über die lokalen Zeiten und multiplizieren mit $\log e$.

Für die Vorplazierung, bei der ein lineares Gleichungssystem gelöst werden muß, können wir Multigrid-Methoden einsetzen, die in Zeit $O(e)$ die Lösung approximieren [7, 12, 13]. ∎

Kommen wir zu den experimentellen Ergebnissen.

4   Die experimentellen Ergebnisse

Die Heuristik wurden außer an einigen vorgegebenen an 200.000 zufälligen Graphen mit zufälligen Kantengewichten getestet. Die Größe der Graphen reichte von 20 bis 100 Knoten und die Kantenanzahl von 100 bis 1000.[2]

Die Überprüfung der Resultate erfolgte mit dem Goldberg/Tarjan Algorithmus [5]. Dieser stand innerhalb des *LEDA*-Systems (Library of Efficient Data Structures and Algorithms) am Lehrstuhl von Kurt Mehlhorn zur Verfügung [9].

Bemerkenswert ist, daß die Ergebnisse in allen Fällen optimal waren. Bisher existiert auch kein Gegenbeispiel. Die Konstruktion eines solchen wird durch die Randbedingung, daß sie (gewichtet) l_2-optimal sein muß, erschwert. Da die Vorplazierung jedoch nur die optimale Einbettung approximiert, ist die Rechengenauigkeit ein 'Unsicherheitsfaktor', der jedoch noch nicht zum Tragen kam. Ob es ein Gegenbeispiel gibt, wenn man von einer optimalen Vorplazierung ausgeht, ist ein offenes Problem.

---

[2]'Zufällig' bedeutet, daß bei gegebener Knoten- und Kantenanzahl die Topologie zufällig gewählt wird.

Danksagungen

Die vorliegende Arbeit entstand im Rahmen des Sonderforschungsbereiches 124, Teilprojekt B1, am Lehrstuhl von Prof. Hotz und floß als eine Anwendungsmöglichkeit der l_p-optimalen Einbettungen in meine Dissertation ein.

Prof. Hotz gab mir die Möglichkeit, an seinem Lehrstuhl zu promovieren, was zuerst einmal zweierlei umfaßt: erstens die wirtschaftliche Grundlage, die er mir durch die Mitarbeit am Sonderforschungsbereich verschaffte, und zweitens seine wissenschaftliche Unterstützung bei der Entwicklung der Arbeit. Was ich aber als mindestens ebenso wichtig empfand, waren seine Geduld und Verständnis in dieser Zeit. Er war immer offen für neue Ideen und gab mir den notwendigen Freiraum, diese zu entwickeln.

Ihm danke ich sehr dafür.

Mein Dank gilt auch Elmar Schömer und Uwe Spanier, die zum Gelingen dieser Arbeit wesentlich beigetragen haben.

Literaturverzeichnis

[1] U. Becker-Groh. *Optimale Einbettung von Graphen mit festem Rand.* Master's thesis, Fachbereich Informatik, Universität des Saarlandes, Im Stadtwald, W–6600 Saarbrücken 11, FRG, 1983.

[2] B. Becker and G. Hotz. On the optimal layout of planar graphs with fixed boundary. *SIAM Journal on Computing*, 16(5), October 1987.

[3] B. Becker and H.G. Osthof. Layouts with wires of balanced length. *Information and Computation*, 73(1):45–58, April 1987.

[4] J. Cheriyan, T. Hagerup, and K. Mehlhorn. Can a maximum flow be computed in $o(nm)$ time ? In *Proceedings of the 17th International Colloqium on Automata, Languages and Programming (ICALP90)*, pages 235–248, Springer, 1990.

[5] A.V. Goldberg and R.E. Tarjan. A new approach to the maximum flow problem. In *Proceedings of the 18th Annual ACM Symposium on Theory of Computing*, pages 136–146, 1986.

[6] A.V. Goldberg, E. Tardos, and R.E. Tarjan. *Network Flow Algorithms.* Technical Report 860, School of Operations Research and Industrial Engineering, Cornell University, Ithaca, NY 14853-7501, September 1989.

[7] W. Hackbusch and eds. U. Trottenberg. Multigrid methods. In *Proceedings of the conference held at Köln-Porz at November, 23-27*, Springer Verlag, 1981.

[8] K. Mehlhorn. *Data Structures and Algorithms 1: Sorting and Searching. EATCS Monographs on Theoretical Computer Science*, Springer-Verlag, Berlin Heidelberg New York Tokio, 1984.

[9] K. Mehlhorn and St. Näher. A library of efficient data types and algorithms. In *Proceedings of the Annual Symposium on Mathematical Foundations of Computer Science (MFCS89)*, pages 88–106, Springer, 1989.

[10] H.G. Osthof. *Optimale Grapheinbettungen und ihre Anwendungen.* PhD thesis, Fachbereich Informatik, Universität des Saarlandes, Im Stadtwald, W–6600 Saarbrücken 11, FRG, 1990. 173 Seiten.

[11] C.H. Papadimitriou and K. Steiglitz. *Combinatorial Optimization: Algorithms and Complexity.* Prentice Hall, Inc., Englewood Cliffs, New Jersey, 1982.

[12] J. Ruge and K. Stüben. *Efficient solution of finite difference and finite elemente equation by Algebraic Multigrid (AMG).* Technical Report 89, GMD, Bonn.

[13] K. Stüben. *Algebraic Multigrid (AMG): Experiences and Comparisons.* Technical Report 23, GMD, Bonn, March 1983.

Computation of the Boolean Matrix–Vector, AND/OR–Produkt in Average Time O(m + nlnn)

C.P. Schnorr

Universität Frankfurt
Fachbereich Mathematik/Informatik
Germany

Abstract

Let $\mathbf{A} = [a_{i,j}]_{1 \leq i \leq n}$ be a boolean matrix and let $\mathbf{b} = (b_1, \ldots, b_m)^\mathsf{T}$ be a boolean vector. The boolean AND/OR–product $\mathbf{c} = \mathbf{A}\mathbf{b}$, $\mathbf{c} = (c_1, \ldots, c_n)^\mathsf{T}$ is given by $c_i = \bigvee_{j=1,\ldots,m} (a_{i,j} \wedge b_j)$ for $i = 1, \ldots, m$. We consider random boolean $n \times m$ matrices $\mathbf{A}$ where the probability for $\mathbf{A}$ is preserved by arbitrary permutations of the elements in each column. We describe an algorithm that first preprocesses the matrix $\mathbf{A}$ using $O(nm)$ steps and then computes the product $\mathbf{A}\mathbf{b}$ using an expected number of $O(m + n \ln n)$ steps. As a consequence the boolean matrix product $\mathbf{C} = \mathbf{A}\mathbf{B}$ of the random boolean $n \times m$–matrix $\mathbf{A}$ and an arbitrary boolean $m \times k$ matrix $\mathbf{B}$ can be computed in average time $O(nm + km + kn \ln n)$.

1   Introduction

The boolean matrix AND/OR–product is closely related to the problem of computing the transitive closure of a digraph. Let $\mathbf{A}$ be the adjacency matrix of a digraph then $\bigvee_{i \geq 0} \mathbf{A}^i$ is the adjacency matrix of the transitive closure graph. The transitive closure of a digraph can be computed in linear expected time, see Schnorr (1978), Karp (1990). This time bound holds for all graph distributions where the probability only depends on the number of edges and the number of vertices of the graph.

A similar time bound for the boolean matrix AND/OR–product is not known. For the particular case that all boolean $n \times n$ matrices have the same probability P.E.O'Neil and E.J.O'Neil (1973) proposed an algorithm with expected time $O(n^2)$. But their algorithm may have expected time $O(n^{2.5})$ for matrix distributions for which the matrix entries are mutually independent and the probability of a 1–entry is $1/\sqrt{n}$. Our results hold for an arbitrary probability distribution on boolean $n \times m$–matrices with the property that the probability of the matrix $\mathbf{A}$ only depends on the vector $(\#_1, \ldots, \#_m)$ where $\#_i$ is the number of 1–entries in column i of $\mathbf{A}$. Equivalently the probability of matrix $\mathbf{A}$ is preserved by arbitrary permutations of the entries of each column. For these distributions the matrix entries in distinct columns are mutually independent but entries in the same column may depend on each other.

We compute the boolean matrix AND/OR–product of a random boolean $n \times m$–matrix $\mathbf{A}$ for such a distribution with an arbitrary fixed boolean $m \times k$–matrix $\mathbf{B}$ in average time $O(nm + km + kn \ln n)$. Our time analysis is valid for an implementation of the algorithm on storage modification machines as proposed by Schönhage. These machines are equivalent to RAM machines with unit costs and addition/subtraction by 1. Our algorithm closely follows the transitive closure algorithm in Schnorr (1978). We use a counting variable to determine the point where the algorithm switches from working with column lists to working with row lists.

2   Algorithm for the boolean matrix–vector AND/OR–product

```
INPUT      A = [a_{i,j}]_{1≤i≤n, 1≤j≤m} ∈ M_{n,m}({0,1})
           b = (b_1,...,b_m)^T ∈ M_{m,1}({0,1})
```

1. (preprocessing the matrix $\mathbf{A}$)

   set up ordered lists for

   $A_j^c := \{i \mid a_{i,j} = 1\}$ for $j = 1, \ldots, m$,

   $A_i^r := \{j \mid a_{i,j} = 1\}$ for $i = 1, \ldots, n$.

2. $z := 0$, $c_1 := c_2 := \cdots := c_n = 0$.

   ```
   FOR  j = 1,...,n  DO
   ```

```
      IF  b_j = 1  THEN DO
      FOR  all  i ∈ A_j^c  DO  [c_i := 1,  z := z + 1]
      END  for i  END if
      IF  z > n ln n  THEN GOTO  4.
   END  for j
```

3. OUTPUT $(c_1, \ldots, c_n)$ and terminate.

4. FOR all i with $c_i = 0$ DO
   FOR all $j \in A_i^r$ DO
   IF $a_{i,j} = 1$ and $b_j = 1$ THEN $c_i := 1$

OUTPUT $\mathbf{c} = (c_1, \ldots, c_n)^\top$.

3 Correctness of the algorithm

We show that the algorithm correctly computes the AND/OR–product $\mathbf{Ab}$. If the algorithm terminates in Step 3 then all the column lists A_j^c representing the columns j with $b_j = 1$ have been exhausted. Then we have $\mathbf{c} = \mathbf{Ab}$. If the algorithm passes Step 4 then all row lists A_i^r with $c_i = 0$ are checked in Step 4 on whether there exists j with $a_{i,j} \wedge b_j = 1$.

4 Time analysis of the algorithm

Theorem 1 *Let $\mathbf{A}$ be a random boolean $n \times m$–matrix for a probability distribution having the property that the probability for $\mathbf{A}$ is preserved by arbitrary permutations of the entries in each column. Then the preprocessing step 1 has time bound $O(nm)$ and Steps 2 – 4 run in expected time $O(m + n \ln n)$.*

Proof: The claim about the preprocessing step is trivial. If the algorithm terminates in Step 3 then Steps 2 – 4 obviously run in expected time $O(n \ln n)$. It remains to consider the case that the algorithm enters Step 4. Upon entry of Step 4 we have that $z < n + n \ln n$. This implies that Step 2 has expected time $O(n \ln n)$.

We next show that upon entry of Step 4 the expected value of $\#\{i \,|\, c_i = 0\}$ is at most 1. This implies that Step 4 runs in expected time $O(m)$ and thus proves the theorem. For every fixed i we have

$$\text{prob}\,[i \notin A_j^c] \;=\; (1 - \frac{1}{n})(1 - \frac{1}{n-1}) \cdots (1 - \frac{1}{n - \#A_j^c})$$

$$\leq\; (1 - \frac{1}{n})^{\#A_j^c}.$$

Therefore the expected value of $\#\{i \mid c_i = 0\}$ is at most

$$n(1 - \frac{1}{n})^z \leq n(1 - \frac{1}{n})^{n \ln n} \leq ne^{-\ln n} = 1.$$

∎

5 Computing the boolean matrix AND/OR–product AB

In order to compute the boolean matrix AND/OR–product **AB** we preprocess the matrix **A** as in Step 1 and we pass Step 2 – 4 for every column vector of matrix **B**. Then Theorem 1 immediately implies the following theorem.

Theorem 2 *Let* **A** *be a random boolean* $n \times m$–*matrix for a probability distribution having the property that the probability for* **A** *is preserved by arbitrary permutations of the entries in each column. Then the AND/OR–product* **AB** *with an arbitrary boolean* $m \times k$–*matrix can be computed in expected time* $O(mn + mk + kn \ln n)$.

Bibliography

[1] R.M. Karp: The Transitive Closure of a Random Digraph. Random Structures and Algorithms. Vol. 1 (1990), pp. 73 – 94.

[2] P.E. O'Neil and E.J. O'Neil: A Fast Expected Time Algorithm for Boolean Matrix Multiplication and Transitive Closure. Combinatorial Algorithms, Ed.: R. Rustin. Algorithmic Press, New York 1973, pp. 59 – 68.

[3] C.P. Schnorr: An Algorithm for Transitive Closure with Linear Expected Time. Siam J. Computing 7 (1978) pp. 127 – 133.

Durch kinematische Szenen erzeugte topologische Räume

Jürgen Sellen

Fachbereich 14 – Informatik
Universität des Saarlandes
Germany

Zusammenfassung

Wir untersuchen die topologische Struktur des Freiraums mechanischer Gebilde und gehen auf die Bedeutung der Fragestellung für den Bereich des qualitativen Argumentierens ein. Topologische Invarianten wie die Fundamentalgruppe sind zwar berechenbar, können jedoch nicht zur Klassifikation beliebiger Mechanismen herangezogen werden. Wir zeigen, daß jede endlich präsentierbare Gruppe als Fundamentalgruppe einer real nachbaubaren, ebenen Mechanik auftreten kann und gehen auf die daraus resultierenden Nichtentscheidbarkeits-Aussagen bzgl. Interpretierbarkeit der Fundamentalgruppe, Homotopie von Wegen im Freiraum und Klassifizierbarkeit des Freiraums nach seinem Homotopietyp ein.

1   Motivation

Im Zuge der zunehmenden Automatisierung der Industriegesellschaft, bei der längst nicht mehr nur stupide Handgriffe von Robotern ausgeführt werden, sondern ganze Fabrikationsprozesse ohne menschliches Zutun ablaufen, hat auch die Robotik ständig an Bedeutung gewonnen. Aktuelle Fragestellungen betreffen dabei verschiedene Gebiete der Informatik, wobei die Künstliche Intelligenz immer häufiger in den Vordergrund tritt : die "denkende Fabrik" der Zukunft sollte z.B. auf unvorhergesehene Ereignisse reagieren und Fehler im Produktionsablauf diagnostizieren können.

Arbeiten aus dem Bereich des qualitativen Argumentierens befassen sich mit dem Problem der Beschreibung komplizierter physikalischer Vorgänge in möglichst einfacher Form, so daß wesentliche Eigenschaften durch die Modellierung nicht verloren gehen. Die qualitative Kinematik beschäftigt sich mit der Darstellung kinematischer Zusammenhänge zwischen Teilen eines Mechanismus. Die existierenden Arbeiten auf diesem Gebiet verfolgen meist das Ziel, mittels einer adäquaten Darstellung der erlaubten Konfigurationen mögliche Bewegungsabläufe zu bestimmen bzw. vorherzusagen (vgl. [3]). Dabei beschränkt man sich auf die Untersuchung einfacher kinematischer Ketten in einem zugrundeliegenden 2–dimensionalen physikalischen Raum.

Von zentraler Bedeutung für den Themenkreis der Robotik ist der Begriff des Konfigurationsraums. Unter dem Konfigurationsraum einer kinematischen Szene versteht man den Raum, der durch sämtliche Parameter aufgespannt wird, welche die Lage der beweglichen Teile der Szene umkehrbar eindeutig beschreiben. Zu einem festen Zeitpunkt wird die Position aller Teile der Szene also durch einen Punkt des Konfigurationsraums beschrieben, den man entsprechend als Konfiguration bezeichnet.

Wählen wir als Beispiel zwei drehbare Zahnräder in der Ebene, so ist der Konfigurationsraum dieser Szene gerade ein Torus:

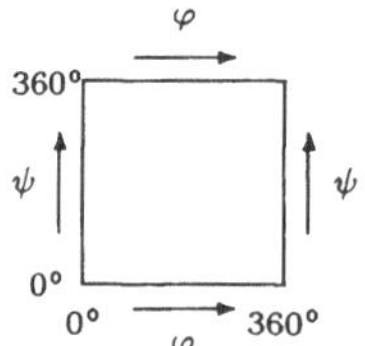

Die Drehwinkel φ und ψ zweier Zahnräder spannen ein Rechteck auf. Durch Identifizierung gegenüberliegender Seiten entsteht ein Torus.

Da die Teile einer Szene sich in ihrer Bewegungsfreiheit gegenseitig behindern können, werden die Konfigurationen unterteilt in erlaubte und nicht erlaubte, d.h. physikalisch sinnlose Konfigurationen. Entsprechend kann der Konfigurationsraum unterteilt werden in einen sogenannten Freiraum und einen blockierten Teil.

Greifen in unserem Beispiel die Zahnräder so ineinander, daß die Drehung eines Rades eine Drehung des anderen erzwingt, so besteht der Freiraum aus mehreren "Bändern", die sich spiralförmig um den Torus winden. Betrachten wir die Zusammenhangskomponente, in der sich ein gegebener Anfangszustand befindet, so handelt es sich bei dem Freiraum also im wesentlichen um eine Kreislinie, die einer Volldrehung beider Zahnräder entspricht.

Die Grenze zwischen Freiraum und blockiertem Teil wird durch diejenigen Konfigurationen gebildet, in denen sich mindestens 2 Objekte der Szene gegenseitig berühren. Die dabei ausgeübten Bewegungszwänge bilden die Grundlage der Arbeit [3].

Die auf einer reinen Simulation von Bewegungsabläufen basierende Methode des qualitativen Argumentierens eignet sich zur rechnergestützten Analyse einfacher kinematischer Ketten. Stellen wir uns hingegen als Aufgabe, die Funktionsweise eines komplexeren Mechanismus zu bestimmen, ohne vorherige Kenntnisse seine Arbeitsweise betreffend zu besitzen, so müssen wir allgemeinere Methoden entwickeln.

Zur Illustration dieser Überlegungen betrachten wir ein konkretes Beispiel. Wir nehmen also an, die Konstruktionsskizze einer Maschine wie beispielsweise eines Automotors vorliegen zu haben. Wissen wir, daß die Kurbelwelle als zentrales Teil des Motors die Stellung aller anderen Motorteile bestimmt, so können wir durch Simulation einer durch die Drehung der Kurbelwelle erzeugten Bewegung ein qualitatives Bild von der Arbeitsweise des Motors gewinnen. Ist uns die Bedeutung der Kurbelwelle jedoch unbekannt, fehlt uns ein Ansatzpunkt für die Simulation : die Bewegung beispielsweise eines einzelnen Ventils reicht nicht aus, um den Motor zu charakterisieren.
Auch kann es sein, daß die Bewegungsmöglichkeiten von mehreren Parametern abhängen. Dies ist beim Autoantrieb z.B. dann der Fall, wenn wir durch Gangschaltung und Kupplung Einfluß auf die Funktion des Motors ausüben können. Als einfachstes Beispiel einer Kupplung können wir uns zwei Zahnräder vorstellen, die ineinander geschoben werden können. Die Topologie des Freiraums liefert uns 3 qualitativ unterschiedliche Verschiebungsbereiche:

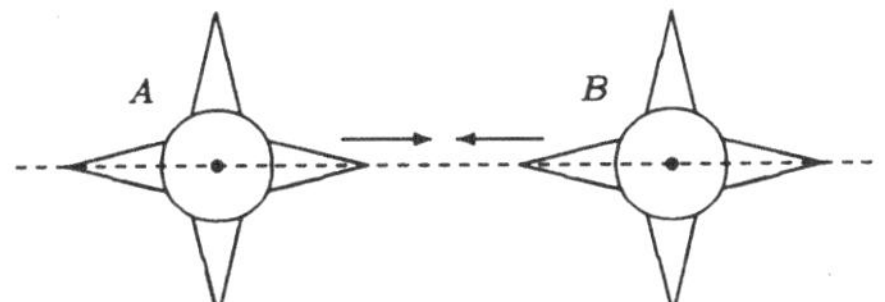

Kann es zu keiner Kollision der Zahnräder kommen, so ist der Freiraum ein Torus. Bei einer vollständigen "Kopplung" entspricht der Freiraum einer Kreislinie.

Zwischen diesen Verschiebungsbereichen gibt es Stellungen, so daß die Zähne zwar miteinander kollidieren können, eine Drehung des einen Rades jedoch nicht eine Drehung des anderen erzwingt. Der Freiraum entspricht in diesem Fall einem Torus mit "Löchern", die durch Kollision der Zahnspitzen entstehen. Diese Situation können wir als "Schleifen" oder "Reiben" der Kupplung interpretieren.

Wir erkennen also, daß die "ungefähre" Gestalt des Freiraums wesentliche Aussagen über die Bewegungsmöglichkeiten und damit die Funktion einer Maschine erlaubt. Die Frage nach einer Klassifikation des Freiraums führt in das Gebiet der Topologie. Nachdem wir in einer kurzen mathematischen Einleitung auf Standardcharakterisierungen wie Homotopietyp, Homöomorphietyp und Fundamentalgruppe eingegangen sind, werden wir Fragen nach Berechenbarkeit, Komplexität und Auswertbarkeit topologischer Eigenschaften behandeln.

2 Mathematische Grundlagen

Die mechanischen Szenen, die wir betrachten wollen, bestehen aus starren Objekten in einem zugrundeliegenden d–dimensionalen euklidischen Raum. Die Objekte besitzen eine Beschreibung als semialgebraische, zusammenhängende Teilmengen des $\mathbf{R}^d$, gegeben durch boolesche Ausdrücke über polynomiellen Ungleichungen der Form $p(x_1, \ldots, x_d) > 0$ oder $p(x_1, \ldots, x_d) \geq 0$ mit ganzzahligen Koeffizienten. Zur Vereinfachung späterer Ausführungen erlauben wir also insbesondere, daß Objekte keine abgeschlossenen Punktmengen sein müssen, sondern "offene" und "abgeschlossene" Seiten besitzen dürfen.

Als mögliche Bewegungen eines Objekts betrachten wir die euklidische Gruppe der starren Transformationen von $\mathbf{R}^d$ bzw. Untergruppen hiervon. Sind für ein Objekt keine Bewegungen erlaubt, sprechen wir auch von einem Hindernis. Eine Kollektion von derart definierten Objekten bezeichnen wir als *kinematische Szene $\mathcal{S}$*.

Der *Konfigurationsraum $\mathcal{K}(\mathcal{S})$* einer Szene $\mathcal{S}$ mit m gegebenen Objekten entspricht also dem Raum $(\mathbf{R}^d \times SO(d))^m$ bzw. einer einfachen Teilmenge davon und kann über die Einträge der Rotationsmatrizen aus der orthogonalen Gruppe $SO(d)$ als semialgebraische Teilmannigfaltigkeit in den $\mathbf{R}^{m(d+d^2)}$ eingebettet werden.

Der *Freiraum $\mathcal{F}(\mathcal{S})$* besteht aus denjenigen Konfigurationen der Szene, in denen sich die Objekte nicht schneiden, d.h. paarweise disjunkte Punktmengen des $\mathbf{R}^d$ überdecken (wir benutzen diese Bezeichnung vereinfachend auch dann, wenn wir uns nur auf eine Zusammenhangskomponente beziehen). Damit ist auch der Freiraum ein semialgebraischer Teilraum des $\mathbf{R}^n$.

Da im $\mathbf{R}^n$ alle Topologien, die auf einer von einer Norm induzierten Metrik basieren, gleich sind, können wir uns bei unseren Betrachtungen also auf den $\mathbf{R}^n$ mit der von der euklidischen Metrik erzeugten Topologie zurückziehen.
Für diesen Spezialfall topologischer Räume erinnern wir an einige häufig benötigte Grundbegriffe.

Definition 1 *Zwei topologische Räume X und Y heißen* homöomorph, *in Zeichen $X \approx Y$, falls es eine bijektive stetige Abbildung (Homöomorphismus) zwischen ihnen gibt, so daß auch die Umkehrfunktion stetig ist.*

Wenn man also einen Teilraum des $\mathbf{R}^n$ verschiebt, dreht, spiegelt, streckt oder staucht, erhält man einen homöomorphen Teilraum. Dies gilt jedoch nicht, wenn man einen Raum auf einen niedrigerdimensionalen Teilraum deformiert. Der Begriff der Homotopie liefert eine schwächere Einteilung topologischer Räume in Klassen:

Definition 2 *Zwei Räume X und Y heißen* homotopieäquivalent *oder vom gleichen* Homotopietyp, *wenn es einen Raum Z gibt, der X und Y als strenge Deformationsretrakte enthält.*

Sind die Freiräume zweier mechanischer Szenen homöomorph, so können wir anhand des Homöomorphismus Bewegungen des einen Mechanismus eindeutig in Bewegungen des anderen übersetzen und umgekehrt. Planungsaufgaben lassen sich so von einer Maschine auf eine andere übertragen. Eng verbunden mit der Frage nach dem Homöomorphietyp oder dem Homotopietyp ist also auch die Frage, ob wir zu jedem Mechanismus eine einfache Normalform konstruieren können, so daß es möglich ist, sich bei der Lösung von Problemen auf diese Normalform zurückzuziehen.

Die Frage, wie man unterschiedliche Bewegungsmöglichkeiten eines Mechanismus charakterisieren bzw. allgemein Wege im Freiraum klassifizieren kann, führt auf den Begriff der Fundamentalgruppe, die gleichzeitig eine nützliche Invariante zur Lösung des Homöomorphie- und des Homotopieproblems darstellt.

Wir betrachten die Menge $\Omega(x)$ der Schleifen an einem festen Punkt x mit der Komposition, d.h. dem "Aneinanderhängen" von Wegen, als Verknüpfung. Sehen wir zwei Wege α und β als äquivalent an, wenn sie stetig ineinander deformierbar (*homotop*) sind, so bildet der Quotient $\Omega(X)/\sim$ eine Gruppe:

Definition 3 *Die Gruppe* $\pi_1(X, x) := \Omega(x)/\sim$ *heißt* Fundamentalgruppe *oder* 1.Homotopiegruppe *von X zum Basispunkt x.*

Ist X wegzusammenhängend, so ist die Fundamentalgruppe unabhängig vom Basispunkt. Sprechen wir von der Fundamentalgruppe $\pi_1(\mathcal{S})$ einer kinematischen Szene $\mathcal{S}$, so meinen wir die Fundamentalgruppe der Zusammenhangskomponente ihres Freiraums, in der der implizit gegebene Anfangszustand liegt.

Jede Gruppe G ist homomorphes Bild einer freien Gruppe F und kann daher mit einem Quotienten F/N identifiziert werden. Ist X ein freies Erzeugendensystem von F und N die normale Hülle einer Menge $R \subseteq F$, so nennen wir das Paar $\mu = (X, R)$ eine *Präsentation* von $G = G(\mu)$ oder auch eine Darstellung von G durch Erzeugende und Relationen. Sind X und R beide endlich wählbar, so heißt G *endlich präsentierbar*. Die Elemente von R definieren dabei die Relationen in F/N, denn dem Element $x_1^{\epsilon_1} \cdot \ldots \cdot x_k^{\epsilon_k}$ aus R entspricht die Relation $x_1^{\epsilon_1} \cdot \ldots \cdot x_k^{\epsilon_k} = 1$ in F/N. Entsprechend schreiben wir (X, R) anschaulicher oft in der Form $< x_1, \ldots, x_m | r_1, \ldots, r_n >$, wobei die x_i die Elemente des freien Erzeugendensystems und die r_i die durch R definierten Relationen darstellen.

Die Erzeugenden der Fundamentalgruppe bieten eine Übersicht über die wesentlich unterschiedlichen zyklischen Bewegungen eines Mechanismus, Relationen beschreiben den Grad der gegenseitigen Störung der Einzelteile.

Die Räume, die als Freiräume kinematischer Szenen auftreten können, haben die angenehme Eigenschaft, daß sie in überschaubarer Weise aus einfachen Teilen zusammengesetzt sind:

Definition 4 *Ein topologischer Raum heißt* k–Zelle, *wenn er zu* $\mathbf{R}^k$ *homöomorph ist. Unter einer* Zellenzerlegung *eines Raumes* $X \subseteq \mathbf{R}^n$ *verstehen wir eine endliche Menge* $\mathcal{Z}$ *von Zellen, so daß* X *gerade die disjunkte Vereinigung der Zellen aus* $\mathcal{Z}$ *ist. Sind die Zellen semialgebraische Mengen, so heißt die Zellenzerlegung algebraisch.*
Der Teilraum $X^k = \bigcup\{e \in \mathcal{Z} \; ; \; dim(e) \leq k\}$ *heißt* k–dimensionales Gerüst *von* X.
Gibt es zu jeder k–Zelle $e \in \mathcal{Z}$ *eine stetige Abbildung* $\varphi_e : [0, 1]^k \to X$, charakteristische Abbildung *genannt, welche* $(0, 1)^k$ *homöomorph auf* e *und den Rand* $\partial[0, 1]^k$ *in das* (k–1)–*dimensionale Gerüst* X^{k-1} *abbildet, so heißt* X *auch* endlicher CW–Komplex.

Um zu entscheiden, ob es einen Weg zwischen zwei Punkten eines CW–Komplexes gibt, genügt es, das 1–Gerüst zu kennen, da ein CW–Komplex X genau dann wegzusammenhängend ist, wenn dies auch für sein 1–Gerüst X^1 gilt. Ähnlich hierzu wird die Fundamentalgruppe eines CW–Komplexes eindeutig durch dessen 2–Gerüst bestimmt.

Diese Eigenschaft ausnutzend werden wir im nächsten Abschnitt einen Algorithmus angeben, der zu einer gegebenen kinematischen Szene eine endliche Präsentation ihrer Fundamentalgruppe bestimmt.

3　Berechnung von Fundamentalgruppen

Der erste Schritt zur Berechnung der Fundamentalgruppe einer beliebigen kinematischen Szene wie auch zur Lösung des allgemeinen Motion Planning Problems besteht in der Berechnung einer algebraischen Zellenzerlegung des durch polynomielle Ungleichungen beschriebenen Freiraums. Hierzu kann der klassische Algorithmus von Collins verwendet werden, der durch sukzessive Projektionen eine "zylindrische" Zellenzerlegung generiert :

Definition 5 *Eine Zellenzerlegung $\mathcal{Z}$ heißt invariant unter einer Menge P von Polynomen, wenn für jedes Polynom $p \in P$ und jede Zelle $C \in \mathcal{Z}$ das Vorzeichen $signum(p(x)) \in \{-1, 0, +1\}$ konstant ist für alle Punkte $x \in C$.*

Satz 1 (G.E. Collins) *Es gibt einen Algorithmus, der zu einer gegebenen Menge P von Polynomen über n Unbestimmten und mit ganzzahligen Koeffizienten eine P-invariante algebraische Zellenzerlegung $\mathcal{Z}$ von $\mathbf{R}^n$ konstruiert. Der Algorithmus liefert zu jeder Zelle $C \in \mathcal{Z}$ einen Testpunkt $p(C) \in C$ mit algebraischen Koordinaten. Die Laufzeit ist doppelt exponentiell in n, genauer $O(m^{2^{O(n)}})$, wobei m die Komplexität von P angibt.*

Über die Testpunkte können wir die zum Freiraum der Szene gehörige Zellenzerlegung $\mathcal{Z}' \subseteq \mathcal{Z}$ bestimmen, die i.a. jedoch weder abgeschlossen noch beschränkt ist. Durch eine Modifikation des durch $\mathcal{Z}'$ definierten Raumes innerhalb seines Homotopietyps läßt sich dies jedoch ohne Mühe erreichen. Damit erhalten wir einen zum Freiraum der Szene homotopieäquivalenten CW–Komplex X.

Mit dem Verfahren von Poincaré können wir die Fundamentalgruppe dieses Komplexes und damit der Szene bestimmen:

1. Fasse das 1–Gerüst X^1 als Graph auf und bestimme einen aufspannenden Baum B dieses Graphen. Orientiere die Kanten aus $X^1 \setminus B$ ("Sehnen") beliebig und markiere sie mit Symbolen $a_1, \ldots, a_k$, den Erzeugenden der zu bestimmenden Fundamentalgruppe.

2. Führe für jede Fläche F des 2–Gerüsts X^2 folgende Aktionen durch :
   Es sei γ ein beliebiger geschlossener Weg auf ∂F um F herum. Es seien

$a_{i_1}, \ldots, a_{i_k}$ in dieser Reihenfolge die Markierungen der von γ betretenen Sehnen; $\varepsilon_l = \pm 1$ gebe die Richtung an, in der die Sehne a_l bzgl. ihrer Orientierung "begangen" wurde.

Dann definiert $\gamma_F = a_{i_1}^{\varepsilon_1} \cdot \ldots \cdot a_{i_k}^{\varepsilon_k}$ eine Relation der zu bestimmenden Fundamentalgruppe.

Die Fundamentalgruppe ergibt sich also zu $\pi_1(\mathcal{S}) = \pi_1(X) = <\ a_1, \ldots, a_k \mid \gamma_F = 1 \ \forall F \in X^2 >$, und wir erhalten:

Satz 2 *Es gibt einen Algorithmus, der eine endliche Präsentation der Fundamentalgruppe $\pi_1(\mathcal{S})$ einer kinematischen Szene $\mathcal{S}$ zu bestimmen vermag. Die Laufzeit ist dabei doppelt exponentiell in der Dimension des Freiraums.*

Die hohe Laufzeit des Algorithmus legt die Frage nach der möglichen Komplexität der Fundamentalgruppe nahe: wieviele Erzeugende kann die Fundamentalgruppe einer Szene in Abhängigkeit von ihrer Komplexität besitzen?

4 Komplexität

In diesem Abschnitt wollen wir ebene Szenen aus translatorisch verschiebbaren Rechtecken und rechtwinkligen Hindernissen untersuchen. Zunächst fragen wir nach dem Aussehen der Fundamentalgruppe, wenn die Szene aus n verschiebbaren Quadraten in der unbeschränkten Ebene besteht.

Betrachten wir drei Quadrate A,B und C, so ist der Freiraum homotopieäquivalent zum $\mathbf{R}^4$, aus dem 3 sich in einem Punkt schneidende Ebenen entfernt wurden. Wir erhalten 3 Erzeugende, die wir als Umkreisung von C durch A, von C durch B und von B durch A bei jeweils fester Position der unbeteiligten 2 Objekte interpretieren können. In diesem Fall treten bereits Relationen auf, wie die folgende Skizze verdeutlicht :

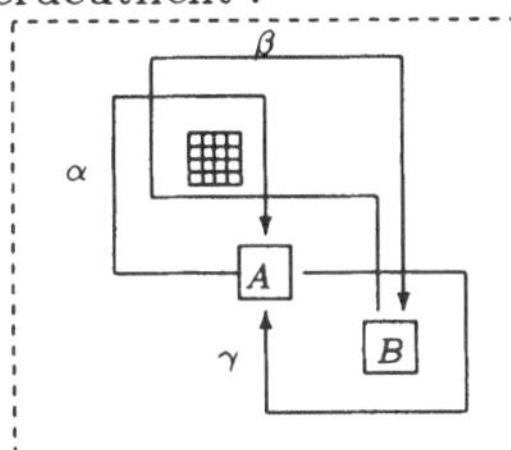

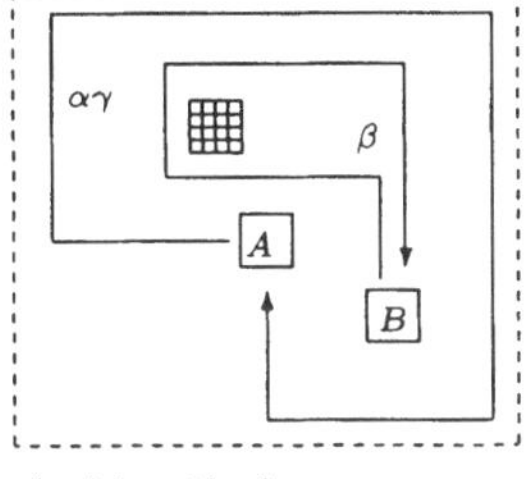

Mit Hilfe dieser geometrischen Interpretation kann man über den Satz von Seifert/van Kampen zeigen, daß die Fundamentalgruppe bei 3 Objekten die Darstellung

$$(\alpha\gamma)\beta = \beta(\alpha\gamma)$$

$\pi_1(\mathcal{S}) =<\ \alpha, \beta, \gamma \mid (\alpha\gamma)\beta = \beta(\alpha\gamma), \ (\gamma\beta)\alpha = \alpha(\gamma\beta) >$ besitzt.

Im Falle n beweglicher Quadrate erhalten wir $\frac{1}{2}n(n-1)$ Erzeugende und ähnliche "Vertauschbarkeitsrelationen".

Was passiert nun mit der Fundamentalgruppe, wenn wir einige der Quadrate zu Hindernissen machen ? Im Freiraum geht man beim "Befestigen" eines Quadrates auf einen linearen Teilraum über, dessen Dimensionszahl um 2 erniedrigt ist. Die folgende

Szene ist ein Beispiel dafür, daß sich die Anzahl der Erzeugenden der Fundamentalgruppe bei solchen Übergängen sukzessive in jedem Schritt mehr als verdoppeln kann
(dick gezogene Linien entsprechen Hindernissen) :

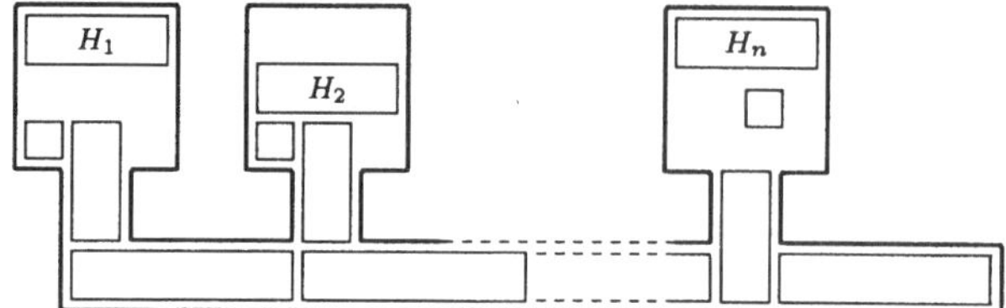

Befestigen wir k der Objekte $H_1, \ldots, H_n$ in ihrer unteren Position, so ist der
entstehende Freiraum homotopieäquivalent zu dem 1–Gerüst eines n–dimensionalen
Würfels. Nach der Eulerschen Formel ist die Fundamentalgruppe eines zusammen-
hängenden Graphen mit α_0 Ecken und α_1 Kanten die freie Gruppe mit $\alpha_1 - \alpha_0 + 1$
Erzeugenden. In unserem Fall erhalten wir damit $k2^{k-1} - 2^k + 1 = (k - 2)2^{k-1} + 1$
Erzeugende für $k > 0$. In den Fällen $k = 0$ und $k = 1$ ist der Freiraum einfach zu-
sammenhängend, d.h. alle Bewegungen sind ineinander deformierbar. Ein genauerer
Beweis kann in ähnlicher Weise wie in unserer nächsten Fallstudie erbracht werden.

Die größte denkbare Einschränkung des Bewegungsspielraums von n Quadraten
liefert eine Szene, die dem "Schiebe-Puzzle" entspricht :

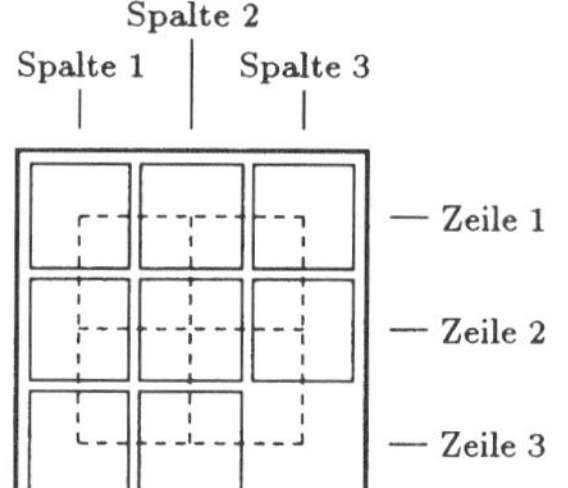

Die Szene $\mathcal{S}_{(m)}$ besteht aus $m^2 - 1$ Quadraten der Kantenlänge 1, die innerhalb eines
großen Quadrates mit Kantenlänge m angeordnet sind, so daß also jedes Quadrat eine
Stellung auf einem $m \times m$–Gitter einnehmen
muß.

Diejenigen Stellungen, bei denen sich jedes der $m^2 - 1$ Quadrate auf einem der
Kreuzungspunkte dieses Gitters befindet, bezeichnen wir als "Permutations-Konfigurationen". Die Anzahl der von einer Startposition aus erreichbaren Permutations-
Konfigurationen ist gerade die Hälfte der Anzahl aller Permutations-Konfigurationen,
also $m^2!/2$.

Der Freiraum $\mathcal{F}(\mathcal{S}_{(m)})$ besteht aus $(m-1)$–Simplices, welche aus denjenigen Konfigurationen gebildet werden, bei denen $m-1$ Quadrate sich irgendwo auf Zeile (Spalte)
i des Gitters befinden und die restlichen Quadrate auf die Gitterkreuzungen der Zeilen
(Spalten) $1, \ldots, i-1$ und $i+1, \ldots, m$ verteilt sind. Wir sprechen von einem "Zeilen-
(Spalten-) Cluster", wenn alle Quadrate in der Szene Positionen einnehmen, die sich
auf einer Zeile (Spalte) des Gitters befinden.

Ein Zeilen- bzw. Spalten-Cluster ist mithin die konvexe Hülle von m Permutations-
Konfigurationen, die durch Verschieben der Quadrate einer Zeile (Spalte) auseinander
hervorgehen können. Dabei ist jeder Zeilen-(Spalten-) Cluster mit m Spalten-(Zeilen-)
Clustern benachbart:

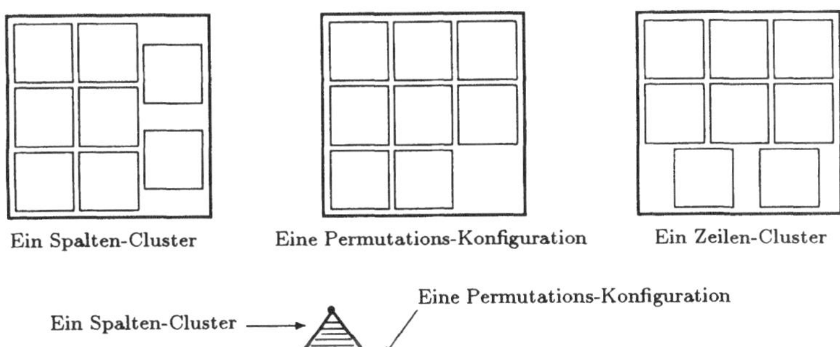

Man erkennt leicht, daß der Freiraum $\mathcal{F}(\mathcal{S}_{(m)})$ homotopieäquivalent ist zu einem Graphen, dessen Ecken gerade die Zeilen- und Spalten-Cluster der Szene und dessen Kanten die Permutations-Konfigurationen sind. Die Anzahl der Zeilen- bzw. Spalten-Cluster beträgt jeweils $m^2!/(2m)$ und insgesamt also $m^2!/m$. Mit der Eulerschen Formel erhalten wir für die Anzahl der Erzeugenden der Fundamentalgruppe von $\mathcal{S}_{(m)}$

$$\alpha_1 - \alpha_0 + 1 = \frac{m^2!}{2} - \frac{m^2!}{m} + 1 > (m^2 - 1)!$$

Satz 3 *Die Anzahl der Erzeugenden der Fundamentalgruppe einer Szene kann in der Größenordnung Fakultät der Anzahl der Objekte wachsen.*

Da das Schiebe-Puzzle intuitiv komplizierter ist als der unbeschränkte Fall liefert die Topologie hier also ein besseres Maß für den "Schwierigkeitsgrad" einer Szene als die geometrische Komplexität des Freiraums, die im Falle des Puzzles nicht höher ist als im Falle freier Beweglichkeit.

5 Allgemeinheit von Freiräumen

In den vorhergehenden Abschnitten haben wir gesehen, daß wir, wenn auch mit großem zeitlichen Aufwand, eine Präsentation der Fundamentalgruppe einer kinematischen Szene berechnen können. Eine andere Frage freilich ist, ob wir diese Präsentation auch interpretieren, also z.B. algebraische Eigenschaften der Gruppe bestimmen oder diese zu Klassifikationszwecken heranziehen können.

Um diese Frage wie auch diejenige nach der Möglichkeit einer Klassifikation kinematischer Szenen nach ihrer topologischen Struktur beantworten zu können, stellen wir die Frage, wie allgemein die Freiräume in diesem Sinne sind. Können wir mit real nachbaubaren mechanischen Gebilden, also kinematischen Szenen in einem höchstens 3-dimensionalen Raum, "beliebig komplizierte" Freiräume erhalten ?

Nach einem Satz von Markov kann man zu jeder Gruppenpräsentation einen 4-dimensionalen Polyeder konstruieren, der die gegebene Gruppe als Fundamentalgruppe besitzt. Mit dieser Konstruktion lassen sich nicht nur die bekannten Nichtentscheidbarkeitsaussagen für Gruppen (Wortproblem, Isomorphieproblem, etc.) auf Fundamentalgruppen übertragen, sondern auch die Nichtentscheidbarkeit des Homöomorphie- und Homotopieproblems für Polyeder zeigen. Wir fragen also, ob wir zu jedem Polyeder eine einfache kinematische Szene konstruieren können, deren Freiraum homöomorph oder homotopieäquivalent zu diesem Polyeder ist.

Bei der Erzeugung von Freiräumen beschränken wir uns zunächst auf einfach zu handhabende Teilräume des $\mathbf{R}^n$, nämlich auf solche, die aus *achsparallelen Quadern* zusammengesetzt sind. Unter "achsparallel" verstehen wir, daß die Kanten des 1-Gerüsts parallel zu den Koordinatenachsen verlaufen. Den offenen Kern eines Quaders bezeichnen wir auch als offenen Quader. Fassen wir Quader als Simplices auf, können wir analog zu Simplizialkomplexen auch *Quaderkomplexe* definieren, die eine zur Realisierung als Freiraum nützliche Zerlegungseigenschaft besitzen:

Im folgenden bezeichnen wir mit p_n die Projektion

$$
\begin{aligned}
p_n : \qquad \mathbf{R}^n \quad &\to \quad \mathbf{R}^{n-1} \\
(x_1,\ldots,x_n) \quad &\mapsto \quad (x_1,\ldots,x_{n-1})
\end{aligned}
$$

Lemma 1 *Es sei X eine endliche, disjunkte Vereinigung offener, achsparalleler Quader in $[0,1]^n$, $n > 1$.*
Dann gibt es eine Zerlegung $[0,1] = (I_1, I_2, \ldots, I_l)$ von $[0,1]$ in l Intervalle, so daß für jedes Intervall I_j der Zerlegung gilt:
a) $p_n(\ ([0,1]^{n-1} \times \{x_1\}) \cap X\ ) = p_n(\ ([0,1]^{n-1} \times \{x_2\}) \cap X\ )\ \ \forall\ x_1, x_2 \in I_j$,
b) $p_n(\ ([0,1]^{n-1} \times \{x\}) \cap X\ )$ ist für $x \in I_j$ beliebig selbst wieder eine disjunkte Vereinigung offener, achsparalleler Quader in $[0,1]^{n-1}$.

Das Lemma besagt, daß wir $X \subseteq \mathbf{R}^n$ zerlegen können in $X = X_1 \times I_1 \dot\cup \ldots \dot\cup X_l \times I_l$, wobei $X_1, \ldots, X_l$ wie X selbst wieder disjunkte Vereinigungen offener, achsparalleler Quader in $\mathbf{R}^{n-1}$ sind. Dies wird uns ermöglichen, eine Szene für einen zu realisierenden n-dimensionalen Freiraum aus Szenen mit (n-1)-dimensionalen Freiräumen zusammenzusetzen.

Mit Hilfe von Lemma 1 definieren wir rekursiv einen Zerlegungsbaum $Z(X)$ für $X \subseteq [0,1]^n$, der X allein durch eine Menge von Intervallen beschreibt:

Definition 6 *Sei $X \subseteq [0,1]^n$ eine endliche Vereinigung offener, achsparalleler Quader. Im Falle $n = 1$ sei $(I_1, \ldots, I_l)$ eine Zerlegung von $[0,1]$ mit $I_j \cap X = I_j$ oder $I_j \cap X = \emptyset$, $1 \leq j \leq l$; für $n > 1$ sei $(I_1, \ldots, I_l)$ wie in Lemma 1. Wir definieren $Z(X)$ als partielle Abbildung von der Menge aller endlichen Tupel $\mathbf{N}^*$ ("Selektoren") in die Menge der Teilintervalle von $[0,1]$:*

$$
\begin{aligned}
X^{[j]} &:= p_n(\ ([0,1]^{n-1} \times \{x_j\}) \cap X\ )\ ;\ x_j \in I_j \text{ beliebig},\ 1 \leq j \leq l, \\
Z(X)[j] &:= I_j\ ;\ 1 \leq j \leq l, \\
Z(X)[j, i_1, \ldots, i_r] &:= Z(X^{[j]})[i_1, \ldots, i_r]\ ,\ 1 \leq j \leq l\ ,\ r < n.
\end{aligned}
$$

Beispiel :

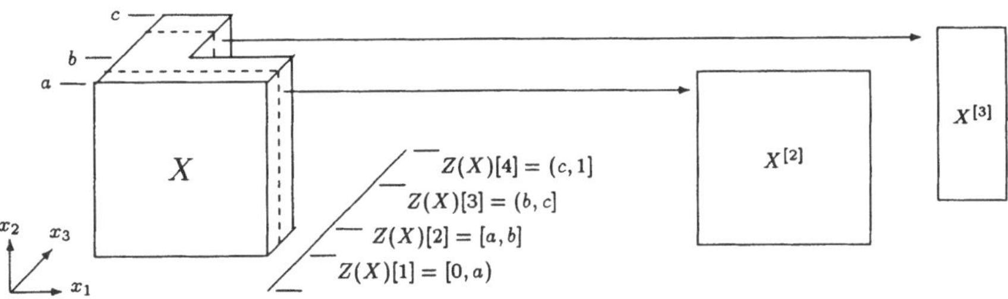

Lemma 2 *Die Zerlegungsintervalle können so gewählt werden, daß $Z(X)[i_1, \ldots, i_r]$ mit $r \leq n-1$ entweder undefiniert, ein offenes oder ein zu einem Punkt degeneriertes Intervall ist. Ist $Z(X)$ eine entsprechend gewählte Zerlegung eines Quaderkomplexes X, $Z(X)[i_1, \ldots, i_r] = \{a\}$ und $Z(X)[i_1, \ldots, i_r \pm 1]$ definiert, so ist $Z(X)[i_1, \ldots, i_r \pm 1]$ ein offenes Intervall und $X^{[i_1, \ldots, i_r \pm 1]} \subseteq X^{[i_1, \ldots, i_r]}$.*

Wir erläutern unsere Vorgehensweise zur Konstruktion von Szenen zu gegebenen Quaderkomplexen an einem ausführlichen Beispiel. Unsere Aufgabe soll sein, die Oberfläche des 3-dimensionalen Einheitswürfels als Freiraum einer kinematischen Szene zu erzeugen:

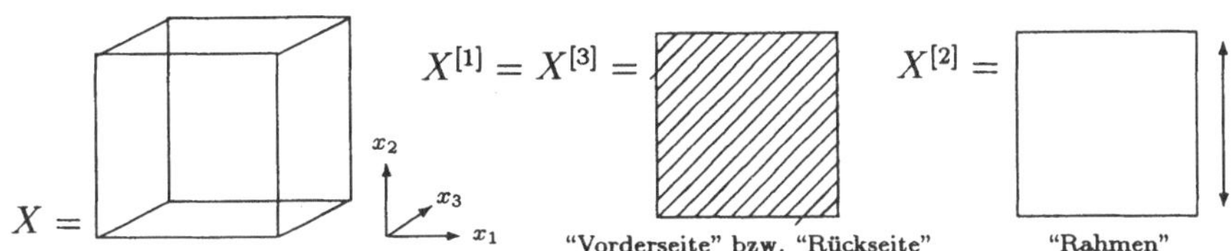

Fassen wir die Zerlegung $Z(X)$ als Baum auf, in dem Zerlegungsintervalle die Knoten und Selektoren $[i_1, \ldots, i_k]$ Pfade zu Intervallen $Z(X)[i_1, \ldots, i_k]$ darstellen, so können wir eine mögliche Zerlegung für unser Beispiel wie folgt angeben:

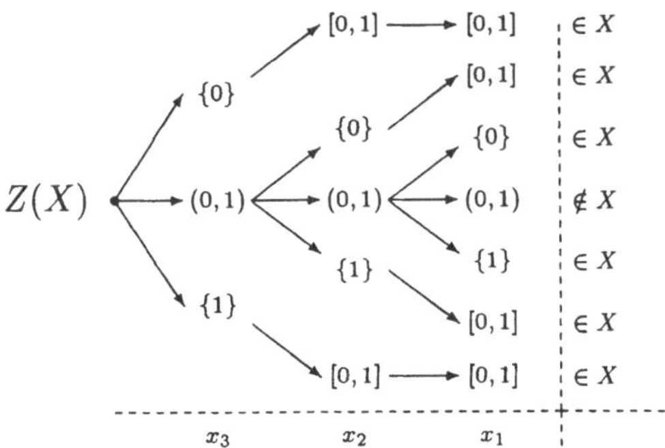

Wir wenden uns zunächst der Frage zu, wie wir einen 1–dimensionalen Freiraum wie z.B. $X^{[2,1]} = [0,1]$ oder $X^{[2,2]} = \{0\} \cup \{1\}$ erzeugen können.

Hierzu konstruieren wir einen in x–Richtung in der Ebene verschiebbaren Balken B, der eine Formung besitzt, durch die wir die freie Beweglichkeit von B mittels Hindernissen auf die gewünschten Intervalle einschränken können :

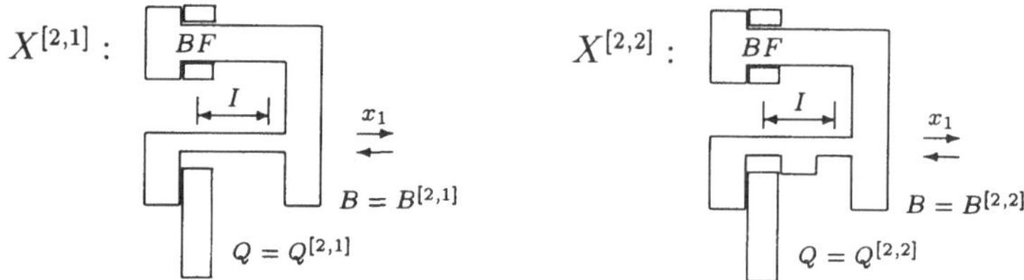

Durch ein Führungsteil BF des Balkens B wird die Beweglichkeit von B mittels zweier Hindernisse auf die x–Richtung beschränkt. Das Hindernis Q beschränkt seinerseits die Beweglichkeit von B in x–Richtung weiter auf das Intervall $X^{[2,1]}$ bzw. auf die diskreten Punkte $X^{[2,2]}$, wenn wir die Stellung von B entsprechend in x–Richtung parametrisieren.

Durch die Verwendung offener und abgeschlossener Seiten können wir erreichen, daß durch eine Bewegung von Q um einen beliebig kleinen Betrag nach unten die Bewegungsfreiheit von B auf das gesamte Intervall $[0,1]$ erweitert wird (die Oberseite des Kolbens und die Unterseite der Formung des Balkens müssen dazu beide abgeschlossen sein). Indem wir Q also zu einem in y–Richtung beweglichen Kolben machen, können wir durch die Stellung von Q Einfluß nehmen auf die Beweglichkeit von B.

Desweiteren können wir die Stellung mehrerer Kolben $Q_1, \ldots, Q_l$ über einen einzelnen Parameter, nämlich die Stellung eines einzelnen Objekts in einer Koordinatenrichtung, beschreiben. Das entsprechende Objekt bezeichnen wir als Schieberegler SR. Wollen wir in unserem Beispiel den Freiraum $X^{[2]}$ erzeugen, so setzen wir zunächst die Szenen für die Freiräume $X^{[2,1]}$, $X^{[2,2]}$ und $X^{[2,3]}$, bestehend aus den Balken $B^{[2,1]}$, $B^{[2,2]}$, $B^{[2,3]}$ und den Kolben $Q^{[2,1]}$, $Q^{[2,2]}$, $Q^{[2,3]}$, nebeneinander und verbinden die drei Balken zu einem Balken $B^{[2]}$. Die Kolben $Q^{[2,i]}$ erweitern wir zu Objekten $QU^{[2,i]}$, deren Stellung sich von der Stellung eines Schiebereglers beeinflussen läßt. Wir wollen einen Schieberegler $SR^{[2]}$ konstruieren, der dafür sorgt, daß der Kolben $QU^{[2,i]}$ *genau dann* "oben" ist und die Bewegungsfreiheit von $B^{[2,i]}$ beeinflußt, wenn die Stellung des Schiebereglers in x–Richtung bei entsprechender Parametrisierung aus dem Intervall $Z(X)[2,i]$ ist (Parameter x_2). Wir versuchen folgende Konstruktion einer Szene $\mathcal{S}^{[2]}$ zur Realisierung von $X^{[2]}$:

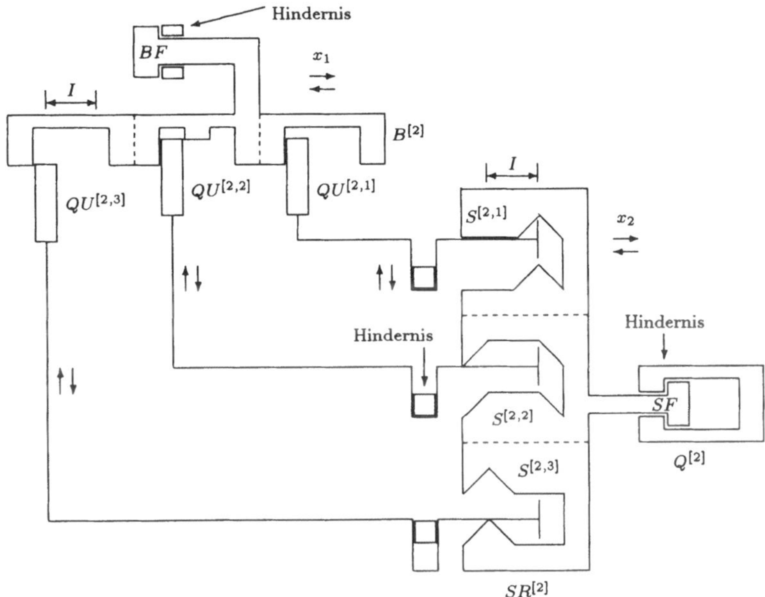

Die Szene erfüllt unsere Anforderungen nicht vollständig: der Kolben $QU^{[2,1]}$ sollte als einziger in die Formung seines Teilbalkens $B^{[2,1]}$ greifen, stattdessen greift zusätzlich $QU^{[2,2]}$.

Es handelt sich hierbei um eine grundsätzliche Schwierigkeit: in unserer Konstruktion können wir nicht verhindern, daß an der Grenze $\overline{I_1} \cap \overline{I_2}$ zweier Intervalle $I_1 = Z(X)[i_1, \ldots, i_r]$ und $I_2 = Z(X)[i_1, \ldots, i_r + 1]$ sowohl die Kolben der Teilszene zu $X^{[i_1, \ldots, i_r]}$ als auch die der Teilszene zu $X^{[i_1, \ldots, i_r+1]}$ greifen, so daß an den Intervallgrenzen der Schnitt beider Räume die erlaubten Konfigurationen beschreibt. Kann die Zerlegung $Z(X)$ so gewählt werden, daß kein Zerlegungsintervall zu einem Punkt degeneriert ist, so ändert sich der Homotopietyp nicht, d.h. der erzeugte Freiraum ist homotopieäquivalent zu X. Aber auch andernfalls können wir uns behelfen, indem wir X zu einem homotopieäquivalenten Raum Y vergrößern, den wir als Freiraum realisieren können.

Wir gehen dazu von einer Zerlegung $Z(X)$ eines achsparallelen Quaderkomplexes X gemäß Lemma 2 aus und "verdicken" den Raum X sukzessive entlang der n–ten bis entlang der 2–ten Koordinate so, daß wir die zu Punkten degenerierten Zerlegungsintervalle durch offene Intervalle ersetzen können:

$$\ldots \qquad \{a_1\} \qquad , \qquad (a_1, a_2) \qquad , \qquad \{a_2\} \qquad \ldots$$
$$\to \quad \ldots \ (a_1 - \varepsilon, a_1 + \varepsilon) \quad , \quad [a_1 + \varepsilon, a_2 - \varepsilon] \quad , \quad (a_2 - \varepsilon, a_2 + \varepsilon) \ \ldots$$

Wir wählen ε kleiner als die halbe Länge des kleinsten offenen Intervalls der Zerlegung $Z(X)$ und definieren damit

$$X'_{(i)} := \{ \, x + (0, \ldots, \overset{\to}{0}, \overset{i}{\delta}, \overset{\gets}{0}, \ldots, 0) \, ; \, x \in X' \, , \, -\varepsilon < \delta < \varepsilon \, \} \cap [0,1]^n,$$
$$Y := (\ldots ((X_{(n)})_{(n-1)}) \ldots)_{(2)}.$$

In unserem Beispiel erhalten wir durch obiges Verfahren mit $\varepsilon = 0.1$:

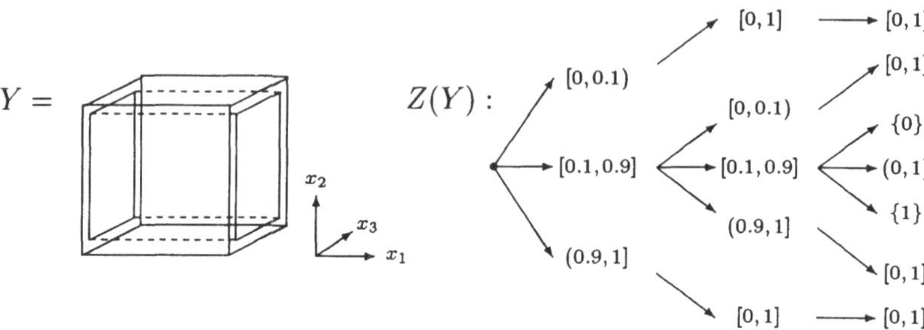

Natürlich ist X Deformationsretrakt von $X_{(i)}$ und Y somit homotopieäquivalent zu X. Führen wir unsere Konstruktion einer Szene nun für Y anstatt für X durch, so daß für jedes Zerlegungsintervall $Z(Y)[s]$, s Selektor mit Länge $< n$, der entsprechende Kolben $QU^{[s]}$ während des abgeschlossenen Intervalls $\overline{Z(Y)[s]}$ greift, so erhalten wir als Freiraum (bis auf Homöomorphie) gerade Y. Dies liegt an der Eigenschaft $X^{[i_1,\dots,i_r\pm 1]} \subseteq X^{[i_1,\dots,i_r]}$ für $Z(X)[i_1,\dots,i_r] = \{a\}$ aus Lemma 2, die sich entsprechend auf die offenen Intervalle aus $Z(Y)$ übertragen läßt und dafür sorgt, daß sich an der Intervallgrenze $\overline{Z(Y)[i_1,\dots,i_r \pm 1]} \cap \overline{Z(Y)[i_1,\dots,i_r]}$ jeweils der zum abgeschlossenen Zerlegungsintervall gehörige Raum $X^{[i_1,\dots,i_r\pm 1]}$ "durchsetzt".

Im letzten Schritt unseres Beispiels wenden wir uns der Konstruktion einer Szene $\mathcal{S}$ für ganz Y zu. Wir nehmen an, wir hätten im vorangegangenen Schritt nicht nur eine Szene $\mathcal{S}^{[2]}$ für $Y^{[2]}$, sondern auch Szenen $\mathcal{S}^{[1]}$ für $Y^{[1]}$ und $\mathcal{S}^{[3]}$ für $Y^{[3]}$ erzeugt. Wir gehen wieder so vor, daß wir mit Hilfe eines Schiebereglers SR zu jedem Parameter aus $[0, 1]$, der die Stellung von SR in x-Richtung beschreibt, bis auf Intervallgrenzen genau eine der Szenen $\mathcal{S}^{[1]}$, $\mathcal{S}^{[2]}$ bzw. $\mathcal{S}^{[3]}$ "aktivieren".
Zu diesem Zweck machen wir die Hindernisse $Q^{[i]}$, welche die Stellung von $SR^{[i]}$ in y-Richtung festlegen, beweglich und steuern die Bewegung mit Hilfe des Schiebereglers SR. Die Hindernisse $Q^{[i]}$ ersetzen wir hierzu durch die Objekte $QU^{[i]}$. $QU^{[i]}$ besitzt dabei genau eine Stellung, in welcher der Regler $SR^{[i]}$ die erlaubten Stellungen der Objekte zu den Parameterwerten $(x_1, x_2) \in [0, 1] \times [0, 1]$ einzuschränken vermag, d.h. in der die Kolben $QU^{[i,j]}$ in die Formung von $B^{[i,j]}$ greifen. Durch Bewegung von $QU^{[i]}$ um einen beliebig kleinen Betrag nach unten werden die Kolben $QU^{[i,j]}$ "freigegeben", d.h. behindern die freie Bewegung des Balkens B in x-Richtung nicht mehr.

Wie im vorhergehenden Schritt müssen wir auch hier die Balken $B^{[1]}$, $B^{[2]}$ und $B^{[3]}$ der Szenen $\mathcal{S}^{[1]}$, $\mathcal{S}^{[2]}$ und $\mathcal{S}^{[3]}$ aneinanderkleben. Um die erlaubten Konfigurationen aus $\mathcal{F}(\mathcal{S})$ wirklich nur durch 3 Parameter beschreiben zu können, müssen wir zusätzlich $SR^{[1]}$, $SR^{[2]}$ und $SR^{[3]}$ in geeigneter Weise "lose" miteinander verbinden, so daß die Stellung eines der Regler in x-Richtung die Stellung der beiden anderen in x-Richtung eindeutig bestimmt, die Stellung der Regler in y-Richtung jedoch eindeutig von SR vorgeschrieben wird. Dies geschieht durch Verbindungsteile der Regler $SR^{[i]}$, die senkrecht ineinander geschoben werden können.

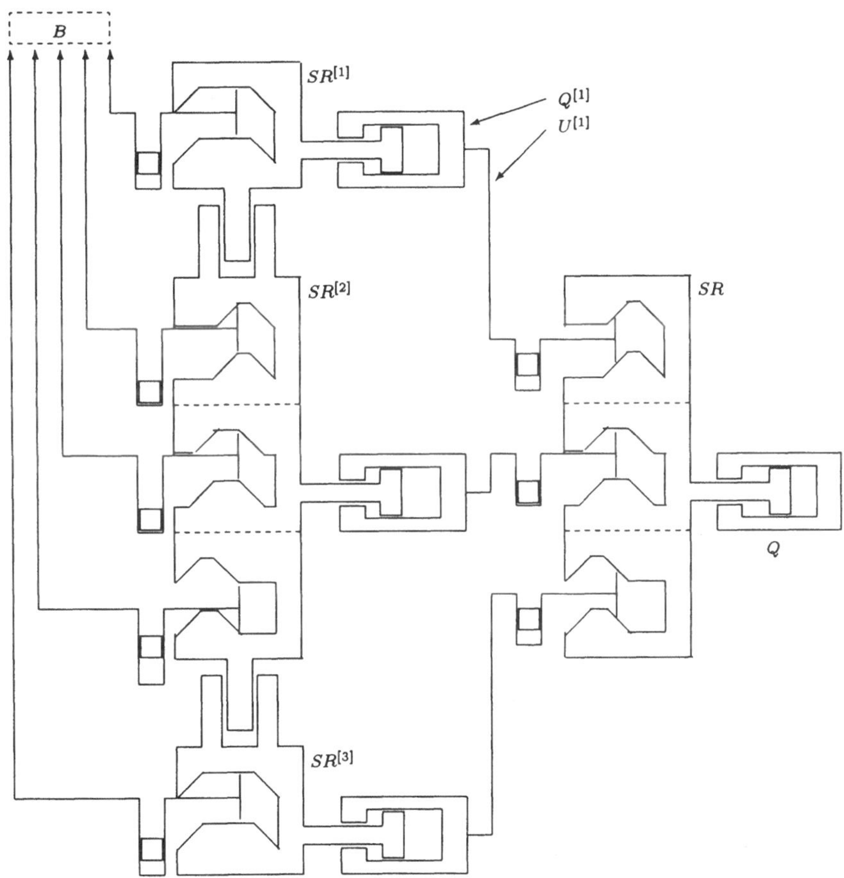

Man erkennt, daß das aufgezeigte Verfahren auf beliebige Dimensionen fortgesetzt werden kann. Die dabei entstehende Szene gleicht dem Zerlegungsbaum insofern, als jedem Intervall $Z(Y)[i_1, \ldots, i_r]$ mit $1 \leq r < n$ ein Schiebereglerteil $S^{[i_1, \ldots, i_r]}$ entspricht. Die Dimensionierung der Objekte muß abhängig von der Dimension n geschehen, da die einzuplanenden Bewegungen der Schieberegler und Kolben in y-Richtung von der entsprechenden "Zerlegungstiefe" abhängen.

Die Stellung der Schieberegler $SR^{[i_1, \ldots, i_r]}$ in "Zerlegungstiefe" $0 \leq r < n-1$ in x-Richtung entspricht dem Parameter x_{n-r}, die Stellung des Balkens B in x-Richtung entspricht dem Parameter x_1. Da die Stellungen der anderen Teile der Szene durch die Stellungen dieser Objekte eindeutig bestimmt werden, ist der Freiraum der so konstruierten Szene homöomorph zu Y bzw. homotopieäquivalent zu X. Wir erhalten als Ergebnis:

Satz 4 *Es sei $X \subseteq [0,1]^n$ ein zusammenhängender, achsparalleler Quaderkomplex. Dann gibt es eine ebene Szene S bestehend aus Hindernissen und translatorisch frei beweglichen Objekten, deren Freiraum $\mathcal{F}(S)$ homotopieäquivalent ist zu X.*

Die in der Verwendung offener und abgeschlossener Seiten beruhende Idealisierung
wird nur benötigt, um bei der Beschreibung der Wechselwirkung zwischen Balken und
Kolben eine Kollision in einer diskreten Position der Kolben erreichen zu können. Man
kommt auch ohne diese Idealisierung aus, indem man die Kolben zum Erreichen ihrer
oberen Position um einen kleinen Betrag in die Ausschnitte der Balken "eintauchen"
läßt, wobei der Homotopietyp des Freiraums unverändert bleibt.

6   Nicht entscheidbare Probleme

Um die Nichtentscheidbarkeit des Homöomorphieproblems bzw. des Homotopiepro-
blems für triangulierbare, kompakte 4–Mannigfaltigkeiten (vgl. [5], [2]) mit Hilfe von
Satz 4 auf Freiräume kinematischer Szenen übertragen zu können, schließen wir unsere
Argumentationskette durch folgendes Lemma:

Lemma 3 *Zu jedem Polyeder $P \subseteq [0,1]^m$ kann man einen zu P homotopieäquivalenten, achsparallelen Quaderkomplex $X \subseteq [0,1]^m$ konstruieren.*

Wir können die bereits motivierten Klassifikationsprobleme damit wie folgt beantworten:

Satz 5 *Es gibt einen Algorithmus, der zu einer gegebenen, endlichen Präsentation
$\mu = (X, R)$ eine ebene, kinematische Szene S bestehend aus polygonalen Hindernissen
und translatorisch frei beweglichen Objekten konstruiert, deren Fundamentalgruppe
isomorph ist zu $G(\mu)$. Desweiteren gilt:*

1. *Es ist nicht entscheidbar, ob die Fundamentalgruppe einer beliebig gegebenen,
   kinematischen Szene isomorph ist zu einer gegebenen, endlich präsentierbaren
   Gruppe.*

2. *Es ist nicht entscheidbar, ob die Freiräume zweier beliebig gegebener, kinematischer Szenen homotopieäquivalent sind.*

3. *Es ist nicht entscheidbar, ob ein gegebener Weg im Freiraum einer beliebigen
   kinematischen Szene nullhomotop ist.*

Unter Ausnutzung der Tatsache, daß man sich auf 4–dimensionale Mannigfaltigkeiten beschränken kann, sollte es möglich sein, auch den Beweis für die Nichtentscheidbarkeit des Homöomorphieproblems für kinematische Szenen zu erbringen.

Die Nichtentscheidbarkeit des Wegeproblems impliziert u.a., daß es keinen Algorithmus geben kann, der einen Weg bzgl. einer im Freiraum gegebenen Metrik nach
seiner Länge durch Deformation zu optimieren vermag.

7   Ausblick

Wir haben gesehen, daß das Wissen um topologische Eigenschaften einen wichtigen Beitrag zum qualitativen Argumentieren leisten kann. Dabei erweisen sich jedoch bereits einfache kinematische Mechanismen als zu komplex, um ohne Hintergrundwissen aussagefähige topologische Invarianten bestimmen zu können.

Dennoch sollten die Nichtentscheidbarkeitsergebnisse dieser Arbeit im Hinblick auf "topological reasoning" nicht entmutigen, weil hier wie in vielen Gebieten der KI das Zusammenwirken verschiedener Methoden und die geschickte Auswertung vorhandenen Wissens zu nutzbaren Tools führen könnte, deren mögliches Einsatzgebiet in der Analyse oder Diagnose von mechanischen Gebilden, Fertigungsprozessen o.ä. liegen könnte. Hierbei mag sich als nützlich erweisen, daß die Nichtentscheidbarkeit erst bei Freiräumen mit Dimension ≥ 4 auftritt.

In dem Abschnitt über Komplexität haben wir gesehen, daß bereits für den sehr eingeschränkten Fall verschiebbarer Rechtecke und rechteckiger Hindernisse in der Ebene die Frage nach Normalformen oder Entscheidbarkeit der Struktur der Fundamentalgruppe nicht trivial ist. Angesichts einer noch immer fehlenden, umfassenden Theorie über Gruppen mit entscheidbarer Struktur ist die Untersuchung einfacher Spezialfälle kinematischer Szenen auch vom Standpunkt der Mathematik aus als interessant anzusehen.

8   Danksagung

Die vorliegende Arbeit enthält wesentliche Teile meiner gleichnamigen Dissertation nach einem Thema von Herrn Professor Hotz. Neben den vielen Kollegen, die mir in Diskussionen weitergeholfen haben, gilt mein besonderer Dank Herrn Professor Hotz, der mir in zahlreichen Gesprächen hilfreich zur Seite stand.

Literaturverzeichnis

[1] Daniel R. Baker : "Some Topological Problems in Robotics", *The Mathematical Intelligencer*, Vol. 12, No. 1, Springer-Verlag New York 1990

[2] W. Boone, W. Haken, V. Poénaru : "On Recursively Unsolvable Problems in Topology and Their Classification", *Contributions to Mathematical Logic*, edited by H. Schmidt, K. Schutte and H.-J. Thiele, North-Holland Pub. Co., 1968, pp. 37-74

[3] Boi Faltings : "Qualitative Kinematics in Mechanisms", *Knowledge Representation 1989*, pp. 436-442

[4] R.C. Lyndon, P.E. Schupp : *Combinatorial Group Theory*, Ergebnisse Math. Grenzgebiete 89, Springer-Verlag New York Heidelberg Berlin 1980

[5] A.A. Markov : "The Problem of Homeomorphy", *Proceedings of International Congress of Mathematicians*, 1958, pp. 300-306 (in Russian)

[6] W.S. Massey : *Algebraic Topology: An Introduction*, Grad. Texts in Math. 56, Springer-Verlag New York Heidelberg Berlin 1967

[7] Michael O. Rabin : "Recursive Unsolvability of Group Theoretic Problems", *Annals of Mathematics*, Vol. 67, No. 1, January 1958, pp. 172-194

[8] J.T. Schwartz, M. Sharir : " Algorithmic Motion Planning in Robotics", *Handbook of Theoretical Computer Science Volume A*, edited by Jan van Leeuwen, Elsevier 1990

[9] R. Stöcker, H. Zieschang : *Algebraische Topologie*, B.G. Teubner Stuttgart 1988

[10] Chee-Keng Yap : "Algorithmic Motion Planning", *Algorithmic and Geometric Aspects of Robotics Volume 1*, edited by J.T. Schwartz, C.-K. Yap, Lawrence Erlbaum Associates 1987

Bemerkungen zum Schätzen von Bayesschen Diskriminantenfunktionen

Hans Ulrich Simon

FB Informatik
Universität Dortmund
Germany

Zusammenfassung

Bei der Approximation von Bayesschen Diskriminantenfunktionen zum Zwecke der automatischen Objektklassifikation ist es üblich, unbekannte Parameter durch empirische Schätzungen zu ersetzen. Wir entwickeln in diesem Beitrag eine ausgefeiltere Schätzmethode mit dem Ziel, auf effiziente Weise und mit (beweisbar) hoher Zuverlässigkeit, zu fast optimalen Diskriminantenfunktionen zu gelangen. Wir wenden dabei Grundgedanken des sogenannten pac–Lernens (pac = probably approximately correct) auf die Bayessche Entscheidungstheorie an.

1 Einleitung

Die Unterscheidung zwischen verschiedenen Objekttypen anhand von Merkmalsvektoren ist ein klassisches Problem der Mustererkennung. Die Merkmalsvektoren könnten zum Beispiel die Pixel handgeschriebener Zeichen sein:

$$\mathcal{I}\ \mathcal{P}\ \mathcal{U}\ \mathcal{A}^1$$

Diskriminantenfunktionen verwandeln die Merkmalsvektoren in reelle Zahlen. Jeder Objekttyp besitzt seine private Diskriminantenfunktion. Diese soll zum Ausdruck geben, wie stark wir daran glauben, daß der betreffende Objekttyp vorliegt. Da es mehrdeutige Merkmalsvektoren gibt, sind gelegentliche Klassifikationsfehler unvermeidlich. Wir dürfen jedoch annehmen, daß die statistischen Dichtefunktionen $p(x|T)$ auf den Merkmalsvektoren x bei gegebenem Objekttyp T sich für verschiedene Objekttypen (ordentlich!) voneinander unterscheiden (wir wollen schließlich Klassifikatoren und keine Hasardeure implementieren). Bayessche Diskriminantenfunktionen sind solche, die eine minimale Fehlerrate realisieren. Zum Beispiel:

$$f(x) = f_T(x) = p(x|T) \cdot \mathrm{pr}(T) = \mathrm{pr}(T|x) \cdot p(x)$$

oder

$$g(x) = g_T(x) = \mathrm{pr}(T|x).$$

Hierbei bezeichnet $\mathrm{pr}(\cdots)$ die (bedingte) Wahrscheinlichkeit von T (bei gegebenem x) und

$$p(x) = \sum_T p(x|T) \cdot \mathrm{pr}(T)$$

die zusammengesetzte Dichtefunktion auf den Merkmalsvektoren. Die erzielte minimale Fehlerrate beträgt

$$\int_x \min_T p(x|T) \cdot \mathrm{pr}(T) = \int_x \min_T \mathrm{pr}(T|x) \cdot p(x).$$

Die bisherigen Ausführungen entsprechen einem kontinuierlichen Merkmalsraum. Im diskreten Fall verwandeln sich alle Dichtefunktionen in Wahrscheinlichkeitsverteilungen und alle Integrale in Summen.

Bei unbekannten Dichtefunktionen ist es naheliegend mit empirischen Schätzformeln $\tilde{h}_T(x)$ für die Bayesschen Diskriminantenfunktionen $h_T(x)$ zu arbeiten. Die Grundidee dabei ist die folgende:

Wenn für festes x und verschiedene T die Werte $h_T(x)$ weit auseinanderliegen, so wird eine gute Schätzung dieser Werte ihre Ordnungsbeziehung nicht verdrehen. Für solche x ist auch die empirische Entscheidung optimal. Liegen die Werte von $h_T(x)$ nahe beisammen, wird die empirische Entscheidung potentiell suboptimal. Der auf x erzeugte Extrafehler ist dann aber nicht allzu groß.

Man beachte aber, daß es nicht ausreicht $p(x|T)$ oder $\mathrm{pr}(T)$ in einem additiven Sinn gut zu schätzen (s. obige Formel für die minimale Fehlerrate: Kleine Extrafehler im

---

[1]Wort für Großer Zauberer in der Sprache der Zulu

Integranden akkumulieren beim Integrieren zu einem unkontrollierbaren Fehlerwert). Wir benötigen Schätzungen, die bis auf einen Faktor von $1 + \mu$ genau sind. Dabei ist μ ein adjustierbarer Genauigkeitsparameter. Die erforderliche Stichprobengröße soll aber nur polynomiell von $1/\mu$ abhängen.

Die Notwendigkeit multiplikativ guter Schätzungen produziert ein Folgeproblem. Werte, die nahe bei 0 sind, lassen sich auf direktem Weg nicht effizient bis auf einen Faktor von $1 + \mu$ schätzen. Liegt $\mathrm{pr}(T)$ nahe bei 0 (dies kann statistisch effizient getestet werden), so eliminieren wir T aus der Menge der relevanten Objekttypen (dies kostet nur einen kleinen Extrafehler). Für die Dichtefunktion $p(x|T)$ ist die Verfahrensweise komplizierter. In den folgenden beiden Abschnitten wollen wir zwei Grundideen skizzieren:

1. Empirische Schätzungen der Dichtefunktionen konvergieren auf 'worstcase'–Merkmalsvektoren zu langsam. Für einige interessante Verteilungsklassen, ist die totale Wahrscheinlichkeit solcher 'schlechter' Merkmalsvektoren aber (adjustierbar) klein und das Prädikat 'schlecht' effizient berechenbar. In diesem Fall haben wir wieder gewonnenes Spiel: die Objekttypen, für die der Merkmalsvektor schlecht ist, können ohne großen Extrafehler ignoriert werden. Für die anderen Objekttypen ist die Konvergenzgeschwindigkeit der empirischen Schätzungen ausreichend.

2. Neuronale Implementierungen dieser Klassifikationsstrategie können mit Hilfe eines Tricks ohne Extralogik für das Prädikat 'schlecht' auskommen. Die Grundidee ist, die empirischen Schätzungen durch raffiniertere Werte zu ersetzen, so daß schlechte Merkmalsvektoren durch winzige Werte der geschätzten Dichtefunktion diskreditiert werden.

2 Schnelle Konvergenz auf guten Vektoren

Wir illustrieren das Prinzip am einfachsten denkbaren Beispiel. Sei b eine binäre Zufallsvariable und p die Wahrscheinlichkeit für $b = 1$. Falls p nahe bei 0 liegt, kann (Chernoff–Schranken) eine empirische Schätzung bis auf Faktor $1 + \mu$ nicht effizient produziert werden. Wenn wir statt p die Formel p^b betrachten, verändert sich die Sachlage zu unseren Gunsten: Falls Parameter p nahe bei 0 liegt, so hat b fast immer den Wert 0 und daher p^b fast immer den Wert 1. Falls p von 0 separiert ist, ist die empirische Schätzung $\tilde{b}$ gut, und wir brauchen den Fall $b = 1$ nicht zu fürchten. Daher ist in beiden Fällen $p^{\tilde{b}}$ (bis auf unwahrscheinliche und effizient erkennbare Ausnahmefälle) eine befriedigende Schätzung für p^b.

Das wichtige primitive Grundmuster der Formel p^b ist: 'Wahrscheinlichkeit eines Ereignisses hoch die charakteristische Funktion des Ereignisses'. Formeln, die diesem Grundmuster genügen, lassen sich in unserem Sinne effizient schätzen. Es ist nicht schwer zu zeigen, daß Produkte gut schätzbarer Formeln (und auch Quotienten unter gewissen Zusatzbedingungen) wieder gut schätzbar sind. Zum Beispiel ist die

Verteilung unabhängiger binärer Merkmale

$$P(x) = \prod_{i=1}^{n} p_i^{x_i} \cdot (1 - p_i)^{1-x_i}$$

ein Produkt aus n Faktoren des primitiven Grundmusters. Mit etwas mehr Aufwand lassen sich auch Markov–Prozesse oder Prozesse, die nichttriviale Korrelationen einer beliebigen (aber konstanten) Ordnung aufweisen, auf diese Weise beschreiben. Wir gehen darauf im Rahmen dieser Übersicht nicht näher ein.

3   Effiziente Neuronale Implementierungen

Wir illustrieren das Grundprinzip am einfachen Beispiel der obenerwähnten Verteilung unabhängiger binärer Merkmale. Um schlechte binäre Vektoren zu idenfizieren, würde man

1. eine Toleranzschwelle ϵ für den von ihnen induzierten Extrairrtum festlegen,

2. diese Toleranz auf die n potentiellen Ausnahmemengen der Form $x_i = 0, 1$ $(i = 1, \ldots, n)$ verstreuen; beachte, daß $x_i = 0$ und $x_i = 1$ nicht simultan Ausnahmemengen sind (daher n statt $2n$),

3. unter Benutzung der Chernoff–Schranken testen, ob wir glauben, daß $p_i < \epsilon/n$ (falls $\tilde{p}_i < \epsilon/(2n)$) oder $p_i \geq \epsilon/(4n)$ (falls $\tilde{p}_i \geq \epsilon/(2n)$). Im letzteren Fall können effiziente empirische Schätzungen produziert werden. Im ersteren Fall sind Vektoren mit $x_i = 1$ schlecht (aber extrem unwahrscheinlich; der Fall schlechter Vektoren mit $x_i = 0$ und Wahrscheinlichkeit $1 - p_i$ ist symmetrisch),

4. das erhaltene 'schlecht'–Prädikat und die erhaltene empirische Schätzformel ausgeben.

Durch Nachrechnen sieht man, daß die geschätzte Wahrscheinlichkeit für gute Vektoren mindestens $\tau = (\epsilon/(2n))^n$ beträgt. Das angekündigte Diskreditieren schlechter Vektoren gelingt nun leicht, indem wir kleine Wahrscheinlichkeiten wie $\tilde{p}_i < \epsilon/(2n)$ mit dem sehr kleinen Wert $\tau/2$ aktualisieren. Beachte, daß die Schätzwerte guter Vektoren darunter invariant sind. Mit diesem kleinen Trick haben wir die Ausnahmelogik eliminiert und brauchen nur noch die geschätzten Wahrscheinlichkeiten eines Merkmalsvektors für 2 Objekttypen miteinander zu vergleichen. Durch Logarithmieren der Verteilungsformel entsteht eine lineare Funktion in den x_i und die fast–optimale Bayes–Entscheidung kann auf einem einzigen Neuron implementiert werden. Analoge Tricks für Markov–Prozesse führen zu neuronalen Netzen mit einem 'hidden layer' und $n - 1$ 'hidden units'. Unter Beachtung von Korrelationen bis zur Ordnung k benötigt das betreffende neuronale Netz einen 'hidden layer' mit

$$(n + 1 - k) \cdot 2^k - n - 1$$

'hidden units'. Auf die Herleitung dieser letzten Ergebnisse gehen wir im Rahmen dieser Übersicht nicht ein.

4 Anwendungen und Literaturhinweise

Abschließend möchten wir der Hoffnung Ausdruck verleihen, daß die Ergebnisse für Verteilungen mit nichttrivialen Korrelationen konstanter Ordnung eine wichtige Rolle bei der automatischen Zeichenerkenung spielen werden. Diese Hoffnung gründet sich auf die Idee, daß die Frage, ob an einer konkreten Bildposition ein Pixel steht, sich statistisch in hohem Maße aus der lokalen Umgebung erklären läßt.[2] Eine genauere Überprüfung dieser Idee erfolgt erst in näherer Zukunft.

Genauere Beschreibungen von Problemen der Mustererkennung und ihren klassischen Lösungen finden sich zum Beispiel in dem Buch von Duda und Hart 'Pattern Matching and Scene Analysis' (s. [2]). Die hier skizzierten Ideen können im Detail nachgelesen werden in den Arbeiten [3] und [1]. In diesen Arbeiten werden auch weitere Querverbindungen beschrieben zum Lernen probabilistischer Konzepte nach dem Modell von Valiant (s. [6]) und seinen Varianten gemäß den Arbeiten von Haussler (s. [4]) sowie von Kearns und Schapire (s. [5]).

Literaturverzeichnis

[1] Svetlana Annulova, Jorge Cuellar, Klaus Uwe Höffgen, and Hans Ulrich Simon. Probably almost optimal neural classifiers. In preparation.

[2] Richard O. Duda and Peter E. Hart. *Pattern Classification and Scene Analysis.* Wiley–Interscience. John Wiley & Sons, New York, 1973.

[3] Paul Fischer, Stefan Pölt, and Hans Ulrich Simon. Probably almost Bayes decisions. In *Proceedings of the 4th Annual Workshop on Computational Learning Theory*, pages 88–94, 1991.

[4] David Haussler. Generalizing the pac model: Sample size bounds from metric-dimension based uniform convergence results. In *Proceedings of the 30'th Annual Symposium on the Foundations of Computer Science*, pages 40–46, 1989.

[5] Michael J. Kearns and Robert E. Schapire. Efficient distribution–free learning of probabilistic concepts. In *Proceedings of the 31'th Annual Symposium on the Foundations of Computer Science*, pages 382–392, 1990.

[6] Leslie G. Valiant. A theory of the learnable. *Communications of the ACM*, 27(11):1134–1142, 1984.

---

[2]Dies steht nicht im Widerspruch zum subjektiven Eindruck einer ganzheitlichen Bilderkennung da diese lokalen Korrelationen sich transitiv zu längeren Wirkungsketten fortpflanzen.

Residuation and Guarded Rules for Constraint Logic Programming

Gert Smolka

German Research Center for Artificial Intelligence and
Universität des Saarlandes
Stuhlsatzenhausweg 3, 6600 Saarbrücken 11, Germany
smolka@dfki.uni-sb.de

Abstract

A major difficulty with logic programming is combinatorial explosion: since goals are solved with possibly indeterminate (i.e., branching) reductions, the resulting search trees may grow wildly. Constraint logic programming systems try to avoid combinatorial explosion by building in strong determinate (i.e., non-branching) reduction in the form of constraint simplification. In this paper we present two concepts, residuation and guarded rules, for further strengthening determinate reduction. Both concepts apply to constraint logic programming in general and yield an operational semantics that coincides with the declarative semantics. Residuation is a control strategy giving priority to determinate reductions. Guarded rules are logical consequences of programs adding otherwise unavailable determinate reductions.

1   Introduction

A major difficulty with logic programming is combinatorial explosion: since goals are solved with possibly indeterminate (i.e., branching) reductions, the resulting search trees may grow wildly. Constraint logic programming systems [5, 12, 7] try to avoid combinatorial explosion by building in strong determinate (i.e., non-branching) reduction in the form of constraint simplification. In this paper we present two concepts, residuation and guarded rules, for further strengthening determinate reduction. Both concepts apply to constraint logic programming in general and yield an operational semantics that coincides with the declarative semantics.

1.1   Residuation

Residuation[1] is a control strategy for constraint logic programming meant to replace the rigid depth first strategy of Prolog, which amounts to eager generation of usually wrong assumptions. Residuation makes determinate reduction the rule and indeterminate reduction the exception that must be requested explicitly by declaring relations as generating. Given a goal, an atom is called *determinate* if reduction with all but possibly one clause defining the atom immediately fails due to constraint simplification. *Residuation* is now the following control strategy:

- given a goal that contains determinate atoms, a determinate atom must be reduced

- given a goal that contains no determinate atoms, an atom whose relation is declared as *generating* must be reduced.

Thus the user controls which atoms can reduce indeterminately by declaring relations as generating. If no relation is declared generating, indeterminate reduction cannot occur. Even with generating relations, indeterminate reduction can only occur if determinate reduction is not possible. A relation is called *residuating* if it is not declared generating. Given a goal, an atom is called *residuated* if it is not determinate and its relation is residuating. An important feature of the residuation strategy is that goals whose atoms are all residuated are taken as answers. Often such complex answers are fine as they are. For instance, if **length** is a length predicate for lists, the goal

$$\exists N \ (\mathsf{length}(L, N) \wedge N \leq 47)$$

("L is a list with at most 47 elements") is a perfect answer. If the user is not satisfied with a complex answer, he can request indeterminate reduction of a residuated atom.

Residuation is similar to the control strategy of the Andorra model [8, 9], with the difference that residuation performs indeterminate reduction only on atoms whose relation is explicitly declared as generating. The philosophy behind residuation is

---

[1] The term residuation was coined by Hassan Aït-Kaci [1] for delaying control schemes.

that for most relations indeterminate reduction simply does not make sense, and that complex answers are often appropriate.

In the examples of this paper we will assume a constraint system with trees and linear integer arithmetic.

A length relation for lists can be defined as follows (constraints are written in *italic font*):

$$\text{length}(\text{L}, \text{N}) \quad \leftrightarrow \quad L = nil \wedge N = 0$$
$$\vee \quad \exists \text{H}, \text{R}, \text{M} \; (L = H.R \wedge N > 0 \wedge M = N - 1$$
$$\wedge \; \text{length}(\text{R}, \text{M})).$$

Instead of the conventional definite clause syntax we use definite equivalences, which make more explicit that the relation on the left hand side is in fact defined (we are committed to least model semantics).[2]

Now, given a goal whose constraint is φ, an atom $\text{length}(\text{L}, \text{N})$ in this goal is determinate if either the constraint $\varphi \wedge L = nil \wedge N = 0$ simplifies to $\bot$, or the constraint $\varphi \wedge \exists H, R \; (L = H.R \wedge N > 0)$ simplifies to $\bot$, where $\bot$ is the canonical unsatisfiable constraint. Assuming a sufficiently powerful constraint simplifier, the goal $\text{length}(\text{X}, \text{N}) \wedge N \geq 2$ reduces in two steps determinately to the goal

$$\exists \text{Y}, \text{Z}, \text{U}, \text{M} \; (X = Y.Z.U \wedge M = N - 2 \wedge M \geq 0 \wedge \text{length}(\text{U}, \text{M})),$$

which is an answer if the relation length is residuating. In any case, it would not make sense to reduce this goal further.

Residuation is a simple and powerful alternative to delay primitives such as the delay annotations of IC-Prolog [4], the freeze construct of Prolog II [6], or the wait declarations of MU-Prolog [15]. Major advantages of residuation over these delay primitives are:

- residuation applies to every constraint system (rather than to tree systems only)

- no annotations in clauses are needed—the programmer only decides which relations should be generating

- residuation is much more flexible—even if all relations are declared generating the search space is considerably pruned since determinate reductions are performed first.

An idealized method for solving problems with residuation splits the problem solver in a propagating and a generating part:

- a predicate $\text{propagate}(\text{S})$ that holds if and only if S is a solution of the problem, and that depends only on residuating relations

---

[2]For the special case of Horn clause programming, the translation from the conventional definite clause syntax to definite equivalences is given by Clark's completion [2].

- a predicate **generate**(S) that defines candidates for (partial) solutions and depends on generating relations.

A problem instance is then given as a query

$$\varphi \wedge \textsf{propagate}(S) \wedge \textsf{generate}(S),$$

where the constraint φ describes the particular problem instance. With residuation $\varphi \wedge \textsf{propagate}(S)$ will reduce determinately to a constraint propagation network consisting of residuated atoms and a shared constraint. In general, the constraint propagation network alone is too weak to exhibit solutions. Thus **generate**(S) is needed to incrementally generate assumptions about the value of the variable S. As soon as an assumption is made, the constraint propagation network will become active since atoms that where residuated before can now fire. Typically, most of the generated assumptions will be invalidated immediately by constraint propagation leading to a failure. To obtain a feasible search space, two things are essential: careful design of the propagation and generation component, and an expressive underlying constraint system.

1.2   Guarded Rules

Guarded rules are logical consequences of the program introducing additional determinate reduction rules. We will see that guarded rules can significantly strengthen the propagation component of a problem solver.

Consider the following definition of list concatenation:

$$\begin{aligned}
\textsf{app}(X, Y, Z) \quad &\leftrightarrow \quad X = \textsf{nil} \wedge Y = Z \\
&\mid \quad X = H.R \wedge Z = H.U \wedge \textsf{app}(R, Y, U).
\end{aligned}$$

It is written in sugared syntax (indicated by writing $\mid$ rather than $\vee$), which suppresses existential quantification of auxiliary variables and allows nesting of constraint terms.

With this definition the goal $\textsf{app}(X, Y, Y)$ does not reduce determinately although it is equivalent to $X = \textsf{nil}$. In fact, the relation **app** satisfies the formula

$$Y = Z \rightarrow (\textsf{app}(X, Y, Z) \leftrightarrow X = \textsf{nil}),$$

which validates the determinate reduction of the atom $\textsf{app}(X, Y, Z)$ to the constraint $X = \textsf{nil}$ if the constraint of the goal entails the "guard" $Y = Z$.

A *guarded rule* is a formula

$$\varphi \rightarrow (A \leftrightarrow G),$$

for convenience written as

$$\varphi \ \square \ A \ \triangleright \ G,$$

where φ is a constraint (called the *guard*), A is an atom, and G is a goal. A guarded rule is *admissible* if it is valid in every model of the declarative semantics (we are committed to least model semantics). Thus admissible guarded rules are redundant as far as the declarative semantics is concerned.

The operational semantics of guarded rules is defined as follows. Given a goal G

$$\exists X(\varphi \wedge A \wedge R)$$

and a guarded rule

$$\psi \ \square \ A \ \triangleright \ G',$$

the goal G can reduce determinately to

$$\exists X(\varphi \wedge G' \wedge R)$$

if the constraint φ entails the constraint ψ, that is, the implication $\varphi \to \psi$ is valid in every model of the constraint system. Note that $\exists X(\varphi \wedge G' \wedge R)$ is logically equivalent to G in all models of the declarative semantics if the guarded rule is admissible. Moreover, $\exists X(\varphi \wedge G' \wedge R)$ is a goal up to constraint simplification and minor syntactic rearrangement.

Two further admissible guarded rules for app are

$$\mathsf{Y = nil} \ \square \ \mathsf{app(X,Y,Z)} \ \triangleright \ \mathsf{X = Z} \wedge \mathsf{list(X)}$$
$$\mathsf{X = Z} \ \square \ \mathsf{app(X,Y,Z)} \ \triangleright \ \mathsf{Y = nil} \wedge \mathsf{list(X)},$$

where the relation list is defined as follows:

$$\mathsf{list(L)} \ \leftrightarrow \ \mathsf{L = nil} \ | \ \mathsf{L = H.R} \wedge \mathsf{list(R)}.$$

Admissible guarded rules are a new concept that must not be confused with the guarded clauses of committed-choice languages such as Concurrent Prolog [16] or Parlog [3]. In these languages guarded clauses are used to define agents, while in our framework relations are defined by definite equivalences and admissible guarded rules are logical consequences of the definitions. Moreover, committed-choice languages usually do not have a declarative semantics. Maher [14] has given a declarative semantics for a strongly restricted class of committed-choice languages, where guards must be mutually exclusive. This is usually not the case for guarded rules, as can be seen in the list concatenation example.

Guarded rules have some similarity with the demon predicates of CHIP [7], but are much more general. First, demon predicates in CHIP are defined by guarded rules only, while in our approach the relation is defined independently by clauses. Second, in CHIP guards are restricted to positive tree patterns. Third, in our approach guarded rules can be given for generating relations, while in CHIP demon predicates are residuating by definition. And last not least, CHIP does not even outline a declarative semantics for demon predicates.

In the presence of guarded rules, an atom in a goal is called determinate if it either is determinate as defined before, or if it can reduce with a guarded rule. Residuation is defined as before, except that it now relies on the stronger notion of determinate atoms.

Residuation with guarded rules yields a surprisingly strong constraint propagation mechanism, which we will illustrate with two further examples.

Consider the following relational definition of the Boolean "and" function:

$$\text{and}(X, Y, Z) \ \leftrightarrow \ X = 1 \wedge Y = Z \wedge \text{bool}(Y)$$
$$| \ \ X = 0 \wedge Z = 0 \wedge \text{bool}(Y)$$

$$\text{bool}(X) \ \leftrightarrow \ X = 1 \ | \ X = 0.$$

First note that the definition of **and** in the presence of residuation already realizes four implicit guarded rules:

$$X \neq 1 \ \square \ \ \text{and}(X, Y, Z) \ \triangleright \ X = 0 \wedge Z = 0 \wedge \text{bool}(Y)$$
$$Y \neq Z \ \square \ \ \text{and}(X, Y, Z) \ \triangleright \ X = 0 \wedge Z = 0 \wedge \text{bool}(Y)$$
$$X \neq 0 \ \square \ \ \text{and}(X, Y, Z) \ \triangleright \ X = 1 \wedge Y = Z \wedge \text{bool}(Y)$$
$$Z \neq 0 \ \square \ \ \text{and}(X, Y, Z) \ \triangleright \ X = 1 \wedge Y = Z \wedge \text{bool}(Y).$$

The second and fourth rule could be optimized since under their guards we have $Y = 1$, but residuation will reduce $\text{bool}(Y)$ anyway to $Y = 1$. By exploiting the symmetry of **and** with respect to its first two arguments we obtain the admissible guarded rules

$$Y \neq 1 \ \square \ \ \text{and}(X, Y, Z) \ \triangleright \ Y = 0 \wedge Z = 0 \wedge \text{bool}(X)$$
$$X \neq Z \ \square \ \ \text{and}(X, Y, Z) \ \triangleright \ X = 1 \wedge Y = 0 \wedge Z = 0$$
$$Y \neq 0 \ \square \ \ \text{and}(X, Y, Z) \ \triangleright \ X = Z \wedge Y = 1 \wedge \text{bool}(X).$$

By adding two further admissible guarded rules

$$X = Y \ \square \ \ \text{and}(X, Y, Z) \ \triangleright \ X = Z \wedge \text{bool}(X)$$
$$X \neq Y \ \square \ \ \text{and}(X, Y, Z) \ \triangleright \ Z = 0 \wedge \text{bool}(X) \wedge \text{bool}(Y),$$

we obtain optimal constraint propagation.

For our next example assume that we want to solve a crossword puzzle. For this task a predicate $s(I, U, J, V)$ is useful that holds if and only if the I's letter of the word U is identical with the J's letter of the word V. This predicate is defined by

$$s(I, U, J, V) \ \leftrightarrow \ I = 1 \wedge U = H.R \wedge \text{at}(J, V, H)$$
$$| \ \ I > 1 \wedge U = H.R \wedge s(I - 1, R, J, V)$$

$$\text{at}(I, U, X) \ \leftrightarrow \ I = 1 \wedge U = X.R$$
$$| \ \ I > 1 \wedge U = H.R \wedge \text{at}(I - 1, R, X).$$

Now the goal $s(2, U, J, V)$ reduces to

$$\exists X, Y, W \ (U = X.Y.W \wedge \text{at}(J, V, Y)),$$

which makes explicit that the word U consists of at least two characters. However, the symmetric goal $s(I, U, 2, V)$ does not reduce determinately. This can be fixed by making the symmetry explicit with the admissible guarded rules

$$J \leq 1 \;\square\; s(I, U, J, V) \;\triangleright\; \exists H, R \; (J = 1 \wedge V = H.R \wedge at(I, U, H))$$
$$J \neq 1 \;\square\; s(I, U, J, V) \;\triangleright\; \exists H, R \; (J > 1 \wedge V = H.R \wedge s(I, U, J - 1, R))$$
$$\neg\exists H, R \; (V = H.R) \;\square\; s(I, U, J, V) \;\triangleright\; \bot.$$

1.3 Nondeclarative Use of Guarded Rules

So far we have only seen admissible guarded rules, that is, guarded rules that were logical consequences of the declarative semantics and whose operational effect was compatible with the declarative semantics. However, the operational semantics obtained by residuation and nonadmissible guarded rules is significantly stronger than what can be captured by classical declarative semantics. In fact, the object-oriented programming techniques developed for Concurrent Prolog [16] become available if determinate atoms are selected for reduction with a fair strategy.

For instance, an agent that reads two input streams X, Y and merges them into one output stream Z can be defined by four nonadmissible guarded rules:

$$X = nil \;\square\; merge(X, Y, Z) \;\triangleright\; Y = Z$$
$$X = H.R \;\square\; merge(X, Y, Z) \;\triangleright\; \exists U \; (Z = H.U \wedge merge(R, Y, U))$$
$$Y = nil \;\square\; merge(X, Y, Z) \;\triangleright\; X = Z$$
$$Y = H.R \;\square\; merge(X, Y, Z) \;\triangleright\; \exists U \; (Z = H.U \wedge merge(X, R, U)).$$

Operationally this merge agent will behave just right: as soon as a message appears on one of the two input streams, it can fire and put the message on the output stream.

It is easy to see that there is no relation **merge** such that the given guarded rules are admissible. For **merge** this could be cured by modeling streams as bags (i.e., lists whose order does not matter) rather than lists, but this would destroy the declarative semantics of most stream consumers.

1.4 Rest of The Paper

The rest of the paper presents a simple and general framework for declarative constraint logic programming with residuation and admissible guarded rules. The complications of Jaffar and Lassez's framework [11] are avoided by not providing for negation as failure.

2 Reduction Systems

The abstract notion of a well-founded reduction system captures important properties of logic programming. It builds on predicate logic in that it takes for granted first-order structures and formulae with the usual connectives and quantifiers. We assume that $\perp$ ("falsity") is a variable-free formula that is invalid in every structure.

A *reduction system* consists of the following:

- a set of formulae called *goals* containing *the trivial goal* $\perp$

- a set of structures called *models* in which the goals are interpreted

- a set of equivalences $G \leftrightarrow G_1 \vee \ldots \vee G_n$ called *reductions* such that:

    - G and $G_1, \ldots, G_n$ are goals, and $G \neq \perp$
    - $G \leftrightarrow G_1 \vee \ldots \vee G_n$ is valid in every model.

A reduction $G \leftrightarrow G_1 \vee \ldots \vee G_n$ *applies* to the goal G and no other goal. Typically, a reduction system contains many reductions with the same left hand side, that is, more than one reduction applies to a goal. A reduction system can be seen as a rewrite system, which allows to rewrite a disjunction of goals into an equivalent disjunction of goals by replacing a goal according to a reduction. The idea is to rewrite until no further reduction applies. The reduction systems corresponding to logic programs are in general nonterminating, that is, there are goals from which infinite rewrite derivations issue.

A reduction system can be separated into a *declarative component* given by its goals and models, and an *operational component* given by its goals and reductions.

We say that a goal G *reduces in one step to* G' and write $G \Rightarrow G'$ if there exists a reduction $G \leftrightarrow G_1 \vee \ldots \vee G_n$ such that $G' = G_i$ for some i. We say that a goal G *reduces to* G' if $G \Rightarrow^* G'$, where $\Rightarrow^*$ is the reflexive and transitive closure of $\Rightarrow$.

An *interpretation* is a pair consisting of a model $\mathcal{A}$ and a variable valuation α into $\mathcal{A}$. A *solution of a goal* G is an interpretation $(\mathcal{A}, \alpha)$ such that G is valid in $\mathcal{A}$ under α. A goal is *satisfiable* if it has at least one solution.

An *answer* is a goal to which no reduction applies. Note that $\perp$ is always an answer (the *trivial answer*). An *answer for a goal* G is an answer G' such that $G \Rightarrow^* G'$. A set of answers for a goal G is *complete* if it contains for every solution σ of G an answer G' such that σ is a solution of G'.

The *computational service* to be provided by a reduction system is *solving of goals*, that is, enumeration of a complete set of answers for a given goal. The declarative component of a reduction system specifies a class of problems, where every goal corresponds to a particular problem instance, and the solutions of the goal are the solutions of the problem instance. The operational component of a reduction system specifies a method for solving problem instances, where solving means to enumerate a complete set of answers.

A reduction system is *well-founded* if there exists a well-founded ordering on pairs of goals and interpretations such that for every reduction $G \leftrightarrow G_1 \vee \ldots \vee G_n$ and every solution σ of G there exist an $i = 1, \ldots, n$ such that $(G, \sigma) > (G_i, \sigma)$ and σ is a solution of G_i. A well-founded reduction system has two important properties:

- every goal has a complete set of answers

- a complete set of answers for a goal G can be enumerated as follows: if no reduction applies to G, then $\{G\}$ is a complete set of answers; otherwise, choose don't care any reduction $G \leftrightarrow G_1 \vee \ldots \vee G_n$ and solve the goals $G_1, \ldots, G_n$ in parallel.

We will see that every Horn clause program yields a well-founded reduction system.

A reduction is *determinate* if its right hand side is a single goal. We say that G *reduces determinately to* G' if G reduces to G' using only determinate reductions. If G reduces determinately to G', then G and G' have exactly the same solutions. A reduction system is *determinate* if it has only determinate reductions. Note that in well-founded and determinate reduction systems there exist no infinite reduction chains $G \Rightarrow G_1 \Rightarrow G_2 \Rightarrow G_3 \cdots$ issuing from a satisfiable goal G.

A reduction system is *terminating* if there exists no infinite chain $G \Rightarrow G_1 \Rightarrow G_2 \Rightarrow G_3 \Rightarrow \cdots$ of reduction steps. Note that a terminating reduction system is always well-founded, but not vice versa. Even a well-founded and determinate reduction system may not terminate on unsatisfiable goals.

3   Constraint Systems

A *constraint system* is a terminating and determinate reduction system whose goals are closed under conjunction, existential quantification, and variable renaming. In a constraint system we call the goals *constraints*, the answers *simplified constraints*, and the process of reducing a constraint to a simplified constraint *constraint simplification*. Note that in a constraint system one can compute for every constraint a simplified constraint. Moreover, if a constraint simplifies to the trivial constraint $\bot$, it must be unsatisfiable. A constraint system is called *complete* if a constraint is unsatisfiable if and only if it simplifies to $\bot$. Thus constraint simplification in a complete constraint system is a decision algorithm for satisfiability of constraints.

The operational component of a constraint system is called a *constraint simplifier*, and the operational component of a complete constraint system is called a *constraint solver*. Our framework for constraint logic programming does not require that the underlying constraint system is complete. Given a set of constraints with the corresponding models, one may prefer in practice an incomplete constraint simplifier since a (tractable) constraint solver may not exist.

Our notion of a constraint system is deliberately very general: every set of formulae with a corresponding class of models can be seen as a constraint system if we provide no reductions and close the formulae under conjunction, existential quantification and variable renaming. Such trivial constraint systems providing no computational service are of course not what we want in practice.

4   Definite Construction

We now introduce definite construction, which is the principle underlying constraint logic programming. We obtain a very simple framework for constraint logic programming with residuation. The two theorems given in this section are consequences of the results in [10].

We assume that a constraint system and a set of *definite relation symbols* are given, where the definite relation symbols take a fixed number of arguments and do not occur in the constraint system.

An *atom* takes the form $r(x_1, \ldots, x_n)$, where r is a definite relation symbol taking n arguments and $x_1, \ldots, x_n$ are pairwise distinct variables. A *definite goal* takes the form

$$\exists X \ (\varphi \wedge R),$$

where X is a possibly empty set of existentially quantified variables, φ is a constraint, and R is a possibly empty conjunction of atoms. Note that the definite goals containing no atoms are exactly the constraints. A *definite equivalence* takes the form

$$A \leftrightarrow G_1 \vee \ldots \vee G_n,$$

where A is an atom and $G_1, \ldots, G_n$ are definite goals called the *clauses* of A. A *definite specification* is a set of definite equivalences containing for every definite relation symbol r exactly one equivalence with r appearing at the left hand side.

In the following we assume that a definite specification is given. Moreover, we assume that φ and ψ range over constraints, A over atoms, R over possibly empty conjunctions of atoms, and G over definite goals. We will construct a reduction system for definite goals by defining definite models (the declarative semantics) and definite reductions (the operational semantics).

For convenience, we will often refer to definite goals simply as goals.

A *definite structure* is a structure that can be obtained from a model of the constraint system by adding interpretations for the definite relation symbols. Definite structures are partially ordered as follows: $\mathcal{A} \leq \mathcal{B}$ iff $\mathcal{A}$ and $\mathcal{B}$ extend the same constraint model and $r^{\mathcal{A}} \subseteq r^{\mathcal{B}}$ for every definite relation symbol r. A *definite quasi-model* is a definite structure that is a model of the definite specification. A *definite model* is a minimal definite quasi-model. The following theorem validates our declarative semantics.

Theorem 1 *For every model of the constraint system there exists exactly one definite model extending it.*

Next we define the operational semantics. We assume that the order in which atoms are written in a definite goal does not matter.

An equivalence $G \leftrightarrow D$ is a *definite reduction* iff the following conditions are satisfied:

- $G = \exists X (\varphi \wedge A \wedge R)$ is a definite goal

- $A \leftrightarrow \bigvee_{i=1}^{n} \exists Y_i \, (\varphi_i \wedge R_i)$ is obtained from a definite equivalence of the definite specification by variable renaming such that only the variables in A are shared with G

- obtain for every clause $\exists Y_i \, (\varphi_i \wedge R_i)$ of the definite equivalence the goal

$$G_i := \begin{cases} \bot & \text{if } \varphi \wedge \varphi_i \text{ simplifies to } \bot \\ \exists X \cup Y_i \, (\psi_i \wedge R_i \wedge R) & \text{if } \varphi \wedge \varphi_i \text{ simplifies to } \psi_i \neq \bot \end{cases}$$

- D is the disjunction of all $G_i \neq \bot$; if all G_i's are $\bot$, then $D = \bot$.

Note that our definition of definite reductions corresponds exactly to SLD-resolution [13] for the special case of Horn clauses.

Given a constraint system $\mathcal{C}$ and a definite specification $\mathcal{D}$ over $\mathcal{C}$, we define $\mathcal{R}(\mathcal{C}, \mathcal{D})$ as the reduction system whose goals are the corresponding definite goals, whose models are the corresponding definite models, and whose reductions are the corresponding definite reductions *together* with the reductions of the constraint system $\mathcal{C}$. It is easy to verify that $\mathcal{R}(\mathcal{C}, \mathcal{D})$ is in fact a reduction system.

Theorem 2 $\mathcal{R}(\mathcal{C}, \mathcal{D})$ *is a well-founded reduction system whose answers are exactly the simplified constraints.*

It is now straightforward to build in *residuation*. We only have to discard unnecessary indeterminate reductions:

- discard all indeterminate reductions for goals that do have determinate reductions

- discard all indeterminate reductions obtained by reduction upon a residuating atom (an atom whose relation is not declared generating).

Let us call the thus obtained reduction system $\mathcal{R}^*(\mathcal{C}, \mathcal{D}, \mathcal{G})$, where $\mathcal{G}$ is the set of generating relation symbols. Clearly, $\mathcal{R}^*(\mathcal{C}, \mathcal{D}, \mathcal{G})$ is still a well-founded reduction system. Moreover, let $\mathcal{R}^*(\mathcal{C}, \mathcal{D})$ be the reduction system $\mathcal{R}^*(\mathcal{C}, \mathcal{D}, \mathcal{G})$ where all definite relations are declared generating. Then $\mathcal{R}^*(\mathcal{C}, \mathcal{D})$ is well-founded and has again exactly the simplified constraints as answers (follows immediately from the above theorem). The important difference between $\mathcal{R}(\mathcal{C}, \mathcal{D})$ and $\mathcal{R}^*(\mathcal{C}, \mathcal{D})$ is that $\mathcal{R}^*(\mathcal{C}, \mathcal{D})$ has significantly smaller search spaces (even for the case of Horn clauses), a fact that has only been realized recently in the Andorra model [8, 9].

5 Guarded Rules

Let a constraint system $\mathcal{C}$ and a definite specification $\mathcal{D}$ over $\mathcal{C}$ be given. A *guarded rule* is a formula

$$\varphi \to (A \leftrightarrow G),$$

where φ is a constraint (called the *guard*), A is an atom, and G is a definite goal. A guarded rule is *admissible* if it is valid in every definite model.

Let $\mathcal{F}$ be a set of admissible guarded rules. Then $G \leftrightarrow G'$ is called a *forward reduction* iff the following conditions are satisfied:

- $G = \exists X(\varphi \wedge A \wedge R)$ is a definite goal

- $\psi \to (A \leftrightarrow \exists Y(\varphi' \wedge R'))$ is obtained from a guarded rule in $\mathcal{F}$ by variable renaming such that only the variables in the atom A are shared with G

- $\varphi \wedge \neg\psi$ is a constraint that simplifies to $\bot$

- $G' = \begin{cases} \bot & \text{if } \varphi \wedge \varphi' \text{ simplifies to } \bot \\ \exists X \cup Y \ (\varphi'' \wedge R' \wedge R) & \text{if } \varphi \wedge \varphi' \text{ simplifies to } \varphi'' \neq \bot. \end{cases}$

If $\varphi \wedge \neg\psi$ simplifies to $\bot$, then φ entails ψ, that is, the implication $\varphi \to \psi$ is valid in every model of the constraint system. Moreover, if the constraint system is complete, then $\varphi \wedge \neg\psi$ simplifies to $\bot$ if and only if φ entails ψ.

The reduction system $\mathcal{R}(\mathcal{C}, \mathcal{D}, \mathcal{F})$ is obtained from $\mathcal{R}(\mathcal{C}, \mathcal{D})$ by adding the forward reductions defined by the admissible guarded rules in $\mathcal{F}$. It is easy to verify that $\mathcal{R}(\mathcal{C}, \mathcal{D}, \mathcal{F})$ is in fact a reduction system, and that every goal of $\mathcal{R}(\mathcal{C}, \mathcal{D}, \mathcal{F})$ has a complete set of answers.

In general, $\mathcal{R}(\mathcal{C}, \mathcal{D}, \mathcal{F})$ is not well-founded; consider, for instance, the admissible guarded rule $\neg\bot \to (A \leftrightarrow A)$. It is the responsibility of the programmer to design the guarded rules in $\mathcal{F}$ such that $\mathcal{R}(\mathcal{C}, \mathcal{D}, \mathcal{F})$ is well-founded. Further research is necessary to find good sufficient conditions for the well-foundedness of $\mathcal{R}(\mathcal{C}, \mathcal{D}, \mathcal{F})$.

Residuation for $\mathcal{R}(\mathcal{C}, \mathcal{D}, \mathcal{F})$ is defined as before.

6 Conclusions

Residuation is a control strategy for CLP meant to replace the rigid depth first strategy of Prolog, which amounts to eager generation of usually wrong assumptions. Residuation makes determinate reduction the rule and indeterminate reduction the exception that must be requested explicitly by declaring relations as generating. Consequently, residuation may produce complex answers containing residuated atoms.

Guarded rules are logical consequences of programs adding otherwise unavailable determinate reductions. Together with residuation guarded rules yield a general and powerful constraint propagation mechanism resulting in drastically smaller search spaces.

Residuation overcomes the strictly sequential computation strategy of Prolog. With residuation every determinate atom can be reduced next, which amounts to multiple threads of computation if a fair selection strategy is used.

The operational semantics of residuation and nonadmissible guarded rules is more expressive than what can be captured by classical declarative semantics. In fact, the object-oriented programming techniques developed for Concurrent Prolog [16] can be expressed.

Topics for further research include: investigation of abstract incrementality properties ensuring efficient implementation if satisfied by constraint simplifiers; design of an abstract machine separating control from constraint simplification; and investigation of parallel reduction strategies.

Acknowledgments. The research reported in this paper was inspired by my collaboration with Hassan Aït-Kaci and Andreas Podelski on the semantics of LIFE. I'm also thankful to Ralf Scheidhauer who contributed several examples.

Bibliography

[1] H. Aït-Kaci and R. Nasr. Integrating logic and functional programming. *Lisp and Symbolic Computation*, 2:51–89, 1989.

[2] K. Clark. Negation as failure. In H. Gallaire and J. Minker, editors, *Logic and Databases*, pages 293–322. Plenum Press, New York, NY, 1978.

[3] K. Clark and S. Gregory. PARLOG: Parallel programming in logic. *ACM Transactions on Programming Languages and Systems*, 8(1):1–49, 1986.

[4] K. L. Clark and F. G. McCabe. The control facilities of IC-PROLOG. In D. Mitchie, editor, *Expert Systems in the Micro-Electronic Age*. Edinburgh University Press, Edinburgh, Scotland, 1979.

[5] A. Colmerauer. An introduction to PROLOG III. *Communications of the ACM*, pages 70–90, July 1990.

[6] A. Colmerauer, H. Kanoui, and M. V. Caneghem. Prolog, theoretical principles and current trends. *Technology and Science of Informatics*, 2(4):255–292, 1983.

[7] M. Dincbas, P. Van Hentenryck, H. Simonis, A. Aggoun, T. Graf, and F. Berthier. The constraint logic programming language CHIP. In *Proceedings of the International Conference on Fifth Generation Computer Systems FGCS-88*, pages 693–702, Tokyo, Japan, Dec. 1988.

[8] S. Haridi. A logic programming language based on the Andorra model. *New Generation Computing*, 7:109–125, 1990.

[9] S. Haridi and S. Janson. Kernel Andorra Prolog and its computation model. In D. Warren and P. Szeredi, editors, *Logic Programming, Proceedings of the 7th International Conference*, pages 31–48, Cambridge, MA, June 1990. The MIT Press.

[10] M. Höhfeld and G. Smolka. Definite relations over constraint languages. LILOG Report 53, IWBS, IBM Deutschland, Postfach 80 08 80, 7000 Stuttgart 80, Germany, Oct. 1988. To appear in the Journal of Logic Programming.

[11] J. Jaffar and J.-L. Lassez. Constraint logic programming. In *Proceedings of the 14th ACM Symposium on Principles of Programming Languages*, pages 111–119, Munich, Germany, Jan. 1987.

[12] J. Jaffar and S. Michaylov. Methodology and implementation of a CLP system. In J.-L. Lassez, editor, *Proceedings of the 4th International Conference on Logic Programming*, Cambridge, MA, 1987. The MIT Press.

[13] J. W. Lloyd. *Foundations of Logic Programming.* Symbolic Computation. Springer-Verlag, Berlin, Germany, 1984.

[14] M. J. Maher. Logic semantics for a class of committed-choice programs. In J.-L. Lassez, editor, *Logic Programming, Proceedings of the Fourth International Conference*, pages 858–876, Cambridge, MA, 1987. The MIT Press.

[15] L. Naish. Automating control for logic programs. *Journal of Logic Programming*, 3:167–183, 1985.

[16] E. Shapiro and A. Takeuchi. Object oriented programming in Concurrent Prolog. *New Generation Computing*, 1:24–48, 1983.

TPNA:
Ein neues Analyseverfahren für
Tandem-Prioritäten-Netze

O. Spaniol, W. Kremer, A. Fasbender
Lehrstuhl für Informatik IV
RWTH Aachen

1. Einleitung

Mehrklassen-Warteschlangennetzwerke mit Produktformlösungen (1), (2) werden benutzt, um Computer und Kommunikationssysteme nachzubilden und zu bewerten (3). Effiziente Berechnungsalgorithmen für Produktformnetze wurden entwickelt (4), (5), (6), (7), mit denen auch größere Netze numerisch ausgewertet werden können. Viele interessante Systemcharakteristiken und Parameter können jedoch in einem Produktformnetz nicht berücksichtigt werden.

Eine wichtige Erweiterung ist die Berücksichtigung von prioritätenorientierten Abarbeitungsdisziplinen in sogenannten Prioritätsnetzen. Gerade bei Kommunikationssystemen in Realzeitumgebungen mit heterogenen Lasten werden Prioritäten besonders wichtig (z.B. Industrieumgebung: MAP (8), (9) oder Inter-Fahrzeug-Kommunikation (10)). Um solche Kommunikationssysteme untersuchen zu können, genügt es Tandem-Prioritäts-Netze (TPN) zu betrachten, bei denen man sich auf offene Auftragsklassen beschränkt. Motiviert wird diese „Einschränkung" durch die Tatsache, daß Kommunikationsarchitekturen, welche auf dem bekannten ISO/OSI-Referenzmodell basieren, mit TPN modelliert werden können. Hierbei wird jede ISO-Ebene durch einen Tandem-Knoten beschrieben.

Durch die Abarbeitung nach Prioritäten geht die Produktformeigenschaft im allgemeinen verloren. Zudem erfordert die exakte Lösung von Prioritätsnetzen einen unvertretbar hohen Rechenaufwand bei der Lösung der globalen Gleichgewichtsgleichungen. Um solche Systeme dennoch analytisch bewerten zu können, muß auf Approximationsmethoden zurückgegriffen werden. Ziel solcher Verfahren ist es, das gewünschte Netzwerk möglichst exakt und mit erträglichem Rechenaufwand zu analysieren. Ferner sind Abschätzungen über die Größe des zu erwartenden Approximationsfehlers notwendig. Für die Berechnung von Prioritätsnetzen sind verschiedene Approximationsverfahren bekannt. Die bekanntesten und

auch effektivsten Verfahren werden nachfolgend vorgestellt und kurz diskutiert. Es zeigt sich jedoch, daß diese Verfahren sehr schlechte Ergebnisse für
Netze mit mehreren Prioritätsknoten liefern.

Daher wird in diesem Beitrag nach der Vorstellung bisher bekannter Verfahren (insbesondere der „*MVA-Priority-Approximation*" **MVA-PA**, (11))
ein neues Approximationsverfahren **TPNA** („*Tandem-Priority-Networks-
Approximation*", siehe auch (12) und (13)) vorgestellt, das speziell für
Tandem-Netze mit der prioritätenorientierten Abarbeitungsstrategie HOL
und mit ausschließlich offenen Klassen entworfen wurde. Als Bedien-
strategie wird das *nicht-unterbrechende* (*nonpreemptive*) „*Head-Of-Line*"
(**HOL**) Verfahren angenommen. Jedoch kann das neu entwickelte Analyseverfahren recht einfach auf die <u>unterbrechende</u> HOL-Disziplin *Preemptive
Resume* (PR) erweitert werden.

TPNA ist erheblich besser als alle uns bekannten Verfahren gleicher Ziel-
richtung, es liefert sogar fast exakte Werte, wenn alle Bedienstationen die
gleiche Bedienrate haben (14).

2. Problemstellung und Modellannahmen

Wie bereits erwähnt, sollen Tandemnetze mit offenen Auftragsklassen
untersucht werden.

> **Definition:** Ein <u>*Tandemnetz*</u> ist ein Warteschlangennnetz, das aus
> nacheinander geschalteten Warteschlangen besteht
> und in dem jeder Kunde eine Warteschlange nur
> maximal einmal passieren kann.

Die Motivation, gerade diesen Typ von Warteschlangennetzen zu untersuchen, liegt in dem Ziel Kommunikationssysteme zu analysieren, die gemäß
dem ISO/OSI-Referenzmodell strukturiert sind und somit aus mehreren
übereinandergeschichteten Ebenen bestehen. Jede dieser Schichten tauscht
über definierte Punkte sogenannte PDU's („Packet Data Units") aus. Eine
Schicht wird durch zwei Warteschlangensysteme modelliert, je eines für
den Empfangs- und den Sendeteil. Die „Server" des Warteschlangensystems
beschreiben das Zeitverhalten der Protokollschichten (Abb. 1).

Das Kommunikationssystem soll Nachrichten unterschiedlicher Prioritäten verarbeiten können. Das bedeutet für unser Modell, daß die Kunden unterschiedlichen Klassen zugeordnet werden müssen. In der Literatur werden diese Klassen überwiegend als geschlossene Kundenklassen modelliert (15), (16). Das heißt, daß die Kunden das System nicht verlassen können und kein neuer Kunde hinzukommen kann. Diese Annahme wird meist aus Vereinfachungsgründen gemacht, da viele approximative Analyseverfahren nur für geschlos-

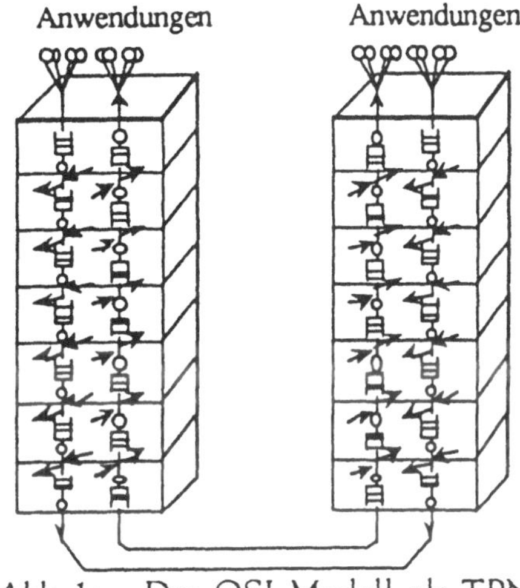

Abb. 1:   Das OSI-Modell als TPN

sene Netze mit einer festen Kundenanzahl anwendbar sind. Für eine realistischere Modellierung ist es jedoch in vielen Fällen zweckmäßiger, zu offenen Kundenklassen überzugehen.

Die Integration mehrerer, möglicherweise parallel ablaufender Anwendungen in einem System führt dazu, daß ein Kommunikationssystem mehrere Anwendungen gegebenenfalls gleichzeitig bedienen muß. Da zudem die Anwendungen häufig verschiedene Anforderungen haben (z.B. Zeitanforderungen), sind den zu übertragenden Nachrichten Prioritäten zuzuordnen, die diesen unterschiedlichen Bedingungen Rechnung tragen. Das TPN-Modell beinhaltet daher Multiklassen-Warteschlangensysteme mit prioritätenorientierten „Server"-Strategien.

Im folgenden wird als „Server"-Strategie das *nichtunterbrechende HOL*-Verfahren gewählt, da unterbrechende HOL-Verfahrens für die Modellierung eines Kommunikationssystems wenig sinnvoll sind. Jeder Knoten habe R offene Auftragsklassen, welche mit 1,...,R nach absteigender Priorität sortiert seien. Jeder Knoten kann somit durch R Eingangswarteschlangen und eine Bedienstation dargestellt werden. Abb. 2 zeigt das entsprechende TPN mit R Prioritätsklassen und N Warteschlangensystemen.

Es wird davon ausgegangen, daß die Warteschlangenpuffer unbegrenzt sind und damit das System verlustfrei ist. Weiter wird angenommen, daß ein Kunde seine Prioritätenklasse im TPN nicht wechselt. Das bedeutet auch, daß in jedem Knoten die gleiche Anzahl von Prioritäten zu verarbeiten sind.

Um das TPN analytisch handhaben zu können, werden die Verteilungen
der Zwischenankunftszeiten in jeder Kundenklasse als exponentialverteilt
angenommen, und zwar sowohl für die Ankünfte im ersten Knoten als auch
für die Ankünfte zwischen den Knoten.

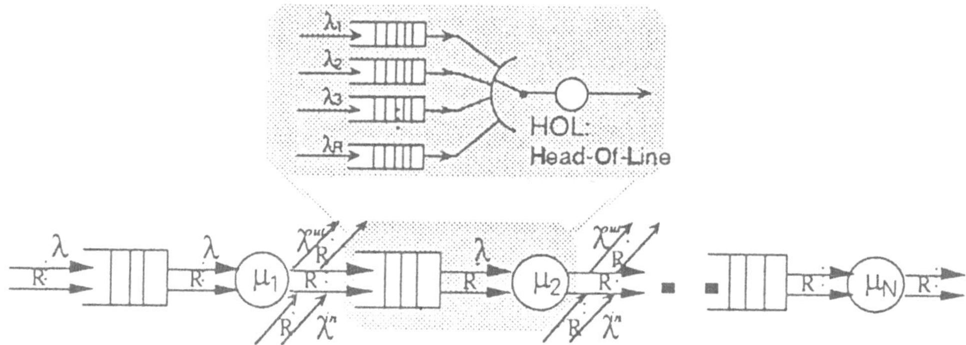

Abb. 2: Das TPN mit N HOL-Bedienstationen und R Prioritätsklassen.

Damit werden fogende Modellannahmen getroffen:

Jeder Knoten im TPN arbeitet nach der Strategie „nonpreemptive HOL".
Das TPN ist ein verlustfreies System.
Das System ist nicht überlastet (Voraussetzung für einen stationären
Zustand).
Ein Kunde ändert seine Klassenzugehörigkeit nicht.
Die Ankünfte der Kunden unterliegen in jedem Knoten und in jeder
Kundenklasse einer Poissonverteilung.

Es werden folgende Schreibweisen eingeführt:

N = Anzahl der Knoten des TPN
R = Anzahl der Kundenklassen im TPN
$\mu_{i,s}$ = mittlere Bedienrate für Kunden der Klasse s in Knoten i
$\rho_{i,s}$ = Auslastung des Knotens i für Kunden der Klasse s

Da im TPN auch „Querströme" (d.h. Input und Output-Ströme zwischen
den Knoten) erlaubt sein sollen, sind folgende Definitionen zweckmäßig:

$\lambda_{i,s}^{in}$ = Zwischenknoten - Ankunftsrate von Kunden der Klasse s
            vor Knoten i
$\lambda_{i,s}^{out}$ = Zwischenknoten - Abgangsrate von Kunden der Klasse s
            vor Knoten i
$\lambda_{i,s}$ = Gesamtankunftsrate von Kunden der Klasse s für Knoten i

Wir gehen davon aus, daß die Zwischenknoten-Ankunfts- und -Abgangs-raten zu den Ankunftsraten des vorigen Knotens addiert werden können, um die Gesamtankunftsrate zu erhalten [1]:

$$\Rightarrow \quad \lambda_{i,s} = \lambda_{i-1,s} + \lambda_{i,s}^{in} - \lambda_{i,s}^{out} := \lambda_{i-1,s} + \Delta\lambda_{i,s}$$

Die Gesamtauslastung des Knotens i wird mit ρ_i bezeichnet, wobei gilt

$$\rho_i = \sum_{s=1}^{R} \rho_{i,s} \qquad \text{mit} \qquad \rho_{i,s} = \frac{\lambda_{i,s}}{\mu_{i,s}}$$

$\mu_i \quad$ = mittlere Bedienrate aller Kundenklassen in Knoten i

$$\Rightarrow \mu_i = \frac{1}{R} \cdot \sum_{s=1}^{R} \mu_{i,s}$$

$\lambda_i \quad$ = Summe der Ankunftsraten aller Kundenklassen in Knoten i

$$\Rightarrow \lambda_i = \sum_{s=1}^{R} \lambda_{i,s}$$

$\bar{k}_{i,s} \quad$ = mittlere Anzahl von Kunden der Klasse s in Knoten i

Für die später vorgestellten Verfahren werden weiterhin benötigt:

$N_{i,s}^{(r)} \quad$ = mittlere Anzahl der Kunden der Klasse s in Knoten i
zum Ankunftszeitpunkt eines ausgezeichneten Kunden
der Klasse r , die vor ihm bedient werden

$M_{i,s}^{(r)} \quad$ = mittlere Anzahl der Kunden der Klasse s in Knoten i
die *nach dem Ankunftszeitpunkt* des ausgezeichneten Kunden
den Knoten betreten , aber vor ihm bedient werden

Gesucht werden die Antwortzeiten (Systemzeiten) des TPN bzw. der einzel-nen Knoten. Folgende Konventionen werden eingeführt:

$W_{i,r} \quad$ = mittlere Wartezeit eines Kunden der Klasse r in Knoten i
$W_{r,0} \quad$ = mittlere Restabarbeitungszeit eines Kunden in Knoten i
$S_{i,r} \quad$ = mittlere Systemzeit eines Kunden der Klasse r in Knoten i
$S_r \quad$ = mittlere Systemzeit eines Kunden der Klasse r
für das gesamte TPN mit N Knoten $\quad \Rightarrow S_r = \sum_{i=1}^{N} S_{i,r} = \hat{S}_{N,r}$

$\hat{S}_{n,r} \quad$ = mittlere Systemzeit dieses Kunden bis einschließlich
Knoten n $\quad \Rightarrow \hat{S}_{n,r} = \sum_{i=1}^{n} S_{i,r}$

---

1 Diese Annahme ist nur dann gültig, wenn alle drei Kundenraten der gleichen Verteilung unterliegen und diese Verteilung ein bestimmtes Aussehen hat (z.B. exponentialverteilt)!

3. Bisherige Verfahren

Der nachfolgende Abschnitt diskutiert kurz bekannte Approximationsverfahren für Prioritätsnetze (11). Dabei interessiert insbesondere die Frage, inwieweit die Verfahren auf das oben beschriebene Tandem-Netz angewendet werden können.

Die *"Reduced-Workrate-Approximation"* (17), (18) und die darauf aufbauende *"Shadow-Approximation"* (19) modellieren die Tatsache, daß ein Klasse-r-Kunde aufgrund der Bevorzugung höherer Prioritäten warten muß, durch eine Reduzierung der Klasse-r-Servicekapazität. Beide sind jedoch in ihrer Anwendung auf Netzwerke mit nur wenigen Prioritätsknoten beschränkt. Bei Netzwerken, die ausschließlich Prioritätsknoten enthalten wie TPN, entspricht der Berechnungsaufwand für die Leistungsgrößen dem einer exakten Analyse. Die *"Composite-Center-Approximation"* faßt von N Knoten, die einen „Central-Server"-Prioritätsknoten enthalten, die $(N-1)$ Nicht-Prioritätsknoten zu einem zusammen. Das resultierende Zwei-Knoten-Netzwerk wird exakt unter Verwendung der globalen Gleichgewichtsbedingungen (20) gelöst. Prinzipiell kann diese Methode auch auf Netzwerke mit mehr als einem prioritätsorientierten Server übertragen werden, die Kosten bei Verwendung der globalen Gleichgewichtsbedingungen steigen jedoch stark an, sodaß dieses Verfahren nicht für die Analyse von TPN geeignet ist. Die *"Heuristic-Aggregation-Method"* (21) kann ebenfalls nur für Netzwerke mit einem Prioritätsknoten effizient benutzt werden, auch bei dieser Methode werden die globalen Gleichgewichtsbedingungen benötigt.

4. Die „MVA-Priority Approximation" (MVA-PA)

Die *„MVA-Priority-Approximation"* (Mittelwertanalyse mit Prioritäts-Approximation, (11)) ist ein Approximationsverfahren für Prioritätsnetzwerke, das auf der von Reiser und Lavensberg (7) zur exakten Analyse geschlossener Warteschlangennetze mit Produktformlösung entwickelten Mittelwertanalyse aufbaut. Diese Approximation ist das bisher wohl am besten geeignete Verfahren zur Analyse von TPN. Daher soll es hier näher vorgestellt und zum Vergleich mit TPNA herangezogen werden.

4.1. Die Mittelwertanalyse für offene Netze

Die Mittelwertanalyse („Mean Value Analysis" MVA) für offene Netze basiert auf zwei Fakten (2):

> Das **Gesetz von Little** (engl. *Little's Result* (22)) besagt:
> Die mittlere Anzahl von Kunden in einem Knoten ist das Produkt von
> Durchsatz und mittlerer Antwortzeit berechnet werden: $\bar{k} = \lambda \cdot S$
> Dieses Gesetz gilt für alle Warteschlangendisziplinen und beliebige
> G/G/m-Systeme im *stationären Zustand.*
> Das **Ankunftstheorem** (engl. *Arrival Theorem* (23), (24)) besagt:
> Die Wahrscheinlichkeit, daß ein Auftrag bei Ankunft an Knoten i den
> Netzwerkzustand $(k_1, ..., k_i, ..., k_N)$ vorfindet, ist für <u>offene</u> Netze gleich
> der Gleichgewichts-Zustandswahrscheinlichkeit $p(k_1, ..., k_i, ..., k_N)$ des
> Netzes.

MVA kann auf gemischte Produktform-Warteschlangennetze (also Netze mit offenen und geschlossenen Klassen) übertragen werden (25). Hierbei gilt jedoch die Einschränkung, daß im Netz nur „single-Server"-Knoten enthalten sein dürfen. Nachfolgend sollen nur offene Netze betrachtet werden. Die mittlere Kundenanzahl in Klasse r kann dabei mit dem Ankunftstheorem wie folgt bestimmt werden (2):

$$\bar{k}_{i,r} = \frac{\rho_{i,r}}{1 - \sum_{s=1}^{R} \rho_{i,s}} = \frac{\rho_{i,r}}{1 - \rho_i}$$

4.2. Die MVA-PA für HOL-Knoten

In einem Prioritätenwarteschlangennetz ist der Ankunftsprozeß eines Knotens i.a. nicht exponentialverteilt (selbst wenn die Bedienzeiten der Knoten exponentialverteilt sind), so daß obige Gleichungen, angewandt auf ein TPN, keine exakten Lösungen liefern. Trotzdem basiert MVA-PA hierauf.

Hierzu bezeichne $\bar{k}_{i,s}^{(r)}$ die mittlere Anzahl von Aufträgen der Klasse s im Knoten i, die ein dort ankommender Klasse-r-Kunde vorfindet. Dabei führt die Beziehung

$$\bar{k}_{i,s}^{(r)} = \lambda_{i,s} \cdot \left(W_{i,s} + \frac{1}{\mu_{i,s}} \right)$$

zu folgender Ausgangsgleichung für MVA-PA:

$$S_{i,r} = \frac{\displaystyle\sum_{s=1}^{r} \frac{\bar{k}_{i,s}^{(r)}}{\mu_{i,s}} + \sum_{s=r+1}^{R} \frac{\rho_{i,s}}{\mu_{i,s}}}{1 - \sum_{s=1}^{r-1} \rho_{i,s}} + \frac{1}{\mu_{i,r}}$$

Die Anwendung des Ankunftstheorems für offene Netze führt zu:

$$\bar{k}_{i,r} = \frac{\rho_{i,r}}{1-\rho_i}$$

Es ergibt sich also für Netze mit ausschließlich offenen Klassen:

$$S_{i,r} = \frac{\dfrac{1}{\mu_{i,r}}\left(1-\sum_{s=1}^{r-1}\rho_{i,s}\right) + \sum_{s=1}^{r-1}\dfrac{\rho_{i,s}}{(1-\rho_i)\mu_{i,s}} + \sum_{s=r+1}^{R}\dfrac{\rho_{i,s}}{\mu_{i,s}}}{1-\sum_{s=1}^{r}\rho_{i,s}}$$

Die Gesamtsystemzeit für einen Klasse-*r*-Kunden in einem TPN mit N Knoten ergibt sich wie folgt:

$$S_r = \sum_{i=1}^{N} S_{i,r}$$

5. Die Analyse eines isolierten HOL-Knotens

Der erste Schritt hin zu TPNA ist die Analyse eines <u>isolierten</u> HOL-Knotens (siehe (26) und (27)). Dazu wird untersucht, in welchem Zustand ein System zum Eintrittszeitpunkt eines ausgezeichneten Kunden ist und was während der Wartezeit dieses Kunden im System passiert.

Zwei Probleme sind mit der Berechnung der Systemzeiten verbunden:

- Berechnung der Restabarbeitungszeit W_0 im Server
- Berechnung der Wartezeiten W_r für jede Klasse r

Aufbauend auf der bekannten *Pollaczek-Khinchin*-Mittelwertformel kann die Restabarbeitungszeit W_0 wie folgt berechnet werden

$$W_0 \overset{M/G/1}{=} \sum_{s=1}^{R}\rho_s \frac{\overline{x_s^2}}{2\cdot \bar{x}_s} = \sum_{s=1}^{R}\frac{\lambda_s \overline{x_s^2}}{2}$$

Für M/M/1- und M/D/1-Systeme gelten somit die Beziehungen:

$$W_0 \begin{cases} \overset{M/M/1}{=} & \sum_{s=1}^{R}\dfrac{\rho_s}{\mu_s} \\[2ex] \overset{M/D/1}{=} & \dfrac{1}{2}\sum_{s=1}^{R}\dfrac{\rho_s}{\mu_s} \end{cases}$$

Die Wartezeiten der einzelnen Klassen bzw. Prioritäten in einem HOL-Knoten können im stationären Zustand über die Warteschlangenlängen der einzelnen Klassen berechnet werden.

Über die Anzahl der Kunden im System erhält man die gesuchte Wartezeit eines ausgezeichneten Klasse-r-Kunden, indem die Kunden berücksichtigt werden, die sich zum Ankunftszeitpunkt des Kunden in den Warteschlangen befinden. Zusätzlich müssen die Kunden berücksichtigt werden, die den ausgezeichneten Kunden während seiner Wartezeit „überholen" und vorher bedient werden. Außerdem ist noch die Restabarbeitungszeit W_0 zu addieren. Die Systemzeit eines Klasse-r-Kunden in einem Knoten kann nun wie folgt berechnet werden (27):

$$\Rightarrow \quad S_r^{\,M/G/1} = \frac{W_0 + \sum_{s=1}^{r} \rho_s W_s}{1 - \sum_{s=1}^{r-1} \rho_s} + \frac{1}{\mu_r}$$

6. Das Verfahren TPNA

Die Tandem-Prioritäten-Netzwerk-Approximation (TPNA) basiert ebenfalls auf der Analyse eines isolierten, nichtunterbrechenden HOL-Prioritätsknoten nach (26) und (27) (siehe oben). Die Idee von TPNA ist, den Ansatz der Analyse eines isolierten Prioritätenknotens für N hintereinander geschaltete Prioritätsknoten zu erweitern. Es wird untersucht, in welchem Zustand sich das Netzwerk befindet, wenn ein ausgezeichneter Kunde das Netz bzw. einen der nachfolgenden Knoten betritt und wenn sich dieser Kunde im „Server" eines Knotens befindet.

Diese Vorgehensweise ist natürlich nur dann sinnvoll, wenn sich das Netz in einem stationären Zustand befindet. In diesem Bericht wird die Frage, ob in einem TPN ein stationärer Zustand vorliegt, nicht weiter erörtert und näherungsweise als gegeben angenommen.

Die gesuchte Systemzeit des TPN wird Knoten für Knoten, beginnend beim ersten, berechnet. Dabei werden die Wartezeiten der vorher berechneten Knoten mit einbezogen. Die Systemzeit eines Klasse-r-Kunden im ersten Knoten wird mit den Gleichungen aus Kapitel 5 ausgewertet.

6.1. Die Systemzeit in einem N-Knoten-TPN

Das Verfahren wird für Tandemnetze mit N Knoten vorgestellt. Für die weitere Vorgehensweise wird ein beliebiger Knoten n betrachtet. Es ergibt sich dabei die Frage: In welchem Zustand befindet sich das TPN, wenn ein ausgezeichneter Klasse-r-Kunde den Knoten n betritt? Um diese Frage zu beantworten, müssen alle Kunden berücksichtigt werden, welche die davor geschalteten Knoten 1 bis (n-1) passieren und bevorzugt bedient werden.

Man erhält somit für die Anzahl der Kunden, die den n-ten Knoten erreichen und vor dem Klasse-r-Kunden bedient werden:

$$\underbrace{\sum_{i=1}^{n-1}\sum_{s=1}^{r-1} M_{i,s}^{(r)}}_{\substack{\text{\# Kunden, die den} \\ \text{ausgezeichneten} \\ \text{Kunden aus Klasse r} \\ \text{überholen}}} + \underbrace{\sum_{i=1}^{n-1}\sum_{s=1}^{r} N_{i,s}^{(r)}}_{\substack{\text{\# Kunden, die in} \\ \text{Knoten 1,..,n-1} \\ \text{bevorzugt bedient} \\ \text{werden}}} + \underbrace{\sum_{i=1}^{n-1} W_{i,0}\cdot\mu_i}_{\substack{\text{\# Kunden, die den} \\ \text{Knoten 1,..,n-1 während} \\ \text{der Restabarbei -} \\ \text{tungszeit passieren}}} + \underbrace{\sum_{i=1}^{n-1}\sum_{s=1}^{R} \frac{\lambda_{i,s}}{\mu_{i,r}}}_{\substack{\text{zusätzliche Kunden ,} \\ \text{die der Flußer-} \\ \text{haltung dienen}}} + \underbrace{\sum_{i=1}^{n-1} \hat{S}_{i,r} \sum_{s=1}^{r} \Delta\lambda_{i+1,s}}_{\substack{\text{\# Kunden aus den} \\ \text{Querströmen , die} \\ \text{während } \hat{S}_{n\text{-}1,r} \text{ den} \\ \text{Knoten n erreichen}}}$$

Alle diese Kunden müssen den n-ten Knoten mit der jeweiligen Bedienzeit des n-ten Knotens durchlaufen. Man erhält dann, durch Multiplikation dieser Summe mit der mittleren Bedienzeit des n-ten Knotens, die erwartete Zeit, die diese Kunden den n-ten Knoten belegt halten noch bevor bzw. während der ausgezeichnete Kunde den n-ten Knoten betritt:

$$\sum_{i=1}^{n-1}\sum_{s=1}^{r-1} \frac{M_{i,s}^{(r)}}{\mu_{n,s}} + \sum_{i=1}^{n}\sum_{s=1}^{r} \frac{N_{i,s}^{(r)}}{\mu_{n,s}} + \sum_{i=1}^{n} \frac{W_{i0}\cdot\mu_i}{\mu_n} + \sum_{i=1}^{n-1}\sum_{s=1}^{R} \frac{\lambda_{i,s}}{\mu_{n,s}\mu_{i,r}} + \sum_{i=1}^{n-1} \hat{S}_{i,r} \sum_{s=1}^{r} \frac{\Delta\lambda_{i+1,s}}{\mu_{n,s}}$$

Der ausgezeichnete Klasse-r-Kunde hat allerdings zu diesem Zeitpunkt selbst die Zeit $\hat{S}_{n-1,r}$ verbraucht.Dadurch verringert sich die restliche, noch zu wartende Zeit. Diese Restzeit wird als „*Residual Delay*" bezeichnet:

$$T_{nr} = \left[\sum_{i=1}^{n-1}\sum_{s=1}^{r-1} \frac{M_{is}^{(r)}}{\mu_{ns}} + \sum_{i=1}^{n}\sum_{s=1}^{r} \frac{N_{is}^{(r)}}{\mu_{ns}} + \sum_{i=1}^{n} \frac{W_{i0}\cdot\overline{\mu}_i}{\overline{\mu}_n} \right.$$
$$\left. + \sum_{i=1}^{n-1}\sum_{s=1}^{R} \frac{\lambda_{is}}{\mu_{ns}\mu_{ir}} + \sum_{i=1}^{n-1} \hat{S}_{ir} \sum_{s=1}^{r} \frac{\Delta\lambda_{(i+1)s}}{\mu_{ns}} \right] - \hat{S}_{(n-1)r}$$

Diese Restzeit kann natürlich nicht negativ werden. Wenn zum Beispiel der „Server" des zweiten Knotens wesentlich schneller ist als der des ersten Knotens und keine zusätzlichen Kunden hinzu kommen, so wird voraussichtlich der Warteraum des zweiten Knotens bei Ankunft des ausgezeichneten Kunden leer sein. Daher muß für ein noch zu bestimmendes Minimum folgende Bedingung erfüllt sein:

$$\oplus \qquad T_{n,r} \geq T_{n,r}^{\min} \qquad \forall\, r = \{1...R\}$$

Für das Minimum sind nur noch die Kunden zu berücksichtigen, die der Flußerhaltung im $(n\text{-}1)$-ten Knoten dienen (näheres in (12), (13)). Der minimale „Residual Delay" für den zweiten Knoten beträgt dann:

$$T_{n,r}^{\min} := \sum_{s=1}^{R} \frac{\lambda_{n-1,s}}{\mu_{n-1,r} \cdot \mu_{n,s}}$$

Somit kann ein rekursiver Term zur Berechnung der Systemzeit angegeben werden, bei dem Bedingung $\oplus$ als Fallunterscheidung berücksichtigt wird:

$$\hat{S}_{n,r} = \begin{cases} \hat{S}_{n-1,r} + T_{n,r} + \sum\limits_{s=1}^{r-1} \dfrac{M_{n,s}^{(r)}}{\mu_{n,s}} + \dfrac{1}{\mu_{n,r}} & , \; T_{n,r} \ge T_{n,r}^{\min} \\[3ex] \hat{S}_{n-1,r} + T_{n,r}^{\min} + \sum\limits_{s=1}^{r-1} \dfrac{M_{n,s}^{(r)}}{\mu_{n,s}} + \dfrac{1}{\mu_{n,r}} & , \; T_{n,r} < T_{n,r}^{\min} \end{cases}$$

Ausgeschrieben und unter Hinzunahme eines empirischen Terms (∇) erhält man folgenden Ausdruck für die Systemzeit (für n>1):

$$\hat{S}_{n,r} = \begin{cases} \hat{S}_{n-1,r} + \left[\sum\limits_{i=1}^{n-1}\sum\limits_{s=1}^{r-1} \dfrac{\lambda_{i,s}W_{i,r}}{\mu_{i,s}} + \sum\limits_{i=1}^{n}\sum\limits_{s=1}^{r} \dfrac{\lambda_{i,s}W_{i,s}}{\mu_{n,s}} + \overbrace{\sum\limits_{i=1}^{n-1} \dfrac{W_{i,0}}{\rho_n}}^{[\nabla]} + W_{n,0} + \sum\limits_{i=1}^{n-1}\sum\limits_{s=1}^{R} \dfrac{\lambda_{i,s}}{\mu_{n,s}\mu_{i,s}} \right. \\[1ex] \qquad \left. + \sum\limits_{i=1}^{n-1} \hat{S}_{i,r} \sum\limits_{s=1}^{r} \dfrac{\Delta\lambda_{i+1,s}}{\mu_{n,s}} \right] - \hat{S}_{n-1,r} + W_{n,r}\sum\limits_{s=1}^{r-1}\rho_{n,s} + \dfrac{1}{\mu_{n,r}} & , \text{wenn } T_{n,r} \ge T_{n,r}^{\min} \\[3ex] \hat{S}_{n-1,r} + \sum\limits_{s=1}^{R} \dfrac{\lambda_{n-1,s}}{\mu_{n-1,r} \cdot \mu_{n,s}} + W_{n,r}\sum\limits_{s=1}^{r-1}\rho_{n,s} + \dfrac{1}{\mu_{nr}} & , \text{wenn } T_{n,r} < T_{n,r}^{\min} \end{cases}$$

Als nächstes muß Bedingung $\oplus$ durch einen handhabbaren Ausdruck ersetzt werden. Hierzu sollen folgende Überlegungen gemacht werden. Der Knoten mit der höchsten Auslastung im TPN wirkt als „Bottleneck". Hat der ausgezeichnete Knoten n eine niedrigere Auslastung als einer der vorigen Knoten 1 bis $(n\text{-}1)$ des TPN, so werden die Kunden in diesem Knoten schneller bearbeitet als in Knoten n. Das bedeutet aber auch, daß zum Eintrittszeitpunkt des Klasse-r-Kunden in Knoten n die Kunden aus $M_{i,s}^{(r)}$ und $N_{i,s}^{(r)}$ diesen Knoten bereits verlassen haben. Dadurch sind diese Kunden nicht mehr zu berücksichtigen, was der Bedingung $T_{n,r} < T_{n,r}^{\min}$ entspricht. Mit diesen Überlegungen ergeben sich folgende Abschätzungen für Bedingung $\oplus$:

$$\left(T_{n,r} \geq T_{n,r}^{\min}\right) \mapsto \left(d\hat{\rho}_{n,r} \geq \max\{\hat{\rho}_{1,r},\ldots,\hat{\rho}_{n-1,r}\}\right)$$

$$\left(T_{n,r} < T_{n,r}^{\min}\right) \mapsto \left(d\hat{\rho}_{n,r} < \max\{\hat{\rho}_{1,r},\ldots,\hat{\rho}_{n-1,r}\}\right)$$

wobei $d\hat{\rho}_{n,r} := \sum_{s=1}^{r}(\rho_{n,s} - \Delta\rho_{n,s})$ und $\hat{\rho}_{n,r} := \sum_{s=1}^{r}\rho_{n,s}$

sowie $\Delta\rho_{n,r} := \dfrac{\Delta\lambda_{n,r}}{\mu_{n,r}}$

Nun kann die rekursive Berechnung der mittleren Wartezeit eines Klasse-r-Kundens in Knoten n wie folgt durchgeführt werden:

$$W_{n,r} = \begin{cases} \dfrac{1}{\sigma_{n,r}} \cdot \left[\sum_{i=1}^{n-1} W_{i,r}\left(\dfrac{\lambda_{i,r}}{\mu_{n,r}} - 1\right) + \sum_{s=1}^{r-1}\left(\rho_{n,s}W_{n,s} + \sum_{i=1}^{n-1}\dfrac{\lambda_{i,s}}{\mu_{n,s}}(W_{i,r} + W_{i,s})\right) \right. \\ \left. \qquad + \sum_{i=1}^{n-1}\dfrac{1}{\mu_{i,r}}\left(\sum_{s=1}^{R}\dfrac{\lambda_{i,s}}{\mu_{n,s}} - 1\right) + \sum_{i=1}^{n-1}\hat{S}_{i,r}\sum_{s=1}^{r}\dfrac{\Delta\lambda_{i+1,s}}{\mu_{n,s}} + \sum_{i=1}^{n-1}\dfrac{W_{i,0}}{\rho_n} + W_{n,0} \right] & , \text{falls } d\hat{\rho}_{n,r} \geq \\ & \qquad \max\{\hat{\rho}_{1,r},\ldots,\hat{\rho}_{n-1,r}\} \\[1em] \dfrac{1}{\sigma_{n,r-1}} \cdot \sum_{s=1}^{R}\dfrac{\lambda_{n-1,s}}{\mu_{n-1,r} \cdot \mu_{n,s}} & , \text{falls } d\hat{\rho}_{n,r} < \\ & \qquad \max\{\hat{\rho}_{1,r},\ldots,\hat{\rho}_{n-1,r}\} \end{cases}$$

Mit der Bewertung des ersten Knotens als isolierten HOL-Knoten und der obigen Gleichung für die restlichen (N-1) Knoten ergibt sich die Gesamtsystemzeit S_r für einen Klasse-r-Kunden in einem N-Knoten TPN wie folgt:

$$S_r = \sum_{i=1}^{N}\left(W_{i,r} + \dfrac{1}{\mu_{i,r}}\right)$$

7. Das Fehlerverhalten von TPNA

Das in Kapitel 6 entwickelte TPNA-Verfahren soll nun auf seine Brauchbarkeit überprüft werden, d.h. es wird untersucht, wie gut es das tatsächliche Verhalten eines TPN beschreibt.

Die exakte Bestimmung der Systemzeiten eines TPN kann entweder durch die Lösung der globalen Gleichgewichtsgleichungen erfolgen, was für offene Netze mit mehreren hundert Kunden sehr aufwendig wird, oder durch Simulationen, die ebenfalls sehr aufwendig sind. Im Rahmen dieser Arbeit beschränken wir uns auf simulative Techniken (13).

Da nicht alle möglichen Konfigurationen eines TPN untersucht werden können, sollen folgende Einschränkungen vorgenommen werden:

Die Bedienraten seien unabhängig von den Kundenklassen:

$$\Rightarrow \quad \mu_{is} = \mu_{ir} = \mu_i \quad \forall\, s, r \in \{1, ..., R\}$$

Die Anzahl der Klassen (Prioritäten) sei $R = 4$.

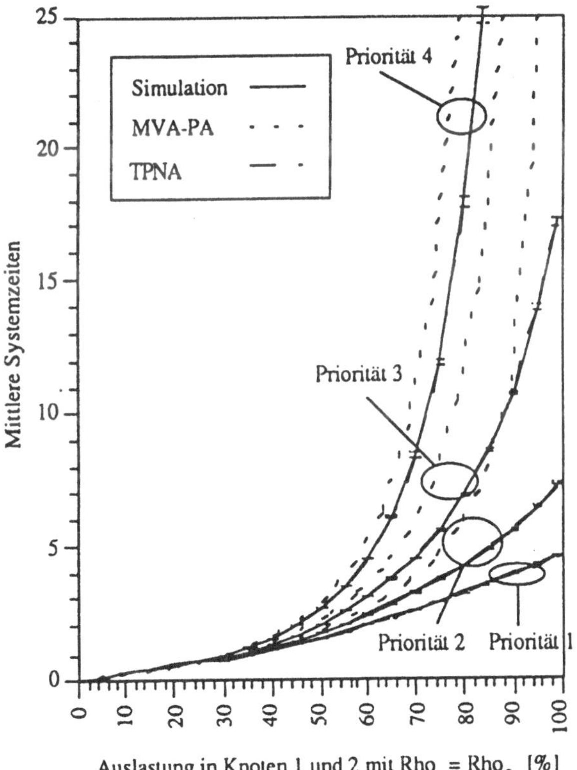

Einen ersten Eindruck zum Fehlerverhalten von MVA-PA und TPNA ermöglicht Abb. 3. Sie zeigt das Verhalten eines 2-Knoten TPN mit gleichen Auslastungen der Knoten ($\rho_1 = \rho_2$). Es werden die Systemzeiten der 4 Prioritäten als Funktion der jeweiligen Auslastung gezeigt.

MVA-PA ist relativ genau für die höchste und niedrigste Prioritätsklasse. Für die anderen Prioritäten wird MVA-PA bei steigenden Lasten immer ungenauer. Da die Systemzeiten bei steigendem ρ_i sehr stark anwachsen, machen sich die Fehler bei MVA-PA besonders bemerkbar. Die TPNA-Kurven stimmen fast exakt mit den Simulationsergebnissen überein.

Abb. 3 Systemzeiten bei 2-Knoten-TPN mit $\rho_1 = \rho_2$.

Ein gutes Maß für die Genauigkeit der Approximationsverfahren ist der sogenannte prozentuale „*Toleranz-Fehler*" Δ_r. Darüber hinaus werden noch die „*maximalen Toleranz-Fehler*" Φ_r bei variierenden „Lastverhältnissen" zwischen den einzelnen Auftragsklassen angegeben. Folgende Einstellungen werden vorgenommen:

$$\Delta_r \quad = \quad \text{prozentualer Toleranzfehler für einen Kunden der Klasse r}$$

$$= \quad \frac{\hat{S}_{N,r}(\vec{\rho}) - SIM_{N,r}(\vec{\rho})}{SIM_{N,r}(\vec{\rho})} \cdot 100 \quad \text{mit } \vec{\rho} = (\rho_1, \rho_2, ..., \rho_N)$$

$$\text{wobei } SIM_{N,r}(\vec{\rho}) = \text{Simulationsergebnisse bei gegebenem } \vec{\rho}$$

$$\Lambda_4 \;=\; \text{Lastverhältnis zwischen den Auftragsklassen}$$

$$\Lambda_4 = \frac{\lambda_1 + \lambda_3}{\lambda_2 + \lambda_4} \text{ mit } \sum\nolimits_{r=1}^{R} \lambda_r = 1$$

$$\Phi_r \;=\; \text{maximaler Toleranzfehler für einen Kunden der Klasse r}$$

$$= \max\left\{ \Delta_r(\vec{\rho}) \mid 0{,}1 \leq \rho_i \leq 0{,}9 \right\}$$

Da Diagramme wie Abb. 3 nur spezielle Lastverhältnisse berücksichtigen können, wird eine Darstellung mit „*Konturdiagrammen*" bevorzugt. Hier können zwei Auslastungen variiert werden. Die Konturdiagramme der Toleranzfehler geben auf der x- und y-Achse verschiedene Auslastungen an, die Toleranzfehler werden wie Höhenlinien eingetragen.

7.1. TPNA im Vergleich mit MVA-PA für 4-Knoten-TPN

Nachfolgend wird ein Tandemnetz mit vier Knoten untersucht. Für alle Server werden exponentialverteilte Bedienzeiten vorausgesetzt, sodaß sowohl MVA-PA als auch TPNA für die Analyse benutzt werden können.

Es werden weiterhin folgende Annahmen gemacht:

$$\lambda_{i,s} = \lambda_{i,r} \quad \forall\, s,r = 1,\dots,4 \text{ und } i = 1,\dots,4$$

d.h. die Ankunftsraten sind in allen Klassen und Knoten gleich.
$$\vec{\rho} = (\rho_{1,2}; \rho_{1,2}; \rho_{3,4}; \rho_{3,4})$$
d.h. die Auslastungen der Knoten 1 und 2 bzw. 3 und 4 werden gleichmäßig variiert.

Durch diese Festlegung der Auslastungen können die Konturdiagramme der Toleranzfehler über die Auslastungen beider Knotenpaare aufgetragen werden.

Für die höchste Prioritätsklasse sind die·Fehlerverhalten beider Verfahren akzeptabel (Abb. 4). Für MVA-PA liegen die Fehler unter 2,4 % und bei TPNA unter 10%. Dieses Verhalten kann damit begründet werden, daß die Kunden höchster Priorität näherungsweise exponentialverteilt die einzelnen Knoten betreten.

Für die Klasse-2-Kunden gilt diese Beobachtung offensichtlich nicht mehr. MVA-PA wird bei Auslastungen > 0,8 merklich schlechter (bis zu 100% Fehler), TPNA dagegen weist nicht mehr als 12 % Fehler auf und bleibt relativ stabil (siehe Abb. 5).

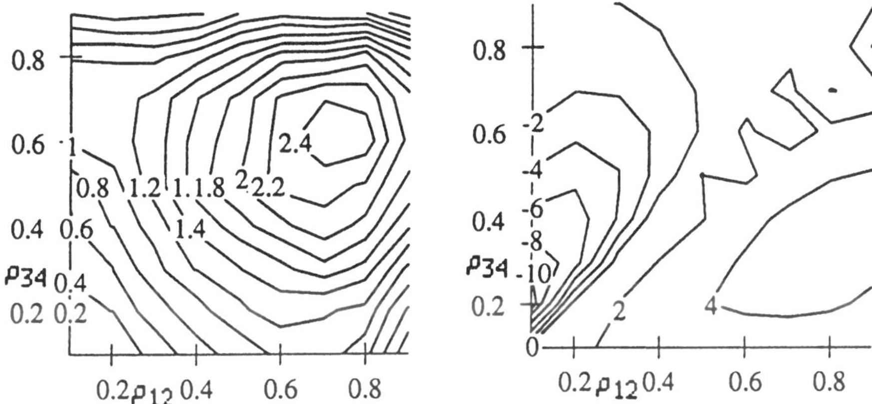

Abb. 4: MVA-PA (links) und TPNA (rechts) für Priorität 1

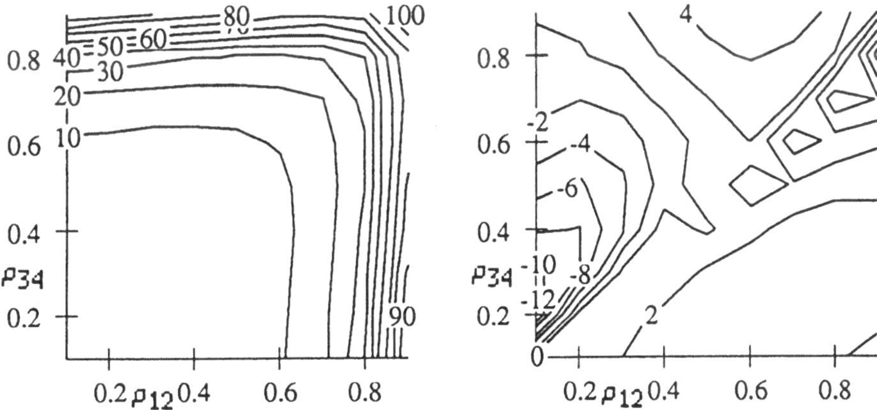

Abb. 5: MVA-PA (links) und TPNA (rechts) für Priorität 2

Für die Klasse-3-Kunden (Abb. 6) und Klasse-4-Kunden (Abb. 7) wird das Fehlerverhalten für MVA-PA ab einer Auslastung von > 0,8 sehr schlecht, wobei Toleranzfehler von bis zu 140 % auftreten.

TPNA dagegen erreicht maximale Toleranzfehler von 25 %. Der Fall $\rho_{1,2} > \rho_{3,4}$ entspricht dem zweiten Teil der TPNA-Gleichung und liefert eine bessere Approximation als der erste Teil der Berechnung. Dies läßt die Vermutung zu, daß die Einführung von Prioritäten weniger Einfluß auf die Systemzeiten der Kunden hat, sobald stark ausgelastete Knoten am Anfang vom TPN liegen.

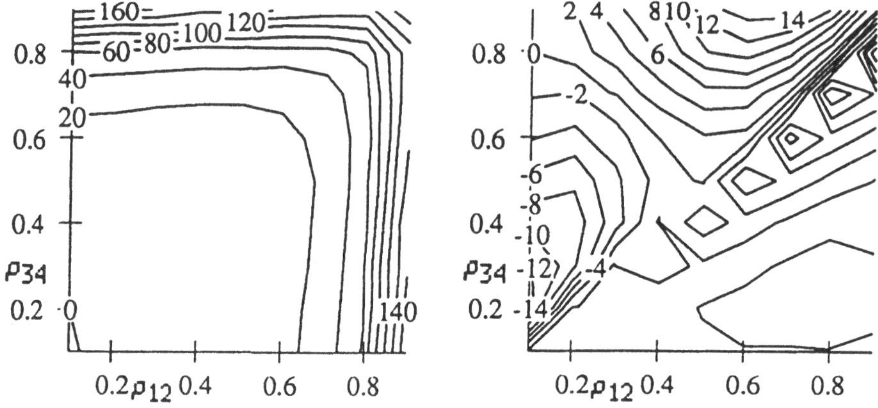

Abb. 6: MVA-PA (links) und TPNA (rechts) für Priorität 3

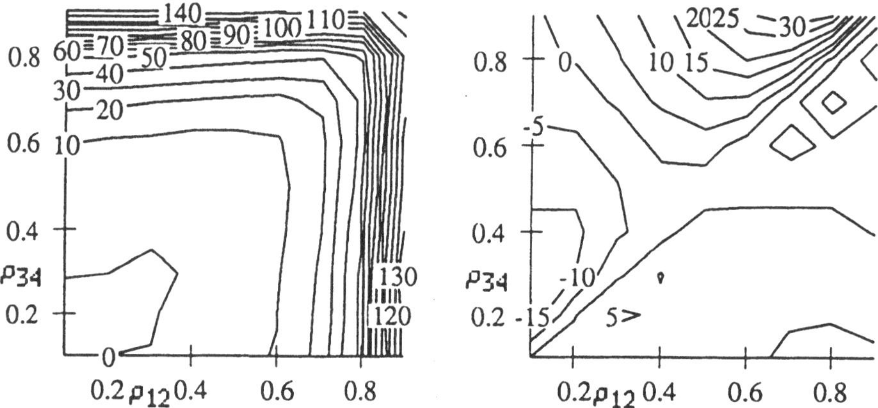

Abb. 7: MVA-PA (links) und TPNA (rechts) für Priorität 4

Die Approximationsgüte beider Verfahren ist für Auslastungen bis zu 70%
vergleichbar. Bei Hochlast ist MVA-PA nicht mehr sinnvoll einsetzbar, da
Fehler von über 140% auftreten.

7.2. Die maximalen Toleranzfehler für 2-Knoten-TPN

In diesem Unterkapitel sollen die maximalen Toleranzfehler Φ_r der
Verfahren MVA-PA und TPNA untersucht werden. In den nachfolgenden
Diagrammen werden die maximalen prozentualen Abweichungen der
Approximationen von den Simulationsergebnissen über die variierenden

Lastverhältnisse Λ_4 in den Prioritätsklassen aufgetragen. Die maximalen Toleranzfehler berechnen sich aus der Maximumbildung eines Kontur-diagramms für das entsprechende Lastverhältnis, wobei jedoch nur Werte aus dem Intervall $0{,}1 \leq \rho \leq 0{,}9$ betrachtet wurden:

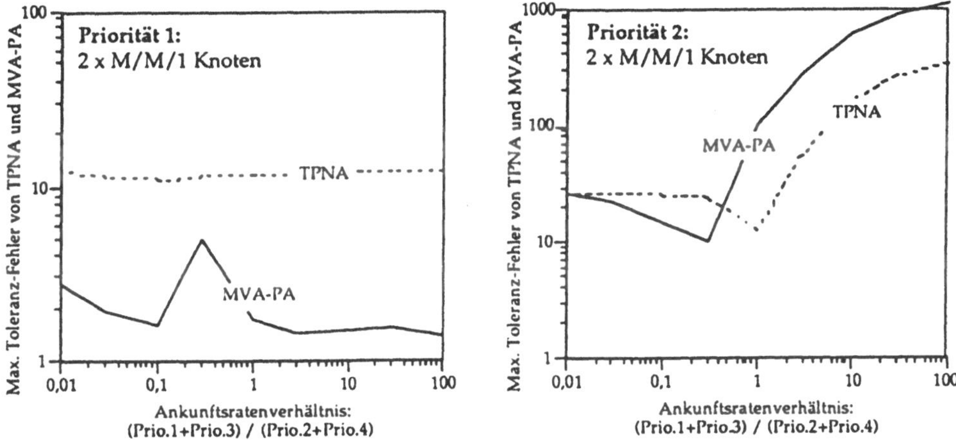

<u>Abb. 8</u>: Maximale Toleranzfehler von Priorität 1 und 2 für 2-Knoten TPN.

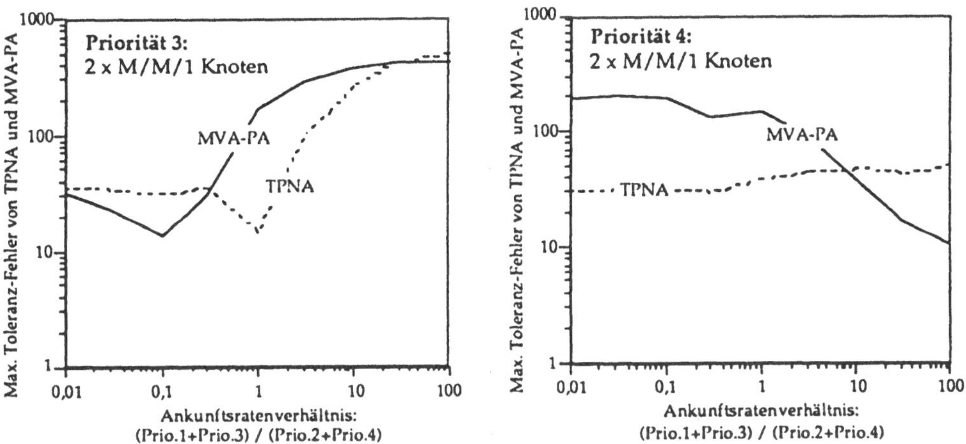

<u>Abb. 9</u>: Maximale Toleranzfehler von Priorität 3 und 4 für 2-Knoten TPN.

Für Klasse 1 sind die maximalen Toleranzfehler bei MVA-PA merklich niedriger als bei TPNA. Allerdings liegen diese Fehler, verglichen mit den unteren Klassen, im noch akzeptablen Bereich von unter 15% (Abb. 8).

Bei den Klasse-2 (Abb. 8) und Klasse-3-Kunden (Abb. 9) lassen sich erstaunliche Parallelitäten im Fehlerverhalten der Verfahren MVA-PA und TPNA feststellen. Beide Verfahren sind recht stabil, solange λ_1 klein ist. Steigt λ_1 an, so erfahren die Kurven einen Knick, worauf die maximalen Fehler an-schließend sehr stark anwachsen. Bei TPNA erfolgt dieser Knick etwas später als bei MVA-PA, wodurch das Fehlerverhalten insgesamt bei TPNA besser ist.

Für Klasse-4-Kunden (Abb. 9) weist TPNA einen fast gleichmäßigen Verlauf der maximalen Toleranzfehler bei $\approx 30\%$ auf. MVA-PA ist jedoch wesentlich schlechter für kleine λ_1 (> 100%) und wird erst ab Λ_4 >10 langsam besser als TPNA.

7.3. Das Fehlerverhalten von TPNA und MVA-PA bei Querströmen

Bisher wurde der Einfluß von Querströmen im TPN noch nicht untersucht. Dies erfolgt jetzt mit einem 3-Knoten-TPN, in welchem ein zusätzlicher Eingangsstrom zwischen Knoten 1 und 2 gelegt wird. Die nachfolgenden Diagramme geben zu jeder Prioritätsklasse einen Ausschnitt eines $d\lambda$-Bereiches an, wobei $d\lambda$ wie folgt definiert ist:

$$d\lambda = \frac{\sum_{s=1}^{R} \lambda_{2,s}^{in}}{\sum_{s=1}^{R} \lambda_{1,s}} \quad \text{im Bereich } 0,5...1,5$$

Da durch den Zusatzstrom der erste Knoten weniger Kunden zu bearbeiten hat und die Konturdiagramme nur von einer „Referenz"-Auslastung abhängig sein sollen, wird folgendes eingesetzt:

$$\rho_i = \begin{cases} \rho - \frac{1}{\mu_2} \cdot \sum_{s=1}^{R} \lambda_{2,s}^{in} & , i = 1 \\ \rho & , i = 2, 3 \end{cases}$$

Durch die Einführung von Querströmen verändert sich das grundsätzliche Fehlerverhalten der beiden Verfahren nur geringfügig. TPNA verbessert sich sogar bei den niedrigeren Prioritäten. Bei MVA-PA setzt sich das schlechte Verhalten unter steigender Auslastung fort, es werden Fehler von über 130 % errreicht. TPNA dagegen erreicht Fehler von maximal 18% (Abb. 10 bis 14).

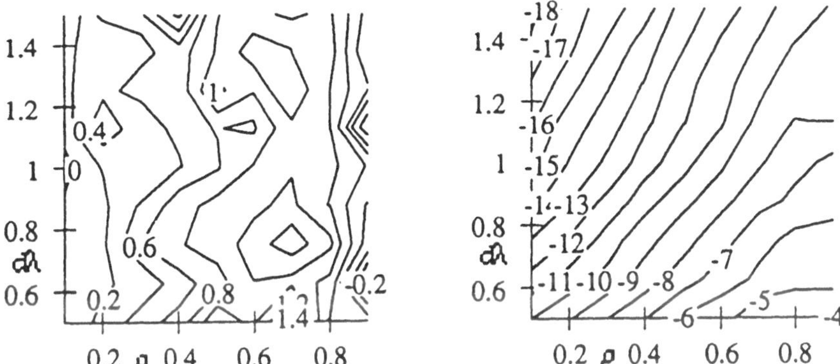

Abb. 10: MVA-PA (links) und TPNA (rechts) für Priorität 1.

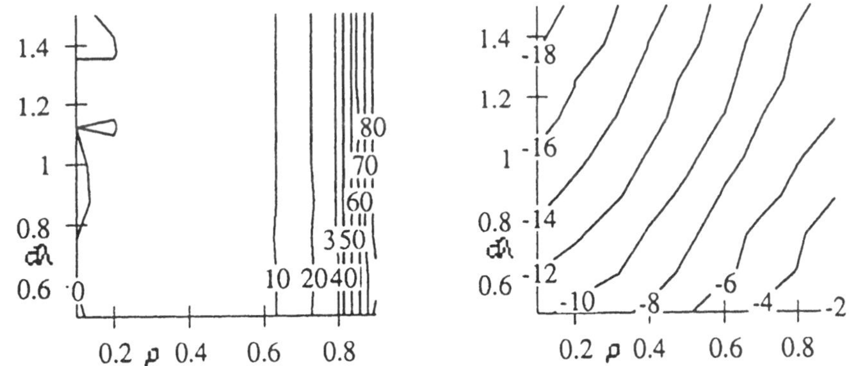

Abb. 11: MVA-PA (links) und TPNA (rechts) für Priorität 2.

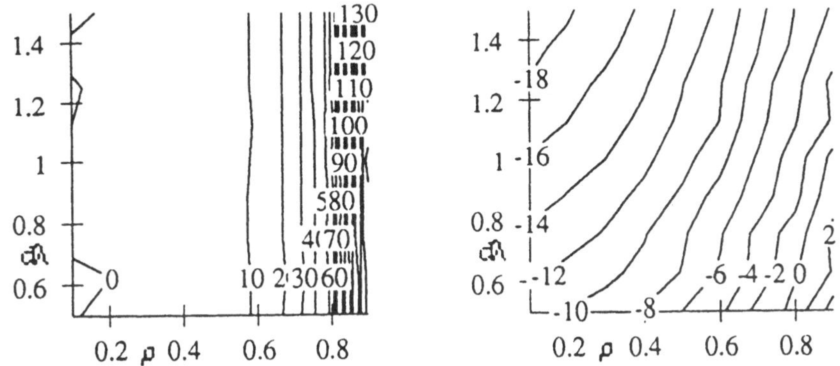

Abb. 12: MVA-PA (links) und TPNA (rechts) für Priorität 3.

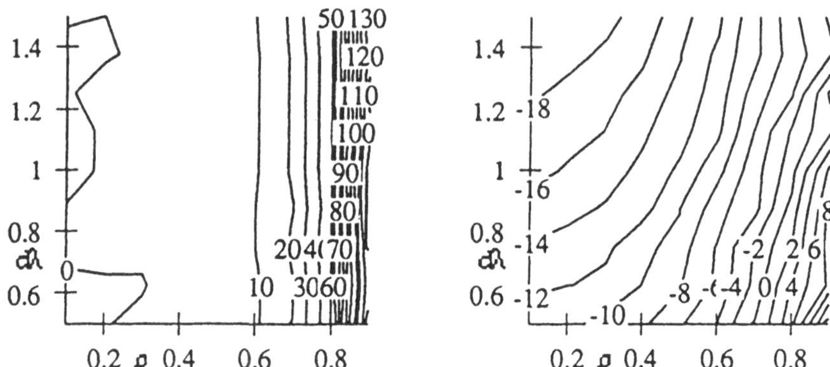

Abb. 13: MVA-PA (links) und TPNA (rechts) für Priorität 4.

In (12) wurden weitere $d\lambda$-Bereiche untersucht (von 0,1 bis 0,5 sowie 10 bis 100). Dabei wurden die hier gemachten Beobachtungen bestätigt.

7.4. Das Fehlerverhalten von TPNA bei Knoten vom Typ M/D/1

Zuletzt soll das Fehlerverhalten des neuen TPNA-Verfahrens bei Verwendung von Knoten mit nicht-exponentiellen Servern untersucht werden. MVA-PA ist hierfür unbrauchbar, daher beschränken wir uns auf TPNA.

Es wird ein Zwei-Knoten-TPN mit deterministischen Servicezeiten untersucht. Dabei soll der Einfluß von Querströmen ausgeklammert werden. Der einzige Unterschied für die Berechnung dieses Netzes nach TPNA liegt in der Bestimmung der Restabarbeitungszeiten.

Somit ergeben sich für das TPN folgende Parametereinstellungen:

$$\lambda_{i,s} = \lambda_{i,r} \quad \forall\, s,r = 1,\dots,4$$

$$\lambda_{i,s} = \lambda_{j,s} \quad \forall\, i,j = 1,2,3$$

d.h. die Ankunftsraten sind in allen Klassen und Knoten gleich.

Wie bereits im vorigen Kapitel werden hierfür Konturdiagramme verwendet, die bei variierenden ρ_1 und ρ_2 die prozentualen Toleranzfehler angeben.

Durch die Verwendung von ·/D/1-Knoten erhöhen sich die mittleren prozentualen Fehler von TPNA nur unwesentlich (Abb. 14 und 15). Die Struktur der Diagramme wird jedoch unregelmäßiger und die <u>maximalen</u> Toleranzfehler wachsen, bedingt durch die fehlende Stationarität des Netzes, auf etwa das Doppelte der Werte bei ·/M/1-Knoten an. Im Bereich $\rho_1 = \rho_2$, also auf der Winkelhalbierenden, wird erneut eine gute Approximation erreicht:

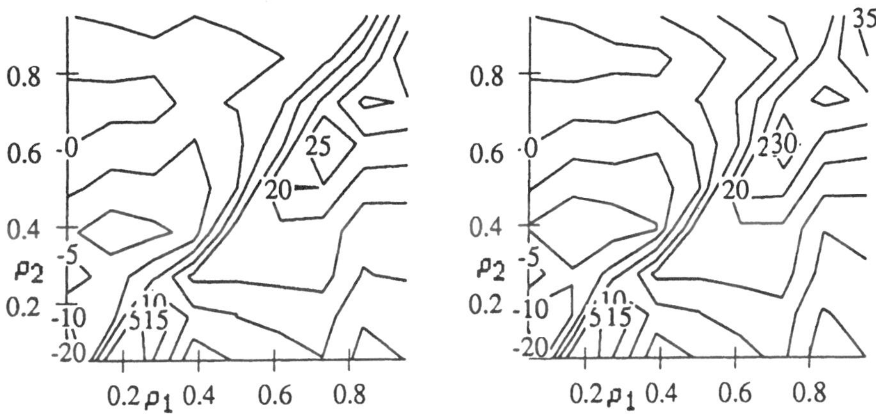

Abb. 14: TPNA für Priorität 1 (links) und Priorität 2 (rechts)

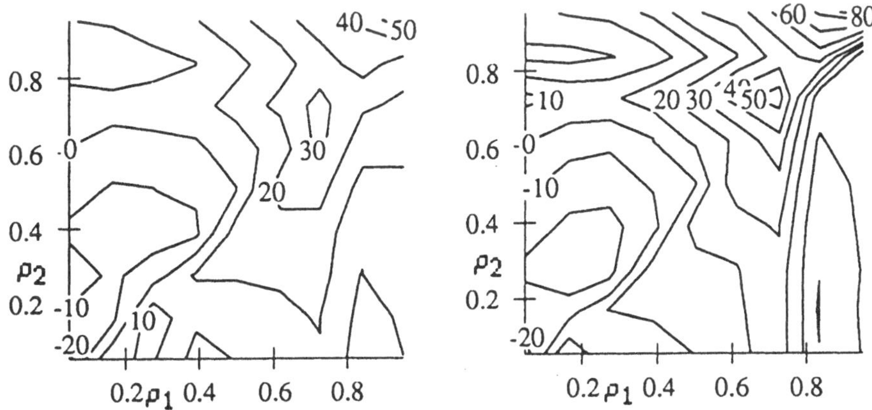

Abb. 15: TPNA für Priorität 3 (links) und Priorität 4 (rechts)

8. Zusammenfassung

Es wurde das neue Approximationsverfahren TPNA vorgestellt, das speziell für offene Tandem-Warteschlangennetze mit mehreren Auftragsklassen und Prioritätenknoten entwickelt wurde. Abfertigungsdisziplin der Knoten ist ein nicht-unterbrechendes HOL-Verfahren.

TPNA basiert auf dem Ansatz der von Cobham und Kleinrock entwickelten

HOL-Analyse für einen isolierten Knoten und wurde für Netzwerke mit beliebiger Knotenanzahl vorgestellt.

Warte- und Systemzeiten des Tandemnetzes werden bei TPNA sukzessive für alle Knoten und Auftragsklassen ausgewertet.

Die Brauchbarkeit von TPNA wurde durch Simulationen untersucht. Das MVA-PA-Verfahren, das bisher für TPN am geeigneten war ((11)), wurde als Vergleich für TPNA herangezogen. Dabei wurden besonders bei höheren Auslastungen (> 0,8) bei MVA-PA sehr hohe Toleranzfehler (> 140%) festgestellt. Dagegen zeigt TPNA ein wesentlich besseres Verhalten und liefert über den gesamten Lastbereich akzeptable Ergebnisse.

Die Einführung von Querströmen verursacht keine nennenswerten Einbußen in der Genauigkeit von TPNA. Bei MVA-PA setzt sich der zuvor beobachtete starke Anstieg der Fehlerraten fort.

Auch für deterministische Bedienstationen kann TPNA eingesetzt werden.

TPNA liefert bei Hochlast (> 80%) wesentlich bessere Ergebnisse als das MVA-PA. Darüberhinaus kann TPNA auch für TPN mit ·/G/1-Knoten mit guter Genauigkeit verwendet werden.

9. Literaturverzeichnis

(1)   F. Baskett, K. Chandy, R. Muntz, P. Palacios.Open, closed and mixed networks of queues with different classes of customers. J. ACM, Vol. 22, No. 2 , April 1975.

(2)   G. Bolch, H. Riedel: *Leistungsbewertung von Rechensystemen*. Verlag B.G.Teubner, Stuttgart 1989.

(3)   G. Graham (Ed.): Special Issue - Queuing network models for computer system performance. ECM Comp. Surv., Vol.20, No.3, Sept. 1987.

(4)   J. Buzen: Computational algorithms for closed queueing networks with exponential servers. Commun. ACM, Vol.16, No.9, Sept. 1973.

(5)   K. Chandy, C. Sauer: Computational algorithms for product form queueing networks. Commun. ACM, Vol.23, No.10, Okt. 1980.

(6)   M. Reiser, H. Kobayashi: Queueing networks with multiple closed chains: Theory and computational algorithms. IBM J.Res. and Devel., Vol.19, No.3, Mai 1975.

(7)   M. Reiser, S. Lavenberg: Mean value analysis of closed multichain queueing networks. J. ACM, Vol.27, No.2, April 1980.

(8)   M. Schümmer: Kommunikationssysteme in Echtzeitumgebungen - Modellierung und Bewertung von Fertigungs- und Intra-Vehikel-Kommunikationssystemen. Dissertation an der RWTH Aachen, Oktober 1990.

(9)   W.T. Strayer, A.C. Weaver: Performance Measurement of data services in MAP, IEEE Network, Vol. 2, No. 3, Mai 1988.

(10)   Wo. Kremer, R. Hager: Feasibility study of a multiplexing strategy in road transport informatics for short range MRNs. ACM/IEEE SAC'91, Kansas City, April 1991.

(11)   R. M. Bryant, A. E. Krzesinski, M. S. Lakshmi, K. M. Chandy: The MVA Priority Approximation. ACM Trans. Comput.Syst., Vol.2, No.4, Nov. 1984.

(12)   Wo. Kremer: Lokale Mobilfunknetze im Straßenverkehr - Entwurf, Dimensionierung und Leistungsbewertung. Dissertation, RWTH Aachen, Lehrstuhl für Informatik IV, 1992.

(13)   A. Fasbender: Leistungsbewertung eines prioritätsgesteuerten Realzeit-Kommunikationssystems. Diplom-Arbeit am Lehrstuhl für Informatik IV der RWTH Aachen, Oktober 1991.

(14)   Wo. Kremer, A. Fasbender: A New Approximation Algorithm for Tandem Networks with Priority Nodes. Aachener Informatik-Berichte, ISSN 0935-3232, Nr. 91-14, 1991.

(15)   A. E. Conway: Performance modeling of multilayered OSI communication architectures. Proc. of Int. Conf. Commun. ICC'89, Boston, June 1989.

(16)   P. Kritzinger: A performance model of the OSI communication architecture, IEEE Trans. Commun. COM-34, No. 6, 1986.

(17)   M. Reiser: Interactive modeling of computer systems. IBM Syst. J., Vol. 15, No. 4, 1976.

(18)   M. Reiser: A queueing network analysis of computer communication networks with window flow control, IEEE Trans. Commun. COM-27, No. 8, August 1979.

(19)   K. Sevcik: Priority scheduling disciplines in queueing network models of computer systems. in Proc. of IFIP Congress `77 in Toronto, North Holland, Amsterdam, 1977.

(20)   C. Sauer, K. Chandy: Approximate analysis of central server models. IBM Journal Res. Devel. 19, Mai 1975.

(21)   D. Neuse, K. M. Chandy: HAM: The heuristic aggregation method for solving general closed network models of computer systems. Perform. Eval. Rev. 11, No. 4, 1982-1983.

(22)   J. D. C. Little: A Proof of the Formular $L = \lambda W$. Oper. Res., Vol.9, No.3, Mai 1961.

(23)   S. S. Lavenberg, M. Reiser: Stationary State Probabilities at Arrival Instants for Closed Queueing Networks with Multiple Types of Customers. J. Appl. Prob., Vol.17, 1980.

(24)   K. Sevcik, I. Mitrani: The Distribution of Queueing Network States at Input and Output Instants. J. of ACM, Vol.28, No.2, April 1981.

(25)   J. Zahorjan, E. Wong: The Solution of Separable Queueing Network Models Using Mean Value Analysis. ACM Sigm.Perform.Eval. Review, Vol.10, No.3, 1981.

(26)   A. Cobham: Priority assignment in waiting line problems. Operations Research, Vol. 2, 1954.

(27)   Kleinrock, L., *Queueing Systems*, Vol. 1 (Theory) and Vol. 2 (Computer Applications), John Wiley & Sons, 1976.

Derivation of High Quality Tests for Large Heterogeneous Circuits: Floating-Point Operations [1]

Uwe Sparmann

Fachbereich Informatik
Universität des Saarlandes,
D-6600 Saarbrücken Germany

Abstract

In this paper the problem of deriving high quality tests for fast combinational floating-point realizations is investigated. Floating-point circuits are heterogeneous, consisting of a large number of regular and irregular modules. Thus, the test strategy applied combines specialized structure based methods and universal test generation. In order to guarantee sufficient controllability and observability of embedded modules small hardware modifications are proposed. As a result we obtain optimal time floating-point circuits for arbitrary operand lengths which can be tested completely with respect to a strong fault model by a minimal number of test patterns.

---

[1]This work has been supported by the 'Leibniz-Preis' awarded to Prof. Dr. G. Hotz in 1987.

1 Introduction

As a result of the advances in VLSI technology hundreds of thousands of transistors may nowadays be packed on a single chip. Because of this miniaturization manufacturing processes are very sensitive to external disturbances, like for example dust particles, which may result in the production of defective chips. Thus, in order to guarantee sufficient quality of the final product, there has to be an extensive test phase where the faulty chips are separated from the good ones. For this test phase input patterns have to be computed which exercise the internal gates of the circuit and completely check them for the most likely defects. Test application and response observation for an internal gate must be done over chip pins. Since the chip pin per gate ratio decreases with increasing miniaturization, the computational costs of test pattern generation have risen dramatically. In order to cut down these costs, design for testability techniques like Scan-Path [9] have been introduced which can be used to reduce the problem of testing a complex sequential circuit to the test development for its combinational parts. As a consequence, we will restrict to the test computation problem for combinational circuits in this paper.

Even for combinational circuits the test generation problem is NP-complete [11]. Thus, *universal test generation algorithms* pay their capability to handle arbitrary circuit structures with worst case exponential running times and despite of their sophisticated heuristics often deliver only poor results for large modules. On the other hand in order to ease the design process, large modules in most cases are designed with a very regular structure. Such structural properties have been used to derive specialized test methods for important circuit types like PLAs [1], iterative logic arrays [7] and tree like structures [5]. These *structure based tests* achieve complete fault coverage with respect to high quality fault models with a minimal number of test vectors. By applying the test theory for iterative logic arrays and tree like structures strong results about the testability of integer arithmetic circuits have been proven [4, 10, 12, 2]. But up to now the problem of how to use these results in order to develop completely testable realizations for the important class of floating-point operations has been unsolved.

In this paper the above problem is solved for fast combinational floating-point adders and multipliers. For brevity only the main ideas will be given, a complete description can be found in [18]. Floating-point circuits are *heterogeneous*, i.e. consist of a great number of different modules which are large and regular or small and irregular[1] in structure. Thus, besides structure based test derivation for the large regular modules, two essential new questions have to be considered: How to apply these tests to embedded modules over the surrounding circuitry? Is it possible to achieve sufficiently good results for the small irregular subcircuits by universal test generation algorithms?[2] Test application to embedded modules is done by utilizing as

---

[1] Note that irregular random logic is difficult to design and, therefore, tends to be small-scale.

[2] For these subcircuits structure based test derivation would be a very troublesome task, since there is no structural regularity to take advantage of.

far as possible the given functionality of the surrounding circuitry. In those cases where this functionality is not sufficient minimal hardware modifications are introduced. For the small irregular subcircuits a technique is developed which eases the job of universal test generation by taking advantage of the designer's knowledge about the interaction of modules.

The paper is structured in three sections: Section 2 outlines a first design of the floating-point circuits only considering points like functionality, speed, area, and ease of design but neglecting testability aspects. It is shown that only insufficient fault coverage can be achieved for these designs in reasonable time limits. The design for testability modifications and the procedure for deriving complete tests for the modified circuits are given in Section 3. To the large regular modules of the mantissa part structure based methods are applied, while for the small exponent part which contains irregular subcircuits universal test generation is utilized. Section 4 summarizes the results. Exact values specifying the design for testability overhead and the size of the complete test set are given for the most important bit widths.

2   Design without consideration of testability

In this section a first design of the floating-point circuits is outlined which does not incorporate testability aspects. As will be clarified by examples, only poor test sets can be generated for these circuits. Hence, testability aspects have to be considered from the very beginning of the design process. Design for testability modifications enabling the construction of high quality test sets will be derived in the next section.

2.1   Design overview

Surely the most important point in the design of floating-point circuits is to select a mathematically well defined and commonly accepted definition of arithmetic for implementation. Thus, the circuits presented here conform to the IEEE-Standard for binary floating-point arithmetic [13]. A specific floating-point word format is fixed by giving the bit lengths m and e for the mantissa and exponent. In order to keep the application range as wide as possible, the ideas presented in this paper are not restricted to the 32- ($m = 24$, $e = 8$) and 64-bit ($m = 53$, $e = 11$) formats of the IEEE-Standard but are valid for any reasonable choice of m and e.

The designs are purely combinational, thus minimizing the latency of a single operation, but naturally the test concept developed here is also applicable to pipelined versions of the floating-point circuits. For all steps (denormalization, addition resp. multiplication, normalization, rounding, overflow/underflow handling) the fastest possible realization was selected. As a result the circuit depth is only of order $\Theta(\log m)$ and thus asymptotically optimal. But not only the speed of single steps was optimized but also the interaction between different steps. Thus, for example rounding and overflow/underflow handling are executed in parallel by an appropriate precomputation on the exponent.

Area requirements were taken into account whenever they did not conflict with speed. As an example consider mantissa multiplication of the floating-point multipliers: A modified Wallace tree [15] is used to sum up the partial products of multiplication in optimal time. In order to cut down the area requirements for this tree without time penalty, modified Booth encoding [17] is utilized to reduce the number of partial products which have to be summed up.

Symbolic layout generators for the floating-point adder and multiplier designs as well as their testable counterparts presented in section 3 were implemented using the hierarchical design system CADIC [3]. The 24-bit ($m = 16$, $e = 8$) version of the testable floating-point adder has been manufactured in $3\mu m$ CMOS gate forest technology at the 'Institut für Mikroelektronik Stuttgart (IMS)' [8].

2.2   Inadequacy of universal test generation algorithms

Let us start by explaining some basic notions: Since completely testing a circuit by applying all possible input combinations is not feasible[3], assumptions have to be made about which faults usually occur in practice. These assumptions are summarized in a *fault model*. For example in the classical (single) *stuck-at* fault model it is assumed that a defect causes exactly one gate input or output to be fixed to either logic '0' or '1'. Since the problem of computing a test input for an arbitrary stuck-at fault is NP-complete, all known universal test generation algorithms are backtracking algorithms with worst case exponential running times. To restrict the overall running time of universal test generation, a limit can be imposed on the maximum number of backtracks allowed for each fault. Thus, the result of such an algorithm in general is only a partial test with respect to the fault model. The percentage of faults which have been processed succesfully, i.e. a test input has been found or they were proven redundant[4], is denoted as *fault coverage*.

Let us now turn to the question whether sufficient fault coverage can be achieved for the floating-point circuits by universal test generation algorithms. Floating-point circuits of different word lengths were generated and test computation for them was performed by applying SOCRATES [19], one of today's best universal test generation programs. Table 1 shows the results of SOCRATES for two floating-point multipliers ($FPM[m, e]$) measured on an APOLLO DN3000. All fault coverages are with respect to the stuck-at fault model.

After a running time of approximately five days, only a fault coverage of 97.68% was achieved for the 32-bit multiplier. To imagine the enormous costs of a further improvement in fault coverage, look at the results in the first and second row for the

---

[3]For a 32-bit multiplier with a running time of 100 nanoseconds such a test would last more than $50,000$ years.

[4]A fault is called *redundant*, iff there exists no test for it, i.e. the correct and faulty circuit compute the same function.

[5]Our version of SOCRATES does not incorporate the features suggested in [16] which might slightly improve the results given here.

circuit $FPM[m,e]$	fault coverage	running time[5]	backtrack limit
$FPM[12,4]$	99.27%	3 h 56 m	1,000
	99.34%	35 h 19 m	10,000
$FPM[24,8]$	97.68%	127 h 8 m	1,000

Table 1: Results of SOCRATES

16-bit floating-point multiplier. Here rising the fault coverage by only 0.07% is payed by an increase in running time of about factor 9.

Since the results for the floating-point adder are similar, we can conclude that even for the simple stuck-at fault model sufficient fault coverage can not be achieved in acceptable time by universal methods. In the next section it will be shown how to derive tests with 100% fault coverage for slightly modified floating-point circuits. For the main part of the circuits even a better fault model than the stuck-at fault model will be considered.

3 Completely testable floating-point circuits

The floating-point circuits consist of two major components. First there is the mantissa part which, since $m \gg e$, dominates the overall size of the circuit. Second we have the exponent part which is small and contains rather irregular subcircuits like the logic for overflow/underflow handling. The large regular modules of the mantissa part will be tested by structure based methods in order to guarantee complete fault coverage with respect to high quality fault models. For the exponent part structure based test derivation would be a very tiresome task because of the rather irregular structure of this subcircuit. Thus, we want to apply universal test generation algorithms here which should deliver complete fault coverage because of the low complexity of this subcircuit. Since the modules are not isolated but embedded into a complex circuit, test application and response observation over the surrounding modules must be guaranteed. In both cases efficient techniques for solving this 'context problem' will be presented.

3.1 Test of mantissa part

For the structure based test of the mantissa part we proceed in three steps: By simple hardware modifications (enlargement of busses) the controllability and observability of internal modules is improved. Thus, test construction for the mantissa part can be nearly reduced to the problem of testing its isolated subcircuits. For some of these circuits structure based tests are already known from previous work, for the remaining modules new tests have to be derived. Finally, in order to minimize test execution time, we will consider in how far module tests can be executed in parallel.

The following paragraphs will explain the application of this procedure to the floating-point adder. For brevity only the main ideas will be given illustrated on a simplified circuit.

Hardware modifications to improve testability

To give reasons for the desired hardware modifications, consider the block diagram of the floating-point adder's mantissa part[6] depicted in figure 1.

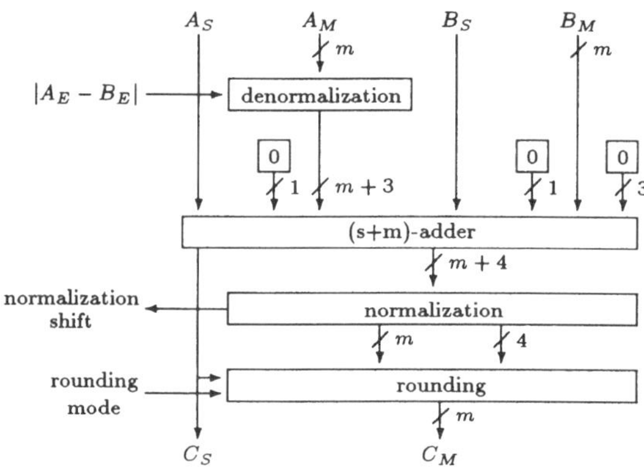

Figure 1: Mantissa part of the floating-point adder

Assume that A is the operand with the smaller exponent which has to be denormalized, i.e. shifted to the right by the exponent difference. Surely this denormalization can not be carried out exactly since then we would have to handle operands of extreme length. Instead, as suggested in [14], we compute a sufficiently precise approximation which can be represented by only $m + 3$ bits. After denormalization the mantissas are added up by a (sign and magnitude)-adder. Here both operands are expanded by a leading zero in order to catch up a possible overflow during addition. The sum is then normalized and finally rounded from the internal precision $m + 4$ to the external precision m.

In principle the modules of the mantissa part are well suited for test application and response observation. The only problem is given by the fact that the maximum internal mantissa precision $m + 4$ is higher than the precision m of the external operands. Thus for example at the (s+m)-adder not all possible values can be applied

---

[6]Note that this block diagram neglects for example mantissa exchange or overflow/underflow handling. The sign, mantissa, and exponent of a floating-point number X is denoted by X_S, X_M, and X_E.

to its inputs and propagation of faulty responses from this module is interfered by the rounding operation which is not injective. Similar observations are true for the remaining modules. To overcome this problem, new primary inputs and outputs were introduced to enlarge all busses of the mantissa part to the maximum internal width $m + 4$ as shown in figure 2.

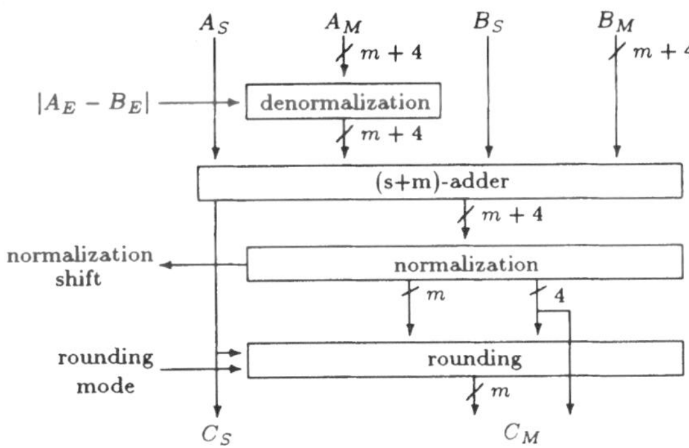

Figure 2: Modified mantissa part

Consider the (s+m)-adder again to see that now any test pattern t can be applied to the embedded module and any difference correct sum (s_c) and faulty sum (s_f) can be observed at the primary outputs:

In order to generate t at the inputs of the (s+m)-adder we simply set the primary inputs of the mantissa part to t and choose the exponents of the two operands equal such that there is no denormalization. Propagation of the difference s_c/s_f is guaranteed by selecting as rounding mode 'round to zero' (truncation). To see that under this condition any difference can be propagated, we have to distinguish two cases. If s_c and s_f differ in the number of leading zeros, the fault will be propagated over the exponent part since different normalization shifts are subtracted from the exponent of the result. In the other case the difference can not be masked by normalization and for rounding mode truncation because of the additional primary outputs the rounding module passes the value of its mantissa inputs identically to its outputs.

Thus any test developed for the isolated (s+m)-adder can also be applied to this module inside the modified floating-point adder. Since the situation for the remaining modules is similar, test construction for the mantissa part of the floating-point adder is nearly[7] reduced to the testing problem for its isolated subcircuits by the above hardware modifications.

---

[7]Some simple testability restrictions remain since only normalized mantissas can be applied to the rounding module.

Structure based test of modules

Let us now turn to the question of how to utilize structural properties of regular circuit families in order to derive efficient test sets for them. The test sets have been constructed with respect to the (single) *cellular* fault model. In this fault model it is assumed that exactly one basic cell of the circuit is faulty and that this fault can vary the cell's output functions in an arbitrary manner as long as they remain combinational. (Note that stuck-at faults are just special cases of cellular faults.) The quality of the test set size as a function of the mantissa length m has been estimated by lower bounds. In all cases the lower and upper bound only differ by a small constant factor[8]. Table 2 shows the dependence between mantissa length and test set size for the modules of figure 2. The results for the (s+m)-adder and rounding are based on the theory developed in [4, 5].

circuit	denormalization	(s+m)-adder	normalization	rounding
test size	$\Theta(m)$	$\Theta(\log m)$	$\Theta(m)$	$\Theta(m)$

Table 2: Dependence between test set size and mantissa length

To illustrate the basic ideas of structure based test derivation, let us consider a simple example. Figure 3 gives the recursive definition of an n-bit leading zeros counter $LZ[n]$ used in the normalization module to determine the shift distance. For simplicity of presentation it is assumed that $n = 2^\alpha$, $\alpha \in \mathbf{N}$, the results for arbitrary $n \in \mathbf{N}$ are nearly the same [20]. The recursion stops for $n = 4$ since $LZ[2]$ is considered to be a basic cell. $MUX[k]$ ($LO[2]$) denotes a multiplexer for busses of width k (2-bit leading ones counter). Since the number of inputs to the left half of $LZ[n]$ is a power of two, addition of the partial results, which is only necessary for an all zeros input to the left half, can be executed in constant time.

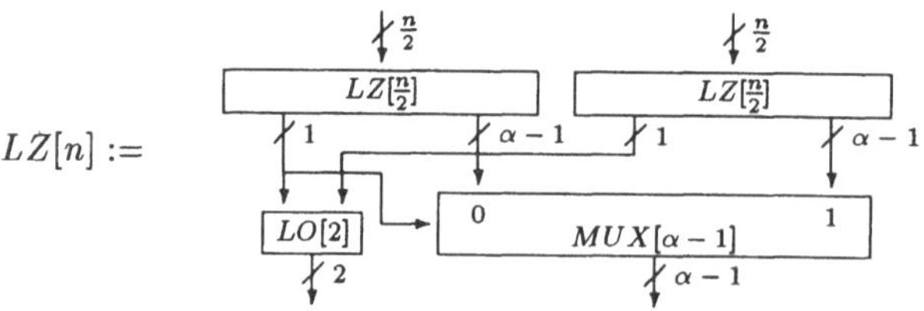

Figure 3: Recursive definition of the leading zeros counter

$LZ[2]$, $LO[2]$, and two input multiplexers will be considered as basic cells which have to be checked for cellular faults. Thus, to each occurence of one of these cells all input combinations must be applied, and any faulty response must be propagated to the primary outputs if possible. A complete test set for $LZ[4]$ is given by the following input patterns:

$$T[4] := \{1101, 1000, 0101, 0100, 0011, 0010, 0001, 0000\}$$

In what follows it will be shown how to inductively construct a test set $T[n]$ for $LZ[n]$, $n > 4$, from a given test $T[\frac{n}{2}]$ for $LZ[\frac{n}{2}]$. The left (right) occurence of $LZ[\frac{n}{2}]$ in $LZ[n]$ will be denoted by $LZ[\frac{n}{2}]_l$ ($LZ[\frac{n}{2}]_r$). The following simple observation gives the restrictions which have to be observed for testing the occurences of $LZ[\frac{n}{2}]$ in $LZ[n]$.

Observation 1 *Let w_l (w_r) be the input combination applied to $LZ[\frac{n}{2}]_l$ ($LZ[\frac{n}{2}]_r$). Assume that exactly one occurence of $LZ[\frac{n}{2}]$ computes a faulty output value, then:*

1. *A faulty output of $LZ[\frac{n}{2}]_r$ is propagated to the primary outputs of $LZ[n]$, iff $w_l = 0^{\frac{n}{2}}$.*

2. *A faulty output of $LZ[\frac{n}{2}]_l$ is not propagated to the primary outputs of $LZ[n]$ iff correct and faulty output of $LZ[\frac{n}{2}]_l$ do not differ in the most significant bit and $w_l = 0^{\frac{n}{2}}$.*

Form part 2 of the above observation it follows that the test $0^{\frac{n}{2}}$ may be devaluated when applied to $LZ[\frac{n}{2}]_l$. To solve this problem, we claim the following property for $T[\frac{n}{2}]$:

(P1) For every fault which can not only be tested by $0^{\frac{n}{2}}$ there exists a test $\neq 0^{\frac{n}{2}}$ in $T[\frac{n}{2}]$.

From observation 1 and property (P1) it follows immediately that:

Lemma 1 *Any test set of the form $\{0^{\frac{n}{2}}t \mid t \in T[\frac{n}{2}]\} \cup \{tw_t \mid t \in T[\frac{n}{2}]\}$, where for $t \in T[\frac{n}{2}]$ $w_t \in \mathbf{B}^{\frac{n}{2}}$ can be chosen arbitrarily, checks the occurences of $LZ[\frac{n}{2}]$ in $LZ[n]$ completely for cellular faults.*

What remains to be done is to test the $LO[2]$ cell and the $\alpha - 1$ multiplexers used to sum up the partial results. As will be shown next, the tests of lemma 1 can also be used for this task. To achieve this goal it is not sufficient to exploit the freedom of choosing the values w_t, but additionally we have to guarantee the existence of appropriate output values during the test of $LZ[\frac{n}{2}]$. Thus, assume that $T[\frac{n}{2}]$ besides (P1) fulfills the property:

(P2) $T[\frac{n}{2}]$ contains six different input patterns $t_1, \ldots, t_6$ such that: t_1 (t_2) (t_3) (t_4) (t_5 and t_6) have exactly $\frac{n}{2}$ ($\frac{n}{2} - 1$) ($\frac{n}{4}$) ($\frac{n}{4} - 1$) (0) leading zeros.

It can be easily verified that $T[4]$ fulfills properties (P1) and (P2). The following lemma shows how to construct $T[n]$ from $T[\frac{n}{2}]$ such that these properties are preserved.

Lemma 2 *Let $T[n] := T[n]_r \cup T[n]_l$, where $T[n]_r := \{0^{\frac{n}{2}}t \mid t \in T[\frac{n}{2}]\}$ and $T[n]_l := \{t0^{\frac{n}{2}} \mid t \in T[\frac{n}{2}] \setminus \{t_1, t_2, t_3, t_4, t_5, t_6\}\} \cup \{t_2 t_1, t_3 t_2, t_4 t_2, t_5 t_1, t_6 t_2\}$. $T[n]$ is a complete test set for $LZ[n]$ which satisfies properties (P1) and (P2).*

Proof: From lemma 1 it follows immediately that $T[n]$ completely tests the occurences of $LZ[\frac{n}{2}]$ in $LZ[n]$. (Note that $t_1 = 0^{\frac{n}{2}}$ need not be applied to $LZ[\frac{n}{2}]_l$ during the test $T[n]_l$ since this input is already applied when testing $LZ[\frac{n}{2}]_r$.) Thus, for proving that $T[n]$ is a complete test set, it only remains to show that every non redundant input combination is applied to the basic cells of the summation logic. This job is done by the input patterns $t_1 t_1$, $t_1 t_2 \in T[n]_r$ and $t_5 t_1$, $t_6 t_2 \in T[n]_l$ for the $LO[2]$ cell. For the multiplexers all input combinations which set the select input s and the input selected for $s = 0$ both to one are redundant. The remaining input combinations are applied by the patterns $t_1 t_6$, $t_1 t_2 \in T[n]_r$ and $t_5 t_1$, $t_6 t_2$, $t_2 t_1$, $t_3 t_2$, $t_4 t_2 \in T[n]_l$.

Since $T[n]$ again satisfies properties (P1) and (P2) if these properties are true for $T[\frac{n}{2}]$, we are done.   ∎

The above construction is optimal since:

Lemma 3 *Let $t(n)$, $n \geq 4$, denote the minimal size of a complete test set for $LZ[n]$, then $|T[n]| = t(n) = \frac{7}{4} \cdot n + 1$.*

Proof: (structural induction)
$T[4]$ has size $8 = \frac{7}{4} \cdot 4 + 1$. To see that this size is optimal consider the $LZ[2]_r$ cell in $LZ[4]$. For a complete test of this cell the inputs 0000, 0001, 0010 and 0011 are necesary. Since these tests only apply two of the six non redundant input combinations to the multiplexer in $LZ[4]$, $t(4) \geq 8$.

Let us assume inductively that the lemma is true for $t(\frac{n}{2})$. For checking $LZ[n]$ both occurences of $LZ[\frac{n}{2}]$ must be tested. By part 1 of observation 1 only one test ($0^{\frac{n}{2}}$) for $LZ[\frac{n}{2}]_l$ can be done in parallel to testing $LZ[\frac{n}{2}]_r$. Thus $t(n) \geq 2 \cdot t(\frac{n}{2}) - 1$ and since $|T[n]| = 2 \cdot |T[\frac{n}{2}]| - 1$ our test is minimal. Its size is given by $|T[n]| = 2 \cdot (\frac{7}{4} \cdot \frac{n}{2} + 1) - 1 = \frac{7}{4} \cdot n + 1$.   ∎

Parallel execution of module tests

The overall test size of the mantissa part is dominated by the modules with linear test complexity, i.e. denormalization shifter, normalization, and rounding (see table 2). Thus, we restrict to examine the possibilities for parallel test execution with respect to these modules. Again the main ideas will only be sketched without giving exact proofs.

Let us start by considering the denormalization shifter. In order to test this module it is sufficient to choose A_M, the exponents, and the rounding mode appropriately. Thus for any test of the denormalization shifter we are free to select the values of A_S, B_S, and B_M. By an appropriate choice of these values an arbitrary test needed for checking the normalization or rounding module can be generated at the outputs of the (s+m)-adder. As a consequence the denormalization shifter can be tested in parallel to these modules.

For normalization and rounding it can be shown that parallel test execution is not possible:

The test size of the normalization (rounding) module is dominated by the leading zeros counter (the incrementer for upward rounding). We already know from observation 1 that for testing the right part $LZ[\frac{n}{2}]_r$ of an n-bit leading zeros counter all inputs to it's left part must be set to zero. Thus, for the test of the leading zeros counter input combinations with strings of leading zeros are necessary. More exactly, it can be easily shown by induction that:

Lemma 4 *Any complete test set for $LZ[n]$, $n = 2^\alpha \geq 4$, must contain $\frac{7}{4} \cdot (n - 4) + 1$ patterns with at least 4 leading zeros.*

Thus nearly all test patterns for the occurence of $LZ[m + 4]$ in the floating-point adder must be of the form $0000w$, $w \in \{0,1\}^m$. Now consider the situation when such a test is applied to the leading zeros counter. The corresponding normalized mantissa $w0000$ has only m significand bits. Thus, there is no upward rounding of the mantissa and no fault of the incrementer inside the rounding module can be propagated.

Since module tests are executed in parallel whenever possible the overall test size is nearly minimal. As an example consider the 32- ($m = 24$, $e = 8$) and 64-bit ($m = 53$, $e = 11$) floating-point adder where the lower bound and the actual test size only differ by factor 2.9 resp. 2.

3.2   Test of exponent part

Because of it's minor size and irregularity test generation for the exponent part should be done by an universal algorithm. But, as becomes clear from the measurements in table 3, while SOCRATES easily copes with the isolated module, the costly treatment of the surrounding circuitry does not allow to generate a complete test for the embedded module in acceptable time. In this section a technique for solving this problem will be presented which combines universal test generation and the designer's knowledge about the global functionality of the circuit.

Consider a module M embedded in a circuit C and let us ask for the influence of C on the testability of M. Clearly this testability depends on the set of input combinations applicable to M in C which will be denoted by $Dom_{M \text{ in } C}$. Additionally, for each value $\alpha \in Dom_{M \text{ in } C}$ we need to know the set $Prop_{M \text{ in } C}(\alpha)$ of all faulty responses of M which can be propagated to the primary outputs of C when α is applied to M. The following definition summarizes all information about C necessary to determine whether a fault inside the embedded module M is testable.

Definition *The* test context *of module M embedded in C is given by*

$$TC_{M \text{ in } C} := \{(\alpha, \beta) | \alpha \in Dom_{M \text{ in } C}, \ \beta \in Prop_{M \text{ in } C}(\alpha)\}.$$

circuit	fault coverage	running time	backtrack limit
$EXP[6]$ isolated	100%	$2\ m$	$1,000$
$EXP[8]$ isolated	100%	$7\ m$	$1,000$
$EXP[6]$ in $FPM[10,6]$	99.29%	$18\ m$	$1,000$
	99.47%	$95\ h$	$1,000,000$
$EXP[8]$ in $FPM[24,8]$	98.64%	$7\ h$	$1,000$

Table 3: Results of SOCRATES for exponent part

Consider again the situation of table 3 where test generation can be easily done for the isolated module M but does not deliver sufficient results when M is embedded in C. Assume that the difficulties in test generation for M in C are only due to the complexity of the circuit context C, but the test context $TC_{M\ \text{in}\ C}$ is simple and can easily be determined by the designer because of his knowledge about the functionality of the circuit. (A trivial example of this situation occurs when arbitrary input combinations can be applied to M and any faulty response can be propagated by setting some control inputs appropriately.) In this situation the procedure illustrated in figure 4 can be used to support the universal test generation algorithm by designer knowledge and thus achieve complete fault coverage.

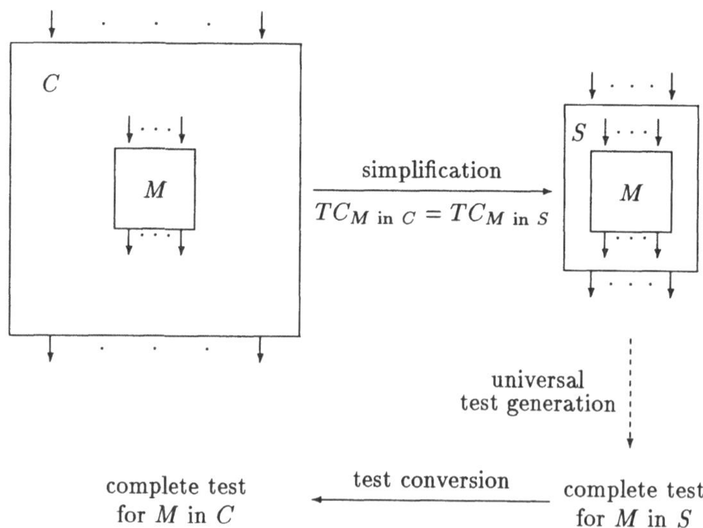

Figure 4: Speeding up universal test generation by circuit simplification

The complex circuit C is replaced by a simplified circuit S. S is chosen such that

$TC_{M\ \text{in}\ C} = TC_{M\ \text{in}\ S}$ and hence any fault of M is testable inside C iff it is testable inside S. Since the gate complexity of the modified circuit is much lower than that of the original one, universal test generation yields a complete test $T_{M\ \text{in}\ S}$ for M embedded in S in acceptable time. From $T_{M\ \text{in}\ S}$ the designer can derive a set $T_{M\ \text{in}\ C}$ of inputs to C which tests the embedded module in an analogous manner i.e. applies the same input combinations to M in C and propagates the same faulty responses.

The above procedure was successfully applied to the exponent parts of the floating-point circuits. As an example of the enormous time savings by this method consider test generation for the 8-bit exponent part of the 32-bit floating-point multiplier. The running time of SOCRATES to compute a complete test set with respect to the simplified circuit was only 10 minutes. Thus, test generation for the embedded module becomes nearly as easy as test generation for the isolated one which could be performed in 7 minutes (see table 3). Surely this has to be payed by an increased effort of the designer who has to perform the two steps of circuit simplification and test conversion. Hence, this method should only be applied in situations where the test context of an embedded module is easy and well understood.

4   Summary and results

In this paper techniques have been presented to solve the test problem for fast combinational floating-point circuits. The procedure proposed combines structure based test methods used for the large modules of the mantissa part and universal test generation applied to the small and irregular exponent part. In order to improve controllability and observability of internal modules design for testability modifications were introduced which mainly consist of additional primary inputs and outputs combined with an enlargement of internal busses. The hardware overhead of these modifications is very low, the depth of the circuits is not increased. The test sets are nearly minimal and check the circuits completely with respect to strong fault models. Concrete results computed in [18] for the two floating-point formats suggested by the IEEE-Standard are summarized in table 4.

circuit	$FPM[24,8]$	$FPM[53,11]$	$FPA[24,8]$	$FPA[53,11]$
gate count	11.672	45.665	4.206	8.806
additional gates	6.6% (3.9%)	3.4% (2%)	3.4%	1.6%
test size	937	1877	508	803

Table 4: Results for some floating-point circuits

The first row of table 4 gives the number of gates for the testable floating-point multipliers ($FPM[m,e]$) and adders ($FPA[m,e]$) (including registers). The overhead introduced by design for testability assuming that the additional primary inputs and outputs are accessed over a Scan-Path is summarized in the second row. Since the overall circuit size grows asymptotically faster than the additional logic introduced for testability purposes, this overhead decreases with increasing operand length. For the floating-point multiplier testability modifications are combined with the introduction of a new operation calculating the exact unrounded product. Since this *double precision multiplication* is not only useful for testability purposes but can also be applied to the fast computation of exact scalar products [14, 6], two values are given in table 4 depending on whether the additional gates needed to implement this operation are counted as overhead or not. The third row lists the actual test sizes of complete tests for the floating-point circuits.

Acknowledgements

First of all I want to express my gratitude to Prof. Dr. G. Hotz for his helpful advice and continuous support and encouragement which made this work possible.

Thanks are also due to R. Drefenstedt, T. Walle, and W. Weber for their engaged work during the implementation of the testable floating-point adder.

Bibliography

[1] V.K. Agarwal. *VLSI Testing*, volume 5 of *Advances in CAD for VLSI*, chapter 3, pages 65–93. North-Holland, 1986. edited by T.W. Williams.

[2] B. Becker and J. Hartmann. Optimal-time multipliers and c-testability. In *Proceedings of the 2nd Annual Symposium on Parallel Algorithms and Architectures*, pages 146–154, 1990.

[3] B. Becker, G. Hotz, R. Kolla, P. Molitor, and H.G. Osthof. Hierarchical design based on a calculus of nets. In *Proceedings of the 24th ACM/IEEE Design Automation Conference (DAC87)*, pages 649–653, June 1987.

[4] B. Becker and U. Sparmann. Regular structures and testing: RCC-adders. In *Proceedings of the 3rd Aegean Workshop on Computing*, pages 288–300, 1988.

[5] B. Becker and U. Sparmann. Computations over finite monoids and their test complexity. *Theoretical Computer Science*, pages 225–250, 1991.

[6] P.R. Capello and W.L. Miranker. Systolic Super Summation. *IEEE Transactions on Computers*, C-37(6):657–677, 1988.

[7] W.T. Cheng and J. H. Patel. Testing in two-dimensional iterative logic arrays. In *Proceedings of the 16th International Symposium on Fault Tolerant Computing Systems*, July 1986.

[8] R. Drefenstedt and T. Walle. Implementierung eines effizient testbaren Gleitkommaaddierers auf einer kommerziellen Sea-of-Gate Struktur. Technical Report 10/1991, SFB 124, Fachbereich Informatik, Universität des Saarlandes, 1991.

[9] E.B. Eichelberger and T.W. Williams. A logic design structure for LSI testability. *Journal of Design Automation and Fault-Tolerant Computation*, 2:165–178, 1978.

[10] J. Ferguson and J.P. Chen. The design of two easily-testable VLSI array multipliers. In *Proceedings of the 6th Symposium on Computer Arithmetic*, pages 2–9, June 1983.

[11] H. Fujiwara and S. Toida. The complexity of fault detection problems for combinational logic circuits. *IEEE Transactions on Computers*, C-31, 1982.

[12] S.J. Hong. An easily testable parallel multiplier. In *18th International Symposium on Fault Tolerant Computing*, 1988.

[13] The Institute of Electrical and Electronics Engineers, Inc. *IEEE Standard for Binary Floating-Point Arithmetic ANSI/IEEE Std 754-1985*, 1985.

[14] U.W. Kulisch and W.L. Miranker. *Computer Arithmetic in Theory and Practice*. Academic Press, 1981.

[15] W.K. Luk and J. Vuillemin. Recursive implementation of optimal time VLSI integer multipliers. In *Proceedings IFIP Congress 83*, pages 155–168, Amsterdam, 1983.

[16] M.H. Schulz and E. Auth. Advanced automatic test pattern generation and redundancy identification techniques. In *18th Symposium on Fault-Tolerant Computing 1988*, June 1988.

[17] O. Spaniol. *Arithmetik in Rechenanlagen*. Teubner Verlag, 1976.

[18] U. Sparmann. *Strukturbasierte Testmethoden für arithmetische Schaltkreise*. PhD thesis, Fachbereich Informatik, Universität des Saarlandes, 1991.

[19] M.H. Schulz, E. Trischler, and T.M. Sarfert. Socrates: A highly efficient automatic test pattern generation system. In *Proceedings of 1987 International Test Conference*, September 1987.

[20] W. Weber. Entwurf und Test einer Familie von Gleitkommaaddierern. Master's thesis, Fachbereich Informatik, Universität des Saarlandes, 1990.

Inductive Theorem Proving by Consistency for First-Order Clauses

Harald Ganzinger, Jürgen Stuber

Max-Planck-Institut für Informatik
Im Stadtwald
6600 Saarbrücken
Germany

Abstract

We show how the method of proof by consistency can be extended to proving properties of the perfect model of a set of first-order clauses with equality. Technically proofs by consistency will be similar to proofs by case analysis over the term structure. As our method also allows to prove sufficient-completeness of function definitions in parallel with proving an inductive theorem we need not distinguish between constructors and defined functions. Our method is linear and refutationally complete with respect to the perfect model, it supports lemmas in a natural way, and it provides for powerful simplification and elimination techniques.

1   Introduction

For proving inductive theorems of equational theories "proof by consistency" is a particularly powerful method. The method has been engineered during the last decade by gradually removing restrictions on the specification side, by reducing the search space for inferences, and by including methods from term rewriting for the simplification and elimination of conjectures. Musser [15] requires the specifications to contain a completely defined equality predicate. During completion inconsistency results in the equation **true** $\approx$ **false**. Huet and Hullot [12] assume a signature to be divided into constructors and defined functions. An equation between constructor terms signals an inconsistency. Jouannaud and Kounalis [13] admit arbitrary convergent rewrite system for presenting a theory. They introduce the notion of *inductive reducibility* to detect inconsistencies. Plaisted [18], among others, has shown that inductive reducibility is decidable for finite unconditional term rewriting systems. Fribourg [9] is the first to notice that not all critical pairs need to be computed for inductive completion. It suffices to consider only linear inferences for selected *complete positions*. Bachmair [1] refines this method to cope with unorientable equations; as a result his method is refutationally complete. His method of proof orderings admits powerful techniqes of simplification and removal of redundant equations without loosing refutation completeness. The latter is essential for verifying nontrivial inductive properties in finite time.

More recently there have been some attempts to extend these techniques to Horn clauses. Orejas [16] places similar restrictions on specifications as Huet and Hullot. Bevers and Lewi [6] build on inductive reducibility, which is a severe restriction as inductive reducibility is in general undecidable for Horn clauses [14].

In this paper we extend the method described by Stuber [20] from Horn clauses to full first-order clauses with equality by adapting the method of [3, 5] for Knuth/Bendix-like completion for first-order clauses. Completion—*saturation up to redundancy*, as we prefer to call this process from now on—serves an important purpose. It produces a representation of a certain minimal model of the given (consistent) first-order theory and allows to prove the validity of ground equations in this model by conditional term rewriting with negation as failure. This distinguished minimal model is called the *perfect model*, and it depends on a given reduction ordering on terms. By inductive theorem proving for first-order theories we mean to prove *validity in the perfect model*, and the method consists in showing that enriching a given theory by a given set of conjectures does not change the perfect model, hence the name *proof by consistency*.

Unlike many other methods of inductive theorem proving [7, 11, 17], our method of proof by consistency does not require that constructors be given explicitly. Moreover we always generate a counterexample if the conjecture is false. In other words, our method is refutationally complete. It also is linear; neither inferences between axioms nor between conjectures have to be computed. The method is rather flexible as it is based on a very general notion of *fair inductive theorem proving derivations* and allows for powerful simplification and elimination techniques. The latter is provided by the notion of redundancy as developed in [3, 5]. In fact we will show that redundancy and inductive validity of clauses are equivalent concepts.

Technically the approach is based on the inference systems for first-order refutation theorem proving presented by Bachmair and Ganzinger [5] and briefly summarized in the appendix.

2   The Method

Clauses are implicitly universally quantified. We make quantifiers explicit and restrict them to generated values by adding a constraint $\mathbf{gnd}(x)$ for every variable x in the clause. We add clauses which define these *type predicates* such that $\models \mathbf{gnd}(t)$ if and only if t is (equivalent to) a ground term of the sort of x. More precisely, for each operator f of arity n a clause

$$\mathbf{gnd}(x_1), \ldots, \mathbf{gnd}(x_n) \to \mathbf{gnd}(f(x_1, \ldots, x_n))$$

is added, and a conjecture $\Gamma \to \Delta$ containing variables $x_1, \ldots, x_n$ becomes

$$\mathbf{gnd}(x_1), \ldots, \mathbf{gnd}(x_n), \Gamma \to \Delta.$$

A clause that is *closed* by explicit quantifiers in this way is valid if and only if it is valid in all generated models (Herbrand models over the given signature.) Validity in all generated models implies validity in the perfect model of a set of clauses, and is a key step towards second-order reasoning.

The perfect model of a set of clauses N is represented, in a sense that will become clear below, by a certain subset N' of N. These clauses define a canonical set R of ground rewrite rules such that the congruence generated by R is the perfect model of N. To prove the inductive validity of a conjecture H, we take its closed version H' and attempt to prove the validity of a set of instances of H' that covers all ground instances of H', assuming that H' is true for all smaller instances. (Here, "smaller" refers to some well-founded ordering on clauses.) The key points of our method are as follows:

(i) The covering set of instances of H' is generated by a narrowing-like process which enumerates the solutions to the antecedent of H' in the perfect model. By closing H we achieve that conjectures are either ground or else have a non-empty antecedent.

(ii) We eliminate an instance of H' if it follows from N and from smaller instances of H'. In this case we call the particular instance of H' *composite*.

(iii) We assume that for a ground instance of H' it is decidable whether or not H' is true in the perfect model. In particular, we assume that N' and R are effectively given in a certain technical sense. This restricts our method, but makes it refutationally complete. If validity of ground clauses were undecidable for a theory N, the problem of inductive theorem proving for N would be hopeless anyway.

(iv) We saturate $N \cup H'$ by applying a positive superposition strategy. To enumerate the solutions of the antecedent of H' we allow to *select* an arbitrary atom A of the antecedent so as to guide the enumeration process to first concentrate on the solutions of A. If A is a type predicate $\mathsf{gnd}(x)$, then the effect is to enumerate all ground substitutions for the variable x in H. It may happen that some type clause

$$C_f = \mathsf{gnd}(x_1), \ldots, \mathsf{gnd}(x_n) \to \mathsf{gnd}(f(x_1, \ldots, x_n))$$

corresponding to some function f itself is an inductive consequence (with respect to N) of some subset B of other type clauses. This is the case if f is a function symbol that is *sufficiently completely* defined relative to the (in some sense more primitive) functions in B. In this case, the C_f needs *not* be superposed on A. In other words, only type clauses for constructors need to be considered for superposition. This optimization is implicitly built into our method as the inductive validity of C_f can be proved in parallel with H. No explicit distinction between constructor symbols and defined symbols is required. Moreover, equalities between constructor terms pose no problem in our framework.

For instance, consider the following specification for natural numbers.

natbase =
    sorts
        nat
    ops
        $0 : \ \to \mathrm{nat}$
        $\mathrm{s} : \mathrm{nat} \to \mathrm{nat}$

The enrichment by type clauses yields

natbase^g = **natbase** +
    gnd : nat
    axioms $\forall \, \mathrm{n} : \mathrm{nat}$
        $\mathsf{gnd}(0)$ (1)
        $\mathsf{gnd}(\mathrm{n}) \to \mathsf{gnd}(\mathrm{s}(\mathrm{n}))$ (2)

Consider the enrichment of the above specification by a definition of $\leq$.

natleq^g = **natbase**^{g} +
    $\leq \ : \mathrm{nat} \times \mathrm{nat}$
    axioms $\forall \, \mathrm{m}, \mathrm{n} : \mathrm{nat}$
        $0 \leq \mathrm{n}$ (3)
        $\mathrm{m} \leq \mathrm{n} \to \mathrm{s}(\mathrm{m}) \leq \mathrm{s}(\mathrm{n})$ (4)

Suppose we would like to prove that $\leq$ is total, i.e. that

$$\to m \leq n, n \leq m,$$

which becomes

$$\mathbf{gnd}(n), \mathbf{gnd}(m) \to m \le n, n \le m$$

after closing, is inductively valid. In this particular case the theory is of Horn clause type so that the perfect model is the initial model.

Whenever a clause is added during a consistency proof, an equation in its antecedent for which solutions are to be enumerated is selected. (The selection will below be indicated by underlining.)

(5) $\underline{\mathbf{gnd}(m)}, \mathbf{gnd}(n) \to m \le n, n \le m$ conjecture

$\quad \mathbf{gnd}(n) \to 0 \le n, n \le 0$ selective resolution (1) on (5)
        composite because of (3)

(6) $\mathbf{gnd}(m), \underline{\mathbf{gnd}(n)} \to s(m) \le n, n \le s(m)$ selective resolution (2) on (5)

$\quad \mathbf{gnd}(m) \to s(m) \le 0, 0 \le s(m)$ selective resolution (1) on (6)
        composite because of (3)

$\quad \mathbf{gnd}(m), \mathbf{gnd}(n) \to s(m) \le s(n), s(n) \le s(m)$ selective resolution (2) on (6)
        composite because of (4) and (5)

We have seen that all clauses that can be enumerated by superposition on selected atoms are composite, i.e. follow from the theory and from smaller instances of the conjecture. For instance, for all ground terms N and M, the clause $C = \mathbf{gnd}(M), \mathbf{gnd}(N) \to s(M) \le s(N), s(N) \le s(M)$ follows from (4) and the instance $D = \mathbf{gnd}(M), \mathbf{gnd}(N) \to M \le N, N \le M$ of (5). D is emdedded in C, hence smaller than C.

The example demonstrates the strong analogy to classical methods of inductive theorem proving. Selecting one of the $\mathbf{gnd}(x)$ corresponds to the selection of an induction variable. Superposition with the type clauses results in a set of new instances representing the different cases to be proved. The elimination of a clause corresponds to an induction step for which the induction hypothesis may be used. Basis of the induction are well-founded orderings on terms which are extended to well-founded orderings on clauses.

For a second example, consider the usual definition of addition for natural numbers.

$\quad \mathbf{natplus} = \mathbf{natbase} +$
$\qquad \mathbf{ops}$
$\qquad\quad + : \mathrm{nat} \times \mathrm{nat} \to \mathrm{nat}$
$\qquad \mathbf{axioms} \quad \forall\, \mathrm{m}, \mathrm{n} : \mathrm{nat}$
$\qquad\quad 0 + \mathrm{n} \approx 0$ (1)
$\qquad\quad s(\mathrm{m}) + \mathrm{n} \approx s(\mathrm{m} + \mathrm{n})$ (2)

Extending the specification by type clauses yields

$\quad \mathbf{natplus}^g = \mathbf{natplus} +$
$\qquad \mathbf{ops}$
$\qquad\quad \mathbf{gnd} : \mathrm{nat}$

$$\textbf{axioms}\quad \forall\, m, n : \text{nat}$$

$$\textbf{gnd}(0) \qquad\qquad\qquad\qquad\qquad\qquad\qquad\qquad (3)$$
$$\textbf{gnd}(m) \rightarrow \textbf{gnd}(s(m)) \qquad\qquad\qquad\qquad\quad (4)$$
$$\underline{\textbf{gnd}(m)}, \textbf{gnd}(n) \rightarrow \textbf{gnd}(m+n) \qquad\qquad\quad (5)$$

We prove that $+$ is a defined operator, that is that (5) is an inductive consequence of (1)–(4). We apply the same method and select the first literal in the antecedent of (5).

$$\textbf{gnd}(n) \rightarrow \textbf{gnd}(0+n) \qquad\qquad \text{selective resolution (3) on (5),}$$
$$\textit{composite} \text{ because of (1) and (3)}$$

$$\textbf{gnd}(m), \textbf{gnd}(n) \rightarrow \textbf{gnd}(s(m)+n) \qquad \text{selective resolution (4) on (5),}$$
$$\textit{composite} \text{ because of (2),(4) and (5)}$$

Clauses which have been proved may be kept and used (as lemmas) for proving compositeness in a subsequent inductive proof. Moreover parallel induction is supported as we allow for arbitrary sets of conjectures to start with.

3 Preliminaries

Equational clauses A *signature* Σ is a set of sorts together with a set of operator declarations $f : s_1, \ldots, s_n \rightarrow s$ over these sorts. $s_1, \ldots, s_n$ is called the *arity*, s the *coarity* of f. A Σ-*term* is a term built according to the operator declarations in Σ, possibly with variables. By a *ground* expression (i.e., a term, equation, formula, etc.) we mean an expression containing no variables. For simplicity we do not allow operator overloading and assume that all sorts are *inhabited*, i.e., admit ground terms. For the moment we will assume a fixed signature Σ. Where necessary, we will use the signature as a prefix, like in Σ-term.

We will define equations and clauses in terms of multisets. A *multiset* over X is an unordered collection with possible duplicate elements of X. Formally a multiset is given as a function M from X to the natural numbers. Intuitively, $M(x)$ specifies the number of occurrences of x in M. An *equation* is an expression $s \approx t$, which we identify with the multiset $\{s, t\}$. A *clause* is a pair of multisets of equations, written $\Gamma \rightarrow \Delta$, where Γ is the *antecedent* and Δ the *succedent*. We usually write Γ_1, Γ_2 and A, Γ instead of $\Gamma_1 \cup \Gamma_2$ and $\Gamma \cup \{A\}$.

A clause represents an implication $A_1 \wedge \cdots \wedge A_m \supset B_1 \vee \cdots \vee B_m$; the empty clause, a contradiction. Clauses of the form $\Gamma, A \rightarrow A, \Delta$ or $\Gamma \rightarrow \Delta, t \approx t$ are called *tautologies*. A *specification* is a set of clauses together with the signature the clauses are defined over.

An inference π is a pair written as

$$\frac{C_1 \ldots C_n}{C}$$

where the *premises* $C_1, \ldots, C_n$ and the *conclusion* C are clauses. An *inference system* $\mathcal{I}$ is a set of inferences. An *instance* of an inference π in $\mathcal{I}$ is any inference in $\mathcal{I}$ with premises $C_1\sigma, \ldots, C_n\sigma$ and conclusion $C\sigma$.

Clause orderings   Any ordering $\succ$ on a set S can be extended to an ordering $\succ_{mul}$ on finite multisets over S as follows: $M \succ_{mul} N$ if (i) $M \neq N$ and (ii) whenever $N(x) > M(x)$ then $M(y) > N(y)$, for some y such that $y \succ x$. If $\succ$ is a total [well-founded] ordering, so is $\succ_{mul}$. Given a set (or multiset) S and an ordering $\succ$ on S, we say that x is *maximal* relative to S if there is no y in S with $y \succ x$; and *strictly maximal* if there is no y in S with $y \succeq x$.

If $\succ$ is an ordering on terms, then the corresponding multiset ordering $\succ_{mul}$ is an ordering on equations, which we denote by $\succ^e$.

We have defined clauses as pairs of multisets of equations. Alternatively, clauses may also be thought of as multisets of *occurrences* of equations. We identify an occurrence of an equation $s \approx t$ in the antecedent of a clause with the multiset (of multisets) $\{\{s, \perp\}, \{t, \perp\}\}$, and an occurrence in the succedent with the multiset $\{\{s\}, \{t\}\}$, where $\perp$ is a new symbol.[1] We identify clauses with finite multisets of occurrences of equations. By $\succ^o$ we denote the twofold multiset ordering $(\succ_{mul})_{mul}$ of $\succ$, which is an ordering on occurrences of equations; by $\succ^c$ we denote the multiset ordering $\succ^o_{mul}$, which is an ordering on clauses. If $\succ$ is a well-founded [total] ordering, so are $\succ^e$, $\succ^o$, and $\succ^c$. From now on we will only consider orderings $\succ$ on terms which are reduction orderings and total on ground terms.

We say that a clause $C = \Gamma \rightarrow s \approx t, \Delta$ is *reductive* for $s \approx t$ if $t \not\succeq s$ and $s \approx t$ is a strictly maximal occurrence of an equation in C. For example, if $s \succ t \succ u$ and $s \succ v$ for every term v occurring in Γ, then $\Gamma \rightarrow s \approx t, s \approx u$ is reductive for $s \approx t$, but $\Gamma, s \approx u \rightarrow s \approx t$ is not. Since the ordering is total on ground terms, a ground clause is reductive if and only if $s \succ t$ and $s \approx t$ is greater than any other occurence of an equation. A nonreductive clause has no reductive ground instances.

Equality Herbrand interpretations   We write $A[s]$ to indicate that A contains s as a subexpression and (ambiguously) denote by $A[t]$ the result of replacing a particular occurrence of s by t. By $A\sigma$ we denote the result of applying the substitution σ to A and call $A\sigma$ an *instance* of A. If $A\sigma$ is ground, we speak of a *ground instance*. Composition of substitutions is denoted by juxtaposition. Thus, if τ and ϱ are substitutions, then $x\tau\varrho = (x\tau)\varrho$, for all variables x.

An *equivalence* is a reflexive, transitive, symmetric binary relation. An equivalence $\sim$ on terms is called a *congruence* if $s \sim t$ implies $u[s] \sim u[t]$, for all terms u, s, and t. If E is a set of ground equations, we denote by E^* the smallest congruence containing E.

By an *(equality Herbrand) interpretation* we mean a congruence on ground terms. An interpretation I is said to *satisfy* a ground clause $\Gamma \rightarrow \Delta$ if either $\Gamma \not\subseteq I$ or else

---

[1]The symbol $\perp$ is not part of the vocabulary of the given first-order language. It is assumed to be minimal with respect to any given ordering. Thus $t \succ \perp$, for all terms t.

$\Delta \cap I \neq \emptyset$. We also say that a ground clause C is *true in I*, if I satisfies C; and that C is *false in I*, otherwise. An interpretation I is said to satisfy a non-ground clause $\Gamma \to \Delta$ if it satisfies all ground instances $\Gamma\sigma \to \Delta\sigma$. An interpretation I is called a (*equality Herbrand*) *model* of N if it satisfies all clauses of N.

A set N of clauses is called *consistent* if it has a model; and *inconsistent* (or *unsatisfiable*), otherwise. We say that N *implies* C, and write $N \models C$, if every model of N satisfies C.

Convergent rewrite systems   A binary relation $\Rightarrow$ on terms is called a *rewrite relation* if $s \Rightarrow t$ implies $u[s\sigma] \Rightarrow u[t\sigma]$, for all terms s, t and u, and substitutions σ. A transitive, well-founded rewrite relation is called a *reduction ordering*. By $\Leftrightarrow$ we denote the symmetric closure of $\Rightarrow$; by $\stackrel{*}{\Rightarrow}$ the transitive, reflexive closure; and by $\stackrel{*}{\Leftrightarrow}$ the symmetric, transitive, reflexive closure. Furthermore, we write $s \Downarrow t$ to indicate that s and t can be rewritten to a common form: $s \stackrel{*}{\Rightarrow} v$ and $t \stackrel{*}{\Rightarrow} v$, for some term v. A rewrite relation $\Rightarrow$ is said to be *Church-Rosser* if the two relations $\stackrel{*}{\Leftrightarrow}$ and $\Downarrow$ are the same.

A set of equations E is called a *rewrite system* with respect to an ordering $\succ$ if we have $s \succ t$ or $t \succ s$, for all equations $s \approx t$ in E. If all equations in E are ground, we speak of a *ground rewrite system*. Equations in E are also called (*rewrite*) *rules*. When we speak of "the rule $s \approx t$" we implicitly assume that $s \succ t$. By $\Rightarrow_{E\succ}$ (or simply $\Rightarrow_E$) we denote the smallest rewrite relation for which $s \Rightarrow_E t$ whenever $s \approx t$ is in E and $s \succ t$. A term s is said to be in *normal form* (with respect to E) if it can not be rewritten by $\Rightarrow_E$, i.e., if there is no term t such that $s \Rightarrow_E t$. A term is also called *irreducible*, if it is in normal form, and *reducible*, otherwise. A rewrite system E is said to be *convergent* if the rewrite relation $\Rightarrow_E$ is well-founded and Church-Rosser. Convergent rewrite systems define unique normal forms.

Predicates   We allow that in addition to function symbols a signature may contain predicate symbols, which will be declared to have a special coarity pred. Thus we also consider expressions $P(t_1, \ldots, t_n)$, where P is some predicate symbol and $t_1, \ldots, t_n$ are terms built from function symbols and variables. We then have equations $s \approx t$ between (non-predicate) terms, called *function equations*, and equations $P(t_1, \ldots, t_n) \approx \mathsf{tt}$, called *predicate equations*, where tt is a distinguished unary predicate symbol that is taken to be minimal in the given reduction ordering $\succ$. For simplicity, we usually abbreviate $P(t_1, \ldots, t_n) \approx \mathsf{tt}$ by $P(t_1, \ldots, t_n)$.

4   The Perfect Model

The proof by consistency method proves properties of a standard model of a specification, which for unconditional equations and horn clauses is the initial model. The initial model can be characterized as the unique minimal (with respect to set inclusion) Herbrand interpretation satisfying N, and for Horn clause specifications it always exists if N is consistent. In the case of first-order clauses more than one minimal model may exist. For instance, if N consists of the single clause $\to p, q$ then both $\{p\}$ and $\{q\}$ are minimal models of N. We will use the ordering $\succ^e$ to single out one of the minimal models as the *perfect model*.

Let $\succ^p = (\succ^e)^{-1}$, then a model I is called *preferable* to J if $J \succ^p_{mul} I$. A *perfect model* (corresponding to $\succ$) is a minimal model with respect to $\succ^p_{mul}$. For instance, if we assume $q \succ p$ then $\{p\} \succ^p_{mul} \{q\}$, i.e., $\{q\}$ is the perfect model. It is important to see that different orderings may yield different perfect models. This is an essential difference to the case of Horn clauses. For general clauses the ordering $\succ$ must be explicitly given in order to uniquely identify the standard model one has in mind.

As $\succ^e$ is well-founded and total, $\succ^p_{mul}$ is a total ordering [19]. Hence there exists at most one perfect model for a set of clauses N. Since $\succ^p_{mul}$ contains $\subseteq$, a perfect model is also minimal.

In the remainder of this section we present methods and techniques for constructing, given a consistent set of clauses and an ordering, the corresponding perfect model. We also explain how to compute in this model. The proofs of the lemmas which justify our techniques may be found in [4] and [5].

4.1 Construction of the Perfect Model

Let N be a set of clauses and $\succ$ be a reduction ordering which is total on ground terms. We shall define an interpretation I for N by means of a convergent rewrite system R. For certain N, I will be the perfect model of N with respect to $\succ$.

First, we use induction on the clause ordering $\succ^c$ to define sets of equations E_C, R_C and I_C, for all ground clauses C over the given signature (not necessarily instances of N). Let C be such a ground clause and suppose that $E_{C'}$, $R_{C'}$ and I'_C have been defined for all ground clauses C' for which $C \succ^c C'$. Then

$$R_C = \bigcup_{C \succ^c C'} E_{C'} \quad \text{and} \quad I_C = R_C^*.$$

Moreover

$$E_C = \{s \approx t\}$$

if C is a ground instance $\Gamma \to \Delta, s \approx t$ of N such that (i) C is reductive for $s \approx t$, (ii) s is irreducible by R_C, (iii) $\Gamma \subseteq I_C$, and (iv) $\Delta \cap I_C = \emptyset$. In that case, we also say that C *produces* the equation (or rule) $s \approx t$. In all other cases, $E_C = \emptyset$. Finally, we define I to be the equality interpretation R^*, where $R = \bigcup_C E_C$ is the set of all equations produced by ground instances of clauses in N.

Instances of N that produce equations are also called *productive*. Note that a productive clause C is false in $I_C = R_C^*$, but true in $(R_C \cup E_C)^*$. The truth value of an equation can be determined by rewriting: $u \approx v \in I$ if and only if $u \Downarrow_R v$. In many cases the truth value of an equation can already be determined by rewriting with R_C. If C is true in I_C then for $D \succeq^c C$ it is also true in I_D and in I.

4.2   Superposition and redundancy

The interpretation I will in general not be a model of N, unless N is closed under sufficiently many applications of certain inference rules. The inference system $\mathcal{S}_S^{\succ}$ which we consider in this paper is the one described in [4, 5] and is also briefly summarized in the appendix. It is based on $\succ$ and on a selection function S. By a *selection function* we mean a mapping S that assigns to each clause C a (possibly empty) multiset of negative occurrences of equations in C. The equations in $S(C)$ are called *selected*. If $S(C) = \emptyset$, then no equation is selected. Selected equations can be arbitrarily chosen and need not be maximal. Selection functions are assumed to be compatible with substitution, i.e. an occurrence of an equation is selected in C if and only if the corresponding occurrence is selected in $C\sigma$, for any substitution σ.

$\mathcal{S}_S^{\succ}$, for short $\mathcal{S}$, if $\succ$ and S are indicated by the context, consists mainly of paramodulation rules which are restricted by ordering constraints derived from $\succ$ or by selection constraints derived from S. Paramodulation affects maximal equations only, unless some atom of the antecedent of a clause is selected. If an equation is selected in (the antecedent of) a clause C, paramodulation on C always occurs into a maximal selected equation of C. An important feature is that clauses for which an equation is selected need *not* be considered for paramodulating into any other clause. They do not directly contribute to the construction of the perfect model. This is made more precise by a notion of redundancy. Redundancy is a key aspect which allows to saturate many nontrivial sets of clauses under $\mathcal{S}$ in a finite number of steps.

A ground clause C is said to be *redundant (in N)* if C is true in I_C. A clause is called redundant in N if all its ground instances are redundant in N. Redundant clauses are true in I. The interpretation I is completely determined by productive clauses, which are non-redundant instances of N.

An inference π from ground clauses is said to be *redundant (in N)* if either one of its premises is redundant in N or else its conclusion is true in I_C, where C is the maximal (the second, if the inference has two premises) premise of π. An inference from arbitrary premises is redundant in N if all its ground instances are redundant in N. We say that N is *saturated* if every ground instance of an inference from premises in N is redundant in N.

Lemma 1 *Let N be a saturated set of clauses. If an instance C of a clause in N contains a selected equation, C is not productive.*

Theorem 1 *Let N be a consistent and saturated set of clauses. Then I is the perfect model of N with respect to $\succ$.*

These two lemmas are the key to our method. The first shows that clauses C in N with selected equations do not contribute to the perfect model. In a saturated set of clauses N, C is therefore an inductive consequence of $N \setminus \{C\}$. The second lemma shows that any consistent set of clauses has a perfect model (with respect to any given complete reduction ordering). In particular, the construction of section

4.1 yields the perfect model, provided N is consistent and saturated. Fair theorem proving derivation are a means to saturate a given set of clauses, though the limit may not be reachable in a finite number of steps. All this gives hints of how to compute in the perfect model, an aspect which is made more precise below.

4.3 Saturation

The next question we address is how to construct a saturated set of clauses. The notion of redundancy is not effectively usable as it is not stable under addition or deletion of clauses.

Let N be a set of clauses and C be a ground clause (not necessarily a ground instance of N). We call C *composite* with respect to N, if there exist ground instances $C_1, \ldots, C_k$ of N such that $C_1, \ldots, C_k \models C$ and $C \succ^c C_j$, for all j with $1 \leq j \leq k$. A non-ground clause is called composite if all its ground instances are composite.

A ground inference π with conclusion B is called *composite* (with respect to N) if either some premise is composite with respect to N, or else there exist ground instances $C_1, \ldots, C_k$ of N such that $C_1, \ldots, C_k \models B$ and $C \succ^c C_j$, for all j with $1 \leq j \leq k$, where C is the maximal premise of π. A non-ground inference is called composite if all its ground instances are composite.

Lemma 2 *For any saturated and consistent set of clauses N compositeness with respect to N implies redundancy with respect to N —for clauses as well as for inferences. Moreover, compositeness is stable under addition of clauses to N and stable under deletion of composite clauses from N.*

A *theorem proving derivation* is a (finite or countably infinite) sequence $N_0, N_1, N_2, \ldots$ of sets of clauses such that either
  (*Deduction*) $N_{i+1} = N_i \cup \{C\}$ and $N_i \models C$, or
  (*Deletion*) $N_{i+1} = N_i \setminus \{C\}$ and C is composite with respect to $N_i \cup \{C\}$.
The set $N_\infty = \bigcup_j \bigcap_{k \geq j} N_k$ is called the *limit* of the derivation. Clauses in N_∞ are called *persisting*.

Deduction adds clauses that logically follow from given clauses; deletion eliminates composite clauses. Simplification can be modeled as a sequence of deduction steps followed by a deletion step.

A theorem proving derivation is called *fair* if every inference in S from premises in N_∞ is composite with respect to $\bigcup_j N_j$.

A fair derivation can be constructed, for instance, by systematically adding conclusions of non-composite inferences in S. As the maximal premise of a ground inference is always greater with respect to $\succ^c$ than its conclusion, the inference becomes composite as soon as the conclusion has been added.

A set of clauses N is called *complete* if all inferences from N are composite with respect to N. A complete set of clauses that does not contain the empty clause is saturated.

Lemma 3 *Let $N = N_0, N_1, N_2, \ldots$ be a fair theorem proving derivation. If N is inconsistent then the empty clause is contained in $\bigcup_j N_j$. Otherwise N and N_∞ are logically equivalent, and N_∞ is complete (and hence saturated).*

4.4 Computing in Perfect Models

The rewrite system R which defines I is canonical, hence constitutes a decision procedure for equality in the perfect model, provided the one-step rewrite relation $\Rightarrow_R$ is computable. This need not be the case in general, not even for finite and complete sets of clauses. However, if the set of clauses is such that matching a ground term against the maximal term of a clause always results in a reductive ground clause, one can use a recursive algorithm to decide the word problem for I.

A clause $C = \Gamma \to \Delta$ is called *universally reductive* if either the succedent Δ is empty, or else Δ can be written as $\Delta', s \approx t$ such that (i) all variables of C also occur in s, (ii) $C\sigma$ is reductive for $s\sigma \approx t\sigma$, for all ground substitutions σ. A set N of clauses is called universally reductive if any clause in N is either universally reductive, or else contains a selected atom.

Lemma 4 *Suppose $\succ$ is decidable, and let N be a saturated, finite and universally reductive set of clauses. Then it is decidable whether a ground equation $s \approx t$ is valid in the perfect model for N.*

Proof: Let $s \approx t$ be such a ground equation. Since R as constructed from N is a convergent rewrite system, it suffices to rewrite s and t to their respective normal forms, and then to check if they are equal. Clauses in N containing selected equations need not be considered, since they do not produce any rewrite rules. The other clauses are universally reductive. By matching the maximal term of all clauses not containing a selected equation against s, we obtain a finite set of reductive ground clauses which could possibly have produced a rewrite rule that can reduce s. Let $C = \Lambda \to \Pi, l \approx r$ be one of the matching instances. We may rewrite l by r if $\Lambda \subseteq I_C$ and $\Pi \cap I_C = \emptyset$. The latter problem is simpler than the problem of reducing s, as I_C is constructed from productive clauses smaller than C.   ■

An obvious consequence of the above lemma is the decidability of validity in the perfect model for ground clauses in the indicated case.

5   Proof by consistency

Inductive validity and redundancy are equivalent concepts. If $C = \Gamma \to \Delta$ is an inductive consequence of a consistent and saturated set N of clauses then the logically equivalent clause $\top \approx \top, \Gamma \to \Delta$, with $\top$ being a new constant of a new sort and maximal with respect to $\succ$, is redundant in N. Conversely, if C is redundant in N is is true in I_C and in I. In this case N and $N \setminus \{C\}$ have the same perfect model. Therefore C is an inductive consequence of $N \setminus \{C\}$. Inductive theorem proving is proving redundancy, and vice versa. Lemmas 2 and 1 are the basis of the technique that we are going propose in this section. In simple cases a conjecture (or some derived clause) can be eliminated by a direct proof of compositeness using specific techniques such contextual rewriting, cf. section 5.5. Otherwise one may attempt to make the clause become redundant in the limit of a saturation process. Closing conjectures by type predicates for their variables is a technical device to translate a non-ground clause into an equivalent one with a non-empty antecedent from which an equation can be selected for superposition. Selecting a type predicate of a variable corresponds to selecting that variable as the induction variable.

5.1 Type Predicates

For each sort s in Σ we add a predicate symbol $\mathbf{gnd}_s$, and for each operator f : $s_1, \ldots, s_n \to s$ in Σ a clause

$$G(f) = \mathbf{gnd}_{s_1}(x_1), \ldots, \mathbf{gnd}_{s_n}(x_n) \to \mathbf{gnd}_s(f(x_1, \ldots, x_n)).$$

$\mathbf{gnd}_s$ is called the *type predicate* for s and $G(f)$ the *type clause* for f. A ground term $t = f(t_1, \ldots, t_n)$ over Σ uniquely determines a ground instance

$$\mathbf{gnd}(t_1), \ldots, \mathbf{gnd}(t_n) \to \mathbf{gnd}(f(t_1, \ldots, t_n))$$

of the clause $G(f)$, which we will denote by $G(t)$. For a signature Σ we define $G(\Sigma)$ as the set of all $G(f)$ where f is an operator in Σ. The union of N and $G(\Sigma)$ will be denoted by N^g, while the extended signature will be denoted by Σ^g. By R^g and I^g we denote the set of rewrite rules and interpretation, respectively, constructed from N^g according to section 4.1. $G(\Sigma)$ encodes the notion of a "ground term", i.e., an atom $\mathbf{gnd}(t)$ is provable if and only if t is equal (modulo N) to a ground term. We assume that the given complete reduction ordering over Σ is arbitrarily extended to a complete reduction ordering over the extended signature Σ^g. Such an extension always exists.

Lemma 5 *Let t be a ground term over Σ of form $f(t_1, \ldots, t_n)$. Then $\mathbf{gnd}(t)$ is true in $(R^g_{G(t)} \cup E^g_{G(t)})^*$ and hence in I^g.*

Proof: Let $C = G(t)$ and $C_i = G(t_i)$. We will use induction on $\succ^c$ for this proof. Since C_i is smaller than C, we may use the induction hypothesis to infer that $\mathbf{gnd}(t_i)$ is true in $(R^g_{C_i} \cup E^g_{C_i})^*$, and hence, in I^g_C. Now suppose that t is reducible by R^g_C, i.e., there exists a term t' such that $t \Rightarrow_{R^g_C} t'$. Then $\mathbf{gnd}(t)$ is already true in $(R^g_{C'} \cup E^g_{C'})^*$, where $C' = G(t')$. We have $C \succ C'$, and conclude that C is true in I^g_C. If on the other hand t is not reducible by R^g_C, C is not true in I^g_C and produces the rule $\mathbf{gnd}(t) \approx \mathbf{tt}$.

■

Lemma 6 *Let N be a saturated set of clauses. Then N^g is also saturated.*

Proof: We have to show that all inferences from clauses in N^g are redundant. Since N is saturated, we only have to consider inferences where at least one premise is not in N. The only nontrivial case is that of a right superposition in which the second premise is in $G(\Sigma)$. If both premises are in $G(\Sigma)$ the inference results in a tautology. If the first premise is in N and the second premise C is in $G(\Sigma)$, any ground instance of the inference has the form

$$\frac{\Lambda \to \Pi, f(s_1, \ldots, s_n) \approx t \qquad \mathbf{gnd}(s_1), \ldots, \mathbf{gnd}(s_n) \to \mathbf{gnd}(f(s_1, \ldots, s_n))}{\Lambda, \mathbf{gnd}(s_1), \ldots, \mathbf{gnd}(s_n) \to \Pi, \mathbf{gnd}(t)}$$

(overlaps inside of the s_i would be in variable positions). t is smaller than $f(s_1, \ldots, s_n)$, hence $\mathbf{gnd}(t)$ is true in I_C^g by lemma 5. Thus the inference is redundant. ∎

Lemma 7 *Let C be a ground instance of a clause in N^g. R_C^g restricted to rules not containing any type predicates is equal to R_C.*

Lemma 8 *Let C be a ground instance of a clause in N. Then C is true in I^g if and only if C is true in I.*

Let $C = \Gamma \to \Delta$ be a Σ-clause and $var(C) = \{x_1, \ldots, x_n\}$. Then we define $G(C)$ to be the Σ^g-clause $\mathbf{gnd}(x_1), \ldots, \mathbf{gnd}(x_n), \Gamma \to \Delta$. $G(C)$ is called the *closed form* of C. For a set of clauses H, we define $G(H)$ as $\{ G(C) \mid C \in H \}$.

Lemma 9 *C is true in I if and only if $G(C)$ is true in I^g.*

Proof: All atoms in Γ and Δ are true in I if and only if they are true in I^g (cf. corollary 8). The additional atoms of the form $\mathbf{gnd}(t)$ are true in I^g (cf. lemma 5). ∎

Let us summarize the contents of this section. If we have a saturated set of clauses N, we can transform it to N^g, which is still saturated. A clause C is valid in I if and only if C is valid in N^g if and only if $G(C)$ is valid in N^g. For inductive theorem proving we may hence we may use N^g in place of N, and we may replace conjectures H by the closed forms $G(H)$.

5.2 Inductive Theorem Proving Derivations

From now on we shall assume to be given a consistent, finite, complete and universally reductive set of clauses N with perfect model I. By an *inductive theorem proving derivation* for N we mean a finite or countably infinite sequence $H_0, H_1, \ldots$ of sets of clauses C such that

(*Deduction*) $H_{i+1} = H_i \cup \{C\}$ and $N \cup H_i \models C$, or

(*Deletion*) $N_{i+1} = N_i \setminus \{C\}$ and C is composite with respect to $N \cup H_i \cup \{C\}$

The set $H_\infty = \bigcup_j \bigcap_{k \geq j} H_k$ is called the *limit* of the derivation. Clauses in H_∞ are called *persisting*.

An inductive theorem proving derivation is called *fair* if every inference by selective equality resolution on a clause in H_∞ and every inference by selective superposition of a clause in N on a clause in H_∞ is composite with respect to $\bigcup_j H_j$.

Given a selection function S, an inductive theorem proving derivation is called *failed*, if there exists a ground clause in $\bigcup_j H_j$ which is false in I (failure with "disproof"), or else there exists a clause in H_∞ for which no equation is selected by S (failure with "don't know").

Theorem 2 *Let $H_0, H_1, \ldots$ be a fair inductive theorem proving derivation for N.*

(i) If the derivation is non-failed then the clauses in H_0 are inductive theorems of N, i.e., valid in I.

(ii) If the derivation fails with "disproof", then H_0 is not valid in I.

Proof: (i) If $H_0, H_1, \ldots$ is fair and non-failed, the sequence $N \cup H_0, N \cup H_1, \ldots$ is a theorem proving derivation (with respect to $\mathcal{S}$). Inferences with premises all in N are composite in N as N is complete. Inferences with at least one premise in H_∞ are composite in $N \cup H'_\infty$ as the derivation is fair and as there is no clause in H_∞ for which no equation is selected by S. Therefore the sequence $N \cup H_0, N \cup H_1, \ldots$ is a fair theorem proving derivation with limit $N \cup H_\infty$, hence $N \cup H_\infty$ is complete. As the clauses in H_∞ have selected equations, they are not productive. Therefore the perfect models of N and $N \cup H_\infty$ (and hence $N \cup H_0$) are identical.

(ii) follows immediately from the soundness of deduction.   ∎

Note that the fairness requirement for inductive derivations does not imply the need for computing any non-linear inferences with premises all in N or with two premises both not in N. To achieve refutation completeness of the method the production of non-ground clauses with an empty antecedent has to be avoided. Using type predicates is a technique to achieve this goal.

5.3 Refutation completeness

Retutation completeness in this context means to avoid failure with "don't know" in inductive theorem proving derivations. For that purpose we can assume without loss of generality that the given theory presentation N includes all type predicates and type clauses, i.e., $N = \tilde{N}^g$, as this does not affect completeness, consistency, finiteness, and universal reductivity. Similarly, we can assume the initial set of conjectures H_0 to be closed. Suppose that we only admit selection functions which always select some type atom of form $\mathbf{gnd}(x)$, x a variable, and only such atoms, if there are any in a given clause. Furthermore assume that in an inductive theorem proving the only deductions one makes are by selective equality resolution on a clause in $\bigcup_j H_j$ or by selective superposition of a clause in N on a clause in $\bigcup_j H_j$. Then any clause in $\bigcup_j H_j$ is closed, hence either is a ground clause, for which validity is decidable, or else contains a selected equation, Therefore failure with "don't know" is impossible.

In practice, however, one would not want to restrict selection functions to always select type predicates only, nor to limit deductions to what is required by fairness.

Otherwise one might not detect situations in which the antecedent is false for certain substitutions. Also simplification, e.g., by demodulation, would not be allowed. Failure with "don't know" will only occur in extreme cases cases anyway, and there are other ways of achieving refutation completeness.

Superposition of a type clause on a selected type atom in a conjecture is a principal kind of inferences to be computed in inductive theorem proving derivations. The more type clauses one can be prove redundant, the less such inferences one needs to consider. This problem will be addressed in the next section.

5.4 Sufficient completeness

A type clause for a function f is redundant, if the function is sufficiently completely defined with respect to the remaining functions. More generally, we consider sub-signatures Σ_B, called *base signatures*, of the given signature Σ. A set of Σ-clauses N (together with an ordering $\succ$) is called *sufficiently complete* with respect to Σ_B if for any Σ-ground term s there exists a Σ_B-ground term t such that $s \approx t$ is true in the perfect model I of N. Again, this property is defined with respect to the perfect model rather than all minimal models. Furthermore, we assume that ground terms over Σ_B are smaller in the reduction ordering than ground terms containing function symbols not in Σ_B.

Lemma 10 N *is sufficiently complete with respect to* Σ_B *if and only if the type clauses* $G(f)$, *for* f *in* $\Sigma \setminus \Sigma_B$, *are redundant.*

Proof: Let R and I be the rewrite system and interpretation, respectively, constructed from the ground instances of N. Moreover, let f be an operator in $\Sigma \setminus \Sigma_B$, and suppose that all ground instances of $G(f) = \mathsf{gnd}(x_1), \ldots, \mathsf{gnd}(x_n) \to \mathsf{gnd}(f(x_1, \ldots, x_n))$ are redundant. We show that any term ground term t with outermost symbol f is reducible by R, which implies sufficient-completeness of N. If $t = f(t_1, \ldots, t_n)$, the redundancy of $C = \mathsf{gnd}(t_1), \ldots, \mathsf{gnd}(t_n) \to \mathsf{gnd}(f(t_1, \ldots, t_n))$ implies that $\mathsf{gnd}(f(t_1, \ldots \ldots, t_n))$ is reducible by R_C. As no type clause smaller than C can produce a rule to reduce $\mathsf{gnd}(f(t_1, \ldots, t_n))$, t must in fact be reducible.

For the converse, suppose there exists a term $t = f(t_1, \ldots, t_n)$ which is not reducible by R, for some operator f in $\Sigma \setminus \Sigma_B$. Consider the ground instance

$$C = \mathsf{gnd}(t_1), \ldots, \mathsf{gnd}(t_n) \to \mathsf{gnd}(f(t_1, \ldots, t_n)),$$

of $G(f)$. $\mathsf{gnd}(f(t_1, \ldots, t_n))$ is not reducible by R_C and the $\mathsf{gnd}(t_i)$ are true in R_C, hence C produces a rule and is not redundant. ∎

Corollary 1 *Let* N *be a saturated set of clauses, and let* S *be a selection function such that each clause in* $G(\Sigma \setminus \Sigma_B)$ *either contains a selected equation if its antecedent is nonempty or else is redundant. Then* N *is sufficiently complete with respect to* Σ_B *if and only if* N^g *is saturated.*

The inductive proof procedure outlined above can be used to prove sufficient-completeness by starting with the theory $N \cup G(\Sigma_B)$ and proving the inductive validity of the clauses in $G(\Sigma \setminus \Sigma_B)$. If this is successful in finite time, the result is a complete presentation $N \cup G(\Sigma_B) \cup G'$ where all clauses in G' have a selected atom. In later inductive proofs no inferences from G', in particular no inferences from any type clause $G(f)$, with f in $\Sigma \setminus \Sigma_B$, need to be computed.

With these remarks, the reader may want to take another look at the proof of sufficient-completeness of addition that we presented in section 2.

Proving the sufficient completeness of a function definition is a particular case in which a lemma is produced that makes subsequent proofs go through or at least makes them more efficient. In general, clauses which have been proved may be kept and used for proving compositeness in a subsequent inductive proof, without the need of superposing such a lemma on some conjecture.

5.5   Proofs of Compositeness

Inductive validity is reduced to proving compositeness of certain clauses that are derived from the conjectures. For the method to be applicable in practice one needs to have powerful methods available for verifying compositeness. Moreover, a failure in proving compositeness often gives an indication of what kind of lemma would be required in order to make the proof go through. It is here that generalization and lemma suggestion techniques should be incorporated. In this paper we shall only scratch the surface by making a few technical remarks on the subject related to orderings and first-order clauses.

In proofs of compositeness, one may assume that instances of the given theory presentation N are smaller (with respect to $\succ^c$) than any "new" clause that is introduced during an inductive theorem proving derivation (including the initial conjectures). Formally this can be justified by assuming that a new clause $\Gamma \to \Delta$ actually represents the logically equivalent clause $\top \approx \top, \Gamma \to \Delta$, where $\top$ is a new constant of a new sort which is the maximal term with respect to $\succ$.

For first-order clauses a technique for obtaining compositeness proofs, called *contextual reductive rewriting* in [2, 3], that has proven to be useful in practice. Let C be a clause and let N be a set of clauses. N_C denotes the set of instances C' of N such that $C \succ^c C'$. Let τ be a skolemizing substitution, i.e., a substitution that replaces variables by new constants. Let $C = \Gamma, u[l\sigma] \approx v \to \Delta$ (or $C = \Gamma \to \Delta, u[l\sigma] \approx v$) be a clause in N. Suppose there exists an instance $D\sigma$ of a clause $D = \Lambda \to \Pi, l \approx r$ in N such that (i) $l\sigma \succ r\sigma$, (ii) $C \succ^c D\sigma$, (iii) $N_C\tau \models \Gamma\tau \to A\tau$ for all equations A in $\Lambda\sigma$ and $N_C\tau \models A\tau \to \Delta\tau$ for all equations A in $\Pi\sigma$. Then C can be contextually rewritten to $\Gamma, u[r\sigma] \approx v \to \Delta$ (or $\Gamma \to \Delta, u[r\sigma] \approx v$). After this the clause C becomes composite in $N \cup \{\Gamma, u[r\sigma] \approx v \to \Delta\}$ and may be eliminated.

We may also do several steps of contextual rewriting in a row; in this case the bound on the complexity is provided by the first clause. If we eventually arrive at a clause that is composite, we have proved that the first clause is composite. Or we may use the method to prove inferences composite; in this case the bound is provided by the maximal premise of the inference.

More liberal ways of rewriting where a term is sometimes replaced by a larger term (as long as the instances of clauses which are involved in the rewriting are still sufficiently small) are suggested by the "rippling-out" method of [8] and can be extended without problems to our framework.

5.6   Another Example

As another example we will prove that $\leq$ is transitive, i.e.,

$$\underline{k \leq m}, m \leq n \to k \leq n \tag{5}$$

By (1)–(4) we refer to the corresponding numbering of axioms in the definition of $\leq$ as presented in section 2. We shall see that in this case we need not even close the conjecture by type predicates.

(6) $m \leq n \to 0 \leq n$ selective resolution (3) on (5), composite because of (3)

(7) $k \leq m, \underline{s(m) \leq n} \to s(k) \leq n$ selective resolution (4) on (5)

(8) $k \leq m, m \leq n \to s(k) \leq s(n)$ selective resolution (4) on (7), composite because of (4) and (5)

6   Conclusion

We have described a method of proof by consistency for first-order clauses with equality. Inductive theorem proving was defined as proving validity in the perfect model of a theory. We have built on methods for saturating sets of clauses and shown that inductive validity and redundancy are equivalent concepts. Selection strategies for superposition provide means to make clauses become redundant in the limit of a successful saturation process. For this idea to always be applicable an explicit closing of clauses by type predicates has been suggested As a side-effect our method allows for proofs of sufficient-completeness of function definitions, a property that is essential in other contexts too, e.g., for hierarchical specifications. We have shown that the concepts of proof by consistency for the purely equational case can be appropriately extended retaining most of their characteristic properties.

We have implemented this method for the restricted case of horn clauses in the CEC system for conditional equational completion [10] and some encouraging, initial practical experience has been made.

Bibliography

[1] Leo Bachmair. Proof by consistency in equational theories. In *Proc. 3rd IEEE Symp. on Logic in Computer Science*, pages 228–233, Edinburgh, July 1988.

[2] Leo Bachmair and Harald Ganzinger. On restrictions of ordered paramodulation with simplification. In *Proc. 10th Int. Conf. on Automated Deduction*, Kaiserslautern, July 1990. Springer LNCS 449.

[3] Leo Bachmair and Harald Ganzinger. Completion of first-order clauses with equality by strict superposition. In *Proc. 2nd Int. Workshop on Conditional and Typed Rewriting Systems*, Montreal, June 1990. Springer LNCS 516.

[4] Leo Bachmair and Harald Ganzinger. Perfect model semantics for logic programs with equality. In *Proc. 8th Int. Conf. on Logic Programming*. MIT Press, 1991.

[5] Leo Bachmair and Harald Ganzinger. Rewrite-based equational theorem proving with selection and simplification. Technical Report MPI-I-91-208, Max-Planck-Institut für Informatik, Saarbrücken, August 1991.

[6] Eddy Bevers and Johan Lewi. Proof by consistency in conditional equational theories. In *Proc. 2nd Int. Workshop on Conditional and Typed Rewriting Systems*, Montreal, June 1990. Springer LNCS 516.

[7] Robert S. Boyer and J. Strother Moore. *A Computational Logic*. Academic Press, New York, 1979.

[8] A. Bundy, F. van Harmelen, A. Smail, and A. Ireland. Extensions to the rippling-out tactic for guiding inductive proofs. In *Proc. 10th Int. Conf. on Automated Deduction*, pages 132–146, Kaiserslautern, July 1990. Springer LNCS 449.

[9] Laurent Fribourg. A strong restriction of the inductive completion procedure. In *Proc. 13th Int. Coll. on Automata, Languages and Programming*, pages 105–115, Rennes, France, July 1986. Springer LNCS 226.

[10] H. Ganzinger and R. Schäfers. System support for modular order-sorted Horn clause specifications. *Proc. 12th Int. Conf. on Software Engineering*, Nice, pages 150–163, 1990.

[11] Stephen J. Garland and John V. Guttag. Inductive methods for reasoning about abstract data types. In *Proc. 15th Annual ACM Symp. on Principles of Programming Languages*, pages 219–228, San Diego, January 1988.

[12] Gérard Huet and Jean-Marie Hullot. Proofs by induction in equational theories with constructors. *Journal of Computer and System Sciences*, 25:239–266, 1982.

[13] Jean-Pierre Jouannaud and Emmanuel Kounalis. Proofs by induction in equational theories without constructors. In *Proc. Symp. on Logic in Computer Science*, pages 358–366, Cambridge, Mass., June 1986.

[14] Stephane Kaplan and Marianne Choquer. On the decidability of quasi-reducibility. *EATCS Bulletin*, 28:32–34, 1986.

[15] David R. Musser. On proving inductive properties of abstract data types. In *Proc. 7th Annual ACM Symp. on Principles of Programming Languages*, pages 154–162, Las Vegas, January 1980.

[16] Fernando Orejas. Theorem proving in conditional-equational theories. Draft.

[17] Peter Padawitz. Inductive expansion: A calculus for verifying and synthesizing functional and logic programs. *Journal of Automated Reasoning*, 7(1):27–103, March 1991.

[18] David A. Plaisted. Semantic confluence tests and completion methods. *Information and Control*, 65:182–215, 1985.

[19] T. C. Przymusinski. On the declarative semantics of deductive databases and logic programs. In J. Minker, editor, *Foundations of Deductive Data Bases and Logic Programming*, pages 193–216. Morgan Kaufmann Publishers, Los Altos, 1988.

[20] Jürgen Stuber. Inductive theorem proving for horn clauses. Master's thesis, Universität Dortmund, April 1991.

Inference rules

The following inference rules are defined with respect to $\succ$ and a selection function S, defining the calculus $S_S^\succ$.

Equality resolution:
$$\frac{\Gamma, u \approx v \rightarrow \Delta}{\Gamma\sigma \rightarrow \Delta\sigma}$$

where σ is a most general unifier of u and v and $u\sigma \approx v\sigma$ is a maximal occurence of an equation in $\Gamma\sigma, u\sigma \approx v\sigma \rightarrow \Delta\sigma$

Ordered factoring:
$$\frac{\Gamma \rightarrow \Delta, A, B}{\Gamma\sigma \rightarrow \Delta\sigma, A\sigma}$$

where σ is a most general unifier of A and B and $A\sigma$ is a maximal occurrence of an equation in $\Gamma\sigma \rightarrow \Delta\sigma, A\sigma, B\sigma$.

Superposition, left:
$$\frac{\Lambda \rightarrow \Pi, l \approx r \quad\quad \Gamma, u[l'] \approx v \rightarrow \Delta}{\Lambda\sigma, \Gamma\sigma, u\sigma[r\sigma] \approx v\sigma \rightarrow \Pi\sigma, \Delta\sigma}$$

where (i) σ is a most general unifier of l and l', (ii) the clause $\Lambda\sigma \to \Pi\sigma, l\sigma \approx r\sigma$ is reductive for $l\sigma \approx r\sigma$ (iii) $v\sigma \not\succeq u\sigma$ and $u\sigma \approx v\sigma$ is a maximal occurrence of an equation in $\Gamma\sigma, u\sigma \approx v\sigma \to \Delta\sigma,$[2] and (iv) l' is not a variable.

Superposition, right:
$$\frac{\Lambda \to \Pi, l \approx r \qquad \Gamma \to \Delta, s[l'] \approx t}{\Lambda\sigma, \Gamma\sigma \to \Pi\sigma, \Delta\sigma, s\sigma[r\sigma] \approx t\sigma}$$

where (i) σ is a most general unifier of l and l', (ii) the clause $\Lambda\sigma \to \Pi\sigma, l\sigma \approx r\sigma$ is reductive for $l\sigma \approx r\sigma$, (iii) the clause $\Gamma\sigma \to \Delta\sigma, s\sigma \approx t\sigma$ is reductive for $s\sigma \approx t\sigma$, and (iv) l' is not a variable.

Merging Paramodulation:
$$\frac{\Gamma \to \Delta, l \approx r \qquad \Lambda \to \Pi, s \approx t[l'], s' \approx t'}{\Gamma\sigma, \Lambda\sigma \to s\sigma \approx t[r]\sigma, s\sigma \approx t'\sigma, \Delta\sigma, \Pi\sigma}$$

where (i) σ is the composition $\tau\varrho$ of a most general unifier τ of l and l', and a most general unifier ϱ of $s\tau$ and $s'\tau$, (ii) the clause $\Gamma\sigma \to \Delta\sigma, l\sigma \approx r\sigma$ is reductive for $l\sigma \approx r\sigma$, (iii) the clause $\Lambda\sigma \to \Pi\sigma, s\sigma \approx t\sigma, s'\sigma \approx t'\sigma$ is reductive for $s\sigma \approx t\sigma$, (iv) $s\tau \succ t\tau$ and $t'\sigma \not\succeq t\sigma$, and (v) l' is not a variable.

The following additional restrictions are imposed: (a) the premises of an inference rule must not share any variables (if necessary, the variables in one premise are renamed); and (b) if C and D are the premises of a paramodulation inference with σ the mgu obtained from superposing C on D, then $C\sigma \not\succeq^c D\sigma$.

The following inference rules are defined with respect to a given selection function S.

Selective resolution:
$$\frac{\Gamma, u \approx v \to \Delta}{\Gamma\sigma \to \Delta\sigma}$$

where σ is a most general unifier of u and v and $u\sigma \approx v\sigma$ is a selected equation in $\Gamma, u \approx v \to \Delta$

Selective superposition:
$$\frac{\Lambda \to \Pi, l \approx r \qquad \Gamma, u[l'] \approx v \to \Delta}{\Lambda\sigma, \Gamma\sigma, u\sigma[r\sigma] \approx v\sigma \to \Pi\sigma, \Delta\sigma}$$

where (i) σ is a most general unifier of l and l', (ii) the clause $C = \Lambda\sigma \to \Pi\sigma, l\sigma \approx r\sigma$ contains no selected equations and $C\sigma$ is reductive for $l\sigma \approx r\sigma$, (iii) $v\sigma \not\succeq u\sigma$ and $u\sigma \approx v\sigma$ is a selected equation in $\Gamma, u \approx v \to \Delta$, and (iv) l' is not a variable.

The inference system $\mathcal{S}_S^\succ$ consists of the above two selective inference rules plus all previous inference rules, with the additional restriction on the latter rules that no premise contain any selected literals.

For predicates we obtain a derived inference rule from the composition of superposition and equality resolution.

Ordered resolution:
$$\frac{\Lambda \to \Pi, P(s_1, \ldots, s_n) \qquad \Gamma, P(t_1, \ldots, t_n), \Gamma \to \Delta}{\Lambda\sigma, \Gamma\sigma \to \Pi\sigma, \Delta\sigma}$$

with the restrictions associated with selective or left superposition.

---

[2] Since we do not require factoring in the antecedent, the equation $u\sigma \approx v\sigma$ may also occur in $\Gamma\sigma$.

Zur Beherrschbarkeit des Entwicklungsprozesses komplexer Software-Systeme

Wolffried Stucky

Andreas Oberweis

Universität Karlsruhe
Institut für Angewandte Informatik
und Formale Beschreibungsverfahren
7500 Karlsruhe
Germany

Zusammenfassung

Es ist bekannt und in der Fachliteratur ebenso wie in Erfahrungsberichten aus der Praxis vielfach dokumentiert, daß die Entwicklung großer Software-Systeme eine komplexe Aufgabe darstellt. Zunächst müssen Anforderungen an das zu entwickelnde System in eindeutiger, widerspruchsfreier und verifizierbarer Form festgelegt werden. Auf der Basis dieser Anforderungen muß dann eine Systemkonzeption entworfen werden, die als Grundlage für die nachfolgende Realisierung dient.

Um die Beherrschbarkeit (d.h. Kontrollierbarkeit und Steuerbarkeit) des Entwicklungsprozesses, an dem auf Auftraggeber- und Entwicklerseite eine Vielzahl von Personen beteiligt ist, zu gewährleisten, werden Vorgehensmodelle vorgeschlagen, die einen generellen Rahmen für die Durchführung der Entwicklungsaktivitäten, die Anfertigung der benötigten Dokumente und die relevanten Querbeziehungen darstellen. Querbeziehungen existieren zwischen Dokumenten, zwischen Aktivitäten sowie zwischen Aktivitäten und Dokumenten.

In diesem Beitrag werden Konzepte für Rechnerunterstützung beim Einsatz eines Vorgehensmodells vorgestellt. Dabei werden insbesondere die folgenden Aspekte berücksichtigt:

- Anpassung eines gegebenen Vorgehensmodells an die Gegebenheiten eines speziellen Entwicklungsprojektes,

- Verwaltung der angefertigten Dokumente (in unterschiedlichen Versionen),

- Überwachung und Steuerung der Entwicklungsaktivitäten,

- Termin- und Kapazitätsplanung,

- Ausnahmebehandlung (z.B. unvorhergesehener Ausfall von Ressourcen),

- Kopplung an Methodentools,

- Auskunfterteilung über aktuellen Projektzustand.

Die Konzepte eignen sich für alle gängigen Vorgehensmodelle und sind unabhängig von den eingesetzten Methoden.

1   Einleitung

Ein *komplexes Software-System* kann im allgemeinen nicht in allen seinen Einzelheiten von einer einzelnen Person verstanden und überschaut werden [40]. Es ist bekannt und in der Fachliteratur ebenso wie in Erfahrungsberichten aus der Praxis vielfach dokumentiert, daß die Entwicklung solcher Systeme eine komplexe Aufgabe darstellt. Zunächst müssen *Anforderungen* an das zu entwickelnde System in *eindeutiger, widerspruchsfreier* und *verifizierbarer* Form festgelegt werden. Auf der Basis dieser Anforderungen muß dann eine *Systemkonzeption* entworfen werden, die als Grundlage für die nachfolgende *Realisierung* dient. Die sich dabei ergebenden Probleme (bzgl. Projektplanung, Ressourceneinsatz, Korrektheitsprüfung etc.) stellen nicht nur eine einfache Vergrößerung derer bei der Entwicklung "kleiner" Systeme dar [40], sondern sind im allgemeinen vollständig anders geartet.

Es sollen hier hauptsächlich solche Software-Entwicklungsprojekte betrachtet werden, in denen ein Software-Haus eine spezifische Anwendung für einen Kunden (*Auftraggeber*) erstellen soll. Einen Spezialfall solcher Projekte stellen unternehmensinterne Entwicklungsprojekte dar, bei denen z.B. Auftraggeber eine Fachabteilung und Entwickler die EDV-Abteilung ist. Die Entwicklung von Standardsoftware für einen unbekannten Kundenkreis wird ausgeklammert.

Auf Auftraggeberseite liegt die gesamte für die Systementwicklung benötigte Information einmal in Form von mehr oder weniger strukturierten vorhandenen Dokumenten vor und zum anderen in Form von (nicht weiter dokumentiertem) Wissen, das verteilt ist auf eine Vielzahl von Personen. Dieser Personenkreis verfügt üblicherweise nicht über weitreichende EDV-Kenntnisse. Auf Entwicklerseite sind ebenfalls eine Vielzahl von Personen beteiligt (z.B. Systemanalytiker, Programmierer, Projektmanager, vgl. [33]), die sich überwiegend zunächst nicht in dem Anwendungsbereich des zu entwickelnden Systems auskennen und sich daher dieses Wissen erst aneignen müssen. *Kommunikationsprobleme* zwischen Auftraggeberseite und Entwicklerseite können diesen Wissenstransfer behindern. Ab einer gewissen Größe der Entwicklergruppe und der Projektgruppe auf Anwenderseite kann es auch innerhalb der beiden Gruppen zu Kommunikationsproblemen kommen, die sich beispielsweise in fehlender Abstimmung untereinander äußern können.

Zu Beginn eines Entwicklungsprojekts müssen verbindliche vertragliche Regelungen zwischen Auftraggeber- und Entwicklerseite getroffen werden (wenn man von "inhouse"-Entwicklungen absieht). Diese Regelungen betreffen einmal inhaltliche Anforderungen an das zu entwickelnde System (*"Was soll das System leisten?"*) und zum anderen Anforderungen an den Entwicklungsprozeß selbst, z.B. bezüglich Einhaltung von Terminen, Lieferung von vordefinierten Zwischenresultaten, Abrechnung, Prüfvorgaben usw.

Die Anforderungen sind jedoch im allgemeinen nicht statisch. Es ist - bei einer üblichen Dauer des Entwicklungsprozesses von der Planung bis zur Realisierung eines komplexen Software-Systems von mehreren Jahren - wahrscheinlich, daß sich die Anforderungen im Laufe des Entwicklungsprozesses ändern, sei es, weil die Umgebung sich ändert (z.B. die organisatorischen Gegebenheiten), sei es, weil sich auf Auftraggeberseite mit fortschreitender Systementwicklung das Verständnis der Anwendung wandelt [12, 40, 3]. Damit ändern sich auch die zu erbringenden Leistungen auf Entwicklerseite, und die vertraglichen Regelungen müssen angepaßt werden[1].

Es stellt sich die Frage, wie der Entwicklungsprozeß in seiner Gesamtheit beherrschbar gemacht werden kann. Beherrschbar wird hier im Sinne von kontrollierbar, überschaubar und steuerbar verstanden. In diesem Zusammenhang ist die Beantwortung folgender Fragen relevant:

- Wie können Kosten- und Zeitaufwand verläßlich geschätzt und kontrolliert werden?

- Wie wirken sich Änderungen in einem Dokument auf andere Dokumente (welche?) aus?

- Wie kann gewährleistet werden, daß das entwickelte Projekt den Anforderungen entspricht?

- Welche Aktivitäten sind in einem gegebenen Zustand des Entwicklungsprozesses (nicht) zulässig?

- Wie können Auswirkungen von Änderungen in vorhandenen Software-Systemen kontrolliert werden?

- Wer ist wofür verantwortlich?

Sowohl die Auftraggeberseite als auch die Entwicklerseite haben ein Interesse an der Beherrschbarkeit des Entwicklungsprozesses.

Häufig läuft die Systementwicklung heute noch so ab, daß der Auftraggeber seine Anforderungen (umgangssprachlich) spezifiziert und die Entwicklerseite nach einer vorgegebenen Zeit das fertige System abliefert. Es fehlen oft Konzepte zur *frühzeitigen* und *projektbegleitenden Einbeziehung* der Auftraggeberseite in den Entwicklungsprozeß ebenso wie zur *Berücksichtigung von geänderten Anforderungen* im Laufe des Entwicklungsprozesses.

Um diese Probleme in den Griff zu bekommen, werden *Vorgehensmodelle* [2, 15] eingesetzt, die einen sowohl von Auftraggeber- als auch von Entwicklerseite akzeptierten Rahmen für die Projektdurchführung darstellen. Vielfach wird bereits von Auftraggeberseite (insbesondere von Seiten der öffentlichen Hand) die Verwendung eines bestimmten Vorgehensmodells vorgeschrieben. Die *erstmalige* Anwendung eines Vorgehensmodells in einem Software-Projekt führt allerdings zunächst einmal zu einem Mehraufwand, da sich alle Projektbeteiligten in die Notation des Vorgehensmodells einarbeiten müssen. Handhabbarkeitsprobleme im Zusammenhang mit den grundlegenden Dokumenten sind bei umfangreichen Vorgehensmodellen fast unvermeidlich.

---

[1]Auf die juristischen Probleme in diesem Zusammenhang kann hier nicht weiter eingegangen werden. Es sei aber auf die Behandlung dieser Aspekte in [13, 20] und der dort angegebenen Literatur verwiesen.

Außerdem müssen projektbegleitend zusätzlich Kontrollen aller durchgeführten Aktivitäten und erstellten Dokumente stattfinden, um Übereinstimmung mit dem Vorgehensmodell sicherzustellen. *Rechnerunterstützung* ist deshalb ab einer bestimmten Projektgröße Voraussetzung für den sinnvollen Einsatz eines Vorgehensmodells.

Im Rahmen dieser Arbeit werden Konzepte zur Rechnerunterstützung bei der Anwendung eines Vorgehensmodells zur Entwicklung komplexer Software-Systeme vorgestellt. Die Konzepte sind auf alle gängigen Vorgehensmodelle (z.B. *Wasserfallmodell*, *Spiralmodell* [6]) übertragbar.

Die Arbeit ist wie folgt gegliedert: Das folgende zweite Kapitel beschreibt allgemeine Aspekte des Einsatzes von Vorgehensmodellen bei der Systementwicklung. Im dritten Kapitel werden Konzepte zur formalen Beschreibung der Struktur des Software-Entwicklungsprozesses beschrieben, im vierten Kapitel wird eine Umgebung vorgestellt, die den Einsatz eines Vorgehensmodells unterstützt. Im fünften Kapitel werden geplante bzw. bereits durchgeführte Implementationsarbeiten beschrieben, das abschließende sechste Kapitel gibt einen Ausblick auf offene Probleme.

2 Vorgehensmodelle

Ein *Vorgehensmodell* ist die Beschreibung des Lebenszyklus eines Software-Produkts (das evtl. eingebettet ist in eine bestimmte Umgebung) in Form von *Aktivitäten* und dazugehörigen *Dokumenten* (vgl. [38]). Es wird festgelegt, in welcher Reihenfolge die Aktivitäten durchgeführt werden können, welche Überschneidungen zulässig sind und welche Bedingungen (z.B. Querbeziehungen betreffend) an die zu erstellenden Dokumente gestellt werden.

Ein Vorgehensmodell legt die allgemeine Struktur des Entwicklungsprozesses fest, eine projektspezifische Anpassung ist im allgemeinen nötig. Diese Anpassung wird als *Tailoring* bezeichnet. Es ist beispielsweise möglich, daß bestimmte Dokumente, die im Vorgehensmodell vorgesehen sind, bereits vorliegen (evtl. mit anderem Namen oder in der Form bzw. im Inhalt geringfügig von den Vorgaben des Vorgehensmodells abweichend). Dann müssen die entsprechenden Aktivitäten zur Erstellung nicht mehr durchgeführt werden. Andererseits kann es auch sein, daß von Auftraggeberseite der Wunsch nach zusätzlichen, im Vorgehensmodell nicht vorgesehenen Dokumenten besteht. In diesem Fall müssen Struktur und Inhalt der zusätzlichen Dokumente ebenso wie die Querbeziehungen zu anderen Dokumenten und die benötigten Aktivitäten genau festgelegt werden. Das Tailoring wird im allgemeinen zu Projektbeginn zwischen Auftraggeber- und Entwicklerseite ausgehandelt und festgelegt. Eventuell ist es allerdings zweckmäßig, bestimmte Punkte zunächst noch offen zu lassen und dafür lediglich allgemeine Entscheidungsregeln vorzusehen.

Generelle Ziele der Verwendung eines Vorgehensmodells sind [10]:

- Verbesserung und Gewährleistung der Softwarequalität.

- Eindämmung der Softwarekosten während des gesamten Software-Lifecycle.

- Verbesserung der Kommunikation zwischen Auftraggeber- und Software-Entwicklerseite.

Als spezielle Vorteile des Einsatzes eines Vorgehensmodells sind zu nennen [14]:
auf Auftraggeberseite:

- Es existiert damit eine Übereinkunft zwischen Auftraggeber- und Entwicklerseite über das Projekt und die Abnahmekriterien für das zu erstellende Endprodukt.

- Wenn sich Angebote unterschiedlicher Softwareentwickler genau an die Vorgaben eines Vorgehensmodells halten, werden die Angebote miteinander vergleichbar.

- Die Abhängigkeit des Auftraggebers vom Software-Entwickler wird verringert, da es bei fest vordefinierten Zwischenergebnissen leichter möglich ist, während eines laufenden Projekts den Entwickler zu wechseln.

auf Software-Entwicklerseite:

- Es existiert eine verläßliche Grundlage für die Abschätzung des Ressourcenverbrauchs (Kosten, Personal (Anzahl und benötigte Ausbildung), Ausrüstung und Zeit).

Es treten allerdings auch zusätzliche Probleme auf: Umfangreiche Vorgehensmodelle, z.B. der *Software-Entwicklungsstandard der Bundeswehr* [10], erfordern eine lange Einarbeitungszeit. Bei einem Umfang der Beschreibung dieses Vorgehensmodells von mehreren Hundert Seiten (dabei sind die zu verwendenden Methoden noch nicht einmal berücksichtigt) sind Handhabbarkeitsprobleme fast unvermeidlich (vgl. auch [35]). Wegen der Vielzahl der Querbeziehungen zwischen den im Rahmen des Entwicklungsprozesses durchzuführenden Aktivitäten erscheint es nicht ausreichend, wenn sich jeder Projektmitarbeiter nur in einem bestimmten Teilgebiet des Vorgehensmodell auskennt. Da nach den Vorgaben eines Vorgehensmodells eine Vielzahl von Dokumenten (Anforderungsdokumente, Entwurfsdokumente, Prüfdokumente, ...) erstellt wird (und zwar im allgemeinen jeweils in mehreren Versionen[2]), ist Rechnerunterstützung sinnvoll und ab einer bestimmten Projektgröße sogar unverzichtbar.

3 Formale Beschreibung der Struktur des Software-Entwicklungsprozesses

Die meisten Standard-Vorgehensmodelle basieren auf Varianten des sogenannten *Wasserfallmodells* [12]. Dabei wird der gesamte Software-Entwicklungsprozeß in Phasen

---

[2]In [45] werden einige reale Fallbeispiele aus der Praxis zur Anforderungsspezifikation beschrieben. Es wird angegeben, daß im allgemeinen die Anforderungsdokumente nach Fertigstellung durch die Systementwicklerseite zweimal der Auftraggeberseite in einem sog. *feedback meeting* zur Stellungnahme vorgelegt und bei Bedarf geändert werden mußten, bis die Auftraggeberseite die Dokumente als ausreichend genau und korrekt akzeptiert hat.

eingeteilt, die aus einer vordefinierten Menge von Aktivitäten bestehen, wobei die jeweils erzeugten Dokumente einer Phase die Eingabe für die nachfolgende Phase darstellen. Wegen der fehlenden Flexibilität und der unzureichenden Einbeziehung der Auftraggeberseite in den Entwicklungsprozeß wird das Wasserfallmodell häufig kritisiert [2]. Es ist aber andererseits unbestritten, daß ein enger und fester methodischer Rahmen für den Entwicklungsprozeß Voraussetzung für eine frühe vertragliche Vereinbarung zwischen Auftraggeber- und Entwicklerseite ist, u.a. beispielsweise auch zur Erstellung von verläßlichen Festpreisangeboten durch die Entwicklerseite.

Bevor das Problem der Rechnerunterstützung für den Einsatz eines Vorgehensmodell bei der Software-Entwicklung gelöst werden kann, muß das Vorgehensmodell selbst in einer (weitgehend) formalen Notation (einem sog. *Metamodell* [44]) beschrieben werden. Dabei müssen die Struktur der Dokumente, die Aktivitäten sowie die Querbeziehungen berücksichtigt werden.

Der Einsatz einer formalen Notation zur Beschreibung eines Vorgehensmodells ermöglicht oft auch das Erkennen von fehlender Genauigkeit (evtl. auch von Inkonsistenzen) in den vorliegenden, natürlichsprachlich abgefaßten Dokumenten zum Vorgehensmodell. Möglicherweise führt die formale Beschreibung daher zu Korrekturen existierender Vorgehensmodelle (vgl. dazu auch [6]).

In [32, 6] wird vorgeschlagen, den Software-Entwicklungsprozeß in einer programmiersprachlichen Notation zu beschreiben, es wird in diesem Zusammenhang von "Software Process Programming" gesprochen. In [21, 23, 44] wird eine logikorientierte Beschreibungssprache verwendet, in der Vor- und Nachbedingungen für das Durchführen von einzelnen Aktivitäten im Rahmen des Software-Entwicklungsprozesses definiert werden. Ein Nachteil dieser rein textuellen Beschreibungssprachen stellt die fehlende Anschaulichkeit im Gegensatz zu graphischen Darstellungsmöglichkeiten dar (siehe [22]).

Zur Beschreibung der Dokumentstrukturen (vgl. auch [4]) schlagen wir daher die Verwendung eines *semantisch-hierarchischen Objektmodells* [9] mit den Strukturierungsmöglichkeiten *Aggregation*, *Generalisierung* und *Gruppierung* vor. Ein Objekt vom Typ *Dokument* ist beispielsweise Aggregation von Objekten der Typen *Einleitung*, *Hauptteil* und *Anhang*. Ein Objekt des Typs *Kapitel* ist (im einfachsten Fall) als Gruppierung von Objekten des Typs *Absatz* gegeben. Schließlich kann der Objekttyp *Entwurfsdokument* als Generalisierung der Objekttypen *Datenentwurf* und *Funktionsentwurf* definiert sein.

In [25] werden die für den Entwicklungsprozeß relevanten Dokumente in einem erweiterten *Entity/Relationship-Modell* beschrieben, das über ähnliche Ausdrucksmöglichkeiten verfügt wie das semantisch-hierarchische Datenmodell. Die Aktivitäten, die zur Erstellung und Änderung dieser Dokumente nötig sind, werden in [25] nicht berücksichtigt.

Wir schlagen vor, diese Aktivitäten und die Querbeziehungen zwischen ihnen als höhere Petri-Netze (Prädikate/Transitionen-Netze [18]) zu beschreiben. Dabei werden Aktivitäten als *Transitionen* und Dokumente als *Marken* in den *Stellen* repräsentiert. Ein Pfeil von einer Stelle zu einer Transition drückt aus, daß die Aktivität beim

Stattfinden ein Dokument des jeweiligen Typs benötigt. Entsprechend drückt ein Pfeil von einer Transition zu einer Stelle aus, daß die entsprechende Aktivität beim Stattfinden ein Dokument des jeweiligen Typs erzeugt.

Es kann zusätzlich für eine Aktivität A eine Zeitdauer eingeführt werden, indem die entsprechende Transition t_A durch eine Start- und eine End-Transition t_{As} bzw. t_{Ae} ersetzt wird. Zusätzlich wird eine Stelle s_A eingeführt, die Eingangsstelle von t_{Ae} und Ausgangsstelle von t_{As} ist. Wenn t_{As} schaltet (d.h. die Aktivität A beginnt), dann wird eine Marke in s_A abgelegt, die ausdrückt, daß A gerade stattfindet. Wenn t_{Ae} schaltet (d.h. A wird beendet), dann wird die Marke aus s_A wiederum entfernt.

Die Einführung eines zusätzlichen *Uhr-* bzw. *Kalendermechanismus* - in Form von speziellen Netzkonstrukten wie etwa in [28] beschrieben - ermöglicht es, *absolute Zeit-dauern* der Art "*mindestens drei, höchstens 6 Stunden*", "*weniger als 2 Wochen*" u.ä. zu modellieren und auch Terminangaben wie "*am 31.12.1991*" oder "*nicht samstags oder sonntags*" auszudrücken.

Die Verwendung von Petri-Netzen bietet folgende Vorteile gegenüber anderen da-tenflußdiagrammartigen Beschreibungstechniken:

- Netze sind leicht verständlich, da es (neben Pfeilen) nur zwei unterschiedliche Symbole (Kreise und Vierecke) gibt.

- Alternativen und Nebenläufigkeiten sind an der Netzstruktur erkennbar.

- Zustände des Entwicklungsprozesses werden durch Marken in den Stellen defi-niert.

- Hierarchiebildung nach formalen Kriterien zur schrittweisen Vorgehensweise beim Entwurf komplexer Netze sowie zur Betrachtung des Entwicklungsprozesses auf unterschiedlichen Genauigkeitsstufen ist möglich.

- Simulation von Abläufen kann einfach durchgeführt werden.

Zusätzliche Anforderungen an die zulässigen Abläufe sowie an die erlaubten Doku-mentstrukturen, die nicht im Netz unmittelbar prozedural beschrieben sind, können deklarativ als *Fakt-Transitionen* bzw. *ausgeschlossene Transitionen* ausgedrückt wer-den [42, 29]. Fakt-Transitionen erlauben beispielsweise die Modellierung von Anfor-derungen der Art "*Wenn Dokument D1 das Kapitel 7 enthält, dann muß Dokument D2 Kapitel 3 enthalten*" oder "*Es ist nicht zulässig (sinnvoll), daß Dokument D1 und D3 gemeinsam existieren*". Ausgeschlossene Transitionen ermöglichen die Beschrei-bung von Anforderungen an Abläufe wie z.B. "*Aktivität A7 muß nach Aktivität A2 stattfinden*" oder "*Wenn Aktivität A3 stattgefunden hat, darf Aktivität A9 nicht mehr stattfinden*".

Mit Fakt-Transitionen und ausgeschlossenen Transitionen ist es auch möglich, *Aus-nahmesituationen* in einem Software-Projekt zu definieren. Eine Ausnahmesituation liegt vor, wenn eine als Fakt-Transition oder ausgeschlossene Transition modellierte Anforderung verletzt wird (z.B. ein Termin wird nicht eingehalten). Für solche Fälle können im Vorgehensmodell spezielle Mechanismen zur Ausnahmebehandlung vorgesehen werden.

Die hier genannten formalen Ausdrucksmöglichkeiten für Zeitaspekte, deklarativ formulierte Systemanforderungen und Exception-Handling sind vollständig integrierbar in existierende Petri-Netzsimulatoren [27, 31]. Andere Ansätze (z.B. [16]) modellieren diese Aspekte als Bedingungs-Stellen in Petri-Netzen, die natürlichsprachliche Beschriftungen tragen. Bei der Simulation mit solchen Netzen ist der Anwender jedoch selbst verantwortlich für die jeweilige Markierung dieser Bedingungs-Stellen. Die Systemdynamik ist hier also nicht vollständig in der Netz-Notation ausgedrückt und kann daher auch nicht formal analysiert werden (beispielsweise auf Redundanz und Widersprüche).

Nicht alle Aktvitäten im Rahmen eines Vorgehensmodell lassen sich allerdings formalisieren. So gibt es manuelle Tätigkeiten oder unstrukturierte Kommunikation, die sich nur unzureichend in einer formalen Notation beschreiben lassen (vgl. etwa [6]). Für solche Aspekte bietet sich ein Übergang zu einer semiformalen Notation mit Kanal/Instanzen-Netzen [34] an, deren Stellen (die Kanäle) und Transitionen (die Instanzen) umgangssprachlich beschriftet sind, um die entsprechenden Dokumente und Aktivitäten zu beschreiben.

4 Werkzeugunterstützung für den Einsatz eines Vorgehensmodells bei der Software-Entwicklung

Es existieren bereits eine ganze Reihe von - auch kommerziell verfügbaren - *Software-Entwicklungswerkzeugen* [5]. Diese sind jedoch im allgemeinen an ein spezielles Vorgehensmodell (üblicherweise eine Variante des Wasserfallmodells) gebunden und unterstützen lediglich bestimmte fest vorgegebene Methoden.

Im folgenden soll ein *Meta-Werkzeug* (ähnlich wie beispielsweise in [25, 41] beschrieben) vorgestellt werden, das an unterschiedliche Vorgehensmodelle angepaßt werden kann und außerdem bezüglich der zu verwendenden Methoden flexibel ist. Es ist nämlich nicht realistisch anzunehmen, daß ein Software-Haus nur *ein* bestimmtes Vorgehensmodell unterstützen muß. - Wenn es selbst auch gewisse Präferenzen haben kann, so muß es sich doch oftmals an Vorgaben von Auftraggeberseite halten. - Mittelfristig werden daran auch Standardisierungsvorhaben wie *EUROMETHOD* [43], einem geplanten einheitlichen EG-weiten Standard der Software-Entwicklung, sicherlich nichts ändern.

Folgende Aspekte sollen unterstützt werden (vgl. auch [35]):

- Durchführung des Tailoring.

- Verwaltung der Dokumente.

- Überwachung und Steuerung der Aktivitäten.

- Auskunft über aktuellen Projektzustand und Projektbeteiligte.

- Termin- und Kapazitätsplanung.

- Unterstützung bei der Ausnahmebehandlung.

- Kopplung an Methodentools.

- Bereitstellung einer Hypermedia-Benutzerschnittstelle.

4.1 Durchführung des Tailoring

Projektspezifisches Tailoring bedeutet einmal, bestimmte Aktivitäten bzw. Dokumente aus dem Vorgehensmodell zu streichen. Wegen der vorgegebenen Querbeziehungen zwischen Aktivitäten und Dokumenten müssen beim Tailoring bestimmte Regeln eingehalten werden.

Möglicherweise kann auf vorliegende Dokumente zurückgegriffen werden, die bestimmte Dokumente des Vorgehensmodells ersetzen. In diesem Fall muß an zentraler Stelle festgehalten werden, daß Dokument *ABC* dem Dokument *XYZ* aus dem Vorgehensmodell entspricht, evtl. mit Hinweis auf vorhandene Abweichungen von den Vorgaben.

Für zusätzlich benötigte Dokumente, die nicht im Vorgehensmodell vorgesehen sind, müssen neben der Struktur die Querbeziehungen zu anderen Dokumenten und Aktivitäten festgelegt werden. Für zusätzliche Aktivitäten müssen die Eingangs- und Ausgangsdokumente bestimmt werden sowie die Beziehungen zu anderen Aktivitäten und die Einordnung in die vorhandenen Abläufe.

Wenn der Software-Entwicklungsprozeß - wie im vorangehenden Kapitel beschrieben - als Petri-Netz modelliert und die Dokumente als Schema nach dem semantisch-hierarchischen Datenmodell beschrieben worden sind, dann führt das Tailoring sowohl zu Änderungen am Netz als auch an dem Dokumenten-Schema. Die Änderungen können mit entsprechenden Editoren durchgeführt werden, die die jeweils relevanten Plausibilitätsbedingungen prüfen.

4.2 Erstellung und Verwaltung der Dokumente

Die Erstellung von Dokumenten kann mit speziellen Editoren unterstützt werden, die die vorgegebene Dokumentstruktur bereits beim Editieren überprüfen. Eine spezielle Mehrbenutzerkontrolle soll es ermöglichen, daß mehrere Personen gleichzeitig im selben Dokument - jedoch in unterschiedlichen Teilen - Änderungen vornehmen können. Es ist in diesem Zusammenhang wichtig, ausreichend "feine" Sperrmöglichkeiten - etwa auf Abschnittebene - zu unterstützen.

Zur Verwaltung der bei der Projektdurchführung erstellten Dokumente gehört insbesondere die Bereitstellung der jeweils aktuellsten Version am Rechner (die Benutzeroberfläche wird in Abschnitt 4.8 skizziert). Alte Versionen müssen archiviert, ein Änderungsdienst muß geführt werden.

Basis der Dokumentenverwaltung kann ein *Repository* bilden, das Versionsmanagement, Mehrbenutzerkontrolle sowie Recovery-Konzepte bereitstellt.

4.3 Überwachung und Steuerung der Aktivitäten

Ausgehend von den zu Beginn des Software-Entwicklungsprozesses bereits vorliegenden Dokumenten werden die im projektspezifischen Vorgehensmodell vorgesehenen Aktivitäten rechnergestützt koordiniert. Dabei werden die vorgegebenen Anforderungen in Form von Querbeziehungen oder vorgegebenen zeitlichen Restriktionen überwacht. Ein ähnliches Konzept wird in [11, 17] beschrieben.

4.4 Auskunft über aktuellen Projektzustand und Projektbeteiligte

Gekoppelt mit der projektbegleitenden Überwachung, Auslösung und Steuerung der Aktivitäten ist eine Auskunftskomponente, die den Projektbeteiligten jederzeit Auskunft über projektrelevante Daten wie den aktuellen Zustand des Projekts erteilen kann. Der *aktuelle Zustand* ist durch die bereits fertiggestellten Dokumente, die bereits abgeschlossenen und die gerade laufenden Aktivitäten bestimmt.

Es ist auch möglich, Informationen darüber zu erhalten, welche Aktivitäten in dem aktuellen Zustand - entsprechend dem projektspezifischen Vorgehensmodell - vorgesehen sind.

Neben diesen, im Laufe des Projekts veränderlichen Angaben können auch die statischen Projektdaten abgefragt werden, z.B. die Zuständigkeit für die Projektplanung oder die Telefonnummer des Verantwortlichen für die Software-Qualitätssicherung. Der Einsatz dieser Auskunftskomponente ersetzt damit weitgehend das Nachschlagen in von Vorgehensmodellen üblicherweise vorgesehenen Projektplänen bzw. Projekthandbüchern (vgl. z.B. [10]).

4.5 Termin- und Kapazitätsplanung

Ausgehend von dem formalen Modell der nach dem projektspezifischen Vorgehensmodell möglichen Abläufe im Entwicklungsprozeß ist es möglich, zu einem gegebenen Zustand unterschiedliche Alternativen für künftige Fortsetzungen des Projektablaufs miteinander zu vergleichen. So können z.B. die Auswirkungen unterschiedlicher Termin- bzw. Kapazitätsvorgaben überprüft werden, indem man geeignete Simulationen durchführt.

4.6 Unterstützung bei der Ausnahmebehandlung

Ausnahmen in einem Software-Entwicklungsprojekt sind Situationen, in denen bestimmte Vorgaben des projektspezifischen Vorgehensmodells oder irgendwelche zusätzlichen Anforderungen zeitlicher Art nicht eingehalten werden können (z.B. bei kurzfristigem Ausfall von Personal, Rechnern, Netzwerk, Datenbank oder Nichteinhalten von Lieferterminen durch Lieferanten). Damit das Projektziel nicht gefährdet wird, müssen Mechanismen zur Behandlung solcher Ausnahmesituationen geplant und bereitgestellt werden.

Da Ausnahmen häufig durch externe Einflüsse bedingt werden, auf die man keinen Einfluß hat, ist es bei der Behandlung solcher Ausnahmesituation im allgemeinen nicht möglich, den Fehlerzustand (ähnlich wie bei Integritätsverletzungen im Datenbankbereich) durch Rücksetzen von Aktivitäten (*Rückwärts-Recovery*) zu beheben. Ziel wird es im allgemeinen vielmehr sein, die Auswirkungen des aufgetretenen Fehlers möglichst gering zu halten. Dies kann einmal dadurch geschehen, daß alternative Abläufe bei Ausfall bestimmter Ressourcen ermittelt werden, um trotzdem bestimmte Resultate zum vorgesehenen Zeitpunkt zu erhalten. Eine Verkürzung der Projektdauer kann auch durch Erhöhung des vorgesehenen Parallelisierungsgrades der Aktivitäten oder durch Beschleunigung einzelner Aktivitäten (evtl. durch Mehreinsatz an Personal) erreicht werden.

Ein in diesem Zusammenhang zu berücksichtigender Aspekt ist die im voraus für die einzelnen Aktivitäten festzulegende Wichtigkeit für das Gesamtprojekt (*Kritikalität*). Die knappen Ressourcen müssen den Aktivitäten unter Berücksichtigung der Kritikalität der Aktivitäten zugeteilt werden.

4.7   Kopplung an Methodentools

Ein Vorgehensmodell legt noch nicht die bei der Durchführung eines konkreten Projekts zu verwendenden Methoden und Werkzeuge - z.B. zur *Datenmodellierung*, zum *Modulentwurf*, zum *Prototyping* - fest. Das hier konzipierte Werkzeug zur Unterstützung des Einsatzes eines Vorgehensmodells bei der Software-Entwicklung soll daher "*offen*" sein für eine Kopplung an vorgegebene Methodentools (vgl. [7, 5]). Dazu gehören beispielsweise *graphische Editoren* (für Datenflußdiagramme, Entity-Relationship-Diagramme, Petri-Netze etc.) sowie *Programm-* und *Datenbank-Generatoren*.

4.8   Bereitstellung einer Hypermedia-Benutzerschnittstelle

Die Entwicklungsumgebung soll über eine *Hypermedia-Benutzerschnittstelle* verfügen, die sowohl Zugriff auf die Dokumente zu dem Vorgehensmodell als auch auf die Dokumente, die im Rahmen des Entwicklungsprozesses erstellt werden, ermöglicht. Existierende Hyper-Techniken [19] können eingesetzt werden, um die vielfältigen Beziehungen zwischen Dokumenten, Aktivitäten, Ressourcen etc. adäquat zu modellieren und dem Benutzer verfügbar zu machen, ohne daß er dazu eine spezielle Abfragesprache erlernen muß. Der Benutzer verfolgt dazu - ausgehend von den gerade betrachteten Daten - *Links* (Verweise), die inhaltliche Zusammenhänge zwischen abgespeicherten Daten repräsentieren und auf inhaltlich nahestehende Daten zeigen.

Das Werkzeug soll schließlich in einer PC-Umgebung (evtl. auf Notebook-Basis) für alle Projektbeteiligten bereitgestellt werden. Dokumente werden auf CD-ROMs abgespeichert und an die Projektbeteiligten verteilt. Dies dient einmal der Eindämmung der Papierflut, die bei einer konsequenten Anwendung von Vorgehensmodellen ansonsten unvermeidlich ist. Zum anderen erleichtert die Verfügbarkeit der Dokumente am Rechner für alle Projektbeteiligten das Suchen von speziellen Informationen sowie die Überwachung von einzuhaltenden Querbeziehungen.

Ein spezielles Sichtenkonzept, wie es aus der Hypertext-Welt bekannt ist, ermöglicht es den einzelnen Projektbeteiligten, Ergänzungen (z.B. Anmerkungen) und auch Kürzungen an den Dokumenten vorzunehmen, die aber nicht sichtbar für die übrigen Projektteilnehmer sind. Änderungen können sowohl die gegebenen Dokumente selbst als auch die Verknüpfungen zwischen den Dokumenten betreffen.

Projektreviews dienen zur Abstimmung der erstellten Dokumente mit den Anforderungen der Auftraggeberseite. Solche Projektreviews bieten sich in Form von Multimedia-Konferenzen [1] an, wenn die räumliche Entfernung zwischen Auftraggeber- und Entwicklerseite regelmäßige Treffen zu aufwendig macht.

5   Implementation

Die in Kapitel 4 vorgestellten Konzepte sollen prototypmäßig implementiert werden. Das zu entwickelnde Werkzeug wird für verschiedene gängige Vorgehensmodelle (Wasserfallmodell, Spiralmodell etc.) einsetzbar sein. Dabei wird zunächst lediglich ein Vorgehensmodell - nämlich das in [10] beschriebene[3] - explizit unterstützt werden. Dieses Vorgehensmodell ist der deutsche Vorschlag für Die Abläufe des Vorgehensmodells werden in Form von Prädikate/Transitionen-Netzen modelliert. Dazu wird ein vorliegender graphischer Editor für Petri-Netze (INCOME-Designer)[4] verwendet, der über eine Schnittstelle zu dem relationalen Datenbanksystem ORACLE[5] verfügt.

Das Tailoring wird mit einer regelbasierten Komponente unterstützt, die in Prolog implementiert wird (vgl. [30]). Regeln werden formuliert, die sich auf Dokumentstrukturen, auf Aktivitäten sowie auf Querbeziehungen (zwischen Dokumenten, zwischen Aktivitäten sowie zwischen Dokumenten und Aktivitäten) beziehen.

Das aufgrund des Tailoring entstandene, den speziellen Projektbedürfnissen angepaßte konkrete Vorgehensmodell in Form eines Prädikate/Transitionen-Netzes zusammen mit einer zusätzlichen Menge von Fakt-Transitionen und ausgeschlossenen Transitionen dient als Grundlage einer Steuerungskomponente für den Software-Entwicklungsprozeß. Dazu verwenden wir einen vorliegenden Simulator für Petri-Netze [27, 31, 36]. Es besteht beispielsweise die Möglichkeit einer vorausschauenden Simulation, um etwa ausgehend von einem gegebenen Ist-Zustand mögliche Alternativen für zukünftige Entwicklungen miteinander zu vergleichen.

Die Steuerungskomponente überwacht Termine, stößt Aktivitäten an und prüft die vorgegebenen Bedingungen bzgl. Querbeziehungen zwischen den Dokumenten und Aktivitäten. Über ein zentrales Entwurfs-Repository wird die Offenheit des

---

[3]Dieses Vorgehensmodell ist der deutsche Vorschlag für EUROMETHOD (Phase 3), den geplanten EG-Standard für die Software-Entwicklung [43].

[4]Produkt der PROMATIS Informatik, Straubenhardt

[5]Produkt der Oracle Corporation, Belmont

Werkzeugs gewährleistet. So wird beispielsweise eine Schnittstelle zu den ORACLE-Case-Werkzeugen[5] (z.B. ein Entity/Relationship-Diagrammeditor) und die INCOME-Tools[4] bereitgestellt. Die methodischen Grundlagen von INCOME wurden am Institut für Angewandte Informatik und Formale Beschreibungsverfahren der Universität Karlsruhe im Rahmen des von der DFG geförderten Projekts "Programmentwurf" entwickelt [24, 26, 37, 39].

6   Ausblick

Ein wichtiger Aspekt künftiger Arbeiten wird es sein, Bewertungsprobleme zu untersuchen. Es stellt sich die Frage, wie Anforderungsspezifikationen, Entwürfe, Software, Dokumentationen, Vorgehensweisen, Methoden und Werkzeuge sowie Projektmanagement zu bewerten sind. Es fehlt noch ein eindeutiges, allgemein akzeptiertes quantifizierbares Qualitätsmaß für solche Aspekte. Der Vergleich unterschiedlicher Vorgehensmodelle und die Untersuchung der Eignung für unterschiedliche Software-Projekte ist aktuelles Forschungsgebiet und wird beispielsweise in [12, 8] behandelt.

Danksagung

Für verschiedene Hinweise und Verbesserungsvorschläge danken wir Peter Jaeschke, Tibor Németh und Volker Sänger.

Literaturverzeichnis

[1] S.R. Ahuja, J.R. Ensor, and D.N. Horn. The RAPPORT multimedia conferencing system. In R.B. Allen, editor, *Proc. Conference on Office Information Systems*, pages 1–8, Palo Alto, 1988.

[2] W.W. Agresti, editor. *New Paradigms for Software Development.* IEEE Computer Society Press, Washington, 1986.

[3] ANSI/IEEE. An american national standard. In M. Dorfman and R.H. Thayer, editors, *Standards, Guidelines and Examples on System and Software Requirements Engineering*, pages 16–38, Los Alamitos, 1990. IEEE Computer Society Press.

[4] V. Ashok, J. Ramanthan, S. Sarkar, and V. Venugopal. Process modeling in software environments. In C. Tully, editor, *Proc. 4th Int. Software Process Workshop*, pages 39–42, Moretonhampstead, 1988.

[5] H. Balzert. *CASE Systeme und Werkzeuge.* Reihe Angewandte Informatik, Bd. 7, BI.-Wiss.-Verlag, Mannheim, 2. Auflage, 1991.

[6] B. Boehm and F. Belz. Applying process programming to the spiral model. In C. Tully, editor, *Proc. 4th Int. Software Process Workshop*, pages 46–56, Moretonhampstead, 1988.

[7] R. Bisiani, F. Lecouat, and V. Ambriola. A tool to coordinate tools. *IEEE Software*, pages 17–25, November 1988.

[8] V.R. Basili and J.D. Musa. The future engineering of software: A management perspective. *IEEE Computer*, pages 90–96, September 1991.

[9] A. Borgida, J. Mylopoulos, and H.K.T. Wong. Generalization/specialization as a basis for software specification. In M.L. Brodie, J. Mylopoulos, and J.W. Schmidt, editors, *On Conceptual Modelling*, pages 87–114, Berlin, Heidelberg, New York, 1984. Springer-Verlag.

[10] Software-Entwicklungsstandard der Bundeswehr, "Vorgehensmodell". Allgemeiner Umdruck. Bundesamt für Wehrtechnik und Beschaffung, Koblenz, Februar 1991.

[11] T.E. Cheatham. Activity coordination programs. In C. Tully, editor, *Proc. 4th Int. Software Process Workshop*, pages 57–60, Moretonhampstead, 1988.

[12] A.M. Davis, E.H. Bersoff, and E.R. Comer. A strategy for comparing alternative software development life cycle models. *IEEE Transactions on Software Engineering*, SE-14(10):1453–1461, Oktober 1988.

[13] Deutsche Gesellschaft für Informationstechnik und Recht e.V. Schlichtung. Informationsschrift Nr. 2, 1989.

[14] M. Dorfman. System and software requirements engineering. In M. Dorfman and R.H. Thayer, editors, *Standards, Guidelines and Examples on System and Software Requirements Engineering*, pages 4–16, Los Alamitos, 1990. IEEE Computer Society Press.

[15] M. Dorfman and R.H. Thayer, editors. *Standards, Guidelines and Examples on System and Software Requirements Engineering*. IEEE Computer Society Press, Los Alamitos, 1990.

[16] V. DeAntonellis and B. Zonta. Modeling events in database applications design. In *Proc. 7th Int. Conference on Very Large Databases VLDB81*, pages 23–31, Cannes, 1981.

[17] C.A. Fritsch and D.L. Perry. A manager/controller for the software development process. In C. Tully, editor, *Proc. 4th Int. Software Process Workshop*, pages 73–75, Moretonhampstead, 1988.

[18] H.-J. Genrich. Predicate/transition nets. In W. Brauer, W. Reisig, and G. Rozenberg, editors, *Advances in Petri nets 86, Vol*, pages 207–247, Berlin, Heidelberg, New York, 1987. Springer-Verlag.

[19] P. Gloor and N. Streitz, editors. *Hypertext und Hypermedia: Von theoretischen Konzepten zu praktischen Anwendungen.* Springer-Verlag, Berlin, Heidelberg, New York, 1990.

[20] D.J. Hildebrand. Die Schlichtungsstelle der Deutschen Gesellschaft für Informationstechnik und Recht (DGIR). *Informatik-Spektrum, Band 12, Heft 3*, pages 162–164, Juni 1989.

[21] P. Hitchcock. The process model of the Aspect ipse. In C. Tully, editor, *Proc. 4th Int. Software Process Workshop*, pages 76–78, Moretonhampstead, 1988.

[22] M.I. Kellner. Representation formalisms for software process modeling. In C. Tully, editor, *Proc. 4th Int. Software Process Workshop*, pages 93–96, Moretonhampstead, 1988.

[23] G.E. Kaiser and P.H. Feiler. An architecture for intelligent assistance in software development. In *Proc. 9th IEEE Int. Conference on Software Engineering*, pages 180–188, Monterey, 1988.

[24] G. Lausen. Grundlagen einer netzorientierten Vorgehensweise für den konzeptuellen Datenbankentwurf. Forschungsbericht 179, Institut für Angewandte Informatik und Formale Beschreibungsverfahren, Universität Karlsruhe (TH), Februar 1987.

[25] A.van Lamsweerde, B. Delcourt, E. Delor, M.C. Schayes, and R. Champagne. Generic lifecycle support in the ALMA environement. *IEEE Transactions on Software Engineering*, SE-14(6):720–741, Juni 1988.

[26] G. Lausen, T. Németh, A. Oberweis, F. Schönthaler, and W. Stucky. The INCOME approach for conceptual modelling and prototyping of information systems. In *CASE89. The First Nordic Conference on Advanced Systems Engineering, Stockholm*, Mai 1989.

[27] Th. Mochel, T. Németh, A. Oberweis, and W. Stucky. Eine offene Simulationsumgebung für Petri-Netze zur Unterstützung des Entwurfs eingebetteter Systeme. In D. Tavangarian, editor, *Proc. 7. Symposium Simulationstechnik Hagen*, pages 510–514, Braunschweig, 1991. Vieweg-Verlag.

[28] A. Oberweis. Zeitstrukturen für Informationssysteme. Dissertation, Universität Mannheim, Fakultät für Mathematik und Informatik, Juli 1990.

[29] A. Oberweis. System simulation with Petri-nets: A new concept combining procedural and declarative system knowledge. In E. Mosekilde, editor, *Proc. European Simulation Multiconference ESM91, Kopenhagen*, pages 59–64, Kopenhagen, 1991.

[30] A. Ohki and K. Ochimizu. Process programming with Prolog. In C. Tully, editor, *Proc. 4th Int. Software Process Workshop*, pages 118–121, Moretonhampstead, 1988.

[31] A. Oberweis, J. Seib, and G. Lausen. PASIPP: Ein Hilfsmittel zur Analyse und Simulation von Prolog-beschrifteten Prädikate/Transitionen-Netzen. "Wirtschaftsinformatik", 33. Jahrgang, Heft 3, Juni 1991.

[32] L. Osterweil. Software processes are software too. In *Proc. 9th IEEE Int. Conference on Software Engineering*, pages 2–13, Monterey, 1988.

[33] D.E. Perry. Problems of scale and process models. In C. Tully, editor, *Proc. 4th Int. Software Process Workshop*, pages 126–128, Moretonhampstead, 1988.

[34] W. Reisig. *Systementwurf mit Netzen*. Springer-Verlag, Berlin, Heidelberg, New York, 1985.

[35] J. Ramanathan and S. Sarkar. Providing customized assistance for software life-cycle approaches. *IEEE Transactions on Software Engineering*, SE-14(6):749–757, Juni 1988.

[36] V. Sänger. Simulation mit deklarativen Systembeschreibungen. Diplomarbeit, Universität Karlsruhe, Institut für Angewandte Informatik und Formale Beschreibungsverfahren, April 1991.

[37] F. Schönthaler. Rapid Prototyping zur Unterstützung des Konzeptuellen Entwurfs von Informationssystemen. Dissertation, Universität Karlsruhe (TH), Institut für Angewandte Informatik und Formale Beschreibungsverfahren, Januar 1989.

[38] F. Schönthaler and T. Németh. *Software-Entwicklungswerkzeuge: Methodische Grundlagen*. B.G. Teubner, Stuttgart, 1990.

[39] W. Stucky, T. Németh, and F. Schönthaler. Modellierung und Simulation verteilter Systeme mit INCOME. In A. Reuter, editor, *Proc. GI-20. Jahrestagung. Informatik auf dem Weg zum Anwender*, Berlin, Heidelberg, New York, 1990. Springer-Verlag.

[40] I. Sommerville. *Software Engineering*. Addison-Wesley, Reading/Massachusetts, 1989.

[41] P.G. Sorenson. The metaview system for many specification environments. *IEEE Software*, März 1988.

[42] K. Voss. Nets in data bases. In W. Brauer, W. Reisig, and G. Rozenberg, editors, *Petri Nets: Applications and Relationships to Other Models of Concurrency, Advances in Petri Nets 1986, II*, pages 234–257, Berlin, Heidelberg, New York, 1987. LNCS 255, Springer-Verlag.

[43] H. Weiler. Euromethod: Ein Standard für die Software-Entwicklung. *Computerwoche*, 11. Oktober 1991.

[44] L.G. Williams. Software process modelling: A behavioural approach. In *Proc. 10th IEEE Int. Conference on Software Engineering*, pages 174–186, Singapore, 1988.

[45] L. Zucconi. Techniques and experiences capturing requirements for several realtime applications. *ACM SIGSOFT Software Engineering Notes*, 14(6):51–54, 1989.

An Intelligent Multimodal Interface[1]

Wolfgang Wahlster

Fachbereich 14 Informatik
Universität Saarbrücken
6600 Saarbrücken
Germany

Abstract

In face-to-face conversation humans frequently use deictic gestures parallel to verbal descriptions for referent identification. Such a multimodal mode of communication is of great importance for intelligent interfaces, as it simplifies and speeds up reference to objects in a visualized application domain. Natural pointing behavior is very flexible, but also possibly ambiguous or vague, so that without a careful analysis of the discourse context of a gesture there would be a high risk of reference failure. The subject of this paper is how the user and discourse model of an intelligent interface influences the comprehension and production of natural language with coordinated pointing, and conversely how multimodal communication influences the user and discourse model. We briefly describe the deixis analyzer of our XTRA system, which handles a variety of tactile gestures, including different granularities, inexact pointing gestures and pars-pro-toto deixis. We show how gestures can be used to shift focus and how focus can be used to disambiguate gestures. Finally, we discuss the impact of the user model on the decision of the presentation planning component, as to whether to use a pointing gesture, a verbal description, or both, for referent identification.

---

[1]This is a condensed and revised version of my paper 'User and Discourse Models for Multimodal Communication', which appears in 'Sullivan, J.W., Tyler, S.W. (eds.) Architectures for Intelligent Interfaces: Elements and Prototypes. Reading: Addison-Wesley 1991.' The research was partially supported by the German Science Foundation (DFG) in its Special Collaborative Programme on AI and Knowledge-Based Systems (SFB 314).

# 1	Introduction

In face-to-face conversation humans frequently use *deictic gestures* (e.g. the index finger points at something) parallel to verbal descriptions for referent identification. Such a *multimodal* mode of communication can improve human interaction with machines, as it simplifies and speeds up reference to objects in a visual world.

The basic technical prerequisites for the *integration of pointing and natural language* are fulfilled (high-resolution bit-mapped displays and window systems for the presentation of visual information, various pointing devices such as mouse, light-pen, joystick and touch-sensitive screens for deictic input, the DataGlove^{TM} or even image sequence analysis systems for gesture recognition). But the remaining problem for artificial intelligence is that explicit meanings must be given to natural pointing behavior in terms of a formal semantics of the visual world.

Unlike the usual semantics of mouse clicks in direct manipulation environments, in human conversation the region at which the user points is not necessarily identical with the region which he intends to refer to. Following the terminology of Clark, we call the region at which the user points *the demonstratum*, the descriptive part of the accompanying noun phrase *the descriptor* (which is optional), and the region which he intends to refer to *the referent* [6]. In conventional systems there exists a simple one-to-one mapping of a demonstratum onto a referent, and the reference resolution process does not depend on the situational context. Moreover, the user is not able to control the granularity of a pointing gesture, since the size of the predefined mouse-sensitive region specifies the granularity.

Compared to that, natural pointing behavior is much more flexible, but also possibly ambiguous or vague. Without a careful analysis of the *discourse context* of a gesture there would be a high risk of reference failure, as a deictic operation does not cause visual feedback from the referent (e.g. inverse video or blinking as in direct manipulation systems).

The subject of this paper is how the user and discourse model of an intelligent interface influences the comprehension and production of natural language with co-ordinated pointing to objects on a graphics display, and conversely how multimodal communication influences the user and discourse model. Before we review previous research on the combination of natural language and pointing and describe some current approaches related to our work let us briefly introduce the basic concepts of user and discourse modeling.

# 2	User Models and Discourse Models

A reason for the current emphasis on user and discourse modeling [20] is the fact that such models are necessary prerequisites in order for a system to be capable of exhibiting a wide range of intelligent and cooperative dialog behavior. Such models are required for identifying the objects which the dialog partner is talking about, for analyzing a non-literal meaning and/or indirect speech acts, and for determining what effects a planned utterance will have on the dialog partner. A cooperative system [19] must certainly take into account the user's goals and plans, his prior knowledge about the domain of discourse, as well as misconceptions a user may possibly have concerning the domain.

We use the following definitions of user and discourse models [21]:

A *user model* is a knowledge source which contains explicit assumptions on all aspects of the user that may be relevant for the dialog behavior of the system.

A *user modeling component* is that part of a dialog system whose function is to

- incrementally build up a user model and to maintain its consistency,

- to store, update and delete entries in it,

- and to supply other components of the system with assumptions about the user.

A *discourse model* is a knowledge source which contains the system's description of the syntax, semantics and pragmatics of a dialog as it proceeds.

A *discourse modeling component* is that part of a dialog system whose function is to

- incrementally build up a discourse model,

- to store and update entries in it,

- and to supply other components of the system with information about the structure and content of previous segments of the dialog.

While it seems commonly agreed upon that a discourse model should contain a syntactic and semantic description of discourse segments, a record of the discourse entities mentioned, the attentional structure of the dialog including a focus space stack, anaphoric links and descriptions of individual utterances on the speech act level, there seem to be many other ingredients needed for a good discourse representation which are not yet worked out in current discourse theory.

An important difference between a discourse model and a user model is that entries in the user model often must be explicitly deleted or updated, whereas in the discourse model entries are never deleted (except for forgetting phenomena). Thus according to our definition above, a belief revision component is an important part of a user modeling component.

3   Related Work on Deictic Input

Although in an intelligent multimodal interface the 'common visual world' of the user and the system could be any graphics or image, most of the projects combining pointing and natural language focus on business forms or geographic maps.

To the best of our knowledge, Carbonell's work on SCHOLAR represents the first attempt to combine natural language and pointing in an intelligent interface

[5]. SCHOLAR, a tutoring system for geography, allowed simple pointing gestures on maps displayed on the terminal screen. NLG [3] also combined natural language and pointing using a touch screen to specify graphics with inputs like (1).

(1) Put a point called A1 here <touch>.

Woods and his coworkers developed an ATN editor and browser, which can be controlled by natural language commands and accompanying pointing gestures at the networks displayed on the screen [23].

In SDMS [2] the user can create and manipulate geometric objects by natural language and coordinated pointing gestures. The first commercially available multi-modal interface combining verbal and non-verbal input was NLMenu [18], where the mouse could be used to rubber band an area on a map in sentences like (2).

(2) Find restaurants, which are located here <pointing> and serve Mexican food.

All approaches to gestural input mentioned so far in our brief review were based on a simple one-to-one mapping of the demonstratum onto a referent and thus have not attacked the central problems of analyzing pointing gestures.

Recently, several research groups have addressed the problems of combining non-verbal and verbal behavior more thoroughly. Several theoretical studies and empirical investigations about the combination of natural language and pointing have been published [7, 9, 12]. Working prototype systems have been described, which explore the use of complex pointing behavior in intelligent interfaces.

For example, the TACTILUS subcomponent (designed and implemented by J. Allgayer) of our XTRA system [10], which we will describe below in more detail, handles a variety of *tactile gestures*, including different granularities, inexact pointing gestures, and *pars-pro-toto deixis*. In the latter case, the user points at an embedded region when actually intending to refer to a superordinated region.

In the DIS-QUE system [22] the user can mix pointing and natural language to refer to student enrollment forms or maps. The deictic interpreter of the T^3 system [17] interacts with a natural language interpreter for the analysis of pointing gestures indicating ship positions on maps and deals also with continuing or repeated deictic input. CUBRICON [11] is yet another system which handles simultaneous input in natural language and pointing to icons on maps, using language to disambiguate pointing and conversely.

While the simultaneous exploitation of both verbal and non-verbal channels provides maximum efficiency, most of the current prototypes do not use truly parallel input techniques, since they combine *typed* natural language and pointing. In these systems the user's hands move frequently back-and-forth from the keyboard to the pointing device. Note, however, that multimodal input makes even natural language interfaces without speech input more acceptable (fewer keystrokes) and that the research on typed language forms the basis for the ultimate speech understanding system.

4 An Intelligent Multimodal Interface to Expert Systems

XTRA (eXpert TRAnslator) is an intelligent multimodal interface to expert systems, which combines natural language, graphics and pointing for input and output. As its name suggests, XTRA is viewed as an intelligent agent, namely a translator who acts as an intermediary between the user and the expert system. XTRA's task is to translate from the high-bandwidth communication with the user into the narrow input/output channel of the interfaces provided by most of the current expert systems.

The present implementation of XTRA provides natural language access to an expert system, which assists the user in filling out a tax form. During the dialog, the relevant page of the tax form is displayed on one window of the screen, so that the user can refer to regions of the form by tactile gestures. As shown in figure 1, there are two

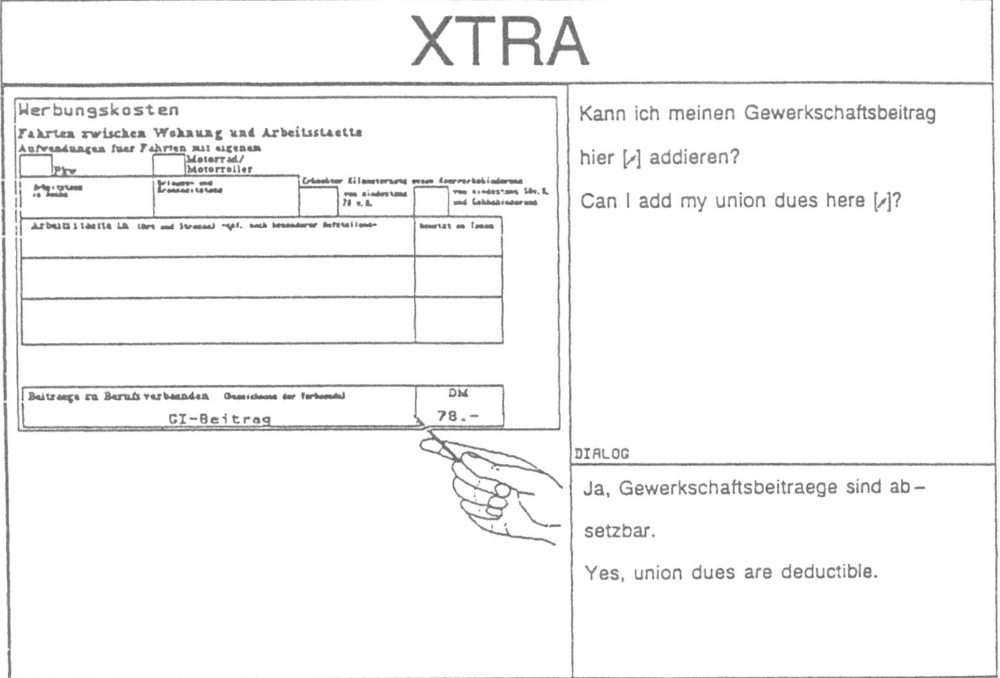

Figure 1: The Combination of Natural Language, Graphics and Pointing in XTRA

other windows on the left part of the display, which contain the natural language input of the user (upper part) and the system's response (lower part). An important aspect of the communicative situation realized in XTRA is that the user and the system share a common visual field - the tax form. As in face-to-face communication, there is no visual feedback after a successful referent identification process. Moreover, there are no predefined 'mouse-sensitive' areas and the forms are not specially designed to simplify gesture analysis. For example, the regions on the form may overlap and there may be several sub-regions embedded in a region of the form.

In addition to the direct interpretation of a gesture, where the demonstratum is simply identical to the referent, TACTILUS provides two other types of interpretation. In a pars-pro-toto interpretation of a gesture the demonstratum is geometrically embedded within the referent. An extreme case of a pars-pro-toto interpretation in the current domain of XTRA is a situation where the user points at an arbitrary part (pars in Latin) of the tax form intending to refer to the form as a whole (pro toto in Latin). Another frequent interpretation of gestures is that the demonstratum is geometrically adjacent to the referent: the user points, for instance, below or to the right of the referent. Reasons for this may be the user's inattentiveness or his attempt to gesture without covering up the data in a field.

The user first chooses the granularity of the intended gesture by selecting the appropriate icon from the pointing mode menu or by pressing a combination of mouse buttons, and then performs a tactile gesture with the pointing device symbolized by the selected mouse cursor. The current implementation supports four pointing modes:

- exact pointing with a pencil

- standard pointing with the index finger

- vague pointing with the entire hand

- encircling regions

The deixis analyzer of XTRA is realized as a *constraint propagation* process on a graph which represents the topology of the tax form. A pointing area of a size corresponding to the intended granularity of the gesture is associated with each available pointing mode. A plausibility value is computed for each referential candidate of a particular pointing gesture according to the ratio of the size of the part covered by the pointing area to the size of the entire region. The result of the propagation process is a list of referential candidates consisting of pairs of region names and plausibility values.

Since pointing is fundamentally ambiguous without the benefit of contextual information, this list often contains many elements. Therefore, TACTILUS uses various other knowledge sources of XTRA (e.g. the semantics of the accompanying verbal description, case frame information, the dialog memory) for the disambiguation of the pointing gesture (see [1] and [10] for further details).

5 The Influence of Pointing Gestures on the Discourse Model

Pointing is not only used for referent identification but also to mark or change the *dialog focus*, i.e. to control or shift *attention* during comprehension. As we noted in section 2, focus is an important notion in a discourse model, since it influences many aspects of language analysis and production. For example, focus can be used to *disambiguate* definite descriptions and anaphora [8].

Figure 2 gives an example of the disambiguation of a definite description using a

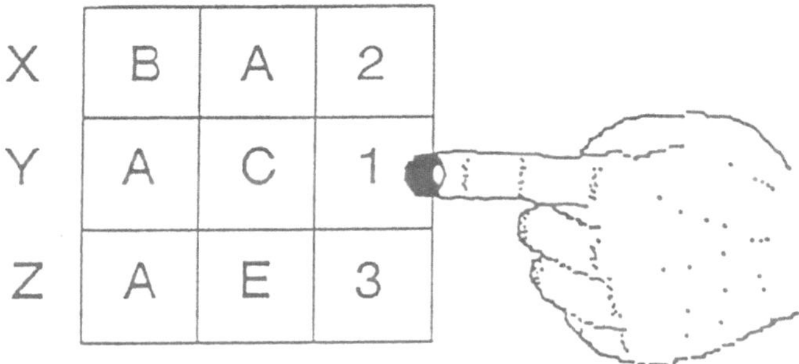

Figure 2: Focusing Gesture Disambiguating the Question 'Why should I delete the 'A''

focusing gesture. Without focus the definite description 'the A' is ambiguous in the given visual context, since there are three objects visible which could be referred to as 'A' (one in each row of the table displayed in figure 2). Together with the pointing gesture at row Y, which marks this row as a part of the immediate focus, the definite description can be disambiguated, since there is only one 'A' in the focused row.

As in the case of gestures for referent identification, the effect of a focusing gesture can also be produced by a *verbal paraphrase*. For the example presented in figure 2, a meta-utterance like 'Now let's discuss the entries in row Y' would have the same effect on the discourse model and help to disambiguate the subsequent definite description.

As we noted earlier, without a discourse context most pointing gestures are ambiguous. In the example above, we have seen that a discourse context can be established not only by verbal information but also by gestures. Thus there is a twofold relation between gestures and focus. Gestures can be used to shift focus and focus can be used to disambiguate gestures.

From this follows that in *simultaneous pointing actions* two communicative functions of pointing can be combined: focus shifting and reference. The following two types of simultaneous pointing can be identified:

- One-handed input:

    - Focusing act: For example, the pencil is put down on the form, so that it points to a particular region on the form.

    - Referential act: A subsequent pointing gesture refers to an object in the marked region

- Two-handed input (see also [4]):

- Focusing act: For example, the index finger of one hand points to a region of the form.

- Referential act: The index finger of the other hand points to an object in the marked region.

Figures 3 and 4 illustrate the use of focusing gestures for the disambiguation of referential gestures. Note that in both situations displayed in figures 3 and 4 the index finger points at the same location on the form and that the utterances combined with these referential gestures are identical. The cases shown in both figures differ only in the location of the pencil which is used for focusing.

Let us explore the processing of these examples in detail. Since the referential gesture with the index finger is relatively inexact, TACTILUS computes a large set of possible referents. The head noun 'numbers' of the verbal description accompanying the pointing gesture imposes two restrictions on this set of possible referents. Since there are only four numbers displayed on the part of the form shown in Figs. 3 and 4, the semantics of the noun restricts the solution space to the power set of {3,4,7,5}, and the plural implies that only sets with at least two elements are considered in this power set. Finally, the position of the index finger on the form makes the interpretations {3,7,5}, {3,4,5}, {3,4,7} and {4,5,7} implausible, so that the resulting set of plausible referential readings becomes {{3,4}, {4,5}, {3,4,7,5}}, where {3,4,7,5} is a typical example of a pars-pro-toto reading.

This means that there remain three possible interpretations before we consider the focusing gesture. It is worth noting that this is one of the cases where the combination of verbal and nonverbal information in one reference act does not lead to an unambiguous reading. Here information from the discourse model helps to disambiguate. In figure 3 the pencil points at the row beginning with 'XYZ', so that this row and all its parts become focused. Now the intersection of the set of plausible referents and the currently focused objects results in the unique interpretation {3,4}. Similarly, in figure 4 the pencil is pointing at the block of columns called 'C3', so that the intersection of the focused elements with the results of the referential analysis is again a unique interpretation, namely {4,5}, but it differs from the set of referents found for the gestural input shown in figure 3. These examples once again emphasize the basic premise of our work, i.e. that pointing gestures must be interpreted in a highly context-sensitive way and that all approaches supposing a one-to-one mapping of the demonstratum onto the referent will fail in complex multimodal interactions.

6   User Modeling for Presentation Planning

As we noted at the outset, an intelligent interface should not only be able to analyze multimodal input, but also to generate multimodal output. The design of XTRA's generator allows the simultaneous production of deictic descriptions and pointing actions [13]. Since an intelligent interface should try to generate cooperative responses,

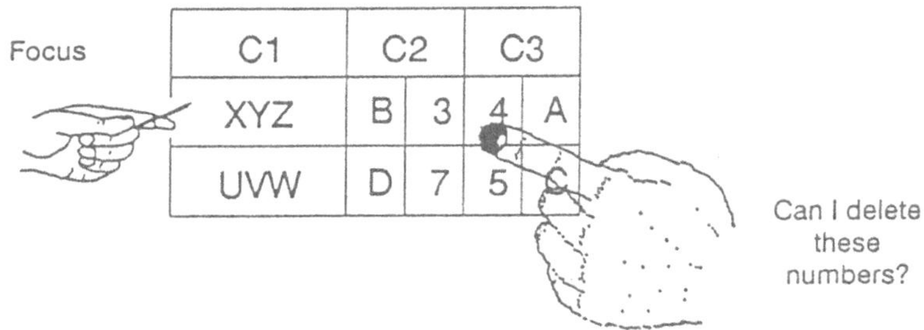

Figure 3: Simultaneous Pointing Gestures

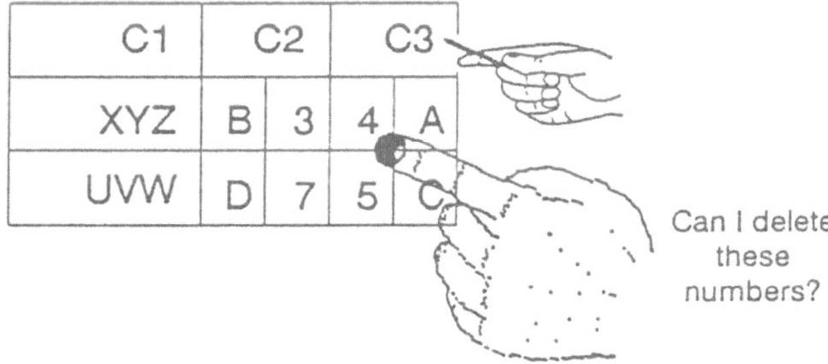

Figure 4: Simultaneous Pointing Gestures with Different Focus

it has to exploit its user model to generate descriptions tailored to users with various levels of expertise.

One important decision which a multimodal presentation planner has to make, is whether to use a pointing gesture or a verbal description for referent identification. Let us explore the impact of the user model on this decision using an example from our tax domain.

Suppose the system knows the concept 'Employee Savings Benefit' and an entry in the user model says that the current dialog partner seems to be unfamiliar with this concept. When the system plans to refer to a field in the tax form, which could be referred to using 'Employee Savings Benefit' as a descriptor, it should not use this technical term but a pointing gesture to the corresponding field. This means that in the conversational context described (3) would be a cooperative response, whereas (4) would be uncooperative.

(3) You can enter that amount here [↗] | in this [↗] field.

(4) You can enter that amount as employee savings benefit.

To summarize that point, if the system knows that a technical term which could be used to refer to a particular part of the tax form visible on the screen is not understandable to the user, it can generate a pointing gesture, possibly accompanied by a mutually known descriptor.

In the following, we discuss a particular method of user modeling, called *anticipation feedback*, which can help the system to select the right granularity of pointing

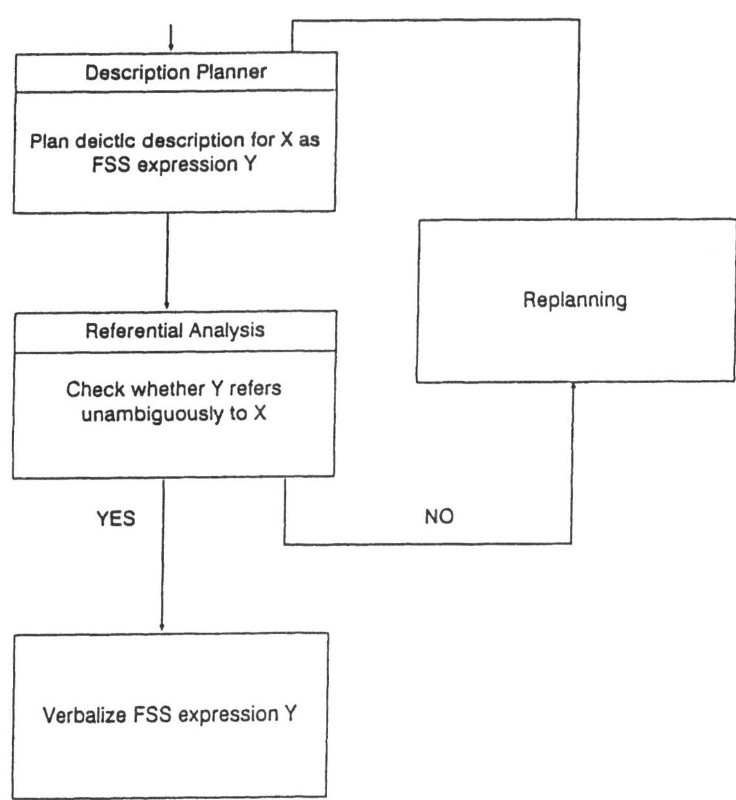

Figure 5: An Anticipation Feedback Loop for Presentation Planning

when generating multimodal output. Anticipation feedback loops involve the use of the system's comprehension capability to simulate the user's interpretation of a communicative act which the system plans to realize [20]. The application of anticipation feedback loops is based on the implicit assumption that the system's comprehension procedures are similar to those of the user. In essence, anticipation on the part of the system means answering a question like (5).

(5) If I had to analyze this communicative act relative to the assumed knowledge of the user, then what would be the effect on me?

If the answer to this question does not match the system's intention in planning the tested utterance, it has to replan its utterance, as in a generate-and-test loop.

Figure 5 shows an extremely simplified version of a multimodal description planning process with an anticipation feedback loop for user modeling. Let us assume that the generator decided to plan a deictic description of an object X, which the systems intends to refer to. The result of the description planning process is a an expression Y of the functional-semantic structure (FSS) together with a planned gesture. The FSS is a surface-oriented semantic representation language used on one of the processing levels of the how-to-say component of XTRA's generator.

Figure 6: Planned Pointing Gesture

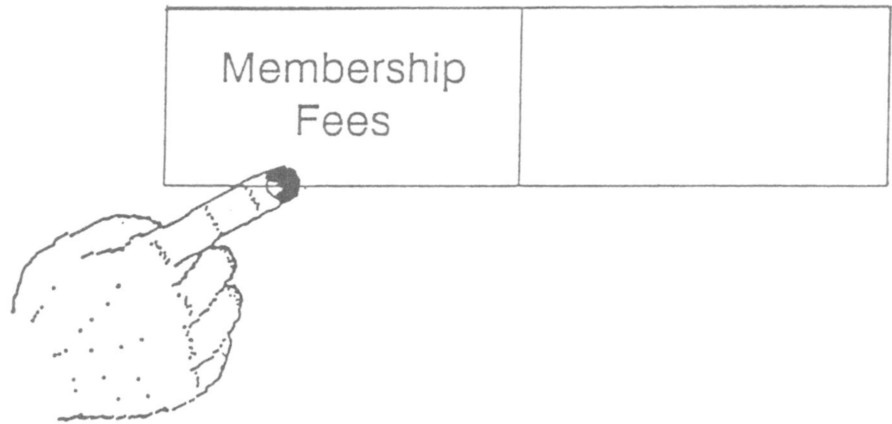

Figure 7: Pointing Gesture after Replanning

This preliminary deictic description is fed back into the system's analysis component, where the referent identification component together with the gesture analyzer TACTILUS try to find the intended discourse object. If the system finds that the planned deictic description refers unambiguously to X, the description is fed into the final transformation process before it is outputted. Otherwise, an alternative FSS and/or pointing gesture has to be found in the next iteration of the feedback process (see figure 5).

Now let us follow the feedback method as it goes through the loop, using a concrete example. Suppose that the system plans to refer to the string 'Membership Fees' in the box shown in figure 6. Also assume that the presentation planner has already decided to generate an utterance like 'Delete this [↗]' together with the pointing gesture shown in figure 6.

For a punctual pointing gesture the system first chooses the pencil as a pointing device. In this case, the exact position of the pencil was selected according to XTRA's default strategy described in [16]: the pencil is below the entry, so that the symbol does not cover it.

When this pointing gesture is fed back into the gesture analyzer of the referent identification component, the set of anticipated reference candidates might be {'Fees', 'e', 'Membership Fees'} containing only elements which can be 'deleted' (the current version of TACTILUS does not deal with characters or substrings of a string). Since the system has detected that the planned gesture is ambiguous, it starts replanning and then selects the index finger icon as a pointing gesture with less granularity (see figure 7). This time, the result of the feedback process is unambiguous, so that the system can finally perform the pointing action.

7   Conclusions

We have shown how the user and discourse model of an intelligent interface influences the comprehension and production of natural language with coordinated pointing to objects on a graphics display, and conversely how multimodal communication influences the user and the discourse model.

First, we described XTRA as an intelligent interface to expert systems, which handles a variety of tactile gestures, including different granularities, inexact pointing and pars-pro-toto deixis, in a domain- and language-independent way. Then we discussed several extensions to the XTRA's deixis analyzer and presented our approach to generating multimodal output.

We showed how gestures can be used to shift focus and focus can be used to disambiguate gestures, so that simultaneous pointing actions combine two communicative functions: focus shifting and reference. We explored the role of user modeling for presentation planning and described how the user model can be exploited to generate multimodal descriptions tailored to the user's level of expertise.

Finally, we discussed anticipation feedback as a particular method of user modeling, which can help the system to select the right granularity of pointing when generating multimodal output.

Bibliography

[1] Allgayer, J. and Reddig, C. 1986. Processing Descriptions containing Words and Gestures - A System Architecture. In Rollinger, C.-R. (ed.), *Proc. of GWAI/ÖGAI 1986*, Berlin, Springer.

[2] Bolt, R.A. 1980. Put-That-there: Voice and Gesture at the Graphics Interface. *Computer Graphics*, 14, pp. 262-270.

[3] Brown, D.C., Kwasny, S.C., Chandrasekaran, B., Sondheimer, N.K. 1979. An

Experimental Graphics System with Natural Language Input. *Computer and Graphics*, 4, pp. 13-22.

[4] Buxton, W. and Myers, B.A. 1986. A Study in Two-Handed Input. *Proc. CHI'86 Human Factors in Computing Systems*, ACM, New York, pp. 321-326.

[5] Carbonell, J.R. 1970. *Mixed-Initiative Man-Computer Dialogues.* BBN Report No. 1971, Bolt, Beranek and Newman, Cambridge, MA.

[6] Clark, H.H., Schreuder, R. and Buttrick, S. 1983. Common Ground and the Understanding of Demonstrative Reference. *Journal of Verbal Learning and Verbal Behavior*, 22, pp. 245-258.

[7] Hayes, P.J. 1986. Steps towards Integrating Natural Language and Graphical Interaction for Knowledge-based Systems. *Proc. of the 7th European Conference on Artificial Intelligence*, Brighton, Great Britain, pp. 436-465.

[8] Grosz, B. 1981. Focusing and Description in Natural Language Dialogues, in Joshi, A., Webber, B., Sag, I. (eds.),*Elements of Discourse Understanding*. New York: Cambridge Univ. Press, pages 84-105.

[9] Hinrichs, E. and Polanyi, L. 1987. Pointing The Way: A Unified Treatment of Referential Gesture in Interactive Discourse. *Papers from the Parasession on Pragmatics and Grammatical Theory at the 22nd Regional Meeting*, Chicago Linguistic Society, Chicago, pp. 298-314.

[10] Kobsa, A., Allgayer, J., Reddig, C., Reithinger, N., Schmauks, D., Harbusch, K. and Wahlster, W. 1986. Combining Deictic Gestures and Natural Language for Referent Indentification. *Proc. of the 11th International Conf. on Computational Linguistics*, Bonn, West Germany, pp. 356-361.

[11] Neal, J.G., Shapiro, S.C. 1988. Intelligent Multi-Media Interface Technology. In *Proc. of the Workshop on Architecures for Intelligent Interfaces: Elements and Prototypes*. Monterey, Ca., pp. 69-91.

[12] Reilly, R. (ed.) 1985. *Communication Failure in Dialogue: Techniques for Detection and Repair*. Deliverable 2, Esprit Project 527, Educational Research Center, St. Patrick's College, Dublin, Ireland.

[13] Reithinger, N. 1987. Generating Referring Expressions and Pointing Gestures. In Kempen, G. (ed.) *Natural Language Generation*, Dordrecht, Kluwer, pp. 71-81.

[14] Retz-Schmidt, G. (1988): Various Views on Spatial Prepositions. In *AI Magazine*, Vol. 9, No. 2, also appeared as: Report No. 33, SFB 314, University of Saarbrücken, Computer Science Department.

[15] Schmauks, D. 1987. Natural and Simulated Pointing. In *Proc. of the 3rd European ACL Conference*, Copenhagen, Danmark, pp. 179-185.

[16] Schmauks, D. and Reithinger, N. 1988. Generating Multimodal Output - Conditions, Advantages and Problems. To appear in *Proc. of the 12th International Conference on Computational Linguistics*, Budapest, Hungary. Also appeared as Report No. 29, SFB 314, Computer Science Department, University of Saarbruecken.

[17] Scragg, G.W. 1987. *Deictic Resolution of Anaphora.* Unpublished paper, Franklin and Marshall College, P.O.Box 3003, Lancaster, PA 17604.

[18] Thompson, C. 1986. Building Menu-Based Natural Language Interfaces. *Texas Engineering Journal*, 3, pp. 140-150.

[19] Wahlster, W. 1984. Cooperative Access Systems. *Future Generation Computer Systems*, 1, pp. 103-111.

[20] Wahlster, W. and Kobsa, A. 1986. Dialog-Based User Models. In Ferrari, G. (ed.) *Proceedings of the IEEE*, 74, 7, pp. 948-960.

[21] Wahlster, W. 1988. Distinguishing User Models from Discourse Models, Report No. 27, SFB 314, Computer Science Department, University of Saarbruecken, Fed. Rep. of Germany, to appear in Kobsa, A. and Wahlster, W. (eds.) *Computational Linguistics*, Special Issue on User Modeling, 1988.

[22] Wetzel, R.P., Hanne, K.H. and Hoepelmann, J.P. 1987. *DIS-QUE: Deictic Interaction System-Query Environment.* LOKI Report KR-GR 5.3/KR-NL 5, Fraunhofer Gesellschaft, IAO, Stuttgart, Fed. Rep. of Germany.

[23] Woods, W.A. et al. 1979. *Research in Natural Language Understanding.* Annual Report, TR 4274, Bolt, Beranek and Newman, Cambridge, MA, USA.

[24] Zimmermann, T.G., Lanier, J., Blouchard, C., Bryson, S. and Harvill, Y. 1987. A Hand Gesture Interface Device. *Proc. CHI'87 Human Factors in Computing Systems*, ACM, New York, pp. 189-192.

Laudatio zum 60. Geburtstag von Prof. Dr. Günter Hotz

Prof. Dr. Otto Spaniol
Lehrstuhl für Informatik IV
RWTH Aachen
Ahornstrasse 55
5100 Aachen
Germany

Lieber Herr Hotz, liebe Familie Hotz, Spektabilitäten, verehrte Festversammlung!

Laudationes (welch gräßliches Wort) mag niemand. Sie bringen im informationstheoretischen Sinne keine Information, denn 'Informationsgewinn' heißt Vermittlung von neuem Wissen bzw. Wegnahme von Unsicherheit, doch über den Werdegang von Günter Hotz ist alles bekannt und über seine Persönlichkeit ist niemand unsicher.

Daher würde eine lexikalische Auflistung des Werdegangs von Günter Hotz nicht mehr bedeuten als eine Eule ins Wappen der Universität Saarbrücken zu tragen.

Es kommt hinzu, daß die mir vom Organisationskomitee dieser Veranstaltung zur Verfügung gestellten Unterlagen zwar umfangreich waren (bzgl. der Seitenzahl), aber nicht über „normale" Verwendbarkeit hinausgingen. Sie enthielten nicht mehr als die üblichen Fakten – und auch diese nur unvollständig, denn bereits der Geburtsort war daraus nicht zu erschließen.

Im Bereich der Logik gibt es das „Paradoxon der unerwarteten Prüfung" oder der „unerwarteten Hinrichtung". Im vorliegenden Fall gab es für mich etwas Ähnliches, nämlich das „Dilemma der unzulässigen Nachfrage". Wen hätte ich um Vermittlung zusätzlicher Angaben bitten sollen?

- Den Jubilar? Das geht aber nun wirklich nicht!

- Einen Kollegen? Das wäre vielleicht denkbar, aber hätte das Risiko der folgenden Gegenfrage aufgeworfen: „Was denn, Sie halten die Laudatio und wissen nicht einmal, daß ...?".

Ich hoffe, daß mein Dilemma verständlich wird.

Was ich demgegenüber vorhabe und wozu ich aufgrund meines Dilemmas gezwungen bin, ist der Versuch, einige markante Punkte von Person und Lebenswerk in einer weniger üblichen Form darzustellen, d.h.:

- Geschehnisse werden weder zeitlich noch bzgl. ihrer scheinbaren oder ihrer realen Wichtigkeit geordnet

- ein Versuch zur Kurzdarstellung der wissenschaftlichen Resultate wird nicht unternommen

- und es ist auch sonst keine Vollständigkeit zu erwarten bzw. angestrebt.

Stattdessen will ich mich auf die Darstellung einiger vielleicht unwichtig erscheinender Details beschränken – auch das aus sehr subjektiver Sicht.

Beginnen möchte ich trotzdem eher konventionell, nämlich mit einer zeitrafferartigen Darstellung einiger Stationen der wissenschaftlichen Laufbahn von Günter Hotz.

Der Lebenslauf

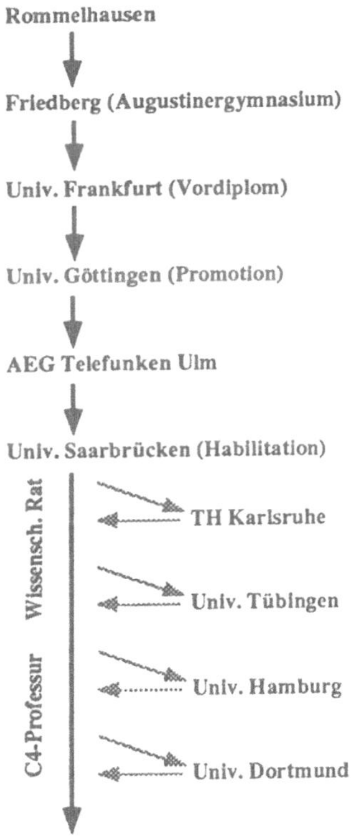

Geboren im oberhessischen Rommelhausen besuchte Günter Hotz das Gymnasium in Friedberg. Er studierte Mathematik und Physik in Frankfurt bis zum Vordiplom und danach in Göttingen.

Nach Diplom (1956) und Promotion (1958) ging er als Entwicklungsingenieur nach Ulm zu AEG-Telefunken. Dort arbeitete er bei Güntsch an der Entwicklung von Rechnern. Danach kam er als Stipendiat der Fritz-Thyssen-Stiftung an das Institut für Angewandte Mathematik der Universität Saarbrücken, welches von Johannes Dörr geleitet wurde.

In Saarbrücken habilitierte er sich im Jahre 1965 mit einer Arbeit zur Algebraisierung des Syntheseproblems von Schaltkreisen, die in zwei Teilen in der Zeitschrift EIK veröffentlicht wurde.

Trotz einer Gastprofessur in Tübingen und trotz mehrerer Angebote bzw. Rufe aus Karlsruhe (1965), Hamburg (1969) und Dortmund (1973) hat er der Universität Saarbrücken bis heute die Treue gehalten.

Während seiner Saarbrücker Tätigkeit bemühte er sich (und er tut es natürlich immer noch) mit großer Intensität und mit ebenso großem Erfolg um die Etablierung der neuen Fachrichtung „Informatik". Er tat dies auch als Gründungsvorsitzender der Gesellschaft für Informatik, als deutscher Delegierter im IFIP-Technical-Committee on Education, als Fachgutachter der Deutschen Forschungsgemeinschaft, als Initiator und Sprecher mehrerer Sonderforschungsbereiche, im Wissenschaftsrat sowie als Mitglied zweier Akademien (nämlich der Akademie für Wissenschaft und Literatur in Mainz sowie der Akademie der Wissenschaften der inzwischen nicht mehr existierenden DDR). Ich werde darauf noch zurückkommen.

Rommelhausen

Friedberg (Augustinergymnasium)

Univ. Frankfurt (Vordiplom)

Univ. Göttingen (Promotion)

AEG Telefunken Ulm

Univ. Saarbrücken (Habilitation)

C4-Professur Wissensch. Rat

TH Karlsruhe

Univ. Tübingen

Univ. Hamburg

Univ. Dortmund

Für sein richtungweisendes Wirken auf dem Gebiet der Informatik erhielt Günter Hotz im Jahre 1989 den Saarländischen Verdienstorden.

Soweit eine sehr kurze Auflistung der Stationen des wissenschaftlichen Lebenslaufs. Aber diese wird der Persönlichkeit von Günter Hotz keineswegs gerecht. Daher soll im folgenden eine Würdigung aufgrund anderer Merkmale versucht werden.

Der Erste

„Die Wirkung nach außen und die Akzeptanz von außen"

Es gibt nur wenige andere Informatiker – wenn überhaupt –, die eine ähnliche Akzeptanz „von außen" (d.h. seitens anderer Fachdisziplinen) und gleichzeitig „nach außen" (durch richtungweisende Veröffentlichungen in Form von Zeitschriftenartikeln, vor allem aber auch durch Monographien, die zu den wenigen Standardwerken in Informatik gehören) gefunden haben wie Günter Hotz.

Seine Wirkung „nach außen" wird nicht zuletzt dadurch sichtbar, daß Günter Hotz kein fremdes Parkett scheut. Man findet seine Veröffentlichungen in Publikationsorganen aus Mathematik und Informatik, aus Physik und Wirtschaftswissenschaften, aus Sprachwissenschaft und Künstlicher Intelligenz. Begeisterte Zustimmung, aber auch Nachdenken bzw. Überdenken haben seine Abhandlungen für die Akademie der Wissenschaften und Literatur ausgelöst.

Die Akzeptanz „von außen" war und ist eine zwangsläufige Konsequenz seines ebenso weitsichtigen wie erfolgreichen Wirkens für die Informatik. In diesem Bereich war Günter Hotz sehr oft „der Erste". Einige Beispiele dafür:

- Er war Gründungsvorsitzender der Gesellschaft für Informatik und sechs Jahre lang Mitglied des Präsidiums der GI.

- Er ist seit 1985 Mitglied der Akademie der Wissenschaften und Literatur des Landes Rheinland-Pfalz.

- Er wurde 1986 erstes auswärtiges Mitglied der Akademie der Wissenschaften der DDR. Dabei bezieht sich die Eigenschaft „erstes auswärtiges Mitglied" nach meiner Kenntnis mindestens auf den Bereich der Informatik, möglicherweise aber auf alle ingenieurwissenschaftlichen und/oder naturwissenschaftlichen Disziplinen. In welcher Form die Mitgliedschaft aufgrund der inzwischen eingetretenen politischen Änderungen fortbesteht, weiß ich nicht, denn zumindest die Eigenschaft der „Auswärtigkeit" kann inzwischen nicht mehr gelten (auch das eine Konsequenz der deutschen Einigung!). Die Ehrung selbst war eine Sensation, und die kritischen Äußerungen von Günter Hotz haben auch einen Beitrag zu den erwähnten politischen Veränderungen geleistet.

- Als erster Informatiker erhielt er den Gottfried-Wilhelm-Leibniz-Preis der Deutschen Forschungsgemeinschaft zusammen mit Kurt Mehlhorn und Wolfgang Paul. Bis heute wurde der Preis an keinen anderen Informatiker vergeben. Das unterstreicht einerseits die herausragende Position von Günter Hotz, andererseits sollte es für uns Informatiker aber auch Ansporn zu Bemühungen sein, wieder einmal bei diesem Preis berücksichtigt zu werden.

- Er war Mitglied des Wissenschaftsrats (von 1987 bis 1989).

- Er erhielt – wie bereits erwähnt – im Jahre 1989 den Saarländischen Verdienstorden.

Das alles sind nur besonders herausgehobene, weil für einen Informatiker erstmalige oder einmalige Ereignisse bzw. Ehrungen.

Der Alchimist   *„Die Suche nach dem Stein der Weisen"*

Wenn man versucht, die Arbeitsweise von Günter Hotz zu charakterisieren, wird man stets die Koexistenz von zwei unterschiedlichen Ansatzpunkten vorfinden:

- ein ebenso originelles wie unkonventionelles Überprüfen aller Möglichkeiten, auch wenn diese zunächst unsinnig oder aussichtslos scheinen

- das ständige Hinterfragen der Gültigkeit von scheinbar „logischen" Gegebenheiten.

Beides zusammen entspricht recht genau dem Alchimieprinzip, denn Alchimisten werden laut Brockhaus durch folgende Eigenschaft gekennzeichnet: „Da sie auf der Suche nach dem Stein der Weisen vor keinem natürlichen Stoff halt machten, gelang ihnen eine Fülle von Entdeckungen".

Die „Suche nach dem Stein der Weisen" (was heute vielleicht mit „grundlegendem Erkenntnisfortschritt" übersetzt werden kann), die Bemühungen um alle möglichen „natürlichen Stoffe" (heute würde man stattdessen „natur- bzw. ingenieurwissenschaftliche Disziplinen" sagen) und schließlich die „Fülle von Entdeckungen" sind nach meiner Auffassung sehr zutreffende Umschreibungen des ersten genannten Arbeitsprinzips von Günter Hotz.

Das zweite Prinzip möchte ich an einer kleinen Anekdote aufzeigen, die von marginaler Bedeutung ist, aber die mir dennoch typisch zu sein scheint. Mich zumindest hat sie außerordentlich beeindruckt und auch zu Nachahmungsversuchen veranlaßt:

Gegen Ende der Sechziger Jahre fand in Saarbrücken eine Tagung der Deutschen Gesellschaft für Operations Research statt. Günter Hotz nahm daran teil, auch Studenten bzw. Mitarbeiter der Angewandten Mathematik bzw. der noch nicht existenten Informatik durften zuhören. Es gab damals noch so wenige Informatiktagungen, daß man sich den Luxus der Teilnahme an fachfremden Veranstaltungen noch erlauben konnte!

Auf der erwähnten DGOR-Tagung gab es einen interessanten Vortrag, ich weiß nicht mehr zu welchem Thema, nur noch daß er von einem Schweizer gehalten wurde (es ist das übliche Schicksal von Vorträgen, daß nur die unwichtigen Dinge in Erinnerung bleiben!). Der Referent behauptete, das von ihm vorgestellte Verfahren sei „optimal". Sein Beitrag fand in der Diskussion zunächst einen deutlich positiven Widerhall, bis Günter Hotz diese Tendenz durch eine simple Feststellung umkehrte – und zwar sagte er, ich weiß es noch wie heute: „Ich kann mir nicht vorstellen, wie Sie die

Optimalität beweisen wollen!" Das war ein zwar nicht beabsichtigter, aber ein dafür umso wirksamerer Blattschuß.

Ich habe mich seither wie viele andere von der Hotzschen Denkweise geprägte „Schüler" um ähnliche – ebenso einfache wie überzeugende und wirksame – Argumentationen bemüht. Es kann festgestellt werden, daß niemand das Vorbild auch nur annähernd erreicht hat.

Der Geometer „Μηδεὶσ ἀγεωμέτρητοσ εἰσίτω"

Leitspruch der philosophischen Schule von Plato war – und so soll es über der Tür seiner Akademie gestanden haben:

„Μηδεὶσ ἀγεωμέτρητοσ εἰσίτω" (Mädeís ageométratos eisíto! Kein Nichtgeometer möge eintreten!). Das Standardlexikon der Antike – der kleine Pauly – bemerkt dazu, daß diese apokryphe Äußerung Plato zugeschrieben werde und zutreffend ausdrücke, daß ein Zugang zur Akademie (und damit zur Weisheit) nur möglich ist für den, der in der Geometrie unterwiesen ist.

Dieser Satz steht zwar heute nicht über der Eingangstür des Lehrstuhls von Günter Hotz, aber er gilt nach wie vor, denn seine Forschungen waren und sind geometrisch geprägt. Er ist Topologe, hat über ein topologisches Thema promoviert – mit einer Arbeit auf dem Gebiet der Knotentheorie –, und er hat die Topologie nie verlassen, auch wenn sich wie bei Informatikern üblich das engere Arbeitsgebiet mehrfach geändert hat. Betrachten Sie die Abbildungen in seinen Veröffentlichungen – etwa in seiner Habilitationsschrift – und Sie verstehen, was ich meine. So wird beispielsweise das von Günter Hotz entworfene Konzept der X-Kategorien am besten verständlich, wenn man es geometrisch deutet bzw. wenn man es auf konkrete Schaltungsprobleme anwendet. In dieser Interpretation erweist es sich dann als außerordentlich gut geeignet, um den Zugang zu Problemen des Schaltungsentwurfs zu erleichtern. Inzwischen finden die von Günter Hotz entwickelten Methoden ständig neue Anwendungen beim Chipdesign und -layout. Dabei werden immer wieder topologische bzw. geometrische Konzepte in die Praxis umgesetzt.

An geometrischen Beispielen demonstriert Günter Hotz auch immer wieder grundlegende offene Probleme von neuen Gebieten, und durch diese – oft verblüffend einfachen – Analogien werden viele Fragestellungen klarer als durch langatmige Formulierungen.

Man nehme etwa sein Essai über künstliche Intelligenz, woraus ich nur ein kleines Detail zitieren möchte: den maschinellen Beweis für die Gleichheit der Basiswinkel im gleichschenkligen Dreieck ohne Zuhilfenahme der Höhe als Hilfslinie. Viele Jahrhunderte lang hatte sich der menschliche Geist auf den ebenso einfachen wie (scheinbar) einzig möglichen Hilfslinienbeweis beschränkt. Erst ein unvoreingenommer Automat konnte dieses starre Denkschema durch eine überraschende Variante erweitern. Diese Beobachtung bewegte Günter Hotz schon lange Jahre vor der Einführung der Künstlichen Intelligenz als wissenschaftlicher Disziplin zum Nachdenken über die Problematik des maschinellen Beweisens. Bereits in seiner ersten Zeitschriftenveröffentlichung ("Ein Satz über Mittellinien"; Archiv der Mathematik, 1959) sind ähnliche Gedankengänge enthalten. Die Möglichkeiten, aber auch die Grenzen der Künstlichen Intelligenz werden auf diese Weise klarer herausgestellt als in vielen Monographien.

Seine Mitarbeiterschar hielt Günter Hotz immer mit einigen geometrischen Problemen beschäftigt. Zwei seiner Fragen, an die ich mich besonders gut erinnere, waren:

- Ein Spiegel vertauscht links und rechts, weshalb vertauscht er nicht oben und unten?

Oder:

- Ist es möglich, einen Zigarrenrauchring durch einen anderen hindurchzublasen?

Diese und andere Denksportaufgaben (z.B. auch über diverse Varianten von Möbiusbändern) stammten mit einiger Sicherheit noch aus seiner Promotionszeit. Seine Dissertation, die er im Jahre 1958 bei K. Reidemeister in Göttingen ablegte, hatte zwar vordergründig betrachtet keinen Einfluß auf die Entwicklung der Informatik, indirekt aber sehr wohl. Eine geometrische Denkweise zieht sich wie ein roter Faden durch alle seine Publikationen. So trägt z.B. die derzeit neueste Veröffentlichung den Titel "On the construction of very large integer multipliers", ein Thema, welches nur unter Zuhilfenahme von geometrischen bzw. topologischen Methoden sinnvoll anzugehen ist.

Der Sponti   *"Der Ulmer Flußspaziergang"*

Der Verlauf einer Karriere wird nur zum kleineren Teil durch kontinuierliche Bemühungen beeinflußt. Bedeutender sind Entscheidungen, welche kurzfristig, also ohne Möglichkeit zum Abwägen aller Alternativen getroffen werden müssen. In Schaltkreisen sind die Verzweigungsstellen einflußreicher als die Drähte, bei Eisenbahnen sind es die im Vergleich zur Gesamtstrecke seltenen Weichen, welche die Wegwahl festlegen.

Auch bei Günter Hotz hingen die Weichenstellungen bzgl. seiner wissenschaftlichen Laufbahn von Zufällen ab – und um einen aus meiner subjektiven Sicht besonders wichtigen Zufall handelt es sich beim "Ulmer Flußspaziergang", dessen Name nicht ganz zufällig an den berühmt gewordenen "Genfer Waldspaziergang" erinnern soll.

Was war da passiert:
Nach seiner Promotion in Göttingen schaute sich Günter Hotz in der Industrie um, weil ihm dort entgegen der üblichen Meinung mehr Freiraum für eigene Kreativität zu bestehen schien als an der Universität, wo er promoviert hatte. Dabei interessierte er sich für eine Tätigkeit bei einem der zwei damals noch großen deutschen Rechnerhersteller, nämlich Siemens und AEG-Telefunken. An beiden Stellen informierte er sich über die Arbeitsbedingungen, und zwar zunächst – was auch eine heimliche Priorität ausdrücken könnte – bei Siemens in München. Die dort zu bearbeitenden Aufgaben

waren interessant, aber: der Vorstellungstag war von einem sprichwörtlichen Sauwetter geprägt. Günter Hotz bekam das, was man im Saarland „einen Moralischen" nennt. Jedenfalls war er nicht zum Abschluß eines Arbeitsvertrags bereit.

Ganz anders in der Ulmer Provinz: Die Sonne schien, es war ein lauer Frühlings- oder Herbsttag. Günter Hotz spazierte nach dem Vorstellungsgespräch in einer der Wetterlage angepaßten Stimmung am Ufer der Donau entlang – und dieses Hochgefühl führte dazu, daß er nach der Promotion zunächst als Entwicklungsingenieur bei AEG-Telefunken in Ulm arbeitete.

Selbst die kühnsten Vorstellungen versagen beim Versuch, sich Aufbau, Ausrichtung und Ansehen der Informatik unter der Bedingung vorzustellen, daß damals in Ulm schlechtes oder aber in München gutes Wetter geherrscht hätte.

Von anderen spontanen Entscheidungen profitierten auch die zahlreichen Mitarbeiter, die das Glück hatten, bei Günter Hotz beschäftigt gewesen zu sein, was im Regelfall eine Gewähr dafür ist, es in kurzer Zeit „zu etwas zu bringen". Als Beleg dafür möchte ich meinen eigenen Werdegang heranziehen: Es kann als gesichert gelten, daß ich meine Beschäftigung als Mitarbeiter bei Günter Hotz den folgenden beiden Ereignissen verdanke:

- Er wurde auf mich aufmerksam, weil ich in seiner Vorlesung immer die Bild-Zeitung las

- und ich wußte zufällig, was die Summe über $\frac{1}{i^2}$ ist, nämlich $\frac{\pi^2}{6}$. Wieso ich das wußte, ist mir noch heute rätselhaft, aber es hat ihn offenbar beeindruckt.

Viele andere Studenten und Mitarbeiter profitierten von ähnlich spontanen Personalentscheidungen. Manchmal riefen diese in deutschen Landen zunächst ein gewisses Kopfschütteln hervor, weil sie allzu riskant zu sein schienen. Im Nachhinein haben sich aber alle als richtig herausgestellt (abgesehen möglicherweise von meiner Person).

Der Wanderer

„Der Saarlandrundwanderweg und Schloß Dagstuhl"

Mit Beginn seines Wirkens in Saarbrücken wurde Günter Hotz nicht nur zum Wahlsaarländer, sondern zum glühenden Verehrer seiner Wahlheimat. Niemand von uns Eingeborenen brachte es je zu einer solch intimen Kennerschaft des Saarlandes, vor allem aber des Saarlandrundwanderwegs – von der Freilichtbühne Gräfinthal bis zu den Baltersweiler Menhir-Imitationen, von den Leitersweiler Buchen bis zur Saarschleife, von Schloß Dagstuhl bis ...

Halt: „Schloß Dagstuhl", dieser Name weckt Assoziationen, die einer Erklärung bedürfen:

Während seiner Wanderungen konnte Günter Hotz nie sein Engagement für die Informatik völlig ablegen. Und es war ihm stets ein Dorn im Auge, daß die Mathematiker in Oberwolfach ein so schönes Tagungszentrum aufgebaut hatten, welches der Informatik aber nur in sehr begrenztem Umfang zur Verfügung stand. Günter Hotz gelang es, Oberwolfach jährlich für eine Woche zu buchen. Er erreichte auch, daß im Laufe der Zeit insgesamt drei Wochen pro Jahr in Oberwolfach für die Informatik vorgehalten werden konnten. Aber diese Kämpfe waren mühsam und wenig ergiebig.

Daher suchte er so lange – natürlich im Saarland – bis er einen in vieler Beziehung ähnlichen Ort in Schloß Dagstuhl gefunden hatte. Die Gründung des „Internationalen Begegnungs- und Forschungszentrums für Informatik" in Dagstuhl ist ohne den Einsatz von Günter Hotz nicht vorstellbar. Dasselbe gilt für das erste Max-Planck-Institut für Informatik – auch wenn diese Einrichtung nicht auf eine Entdeckung anläßlich einer Wanderung zurückgeht, sondern auf die kontinuierlichen und erfolgreichen wissenschaftspolitischen Bemühungen von Günter Hotz – in der Gesellschaft für Informatik, im Wissenschaftsrat und bei vielen anderen Institutionen.

Der Beobachter   *„Der Aha-Effekt"*

Nicht alle negativen Erfahrungen sind schädlich, im Gegenteil: Sie können positive Denkanstöße bewirken, neue Probleme aufdecken und unkonventionelle Lösungen erschließen. Ein gutes Beispiel dafür ist eine Beobachtung, die Günter Hotz während seiner Tätigkeit im Ulmer Forschungslabor machte. Er hatte dort ein Programm zur Schaltkreisminimisierung geschrieben (nebenbei bemerkt fragten wir uns in seinen Vorlesungen immer, was denn der korrekte Begriff sei: Minimierung, Minimisierung, Minimalisierung, Minimation, Minimisation,.... Günter Hotz benutzte sie alle synonym). Zur automatischen Lösung dieses Problems hatte er eine der damals schnellsten Rechenanlagen programmiert. Ebenso überraschend wie prägend war für Günter Hotz die Beobachtung, daß der Rechner für kleine Anzahlen von Eingabevariablen die Lösung sozusagen im Nullkommanichts ausspuckte, daß er aber schon auf geringfügig größere Variablenzahl überhaupt nicht mehr reagierte. Um diesem Effekt auf die Schliche zu kommen, machte Günter Hotz den damals revolutionären Schritt zur Abschätzung des benötigten Rechenaufwands und kam zum Ergebnis, daß die Maschine noch viele Hundert Jahre zu tun haben würde, um die Aufgabe zu lösen. Er hat uns oft erzählt, daß dieses Ereignis ihn veranlaßte,

(a) den Rechner umgehend abzustellen,

(b) die Rückkehr an die Universität in Betracht zu ziehen und

(c) sich den Problemen der Komplexitätstheorie und der effizienten Algorithmen zu widmen.

Damit wurde eine mehr oder weniger zufällige Beobachtung zusammen mit einem kritischen Hinterfragen der dafür verantwortlichen Ursachen zur Geburtsstunde der Komplexitätstheorie (zumindest an der Universität Saarbrücken).

Der Vordenker   *„Theorie und Praxis"*

Die 70'er Jahre – also die Sturm-und-Drang-Zeit der Informatik – waren im Deutschland gekennzeichnet von einem Kampf zweier Welten. Hier 'Bauer/Goos', dort 'Hotz', was sich auf die Autoren der beiden ersten Bücher bezog, die für sich in Anspruch nehmen konnten, so etwas wie Standardwerke zu sein. Obwohl es vielleicht allzu vereinfachend ist, kann man sagen, daß Ausgangspunkt der Münchner Schule eher die Software war und ist, in Saarbrücken dagegen mehr die Hardware bzw. ein vorwiegend struktureller Zugang. Sicher ist, daß beide Fachbücher die Entwicklung der Informatik in Deutschland entscheidend beeinflußt haben.

Der Münchner Zugang war – um es wiederum sehr plakativ darzustellen – geprägt vom Gedanken „How I did it!", die Saarbrücker Seite konzentrierte sich dagegen mehr auf die Frage „How to do it!". Das brachte dann München den Ruf der Praxis und Saarbrücken den der Theorie ein. Beides hat seine Berechtigung. Donald Knuth hat den Unterschied zwischen Theorie und Praxis durch die Übersetzung der entsprechenden griechischen Wortstämme charakterisiert: Theorie kommt von „Sehen" oder „Betrachten", Praxis dagegen bedeutet „Tun".

Man kann heute feststellen, daß die beiden unterschiedlichen Schulen in der Zwischenzeit sich einander angenähert und sich gegenseitig schätzen gelernt haben. Dies war nicht immer so, denn während wir Saarbrücker vor den Münchnern immer eine gehörige Hochachtung und wegen unserer angeblichen oder realen Theoriebezogenheit auch einen ausgeprägten Minderwertigkeitskomplex hatten (ich spreche diesbezüglich gesichert nur für mich, vermute es aber auch von anderen), betrachteten einige Angehörige der Münchner Schule lange Zeit Saarbrücken eher als eine Art von Kuriosum. Ich erinnere mich an ein Gespräch an einem der gemeinsamen Kneipenabende während der GI-Jahrestagung 1973 in Hamburg, wo die These vertreten wurde, die deutsche Informatik könne sich ja vielleicht „ein Saarbrücken" leisten, aber nicht mehrere. Es ist vor allem das Verdienst von Günter Hotz, daß solche Äußerungen seit mindestens 15 Jahren nicht mehr vorstellbar sind.

Der Trainer   *„... und seine Fohlenelf"*

Die „Saarbrücker Informatik-Schule" – die von Günter Hotz aufgebaut und seit jeher entscheidend von ihm geprägt wurde – hat national wie international höchste Anerkennung gefunden. Woran lag das? Die Frage soll durch einen ziemlich gewagten Vergleich beantwortet werden:

Wenn ich eine Analogie zum Fußball arg strapazieren darf (ich habe mir lange überlegt, ob sie zulässig ist oder nicht, habe sie auch mehrfach verworfen, aber jetzt wage ich es doch), dann könnte man sagen, daß die Mannschaft von Günter Hotz mit der Fohlenelf von Borussia Mönchengladbach der 70'er Jahre zu vergleichen war. Genau wie dort wurde manches Mal schwach gespielt, aber wenn sie zur Form auflief (und das passierte ziemlich oft), dann spielte sie jeden Gegner an die Wand. Verantwortlich dafür war die Identifikation und die Begeisterung der Mannschaft. Wir waren von der Qualität unseres Trainers einfach überzeugt. Günter Hotz als Hennes Weisweiler der Informatik, ein kühner oder vielleicht sogar bizarrer Vergleich, aber: Ihr sollt ihn lassen stahn. Und wie im Beispiel wurden und werden die Spieler dieser Fohlentruppe in großer Zahl weggekauft. Ich kenne keinen anderen deutschen Informatiker, der ähnlich viele direkte wissenschaftliche „Nachkommen" aufzuweisen hat wie Günter Hotz. Neben einer sehr großen Zahl von Promotionen und Habilitationen wird dies durch eine zweistellige Zahl von Informatik-Professoren belegt, die bei Günter Hotz ihre ersten akademischen Meriten erhielten. Die Schar der nicht-direkten wissenschaftlichen „Nachkommen" (also derjenigen, die etwa erst nach ihrer Promotion zum Team von Günter Hotz stießen) ist kaum überschaubar und wohl nirgends vollständig aufgelistet. Sie enthält zum Beispiel Kurt Mehlhorn, um nur einen aus dieser Reihe zu nennen.

Die Analogie zum Fußball ist aber aus zwei Gründen fragwürdig bzw. inkorrekt, denn:

- erstens ist seit den frühen Achtziger Jahren die große Zeit von Borussia Mönchengladbach vorbei (ich wohne nicht allzuweit vom Stadion entfernt und leiste mir manchmal das zweifelhafte Vergnügen des entsprechenden Gekickes), während die Saarbrücker Schule ungebrochen produktiv ist

- und zweitens war und ist Günter Hotz nicht nur Trainer, sondern in erster Linie ja auch Spieler, und zwar was für einer! Also, Günter Hotz als Hennes Weisweiler und Günther Netzer in einer Person! Aber damit will ich es bewenden lassen, sonst wird der Vergleich allzu abenteuerlich.

Der Kritiker *„... und seine Kritikerin (nen)"*

Es gab Zeiten (und sie sind noch gar nicht so lange vorbei), in denen es ungewöhnlich war, den Bereich des Saarlandes zu verlassen und in denen von Saabrücken aus gesehen bereits Kaiserslautern als unwirklich ferne und eher gefährliche Welt erschien. Das hat sich heute nicht zuletzt durch Günter Hotz geändert, der ja Initiator und Sprecher des gemeinsam von Saarbrücken und Kaiserslautern getragenen DFG-Sonderforschungsbereichs „VLSI-Entwurfsmethoden und Parallelität" war bzw. ist und der auch dem Sonderforschungsbereich „Künstliche Intelligenz" mit seinem Rat zur Seite steht.

Aber nicht auf diese Tätigkeiten will ich hinaus, sondern auf das Wirken von Günter Hotz in noch weiter östlichen Gefilden. Denn: Günter Hotz war einer der ersten Informatiker, die das Gespräch mit Kollegen suchten, die eine andere politische Einstellung hatten oder zu haben vorgeben mußten. Dies ergab sich fast zwangsläufig als Konsequenz seiner Einladungen zu Hauptvorträgen, die ihn auch in zahlreiche Länder und Orte des früheren Ostblocks führten, vor allem auch in die DDR.

Gar seltsame Dinge wußte Günter Hotz von diesen Reisen zu berichten. So erzählte er beispielsweise von einem Theaterstück in Weimer, Gera oder Jena (für uns waren diese Orte damals völlig ununterscheidbar) – wenn ich mich recht erinnere, hieß es „Der Hahn des Schusters" –, worin zwar versteckte, aber deutliche Systemkritik enthalten gewesen sei.

Wir Mitarbeiter waren hochbeeindruckt vom Engagement unseres Vorgesetzten und über seine eigenen dort vorgetragenen kritischen Äußerungen. Solche Kritik formulierte er nie polemisch, sondern leise, dafür aber penetrant – ja mit einer gewissen oberhessischen Hartköpfigkeit – und deshalb mit besonders hohem Wirkungsgrad. Diese Zähigkeit und Sachkompetenz gleichermaßen waren auch verantwortlich dafür, daß Günter Hotz zum ersten westdeutschen Mitglied der Akademie der Wissenschaften gewählt wurde. Das war – ich erwähnte es bereits – eine wirkliche Sensation.

Kritische und gleichzeitig konstruktive Anregungen wurden und werden Günter Hotz entgegengebracht auch aus einer ganz anderen Richtung, nämlich von Ehefrau und von fünf Töchtern. Seine Frau Roswitha, geb. Trommsdorff, ist Diplommathematikerin und hat vor nicht allzu langer Zeit einen Doktortitel erhalten. Sie versteht wie niemand sonst die originellen und unkonventionellen Ideen Ihres Gatten einzuordnen und gegebenenfalls zu kanalisieren. Ohne diese Rückkopplung wären viele der Ideen von Günter Hotz nicht annähernd so erfolgreich geworden wie sie es heute sind. Auch Diskussionen mit seinen Töchtern, die sich alle für der Informatik eher fernliegende Gebiete entschieden haben, tragen nicht zuletzt wegen dieser Distanz sehr dazu bei, daß die Arbeiten und auch die Arbeitsrichtung von Günter Hotz des öfteren neue Denkanstöße und Umorientierungen erfahren haben.

Der Gönner   *„Ein weiterer Bezug zur Alchimie"*

Große Persönlichkeiten brauchen Gönner, und zwar nicht einmal so sehr für's Finanzielle, auch wenn das keinesfalls unwichtig ist. Nein, eher noch braucht man Gönner zur Schaffung von Freiräumen. Das galt schon zu Zeiten Augusts des Starken (auch diese Analogie wurde uns von Günter Hotz vermittelt): Dieser holte sich Alchimisten ins Land, deren geheimnisvolle Operationen er nicht verstand, die er aber gewähren ließ – im Gegensatz zu anderen Landesfürsten, welche erfolglose Alchimisten einen Kopf kürzer zu machen pflegten. Das Ergebnis dieser Versuche, nämlich Porzellan, entsprach nur sehr bedingt dem Ausgangsziel, nämlich Gold. Der Landesfürst hätte also böse sein dürfen oder müssen. Aber er akzeptierte das zunächst zweitrangig scheinende Resultat dankend, wenngleich etwas knurrig. Die Geschichte hat gezeigt, daß er gut daran tat.

Solch einen Gönner hatte Günter Hotz auch in seiner Startphase in Saarbrücken, nämlich Johannes Dörr. Dieser veranlaßte ihn zum Wechsel nach Saarbrücken – und zwar in den Bereich der Angewandten Mathematik, denn die Fachrichtung Informatik wurde erst später von Günter Hotz mitbegründet (obwohl seine Arbeitsrichtung schon damals als 'informatisch' zu bezeichnen war).

Die Forschungsinteressen von Dörr und Hotz waren sehr unterschiedlich. Kein C4-Professor würde heutzutage derart unterschiedliche Arbeitsrichtungen im gleichen

Institut dulden. Auch für Johannes Dörr wäre es ein Leichtes gewesen, bei der Besetzung einer Wissenschaftlichen Ratsstelle (was heute formal einer C3-Stelle entspräche) den bequemeren Weg zu gehen, nämlich einen Bewerber seiner eigenen Fachrichtung zu favorisieren. Aber das tat er nicht, und wir alle sind ihm zu außerordentlich großem Dank dafür verpflichtet, sondern er entschied sich für Günter Hotz, weil er spürte, daß sich hier etwas Neues und etwas Großes anbahnte. Mehr noch: Er ließ ihn gewähren, auch wenn er insgeheim – das erzählte er uns in stillen Stunden – einige Ideen von Günter Hotz für Träumereien hielt. Aber er hatte eine ungeheure Hochachtung vor wissenschaftlichen Potential, welches unverkennbar von Günter Hotz ausging.

Bitte mißverstehen Sie den folgenden Vergleich nicht als Blasphemie, ich nehme mir die Berechtigung dazu 'nur' wegen der Namensgleichheit der Vornamen: In gewisser Hinsicht war Johannes Dörr so etwas wie ein „Johannes der Täufer" für die Informatik, also ein Wegbereiter für den Verkünder einer neuen Religion bzw. einer neuen Fachrichtung.

Gott sei Dank   *„Göttingen sei Dank, Göttingern sei Dank"*

„Gott sei Dank" wurde Günter Hotz nach seinem Mathematik- und Physikstudium zum Informatiker. Etwas pointierter darf man sogar sagen: „Göttingen sei Dank", daß er zum Informatiker konvertierte. Warum dieses? Nun, die Initialzündung für die Wandlung vom Saulus zum Paulus kam eben aus Göttingen, wenngleich sie in dieser Form nicht beabsichtigt war.

Denn: Im Grunde – und ich bin sicher, daß das Geburtstagskind mir da recht geben wird – liegt ihm die Mathematik immer noch sehr am Herzen (Mathematiker zu sein ist per se noch nichts Ehrenrühriges). Eine Karriere in Göttingen wäre so schlecht nicht gewesen. Wer weiß, was passiert wäre, wenn er nach der Promotion in Göttingen geblieben wäre.

Daß es anders kam, ist dem Umstand zu danken, daß eine einflußreiche Gruppe von Göttinger Mathematikern es nicht ertragen konnte, eine solche Kapazität neben sich erstarken zu sehen. Die Folge davon war, daß Göttingen – oder vielmehr: einige Göttinger Mathematiker – eines ihrer Aushängeschilder nach seiner Promotion vergraulten. Zwar bot man ihm eine Assistentenstelle an, aber der Freiraum war so sehr eingeengt, daß Günter Hotz eine Beschäftigung in der Industrie bevorzugte; danach war seine Hinwendung zur Informatik nicht mehr aufzuhalten. Auslöser dafür waren also einige Göttinger Mathematiker. Die Wissenschaft hat nicht viele Beispiele, wo eine offensichtliche Fehlentscheidung mittel- und langfristig zu solch positiven Konsequenzen geführt hat. Daher: „Göttingen sei Dank" oder genauer: „Göttingern sei Dank".

Zum Abschluß ....

ein kleines Gedicht. Man verzeihe die offensichtliche Anlehnung an Eugen Roth:

Ein Mensch

E in Mensch aus Rommelhausen (Hessen)
I st zum Geburtstag nicht vergessen.
N atürlich gilt dies ohne Frage

M itnichten nur am heut'gen Tage.
E r ist, was ständig uns beweist er,
N och immer unser großer Meister.
S o lautet jedenfalls, punktum,
C laus Volkers Charakteristikum.
H ochgeschätzt haben sein Genie

G ottfried Leibniz und die Akademie.
U nter and'rem sagen Dank
E in Schloß in Dagstuhl und Max-Planck.
N atürlich auch der Fachbereich,
T ausend Studenten tun's ihm gleich.
E r ist uns allen Spielgestalter.
R eal gesagt: Von A(rz) bis W(alter).

H och lebe drum der Jubilar,
O rdensgeschmückt ist er sogar.
T önt mit Trompeten und Fagotts
Z um Sechzigsten von Günter Hotz.